Global Issues in Water Policy

Volume 9

More information about this series at http://www.springer.com/series/8877

Ariel Dinar • Víctor Pochat
José Albiac-Murillo
Editors

Water Pricing Experiences and Innovations

Springer

Editors
Ariel Dinar
School of Public Policy
University of California
Riverside, CA, USA

José Albiac-Murillo
Department of Agricultural Economics
Agrifood Research and Technology Centre
CITA-DGA
Zaragoza, Spain

Víctor Pochat
Víctor Pochat Consulting
Buenos Aires, Argentina

ISSN 2211-0631 ISSN 2211-0658 (electronic)
Global Issues in Water Policy
ISBN 978-3-319-16464-9 ISBN 978-3-319-16465-6 (eBook)
DOI 10.1007/978-3-319-16465-6

Library of Congress Control Number: 2015941020

Springer Cham Heidelberg New York Dordrecht London

Printed on acid-free paper

Springer International Publishing AG Switzerland is part of Springer Science+Business Media (www.springer.com)

To my granddaughter Yael Aviv who, at this stage of her life, doesn't like to drink water. But she will soon.

—Ariel Dinar

To my wife, Marta, for her lifelong support.

—Víctor Pochat

To my wife, Teresa, and my daughter, Silvia.

—José Albiac-Murillo

Contents

Contributors

Anne-Laurence Agenais ACTeon – Innovation, Policy, Environment, Colmar, France

José Albiac-Murillo Unidad Economía Agraria, CITA-DGA, Zaragoza, Spain

Tom Ash Inland Empire Water Utility, Chino, CA, USA

Tim Barr Western Municipal Water Disctrict, Riverside, CA, USA

Bernard Barraqué Agroparistech-Engref, Paris, France

Nir Becker Department of Economics and Management, Faculty of Social and Humanities science, Tel-Hai College, Upper Galilee, Israel

Javier Calatrava Escuela Técnica Superior de Ingeniería Agronómica, Universidad Politécnica de Cartagena, Cartagena, Spain

Yue Che Department of Environmental Science, East China Normal University, MinHang, Shanghai, China

Bethany Cooper Centre for Water Policy and Management, La Trobe University, University Drive, Wodonga, VIC, Australia

Lin Crase Centre for Water Policy and Management, La Trobe University, University Drive, Wodonga, VIC, Australia

Ariel Dinar School of Public Policy, University of California, Riverside, CA, USA

Guillermo Donoso Department of Agricultural Economics, Pontificia Universidad Católica de Chile, Macul, Chile

Diane P. Dupont Department of Economics, Brock University, St. Catharines, ON, Canada

Diego Fernández Econtec Consultores Ltda., Bogotá, Colombia

Marian García-Valiñas Facultad de Economía y Empresa, Departmento de Economía, Universidad de Oviedo, Oviedo, Spain

Alberto Garrido ETS Ingenieros Agrónomos, Madrid, Spain

Raymundo José Santos Garrido Salvador City, Estado da Bahia, Brazil

Faustino Gómez-Sántiz Facultad de Economía, Universidad Michoacana de San Nicolás de Hidalgo (UMSNH), Morelia, Michoacán, Mexico

Francisco González-Gómez Departamento de Economía Aplicada, Facultad de Ciencias Políticas y Sociología, Universidad de Granada, Granada, Spain

Hilda Guerrero-Garcia-Rojas Facultad de Economía, Universidad Michoacana de San Nicolás de Hidalgo (UMSNH), Morelia, Michoacán, Mexico

P. Hellegers Water Resources Management Group, Wageningen, The Netherlands

Francesc Hernández-Sancho Department of Applied Economics II, Faculty of Economics, University of Valencia, Valencia, Spain

Bryan Jenkins Waterways Centre for Freshwater Management, University of Canterbury and Lincoln University, Christchurch, New Zealand

Krishna Reddy Kakumanu International Water Management Institute (IWMI), Patancheru, Andhra Pradesh, India

Sébastien Loubier UMR G-Eau – Irstea, Montpellier cedex 5, France

Ravinder P.S. Malik International Water Management Institute, New Delhi, India

Antonio Massarutto DIES – University of Udine, Udine, Italy

Abel Mejía CAF Latin America Development Bank, Rockville, MD, USA

Maria Molinos-Senante Department of Mathematics for Economics, Faculty of Economics, University of Valencia, Valencia, Spain

Marielle Montginoul UMR G-Eau – Irstea, Montpellier cedex 5, France

Javier Santiago Ortiz Correa Departamento de Economia, Universidade de Brasília, Brasília, DF, Brazil

Kuppannan Palanisami International Water Management Institute, Patancheru, Hyderabad, India

Nicholas Pawsey Centre for Water Policy and Management, La Trobe University, University Drive, Wodonga, VIC, Australia

Víctor Pochat Víctor Pochat Consulting, Buenos Aires, Argentina

Steven Renzetti Department of Economics, Brock University, St. Catharines, ON, Canada

Moises de Andrade Resende Filho Departamento de Economia, Universidade de Brasília, Brasília, DF, Brazil

Daniel Rivera CAF Latin America Development Bank, Ciudad de Panamá, Panamá

J. Refugio Rodríguez-Velázquez Facultad de Economía, Universidad Michoacana de San Nicolás de Hidalgo (UMSNH), Morelia, Michoacán, Mexico

Ramón Sala-Garrido Department of Mathematics for Economics, Faculty of Economics, University of Valencia, Valencia, Spain

Jose Luis Santos ECAPAG Empresa Cantonal de Agua Potable y Alcantarillado de Guayaquil, Guayaquil, Republica del Ecuador

Barbara Schreiner Pegasys, Hatfield, Pretoria, South Africa

Marianne S. Schuerhoff PBL, Netherlands Environmental Assessment Agency, Utrecht, The Netherlands

Zhaoyi Shang Department of Environmental Science, East China Normal University, MinHang, Shanghai, China

Dajun Shen School of Environment and Natural Resources, Renmin University of China, Beijing, China

Jian Shi School of Environment and Natural Resources, Renmin University of China, Beijing, China

Marcelo de Oliveira Torres Departamento de Economia, Universidade de Brasília, Brasília, DF, Brazil

Germán E. Uzcátegui UCAB Universidad Católica Andres Bello, Caracas, Venezuela

Caroline van den Berg The World Bank, Washington, DC, USA

Xudong Yu School of Environment and Natural Resources, Renmin University of China, Beijing, China

Author Bios

Anne-Laurence Agenais is an engineer in agricultural sciences and environmental economics. She is a project engineer specializing in water and environmental economics at the consulting and research company ACTeon. Before that, she was a research assistant at the French National Research Institute of Science and Technology for Environment and Agriculture (Irstea). She worked on the analysis of household water consumption drivers and the study of water and wastewater rate structures in French municipalities. In her research, she has used various methods, e.g., surveys, statistical analysis, and building of the econometrics model.

José Albiac-Murillo is a researcher at the Agrifood Research and Technology Center (CITA-Government of Aragon) and a professor at the University of Zaragoza, Spain. He earned his bachelor's degree in economics at the University of Zaragoza and his master's and doctoral degrees in agricultural economics at the University of Illinois. His research specializes in environmental and natural resource economics, environmental and agricultural policies, water management, water scarcity and droughts, irrigation, groundwater and ecosystems protection, nonpoint pollution, and climate change.

Tom Ash as a University of California Extension Advisor, helped design, implement, and administer the first water budget rate structure in 1991. Tom has assisted a dozen agencies to achieve financial stability and water efficiency, while growing consumer equity and awareness of water use efficiency.

Tim Barr as the deputy director of water resources at Western Municipal Water District, has pioneered water conservation for 25 years. He managed the water rate structure changes and implementation at WMWD. Tim is recognized as a leading innovator in conservation programs design and serves as a board member of the California Urban Water Conservation Council.

Bernard Barraqué a civil engineer and urban planner, is research director in the French National Science Centre (emeritus), focusing on water resources and also public services. His laboratory is CIRED, Centre International de Recherches sur l'Environnement et le Développement, attached to Paris-Tech. He coordinated an important research program on the sustainability of water services in French cities, with an opening on Europe, the United States, and Australia. http://eau3e.hypotheses.org. He edited *Urban Water Conflicts* (UNESCO-IHP/Taylor & Francis 2012).

Nir Becker's main interest is with natural resources and environmental economics. He is currently the dean of the faculty of social sciences and humanities at Tel Hai College. He has published 60 refereed papers. In his research, he utilizes concepts such as optimization techniques, game theory, cost-benefit analysis, and valuation methods. He holds B.Sc. and M.Sc. degrees from Hebrew University in Jerusalem and a Ph.D. from the University of Minnesota. Besides his academic activities, Becker also serves as a consultant to the Nature Protection Agency, the Ministry of the Environment, and was a member of several governmental committees. Becker published a book on water policy in Israel in 2013 (Springer publishing). He has supervised eight Ph.D. students and more than 20 M.Sc. students.

Javier Calatrava is an associate professor of agricultural economics and policy and of rural sociology at Universidad Politécnica de Cartagena (UPCT), Spain. He has conducted extensive research in the field of agricultural and resource economics, mainly focused on the economics and policy of water resources and agricultural soil conservation. He has participated in 15 public research projects and supervised three doctoral dissertations. He has authored more than 40 academic references, including 25 articles in peer-reviewed scientific journals and 15 chapters in international books. He has conducted consultancy work for several Spanish agricultural and environmental administrations, the European Commission, and the OECD.

Yue Che is a professor of environmental management and sustainability at the Ecological and Environmental Sciences School at the East China Normal University. He was also a visiting researcher at the University of Sheffield (December 2013 to December 2014). He received a doctoral degree from the East China Normal University. His research has focused on water resources management, waste management, sustainability assessment, and low carbon planning.

Bethany Cooper is currently a postdoctoral fellow at the Centre for Water Policy and Management. In recognition of the quality of her research, Cooper has received numerous scholarly awards, including a Committee for Economic Development of Australia award. Her Ph.D. and subsequent research has focused on choice behavior and water policy, and she has published several empirical studies in prestigious scholarly journals, such as the *Australian Journal of Agricultural and Resource Economics* and *Water Resources Research*. Cooper has also completed a number of projects funded by major water utilities and regulatory bodies.

Lin Crase is a professor of applied economics and director of the Centre for Water Policy and Management at La Trobe University, Victoria, Australia. Lin holds a Ph.D. in economics and has published over 100 refereed journal articles on resource management issues. He has also edited several books in this field and has undertaken research for the Australian government, the OECD, and numerous state agencies responsible for managing water resources.

Ariel Dinar is a professor of environmental economics and policy at the School of Public Policy, University of California, Riverside. He teaches and conducts research on issues related to water economics, climate change economics, regional cooperation, and international water management.

Guillermo Donoso is an agricultural and natural resource economist, specializing in water economics, water allocation mechanisms, and water management. At present, he is a full professor of agricultural economics in the Department of the Pontificia Universidad Católica de Chile. He has been director of the National Agricultural Research Institute of Chile and held the position of dean of the Facultad de Agronomía e Ingeniería Forestal of the Pontificia Universidad Católica de Chile from 1998 to 2007. He was previously chair of the Agricultural Economics Department between 1995 and 1998. His publications include books such as *Mercados (de derechos) de agua: Experiencias y Propuestas en América del Sur. Serie Recursos Naturales e Infraestructura N° 80. CEPAL. 80 pp.* and *Mercados e Instituciones de Aguas en Bolivia. UDAPE: Unidad de Análisis de Políticas Sociales Económicas, 1998*, 241 pp. His book chapters include "Water markets in Chile: Are they meeting needs?"; "Regulating groundwater use through economic instruments: Current practices, challenges and innovative approaches"; "The evolution of water markets in Chile"; and "Vulnerability and adaptation to climate change in an irrigated agricultural basin in semi arid Chile." He also has publications in *Water Resources Management*, *Agricultural Water Management*, *Water Policy*, *Water Science and Technology*, *Futures Markets*, *Cuadernos de Economía*, *Ciencia e Investigación Agraria*, and *Revista de Derecho Administrativo Económico*.

Diane P. Dupont is a professor in the Economics Department at Brock University and was the recipient of Brock's Distinguished Research and Creative Activity award for 2012. Her research program concentrates upon examining ways to encourage more efficient and sustainable use of water resources, including the use of non-market valuation approaches to value water in relation to perceptions of health risks. She also works on transdisciplinary projects relating to water with researchers in fields such as engineering, hydrology, nursing, geography, and agriculture. She is an associate editor for *Water Resources Research*, *Canadian Journal of Water Resources*, and *Water Resources & Economics*.

Diego Fernández is an economist. He holds an M.A. from Instituto Torcuato Di Tella, Buenos Aires, Argentina; and is conferred with the title of Data Analysis

Specialist by Universidad La Salle, Bogotá, Colombia. He has more than 25 years of experience in water, wastewater, and municipal solid waste services. As director of The Water and Sanitation Regulatory Commission of Colombia, he developed the first regulatory tariff regimes applied for all water, wastewater, and municipal solid waste (MSW) companies (public and private). He had worked in the definition or revision of regulation, institutional reform, and/or water rates methodologies for authorities of Colombia, Ecuador, Bolivia, Dominican Republic, Guatemala, Honduras, Mozambique, Panama, Peru, Uruguay, and Venezuela. He has made tariff studies on water services for public and private companies in many of these countries.

Marian García-Valiñas is associate professor at the School of Economics and Business and member of the Oviedo Efficiency Group (Department of Economics) at the University of Oviedo (Spain). Since she obtained her Ph.D. in 2003, she has specialized in environmental economics, in particular water economics. Among other topics, she has worked on urban water demands characterization, water sector efficiency, and the use of alternative instruments in order to control water consumption, such as prices, low-consumption technologies, and rationing programs. She has published 30 papers in prestigious peer-reviewed journals, such as *Environmental and Resource Economics*, *Urban Studies*, and *Environmental and Development Economics*. She has collaborated intensively with several public and private institutions, such as the Spanish Ministry of Environmental Issues, and several river basin confederations, analyzing urban water demands in some Spanish regions, in order to implement the economic analysis needed for the EU Water Framework Directive (WFD). She has also a significant evidence of international research collaborations in Australia, Canada, and Europe.

Alberto Garrido is a professor in the Department of Agricultural Economics and Social Sciences at Universidad Politécnica de Madrid. He is also deputy director of the Research Centre for the Management of Agricultural and Environmental Risks (CEIGRAM), a research unit of UPM, and of the Water Observatory of the Botin Foundation. He has a B.S. in agricultural engineering from UPM (1989); an M.S. in agricultural economics from the University of California, Davis (1992); and a doctoral degree from UPM (1995). His research primarily focuses on agricultural and natural resource economics and policy and risk management. He has authored more than 150 publications (including 60 articles in peer-reviewed scientific journals) on a wide range of topics, ranging from water economics to food security, climate change, risk management, agricultural insurance, and sustainability assessments. He has consulted for may international, national, and regional institutions and governments, including OECD, Inter-American Development Bank, FAO, European Parliament, European Commission, and the governments of Spain, Chile, Colombia, Suriname, Vietnam, Panama, Nicaragua, and the Netherlands. He has also supervised 10 doctoral theses and 14 master theses.

Raymundo José Santos Garrido received his civil engineering degree in 1970 (Federal University of Bahia) and his master's degree in economics in 1991 (Federal

University of Bahia). He was National Secretary of Water Resources of Brazil from 1999 to 2002. He was general director of the Superintendency of Water Resources, State of Bahia (1991–1994); technical director of the National Secretariat of Water Resources of Brazil (1995–1998); president of the International Network of Basin Organizations—INBO (1998–2000); member of the Global Water Partnership Steering Committee (1999–2001); professor-researcher on microeconomics analysis at the Federal University of Bahia (2003–present); coauthor of the book *Economics of Water Resources*; and author of chapters in several books and of technical articles on water resources management.

Faustino Gómez-Sántiz is an economist at the University of Michoacán and has done his master's in human geography at El Colegio de Michoacán (Colmich). His areas of research are environmental economic assessment, ecological economics, common property resources, and collective action practices and natural resource management in rural communities. He has published and disseminated at international conferences scientific articles on topics such as economic assessment of water, payment for environmental services (hydrologic), and community management mechanisms. He is an independent consultant for public and private companies.

Francisco González-Gómez is a professor in the Department of Applied Economics at the School of Economics and Business and a researcher at the Water Research Institute at the University of Granada (Spain). In recent years, he has specialized in water economics. Among other topics, he has worked on public choice of managing ownership, water tariffs, water sector efficiency, and, more recently, on willingness to pay for improvements in water services. He has published 40 papers in prestigious peer-reviewed journals, such as *Public Choice*, *Public Administration*, *Water Resources Management*, and *Agricultural Water Management*. He is a member of the editorial board of the *International Journal of Water Resources Development*. Most recently, he has collaborated in several public research projects on water economics.

Hilda Guerrero-Garcia-Rojas is a professor and researcher in economics, Universidad Michoacana de San Nicolas de Hidalgo (UMSNH), México. She holds a bachelor's degree in economics (Universidad Michoacana de San Nicolas de Hidalgo, México); a master's degree in applied mathematics and economics (CIDE); a master of arts degree in economics (University of Colorado at Boulder); and a Ph.D. in economic sciences (Université des Sciences Sociales—Toulouse). Her Ph.D. thesis focused on "economic assessment of policy interventions in the water sector; industrial water demand in Mexico: econometric analysis and implications for water management policy." As a consultant, she has been collaborating with the World Bank to coordinate the project "AAA Informing Policy on Water Quality in Mexico" and with the Inter-American Development Bank (IDB) on the research "The Cost of Providing and Willingness to Pay for the Water in the Municipal and Industrial Sectors," among others. Publication topics involving Guerrero include

"Water pricing reforms in Mexico: The case of manufacturing sector"; "Applying water demand: Case for industrial use on Mexico"; "Water in Mexico: Impact of policy interventions in the sector"; and "Water pricing reforms in Mexico: The case of the manufacturing sector, among others." Guerrero's areas of interest are natural resources and environmental economics, integrated water management (IWM), economic instruments applied to natural resources (water management), water price and its reforms, applied economics, environmental economic valuation (contingent valuation), micro-econometrics and climate change in the water sector, and economics of climate change.

P. Hellegers is professor of economics of water and climate change and chair of the Water Resources Management Group of Wageningen University. She worked as a senior water economist for Landbouw Economisch Instituut (Agricultural Economics Institute) (LEI), part of Wageningen UR, between 1994 and 2013, and for the International Water Management Institute (IWMI) in Sri Lanka between 2005 and 2007. She has extensive experience in the economic analysis of water and water allocation issues, nationally and internationally (including Egypt, India, Morocco, and Mozambique).

Francesc Hernández-Sancho is an associate professor at the University of Valencia (UV). He is head of the Water Economics Research Group in UV and director of the Master on Water Management program. He is also a member of the Management Committee of the IWA SG on Statistics and Economics and leader of the IWA Water Economics Working Group. His research focuses on economic feasibility studies of wastewater treatment and reuse projects. He has also worked on water tariffs and the economic efficiency of water utilities.

Bryan Jenkins is a professor of strategic water management at the Waterways Centre for Freshwater Management, a joint center of the University of Canterbury and Lincoln University. The Center provides undergraduate and postgraduate courses in water and natural resources management and coordinates research across the universities in water resource issues. Prior to this appointment, he was chief executive of Environment Canterbury for more than 7 years. Environment Canterbury is the regional council for Canterbury whose responsibilities include water resource management. He was chief executive of the Department of Environmental Protection in Western Australia for 7 years. Before that, he had more than 20 years of experience in resource and environmental management consulting throughout Australia, South East Asia, India, and China. He has a Ph.D. in environmental planning from Stanford University, a master's and first class honors degrees in civil engineering from Adelaide University, and a master of administration degree from Monash University.

Krishna Reddy Kakumanu is a researcher at the International Water Management Institute, Hyderabad, India. He is an agricultural economist and obtained his Ph.D. from Justus-Liebig University, Germany. He is widely experienced on the economics of conjunctive use of water, energy pricing, climate change impact assessment and adaptation in agricultural and water sectors in India, as well as implementing and uptaking the progressive adaptation practices through stakeholder's integration. He has also published various research papers in international journals and books.

Sébastien Loubier has a Ph.D. in environmental and resource economics. He is a research engineer in the French National Research Institute of Science and Technology for Environment and Agriculture (Irstea). Most of his research concerns the effectiveness of regulatory and/or economic instruments for agricultural water management at the river basin level or at the water user associations' level. He is an advisor for the Ministry of Agriculture and the French National Agency for Water and Aquatic Environments (ONEMA) concerning the previous topics and the implementation of the water framework directive (WFD).

Ravinder P. S. Malik is a principal researcher of economics with the International Water Management Institute, based in New Delhi, India. Over a span of more than three decades, he has worked with several organizations, including the World Bank, University of Delhi, and the Afro-Asian Rural Development Organization. He has published extensively on various issues related to water economy and agricultural development.

Antonio Massarutto teaches economic and monetary policy and public economics at the University of Udine. He is research director at IEFE (Center for Research on Economics and Policy of Energy and the Environment) at Bocconi University, Milano. He collaborates on a project base with many other leading research centers in Italy and in Europe. His field of research concerns applied environmental and regulatory economics, particularly in the field of water resources, water services, and waste management.

Abel Mejía is an international water consultant with more than 35 years of experience. Until 2009, he was manager of water and environment units at the World Bank, with responsibility for designing and implementing water infrastructure projects, and leading water policy and strategy at regional and global levels. Mejia is a civil engineer with master's degrees in civil engineering management and industrial engineering from Stanford University. He also did graduate training at Cambridge (UK) and Harvard. Since 2010, he has been consulting for international development banks, UN agencies, and the private sector on water infrastructure projects.

Maria Molinos-Senante has a Ph.D. in local development and territory. She is secretary of the Economics and Statistics Specialist Group of the International Water Association. Her research focus is on economic feasibility studies, including the economic valuation of externalities and sustainability assessment of water utilities. She has carried out research in different countries, such as the United Kingdom, Italy, Argentina, Chile, and Greece.

Marielle Montginoul is research director in economy at the National Research Institute of Science and Technology for Environment and Agriculture (Irstea) at the Joint Research Unit G-Eau. She works on the understanding of water consumption of farmers and on households' behavior, on the manner to reveal these behaviors, and on the type of economic tools to manage water withdrawals. She conducts research on water pricing implemented for urban and agricultural uses, in particular looking at efficiency concerning water savings, budget balance, and social objectives. She mobilizes different types of methods, in particular surveys, experimental economics, and scenario workshops.

Javier Santiago Ortiz Correa holds a B.A. in economics from Universidad EAFIT, an M.A. in international and developmental economics from the University of San Francisco, and a Ph.D. in economics from the University of California, Riverside. He is a visiting assistant professor and a postdoctoral fellow in the Department of Economics at the University of Brasilia.

Kuppannan Palanisami is a principal researcher of economics and policy with the International Water Management Institute, based in Hyderabad. He has worked in several countries on water-related issues. He also served as a consultant to the World Bank, Asian Development Bank, and United Nations. Currently, he is working on climate change impacts, technology transfer, and uptake and policy interfacing with the government.

Nicholas Pawsey is currently a lecturer in accounting at La Trobe Business School and a member of the Centre for Water Policy and Management. Pawsey's principal areas of research expertise include public sector accounting, water prices, and international accounting standards. He has prepared a number of journal articles, book chapters, and funded reports on these issues. Through his work on Victorian water businesses, Pawsey has scrutinized the regulatory and statutory accounting frameworks governing the sector and the infrastructure reporting practices of individual businesses.

Víctor Pochat is an international consultant on water resources planning and management and a professor at the National University of Litoral, Argentina. He earned his civil engineering degree at the National University of Cuyo and his master's degree in engineering at the University of California. He has held first-level positions at national institutions in Argentina, related to planning and implementation of the national water policy, water science and technology, and construction and operation of hydraulic works.

Steven Renzetti is a professor of economics at Brock University. His research is principally concerned with the economics of water resources. He is the author of *The Economics of Water Demands* (Kluwer 2002) and coauthor of *Economics of the Environment and Natural Resources* (Blackwell 2004). His research has been published in leading peer-reviewed journals. He is an associate editor of *Water Resources*

Research and was recently appointed to the IJC's Great Lakes Science Advisory Board. He is also the program director of the Water Economics, Policy and Governance Research Network.

Moisés de Andrade Resende Filho holds a B.S. in agronomy (1993) and an M.S. in agricultural economics (1996) from the Federal University of Viçosa and a Ph.D. in applied economics from the University of Minnesota (2006). He is an associate professor in the Department of Economics at the University of Brasília, where he teaches and does research in agricultural economics and applied microeconomics and econometrics.

Daniel Rivera is director of social projects at CAF—Development Bank of Latin America. He serves as lead water engineer and water projects manager. He has more than 25 years of professional experience in sector governance, strategic planning, financial structuring, project implementation, and private sector participation in Latin America and the Caribbean. Prior to joining CAF, Rivera worked in Colombia as water director and national coordinator for the Colombian Ministries of Environment, Economic Development, and the National Department of Planning, where he led emblematic national water and sanitation development and reform projects. As a consultant, Rivera has conducted water sector–related assignments for the World Bank, USAID, IDB, international consulting firms, and Latin American governments. He has a master's degree in public policy from Duke University and a B.S. in civil engineering.

J. Refugio Rodríguez-Velázquez is a Ph.D. candidate in applied economics at Universidad de Santiago de Compostela, Spain. He has been working at the Universidad Michoacana de San Nicolas de Hidalgo (UMSNH), México, for over 25 years in the faculty of economics. He has a degree in economics and a master's degree in applied economics, the latter obtained through research using the methodology of Data Envelopment Analysis (DEA) for the education sector in Michoacan state. In the last 5 years, he has been working on the design of environmental indicators.

Ramón Sala-Garrido has a Ph.D. in economics and is a professor at the University of Valencia. His research interests include optimization processes applied mainly to solve water scarcity problems and productivity assessment of water companies. He has been involved in several European projects and has published more than 40 papers in international journals, such as *Journal of Hydrology*, *Science of the Total Environment*, and *Environmental Science and Policy*.

Jose Luis Santos is the general manager of the water regulatory agency in Guayaquil (EMAPAG). Santos is an engineer with more than 30 years of experience in planning, engineering, and construction of water infrastructure projects financed by the Inter-American Development Bank and the World Bank in several Latin American countries.

Barbara Schreiner is executive director of the Pegasys Institute, chairperson of the Board of the Water Research Commission, and a director in Pegasys Consulting. She has 20 years of experience in the water and environmental sector, having worked as advisor to Minister Asmal, as a senior official in the Department of Water Affairs, and subsequently in the consulting and research fields locally and internationally. She is an experienced senior manager and consultant with highly developed leadership and communication skills and considerable expertise in issues of policy implementation, monitoring, and evaluation; integrated water resources management; and institutional and organizational development. She has also worked in the field of transboundary water management. She has written extensively on issues of water resources management in developing countries, with a particular focus on issues of gender and poverty. She coedited the book *Transforming Water Management in South Africa—Designing and Implementing a New Policy Framework*. She has worked in a number of SADC countries and on projects that span Africa and sub-Saharan Africa and with the SADC water sector. Her high level of international experience includes being a member of the international Global Water Partnership Steering Committee for several years and a board member of the international CGIAR Water for Food Challenge Programme. She is currently on the board of the International Water Management Institution and the acting chair of the Programme Oversight Panel of the Aquatic Agricultural Systems research of the CGIAR. She is a member of the high-level panel of experts appointed by the UN Committee on Food Security to draft a position paper on water and food and nutrition security. She was chairperson of the international Gender, Water and Development Conference, held in East London, South Africa, in November 2014. In her spare time she writes poetry.

Marianne S. Schuerhoff works at PBL Netherlands Environmental Assessment Agency as a researcher for the TEEB Netherlands program. After she graduated from Wageningen University, she worked as a statistical researcher for the Environmental Accounts at Statistics Netherlands.

Zhaoyi Shang is a Ph.D. student of the Ecological and Environmental Sciences School at the East China Normal University. She received her bachelor's degree from the East China Normal University in 2010. Her research has mainly focused on urban water resource assessment and catchment management.

Dajun Shen is a professor at the School of Environment and Natural Resources, Renmin University of China, since 2011. He has been conducting research on water resources management in China for about 20 years, in terms of water pricing, water rights, management instruments, and policies. He has published 10 books and more than 50 articles.

Jian Shi is a graduate student in the School of Environment and Natural Resources, Renmin University of China. Her research focuses is on economic modeling of water resources.

Marcelo de Oliveira Torres is an economist with a B.S. in economics from the University of Brasília (1993), an M.S. in economics from the University of São Paulo, and a Ph.D. in agricultural and resource economics from the University of California, Davis (2004). He is a professor in the Department of Economics at the University of Brasília, with a research focus on water resources and environmental economics.

Germán E. Uzcátegui is an international consultant in the field of water with more than 40 years of experience. He was director of the postgraduate courses on hydraulic planning at the University of Los Andes, under the academic supervision of Professor Ray Linsley of Stanford University. He was director of the Inter-American Center for Water and Land Integrated Development sponsored by the Organization of American States and president of the National Institute of Sanitary Works of Venezuela. Uzcategui is a civil engineer who did his postgraduate work at Utah State University. Since 1990, he has worked as a consultant in hydraulic projects for public and private institutions.

Caroline van den Berg is a lead water and sanitation specialist, working mainly on the economics of water supply, wastewater, and irrigation water services, while also working on economics of water projects, monitoring, and evaluation and public finance in the water sector. She has extensive experience in the design and implementation of applied research projects and investment loans, with work experience in about 40 countries in Africa, Asia, Latin America, and the Middle East. Prior to joining the World Bank, she was working as a research and project economist in the private sector. She earned her M.A. in macroeconomics from the Erasmus University of Rotterdam (the Netherlands) and her Ph.D. in spatial sciences from the University of Groningen (the Netherlands).

Xudong Yu is a Ph.D. student in the School of Environment and Natural Resources, Renmin University of China. His research focuses on natural resources management.

Chapter 1
Introduction

Ariel Dinar, Víctor Pochat, and José Albiac-Murillo

Abstract This book is to provide credible evidence from water pricing experiences in various countries around the world. The book chapters, written by experts in water pricing from various countries, document the past 10–15 years of water pricing experiences in Australia, Brazil, Canada, China, Colombia, France, India, Israel, Italy, Mexico, The Netherlands, New Zealand, South Africa, and Spain. The chapters on water pricing experiences can be found in Part I of the book. Part II of the book includes several chapters that review innovations in water pricing in various countries, such as new reform mechanisms, achieving social objectives via water pricing, achieving revenue recovery, water use efficiency and customer equity, and charging the poor.

Keywords Water pricing • Urban water pricing • Agricultural water pricing • Economic incentives • Social objectives

1.1 Introduction

The water sector has seen many policy interventions aimed at regulating water consumption by users in the various sectors—irrigation, households, and industry. Among such regulations are various pricing methods that have been implemented with more or less success.

A. Dinar (✉)
School of Public Policy, University of California, 7852 Woodshole Ct., Riverside, CA 92506, USA
e-mail: adinar@ucr.edu

V. Pochat
Víctor Pochat Consulting, La Pampa 2086 – Piso 13, C1428EAJ Buenos Aires, Argentina
e-mail: vpochatm@yahoo.com.ar

J. Albiac-Murillo
Unidad Economía Agraria, CITA-DGA, Avenida de Montañana 930, 50059 Zaragoza, Spain
e-mail: maella@unizar.es

A. Dinar et al. (eds.), *Water Pricing Experiences and Innovations*, Global Issues in Water Policy 9, DOI 10.1007/978-3-319-16465-6_1

Water pricing gained popularity in the 1990s as a policy intervention tool that could be used to affect the environmentally, socially, and economically efficient use of water. While pricing of water was practiced earlier, the attention to water pricing followed the 1992 Fourth Dublin Principle, which was supplemented in the same year by what became known as the 1992 First Rio Principle. These two principles reflect the contrast of beliefs among water experts and policymakers and demonstrate, to a larger extent, the polarization that could and actually does take place in the global dialogue on managing our scarce water resources. While the Dublin Principle views water as an economic good, the Rio Principle suggests, implicitly, that water is a social good and humans are entitled to, at least, a minimal level of quantity and quality of safe water. Such views and beliefs also affect the existing debate over pricing or charging for water by experts representing various disciplines. Such debates also reflect the growing and declining interests in water pricing as a policy tool for water management that took place in the early 2000s, following disappointments from the ability of governments to implement water pricing.

Following this time line, international development agencies promoted, during the early 1990s, water-pricing initiatives and conditioned financial help on having pricing schemes (e.g., cost recovery) as part of the government obligations in the project. In the early 2000s, with increased social resistance to charging for water, the implementation of water pricing witnessed some decline in many countries. However, the European Union Water Framework Directive promoted some form of taxing water users in a way that will reflect the scarcity value of water. Recently, many countries face looming evidence of increased water scarcity due to realization of population growth, climate change, deterioration of water quality, among other issues. As a result, many countries, regions, and agencies turned again to water pricing as a policy intervention that could help in managing the dwindling water supplies.

While all agree that water is a resource/commodity that is essential for life, there is less agreement on the appropriate ways it should be regulated by society. Differences in views of the role governments shall play in water regulation are based on cultures, religions, and political interests (Dinar 2000). Changes in the development of water resources have been seen lately. First, the more ideal sites for construction of reservoirs have been already used, leading to increased cost of water production. Second, awareness led by changes in social priorities in many countries about the environmental impacts of water extraction and use introduced an additional layer of consideration in the allocation of water. And third, increasing competition by various sectors for scarce water resources in the developing world resulting from growing population, and increased economic activity and prosperity led to higher opportunity costs for water use. All of these changes led to a fundamental shift in the understanding of policymakers (Cummings et al. 1996).

Past literature that reviews water pricing efforts in a global setting (comparing experiences and procedures from various countries and comparing efficiency across various water pricing methods) include Dinar et al. (1997), Dinar and Subramanian (1997), Tsur and Dinar (1997), Dinar and Subramanian (1998), Jones (1998), Johansson et al. (2002), Tsur et al. (2004a), Dinar and Mody (2004), Tsur et al.

(2004b), Cornish et al. (2004), and Molle and Berkoff (2007). The overarching conclusion from these and other studies that compare performances of water pricing among countries is that there is no best practice that can be recommended to one country or sector. Water-using sectors in various locations face different situations and needs for pricing approaches. Future scarcity affected by climate change will most likely lead to different water pricing needs than the schemes we know from the past.

This book presents the latest observed pricing experiences from Australia, Brazil, Canada, Chile, China, Colombia, France, India, Israel, Italy, Mexico, The Netherlands, New Zealand, South Africa, and Spain. The chapters review both past and present experiences, as well as debates in the countries regarding future pricing directions. In addition, the book includes several chapters that review various innovations in water pricing in several countries, including new reform mechanisms, achieving social objectives via water pricing, achieving revenue recovery, water use efficiency and customer equity, and schemes for charging the poor. The following sections include a description of the main content of the various chapters.

1.2 Description of Chapters in Part I: Water Pricing Experiences

Chapters in Part I of the book follow a similar format that allows discussion of water pricing in the past, in the present, and debates about future options for the main four sectors: irrigation, urban, industrial, and environment.

The water pricing reforms in Australia are analyzed in Chap. 2 by Crase et al. (2015). The authors indicate that this reform, based in a common national water-pricing framework, has resulted in different approaches to water pricing, not only across states but also across water sectors (urban, rural, and environmental). Water pricing is the responsibility of states, and water is mostly managed by the public sector. Water pricing is a complex issue because of the different types of prices related to prices of end users, water market prices, and prices of water acquired for the environment. Urban water is provided by public utilities, irrigation water is managed by irrigation districts, and water pricing in both sectors is under regulation by different government agencies. There is a shared commitment to full cost recovery by states and water sectors, although this political will is sometimes weakened in the political process, especially during droughts. The authors indicate that the economic regulation of water needs additional research, especially in areas with high water resource variability.

Brazil is a "continental" country with big extremes, ranging from very wet and water-abundant regions to very dry and water-scarce regions. Filho et al. (2015) review in Chap. 3 selected (and representative) water pricing schemes and experiences during the past 15 years in the agricultural, residential, industrial, and environmental sectors in several river basins in Brazil, as was designed by the

Brazilian National Water Agency (ANA). The pricing system is very similar across all river basins under the federal and state jurisdiction, with similar pricing principles, but with technical specifications that vary by basin, such as the value of the coefficients used for computing some of the prices. The authors discuss the strengths and weakness of those pricing systems, and review new approaches for water pricing under consideration in Brazil at present.

Dupont and Renzetti (2015) present in Chap. 4 a critical review of past and current practices related to water pricing in irrigation, residential, and industrial sectors, as well as water pricing related to the provision of environmental services in Canada. They argue that water prices in most sectors have historically been quite low relative to the costs of supply, and relative to international standards. More recently, some provinces, irrigation districts, and municipalities have raised rates to promote conservation and increase supply network's financial sustainability. The chapter concludes by pointing to a number of important emerging issues related to water pricing in Canada.

Chile is a country known for its leadership in water reforms. Chapter 5 by Donoso (2015) summarizes and criticizes the decentralization and pricing reforms in Chile. The chapter focuses on recent pricing experiences in the urban residential and rural sectors. While the pricing and decentralization reforms were proven to be efficient, there are challenges that the water sector still faces, such as increasing extreme climatic events, and a highly informed and organized consumer base. In addition, there are concerns with respect to sustainability of groundwater extraction and deterioration of water-dependent ecosystems, due to over-allocation of water rights. The chapter also reviews Chile's national Rural Potable Water (APR) program, considered to be a successful program and different than the urban water supply services.

China is a vast country with a considerable endowment of water resources. The high growth rate of population and economic activities in recent decades has resulted in severe pressures on the quantity and quality of water resources, especially in Northern and Central China. Che and Shang (2015) review in Chap. 6 the experiences of water pricing in the irrigation, urban, and industrial sectors. In the urban and industrial sectors, there has been a strong increase in water prices, coupled with the introduction of block tariffs to achieve cost recovery. In the irrigation sector, prices have also increased, but they remain too low for cost recovery. The chapter highlights the current problems of water pricing in the context of the strong development and urbanization process, and makes recommendations for further reforms of water pricing. These reforms should address the water scarcity and water quality degradation problems, enhance competition among water supplying companies, develop incentive systems for water savings, and improve legislation and regulations for a better coordination of the institutions responsible for water management.

For a framework in which the majority of the population and economic activities of Colombia are located in areas with low water supply, Fernández (2015) shows in Chap. 7 that historically Colombia has always had a great concern for the prices of public services—including those for water and wastewater—having from 1968 to 1993 a national board that centrally regulated and even set pricing for them for the entire country. With the enactment of a special law in 1994, the door was opened for private participation and the operation of the market for water services, respecting competition rules and a clear economic plan, with definitions related to coverage of costs by tariffs, cross-subsidies and budget contributions.

France has quite an impressive past experience with water pricing. Montginoul et al. (2015) describe in Chap. 8 how the focus of pricing policy progressively shifted from budget balancing (cost recovery) to water conservation and then to social protection over the past 15–20 years. In addition to describing water pricing practices in the urban sectors, the authors also discuss water pricing in the agricultural sector at different scales: large public irrigation schemes, smaller water user associations and individual irrigation systems. The chapter discusses the efficiency of water pricing in urban and irrigation sectors and addresses limitations of water pricing that have to be taken into account by the regulator.

India is a federal state where water is a state issue. Palanisami et al. (2015) explain in Chap. 9 how the public concern about water scarcity has led to renewed focus on the issue of water pricing and cost recovery in India. In this debate on water pricing, the irrigation sector, which accounts for almost 80 % of the total water use but for which water is charged at a fraction of the supply cost was central. The chapter examines several issues at the heart of the water-pricing debate in India: the basis for setting water rates/charges in different sectors; the difference in water prices among different states and uses; the periodicity of revision in water rates and methods of charging; the implications of low water rates—on cost recovery, on inefficiency in allocation and use of water, on availability of finances for maintenance of infrastructure and investments in new supply sources; and cost allocation and subsidy in major irrigation projects.

Due to the high level of scarcity it faces, Israel manages its water using quantitative and pricing regulations. As a semi-arid climate country, efficient water pricing might prove to have much more potential welfare implications. Becker (2015) summarizes in Chap. 10 the theoretical background of the various water pricing policies and reforms that have been recently implemented in Israel. At present, prices of water reflect the true scarcity value of the resource. Since Israel utilizes many of its water sources (treated wastewater, desalinated water, groundwater, storm water, surface water, etc.) there are different pricing schemes for each of the sources and uses, including non-market nature (e.g., in-stream value) and ones that should be based on basin cooperation among Israel's neighbors.

Chapter 11 by Massarutto (2015) analyzes water pricing in the context of water management in Italy. The water governance and regulatory system is presented describing the financial flows in the water sector, and the water-pricing practices are explained for the urban, industrial, and irrigation sectors. The water-pricing experiences indicate that water pricing is driven by financial considerations to achieve the economic viability of companies in the water sector. In some cases, water companies are not able to sustain the needed investments, and water taxes are being considered to provide the additional funding. The use of water pricing as an incentive for water savings and pollution abatement is far from being considered in Italy. At present, regulation and management institutions provide incentive schemes only to influence water investment decisions. The consideration of the environmental and resource costs of water utilization, advanced by the European water legislation, is an opportunity to make use of water pricing as an incentive for a more efficient and sustainable water sector.

Mexican water price structure is set to reflect water availability and its economic value. Guerrero-García-Rojas et al. (2015) explain in Chap. 12 the zonal and sectoral distribution of pricing in Mexico, where water price schemes reflect the level of water scarcity in the various zones, and the tariff differentiation according to sectorial users, such as industry, households, and agriculture. The chapter briefly describes an instrument of environmental policy, which affects pricing water from environmental conservation.

The Netherlands applies a very comprehensive set of regulations that include water quotas and pricing/taxes on the various sources of water used for consumption. Schuerhoff and Hellergers (2015) review in Chap. 13 the various taxes imposed in The Netherlands, which aim to recover costs, trigger adoption of water-saving technologies, or reduce water demand to leave more for environmental purposes. In particular, the authors analyze the 1995 national groundwater tax, which was considered a “win–win green tax,” but was realized to be fiscally inefficient and environmentally ineffective, and revoked in 2012. Further attempts to tax tap water is the national tap water tax, which was increased in 2014. However, this increase will not yield effective results since, on average, only 0.6 % of households’ budget is spent on tap water, so its impact on water demand is minute.

Water pricing in New Zealand is analyzed in Chap. 14 by Jenkins (2015). The reform of the water sector in New Zealand during the 1980s maintained the responsibility of municipalities over urban water supply, but abolished the public involvement over irrigation and hydroelectricity by privatizing irrigation schemes and hydropower generation. The water sector is considered inefficient in making investments for water infrastructure, and also in improving the economic and environmental welfare of society. The chapter analyzes the performance of water pricing in the urban, irrigation and hydroelectric sectors, and the contribution of water pricing to solve the main challenges faced by water resources in the country. These challenges are the quality of drinking water in some locations, the growing water scarcity driven by the expansion of irrigation, and the future impacts of climate change.

Schreiner (2015) shows in Chap. 15 that charging for water use was established in South Africa in different areas and circumstances across the country. Water charging is set usually at the local level, and gradually, over time, a more coherent approach to pricing of water was introduced at the national level for raw water. The pricing of water supplied by municipalities, however, has always been the individual purview of local government. There has thus been a continual evolution of pricing of water across the value chain, until a most recent introduction, which is looking at the application of a charge to discharge of wastewater into a water resource.

Spain is known for its long-standing water pricing practices. However, Spain faces increasing rates of recurrences of droughts, severe water pollution, and expected negative impacts of climate change. Calatrava et al. (2015) suggest in Chap. 16 that water demand management is now one of the pressing issues in the Spanish water policy agenda. The chapter reviews water pricing schemes for the main users in Spain, including irrigated agriculture, residential, and industrial sectors. In addition, the chapter also discusses several controversial issues in the water-pricing debate in Spain. Among such issues is (1) the Water Framework Directive (WFD), which some of its principles are not popular in Spain and may lead to significant consequences in terms of affordability for all water users in Spain; (2) the irrigation water pricing that was put on hold in the 2014 reform of the Common Agricultural Policy (CAP), which allow the federal and regional governments to stick to the status quo of pricing irrigation water. But it has been realized that all future climate change adaptation programs cannot be developed without proper water pricing of the irrigation sector in the country. The chapter introduces also the links between non-water reforms (fiscal and decentralization) and water pricing, which emerge as a significant challenge to improve water resources allocation in Spain.

1.3 Description of Chapters in Part II: Innovations in Water Pricing

China is a very large country with a range of water-scarcity levels. Water-pricing efforts in China have taken both a general, but also a local focus. Shen et al. (2015) describe in Chap. 17 the pricing reform process and analyze the pricing structure, referring to several case studies in China in the past 60 years, with special focus on the process after 1980. China introduced the water resources fee in the 1980s, and the wastewater treatment and collection fee in the late 1990s. In the late 2000s, a comprehensive pricing system was developed. The pricing efforts and changes in prices for various sectors in China are demonstrated in two case studies: Beijing and Shanxi Province.

Barraqué and Montginoul (2015) introduce in Chap. 18 the social dimension of water pricing, which has been less studied compared to efficiency aspects. Social dimensions of water pricing become more acute with the increased trends of privatization of water services in the world. The chapter presents various solutions to support low-income populations facing payment of their bills. Possible solutions include reduced bills for targeted populations (rebates, increasing blocks); supporting the income of targeted populations; reducing bills for all customers and reintroducing taxation as a source of revenue. The chapter discusses the incremental transaction (administrative) costs that arise from "social tariff design" that may offset the benefits to society.

The best evidence for the impact of water pricing on consumer demand is the fact that consumers respond to increases in prices by reducing consumption, which leads to financial instability of the utilities, and political unrest among the customers. Barr and Ash (2015) describe in Chap. 19 the development and implementation of the "water budget-based rates" or, more accurately, the "sustainable rate design" in the Western Municipal Water District in Southern California. The chapter describes the principles by which the utility implemented rate structures that accurately reflect the costs of water and its delivery service, recognize customers by their water use efficiency, and also provide a strong economic signal as to the opportunity costs of water. Given the principle of zero benefits of water utilities, revenue that exceeds cost goes into conservation programs.

Mejia et al. (2015) justify and describe in Chap. 20 a pragmatic approach developed and implemented in the city of Guayaquil, Ecuador, for pricing urban water services (water supply, wastewater, storm water) via a long-term "regulated" concession to a private company. The authors offer this model to other big cities in the developing world. The chapter discusses the pricing adjustment mechanisms to account for inflation, and meet investment and service targets that are updated periodically. The mechanisms used address finance of unexpected shortfalls, government guarantees, meeting poverty, and environmental goals.

Santos Garrido (2015) analyzes in Chap. 21 pricing of bulk groundwater for domestic water supply from the Tucano aquifer in Bahia, Brazil. Using a market equilibrium approach for water demand and supply, and financial balance of public expenditures, the chapter offers bulk water tariff levels that address both water scarcity as well as level of poverty that characterizes the region.

Water-scarcity increases the value of recycled wastewater. However, the cost of treated wastewater and the level of tariffs paid by water users suggest that, in the majority of water reuse projects, the principle of cost recovery is not met. However, water reuse projects generate positive externalities (reduced health risks, improved environmental quality, and reduced competition among sectors), contributing to improved social welfare. Hernandez-Sancho et al. (2015) suggest in Chap. 22 to deal with all water sources rather than with individual sources for pricing purposes. The chapter describes a proposed framework for pricing of reclaimed wastewater in the Valencia region of Spain. The framework calls for a pricing policy to view reclaimed water from an integrated water resource management point of view, addressing the costs and benefits of all sources. A two-part tariff with a combination of a decreasing and increasing rate structure is proposed as a partial solution to improve the cost recovery of water reuse projects.

Van den Berg (2015) examines in Chap. 23 how balancing the principles of managing water as a social and economic good has worked out globally. She explains that while the costs of providing water and wastewater services have been increasing, these cost increases have been matched by increases in tariffs. For many residential water users, that effect has been partially compensated by increases in cross-subsidies. Yet, the combination of higher future costs of water and the high levels of affordability mean that there is scope for balancing the goals of revenue sufficiency and affordability more in favor of the first, as government subsidies will otherwise increase rapidly.

1.4 Conclusion

There are several take-away lessons available from the rich cases presented in this book; however, we are not pretending that these country cases represent the state of the world. The water pricing experiences and innovations presented in the chapters are by no means representative samples of countries in the world, but they cover the main issues and debates being considered by decision-makers in managing basins and water sectors across the world. Water pricing has become a key issue in both the developed and developing world, because the unrelenting investments in water technologies that guarantee human water security require an enormous effort of private and public financing. The book also contains a nice and representative balance among developing countries and developed countries. Water pricing could also be used as an environmental instrument in protecting water resources from excessive depletion and quality degradation, which is pervading in many basins in all regions around the world.

The experiences of water pricing are not cases of "best practice," but rather a good collection of attempts to improve water management. The outcomes from these attempts highlight the challenges of using water pricing and the implementation difficulties that appear in the political process. The lessons learned from these experiences are important for the design and implementation of potential future water pricing initiatives.

Several experiences suggest a shift in water pricing, from dealing mainly with cost recovery or efficiency to addressing social parameters and environmental considerations, as well. The social aspects include the impact of water pricing on impoverished population groups, and on vulnerable economic sectors. Another issue is the social perception that water is not a pure economic good, and should not be treated as a private commodity. The environmental considerations focus on the contribution of water pricing towards the sustainable management of water resources. The rapid growth in population and income requires large water investments and financing that result in further pressures on the quantity and quality of water resources. In these conditions, the water demand response to pricing is not strong enough to reduce water extractions by the urban, industrial, and irrigation sectors.

Several observations suggest that water pricing is no longer a bad word among policymakers and water customers; that incentive-based tariffs are more prominent than just cost recovery objectives, and that there is a shift from efficiency objectives only to social and environmental considerations along the efficiency goals.

We observed from the various chapters that there has been much more use of incentive tariffs in all sectors than reported in the past (Dinar et al. 1997; Dinar and Subramanian 1997, 1998; Tsur and Dinar 1997; Jones 1998; Johansson et al. 2002; Tsur et al. 2004a, b; Dinar and Mody 2004; Cornish et al. 2004; Molle and Berkoff 2007). This reflects the devolution of water management responsibilities from central governments to regional or local administrations, the growing private sector involvement in providing water services, and the financing of large water investments through public-private partnerships. These changes enhance the importance of water pricing, but also call for strong regulation mechanisms that could guarantee the protection of the public interest.

Finally, we observed also that there is a shift (both in developed and developing countries) from cost-recovery principles only to social and environmental considerations of the pricing tariffs. This phenomena reflects the fact that customers are more informed and outspoken, that water service charges become a major budgetary constraint to more than just the poor, but also to mid-income strata of the society, and that norms of environmental conservation have become an integral part of our society preferences. All of these suggest that pricing of water and water services are an acceptable practice that may, with some carful implementation, lead us in the right direction of water conservation and sustainable management.

References

Barr, T., & Ash, T. (2015). Sustainable water rate design at the Western Municipal Water District: The art of revenue recovery, water use efficiency and customer equity (Chapter 19). In A. Dinar, V. Pochat, & J. Albiac (Eds.), *Water pricing experiences and innovations* (pp. 375–394). Dordrecht: Springer.

Barraqué, B., & Montginoul, M. (2015). How to integrate social objectives into water pricing? (Chapter 18). In A. Dinar, V. Pochat, & J. Albiac (Eds.), *Water pricing experiences and innovations* (pp. 361–373). Dordrecht: Springer.

Becker, N. (2015). Water pricing in Israel: Various waters, various neighbors (Chapter 10). In A. Dinar, V. Pochat, & J. Albiac (Eds.), *Water pricing experiences and innovations* (pp. 183–201). Dordrecht: Springer.

Calatrava, J., García-Valiñas, M., Garrido, A., & González Gómez, F. J. (2015). Water pricing in Spain: Following the footsteps of somber climate change projections (Chapter 16). In A. Dinar, V. Pochat, & J. Albiac (Eds.), *Water pricing experiences and innovations* (pp. 315–342). Dordrecht: Springer.

Che, Y., & Shang, Z. (2015). Water pricing in China: Impact of socioeconomic development (Chapter 6). In A. Dinar, V. Pochat, & J. Albiac (Eds.), *Water pricing experiences and innovations* (pp. 97–115). Dordrecht: Springer.

Cornish, G., Bosworth, B., Perry, C., & Burke, J. (2004). *Water charging in irrigated agriculture. An analysis of international experience* (FAO Water Reports 28). Rome: FAO.

Crase, L., Cooper, B., & Pawsey, N. (2015). Water pricing in Australia: Unbundled politics, accounting and water pricing (Chapter 2). In A. Dinar, V. Pochat, & J. Albiac (Eds.), *Water pricing experiences and innovations* (pp. 15–39). Dordrecht: Springer.

Cummings, R., Dinar, A., & Olson, D. (1996, December). *New evaluation procedures for a new generation of water-related projects* (World Bank Technical Paper 349). Washington, DC.: World Bank.

Dinar, A. (Ed.). (2000). *The political economy of water pricing reforms*. New York and Washington DC: Oxford and World Bank.

Dinar, A., & Mody, J. (2004). Irrigation water management policies: Pricing and allocation principles and implementation experiences. *Natural Resources Forum, 28*, 112–122.

Dinar, A., & Subramanian, A. (1997). *Water pricing experiences—an international perspective* (World Bank Technical Paper No. 386). Washington, DC: World Bank.

Dinar, A., & Subramanian, A. (1998). Policy implications from water pricing experiences in various countries. *Water Policy, 1*(2), 239–250.

Dinar, A., Rosegrant, M. W., & Meinzen-Dick, R. (1997, June). *Water allocation mechanisms: Principles and examples* (World Bank Policy Research Working Paper 1779). Washington, DC: World Bank.

Donoso, G. (2015). Water pricing in Chile: Decentralization and market reforms (Chapter 5). In A. Dinar, V. Pochat, & J. Albiac (Eds.), *Water pricing experiences and innovations* (pp. 83–96). Dordrecht: Springer.

Dupont, D., & Renzetti, S. (2015). Water pricing in Canada: Recent developments (Chapter 4). In A. Dinar, V. Pochat, & J. Albiac (Eds.), *Water pricing experiences and innovations* (pp. 63–82). Dordrecht: Springer.

Fernández, D. (2015). Water pricing in Colombia: From bankruptcy to full cost recovery (Chapter 7). In A. Dinar, V. Pochat, & J. Albiac (Eds.), *Water pricing experiences and innovations* (pp. 117–139). Dordrecht: Springer.

Filho, M. A. R., Correa, J. S. O., & Torres, M. O. (2015). Water pricing in Brazil: Successes, failures, and new approaches (Chapter 3). In A. Dinar, V. Pochat, & J. Albiac (Eds.), *Water pricing experiences and innovations* (pp. 41–61). Dordrecht: Springer.

Garido, R. J. S. (2015). Price for domestic water supply: An innovative method developed for the Tucano aquifer in the state of Bahia, Brazil (Chapter 21). In A. Dinar, V. Pochat, & J. Albiac (Eds.), *Water pricing experiences and innovations* (pp. 409–431). Dordrecht: Springer.

Guerrero-García-Rojas, H., Gómez-Sántiz, F., & Rodríguez-Velázquez, J. R. (2015). Water pricing in Mexico: Pricing structures and implications (Chapter 12). In A. Dinar, V. Pochat, & J. Albiac (Eds.), *Water pricing experiences and innovations* (pp. 233–249). Dordrecht: Springer.

Hernandez-Sancho, F., Molinos-Senante, M., & Sala-Garrido, R. (2015). Pricing for reclaimed water in Valencia, Spain: Externalities and cost recovery (Chapter 22). In A. Dinar, V. Pochat, & J. Albiac (Eds.), *Water pricing experiences and innovations* (pp. 433–444). Dordrecht: Springer.

Jenkins, B. (2015). New Zealand water pricing (Chapter 14). In A. Dinar, V. Pochat, & J. Albiac (Eds.), *Water pricing experiences and innovations* (pp. 265–290). Dordrecht: Springer.

Johansson, R. C., Tsur, Y., Roe, T. L., Doukkali, R. M., & Dinar, A. (2002). Pricing and allocation of irrigation water: A review of theory and practice. *Water Policy, 4*(2), 173–199.

Jones, T. (1998). Recent development in the pricing of water service in OECD countries. *Water Policy, 1*, 637–651.

Massarutto, A. (2015). Water pricing in Italy: Beyond full-cost recovery (Chapter 11). In A. Dinar, V. Pochat, & J. Albiac (Eds.), *Water pricing experiences and innovations* (pp. 203–232). Dordrecht: Springer.

Mejia, A., Santos, J. L., & Rivera, D. (2015). Pricing urban water services in the developing world: The case of Guayaquil, Ecuador (Chapter 20). In A. Dinar, V. Pochat, & J. Albiac (Eds.), *Water pricing experiences and innovations* (pp. 395–407). Dordrecht: Springer.

Molle, F., & Berkoff, J. (Eds.). (2007). *Irrigation water pricing: The gap between theory and practice* (Comprehensive Assessment of Water Management in Agriculture, Vol. 4). Wallingford: CABI Publication.

Montginoul, M., Loubier, S., Barraqué, B., & Agenais, A.-L. (2015). Water pricing in France: Toward more incentives to conserve water (Chapter 8). In A. Dinar, V. Pochat, & J. Albiac (Eds.), *Water pricing experiences and innovations* (pp. 141–162). Dordrecht: Springer.

Palanisami, K., Reddy, K., & Malik, R. P. S. (2015). Water pricing experiences in India: Emerging issues (Chapter 9). In A. Dinar, V. Pochat, & J. Albiac (Eds.), *Water pricing experiences and innovations* (pp. 163–182). Dordrecht: Springer.

Schreiner, B. (2015). Water pricing—The case of South Africa (Chapter 15). In A. Dinar, V. Pochat, & J. Albiac (Eds.), *Water pricing experiences and innovations* (pp. 291–313). Dordrecht: Springer.

Schuerhoff, M., & Hellergers, P. (2015). Water pricing in the Netherlands (Chapter 13). In A. Dinar, V. Pochat, & J. Albiac (Eds.), *Water pricing experiences and innovations* (pp. 251–264). Dordrecht: Springer.

Shen, D., Yu, X., & Shi, J. (2015). Introducing new mechanisms into water pricing reforms in China (Chapter 17). In A. Dinar, V. Pochat, & J. Albiac (Eds.), *Water pricing experiences and innovations* (pp. 345–360). Dordrecht: Springer.

Tsur, Y., & Dinar, A. (1997). On the relative efficiency of alternative methods for pricing irrigation water and their implementation. *World Bank Economic Review, 11*(2), 243–262.

Tsur, Y., Roe, T., Doukkali, R., & Dinar, A. (2004a). *Pricing irrigation water: Principles and cases from developing countries*. Washington, DC: Resources for the Future Press.

Tsur, Y., Dinar, A., Doukkali, R., & Roe, T. (2004b). Irrigation water pricing: Policy implications based on international comparison. *Environment and Development Economics, 9*(6), 735–755.

van den Berg, C. (2015). Pricing municipal water and wastewater services in developing countries: Are utilities making progress towards sustainability? (Chapter 23). In A. Dinar, V. Pochat, & J. Albiac (Eds.), *Water pricing experiences and innovations* (pp. 445–464). Dordrecht: Springer.

Part I
Water Pricing Experiences in Selected Countries

Chapter 2
Water Pricing in Australia: Unbundled Politics, Accounting, and Water Pricing

Lin Crase, Nicholas Pawsey, and Bethany Cooper

Abstract This chapter presents a review of water-pricing arrangements in each of Australia's state jurisdictions. The pricing approaches for urban, environmental, and rural (i.e., agricultural) water uses are scrutinized and compared against the ambitions established as part of the National Water Initiative (NWI). While the framework for water pricing in the NWI has been generally deployed in most states, local nuances give rise to quite different price outcomes. Moreover, there is still opportunity for political influences to shape water prices, even though the NWI is committed to full-cost recovery with regulatory oversight that seeks to objectively align costs and prices. We conclude that there remains scope for improvement that would remove artificial differences in the way water is priced for different water users and thus support the distribution of water to its highest values.

Keywords Australia • Urban water pricing • Rural water pricing • Environmental water pricing • Sewage charges

2.1 Introduction

All Australian state jurisdictions agreed on water-pricing reforms in the 1990s and reaffirmed their commitment to the principle of cost-reflective pricing, along with enhanced institutional arrangements for managing water in the early 2000s in the form of the National Water Initiative (NWI). Similarly, in 2010, jurisdictions confirmed their commitment to the NWI pricing principles via the agreement signed by the National Resource Management Ministerial Council (DAFF and DEWHA 2010). Regardless of the apparent enthusiasm for pursuing common goals in water pricing, each state has followed different reform trajectories. Partly, these differences reflect historical institutional arrangements, but hydrology also varies between and within states, and this has also led to discrepancies. The political will to pursue

L. Crase (✉) • N. Pawsey • B. Cooper
Centre for Water Policy and Management, La Trobe University,
University Drive, Wodonga, VIC 3690, Australia
e-mail: l.crase@latrobe.edu.au; n.pawsey@latrobe.edu.au; b.cooper@latrobe.edu.au

A. Dinar et al. (eds.), *Water Pricing Experiences and Innovations*,
Global Issues in Water Policy 9, DOI 10.1007/978-3-319-16465-6_2

efficient water prices has also varied across jurisdictions and over time. For instance, severe water shortage experienced as part of the extended drought in the first decade of the 2000s played an important part in shaping different approaches to water pricing, albeit supposedly within a common national framework. The impacts on water planning and the effect on water prices from these interventions have now resulted in marked divisions between the ways urban, rural, and environmental water users are charged in many jurisdictions.

The Australian experience with water pricing is thus informative, because it offers insights into how a common national water-pricing framework can still give rise to very different outcomes for water users. The experience also highlights the difficulty of "staying the distance" when it comes to pursuing the efficiency goals with which the principle of cost recovery pricing is often associated. More specifically, the experience in Australia shows that even minor differences in regulation or interpretation of accounting standards can be used to pursue a range of noneconomic objectives while seemingly remaining within a national framework based on full-cost recovery (see, e.g., Pawsey and Crase 2013).

This chapter is used to explore water-pricing reforms in Australian jurisdictions. We provide a synoptic overview of water pricing across contrasting states and detail the varying regulatory arrangements, pricing structures, implementation approaches, and implications for water users. We also briefly contrast the existing water-pricing outcomes with the principles that were agreed in the national reforms in the early 2000s and restated in 2010.

The chapter itself comprises three additional parts. In the following section, we provide a broad overview of water pricing at a national level and note significant influences in this context. Section 2.3 is used to detail the status of water pricing in individual jurisdictions. In this section, we consider the pricing arrangements for urban, rural, and environmental water uses separately, in part to highlight differences. The fourth section of the chapter briefly explores some of the themes that transcend jurisdictions and the resulting price outcomes and includes some brief concluding remarks.

2.2 A National Synopsis

Australia is a federation in which constitutional control of water rests with the constituent states. Some changes have occurred with the management of water in the Murray-Darling basin, but in essence, states are responsible for any prices charged to water users. The 1980s and 1990s were a period of dramatic economic reform in Australia. International competitiveness and declining terms of trade were prominent in the minds of national governments. The general policy solutions involved increased openness to trade and a reconsideration of the role of the government generally. A series of competition reforms[1] were ushered in which included the

[1]Competition reform is a generic term that became popular in Australia in the 1980s. The notion involves a broad suite of policy changes ranging from modifications to the way banking regulation might limit international competition through to questions regarding the efficacy of state owner-

privatization of some utility services, such as telephone and in some cases electricity, and greater attention was given to prices and costs in those utilities that remained in public hands (Crase 2009).

Water is still largely managed within the public sector in each state, although the notion of corporatization also features prominently in several jurisdictions. In addition, there has been a marked expansion of contracting and market instruments in an effort to deliver greater efficiency within the sector. The advantages of better aligning prices and costs should not require detailed elucidation here. Nevertheless, in a country often typified as being the "driest on earth," the strengthening of incentives for a more cautious use of the resource and enhanced signaling for investment were seen as clear benefits. In addition, mounting evidence about overextraction in various basins and in some aquifers provided grounds for increased attention to water pricing. To facilitate this, the national reform agenda also included a commitment from states to introduce independent economic regulation as part of price-setting arrangements.

Several additional important national reforms that impact water pricing also warrant mention. First, under the NWI, states agreed that water prices should be based, in part and wherever practical, on volumetric use. This results in clearer signals to end users about the consequences of profligate use. Second, water rights were separated from land, and trade in rights was encouraged. Accordingly, water access and use rights are now regularly exchanged between larger users, such as irrigation farmers, environmental reserve managers, and, to a lesser extent, urban bulk water suppliers. Trade can only occur when there is hydrological connectivity, and administrative and legislative arrangements are in place to support market exchange. Nonetheless, it is important to understand that there are now different "prices" for water in Australia. On the one hand, there are a set of prices that relate to the operation of water markets, which covers trade of allocations (i.e., annual water access), prices that pertain to long-term entitlements (i.e., perpetual access rights), and a range of derivatives.[2] On the other hand, there is a set of water prices paid by end users. These prices relate to delivery services and infrastructure access in the case of irrigators, environmental reserve managers, and bulk urban water suppliers. In the case of residential and most commercial urban users, prices also cover access to the resource itself, since urban water is not generally unbundled and traded by this group.[3] Clearly, however, if charges for water access and use do not accurately reflect costs, then there will be distorting impacts in the water markets that allocate

ship of specific assets. The underling question that drove competition reform was the extent to which market competition might make Australian production uncompetitive in international settings. This is especially important because Australia is a small open economy heavily reliant on trade with the rest of the world to achieve and maintain high living standards.

[2] In the case of the latter, for instance, recently announced changes to federal regulations mean that formalized forward markets are now emerging (see, e.g., WaterFind 2014).

[3] In a small number of instances, urban water trade at the customer level arose during the prolonged drought at the beginning of the century.

bulk supplies. This matter has been addressed elsewhere (see, e.g., Crase et al. 2013b) but remains an important area requiring further analysis.

One final national "complication" in the context of water pricing relates specifically to developments in the Murray-Darling basin.[4] The main driver of policy change in this region has been the broad acceptance that water had been excessively allocated for consumptive use and that more water was needed for environmental purposes. This led to the creation of the Commonwealth Environmental Water Holder, a national agency that is now in command of a large volume of water acquired through a combination of market purchases, infrastructure-for-water swaps, and administrative changes to water entitlements. The water held by the Commonwealth Environmental Water Holder is deployed with the aim of restoring ecological processes within the basin[5] but there are costs associated with managing and monitoring this work. Currently, these costs are shared between the national and relevant state governments, though this remains contentious and the extent to which such costs should be passed to consumptive users is unresolved. More generally, the requirement that water prices should also cover the cost of water planning and management remains a work in progress in most jurisdictions.

2.3 Pricing Reform by Jurisdiction

We now turn to water pricing in different state jurisdictions in Australia. To reiterate, all jurisdictions are signatories to the NWI and have received funds from the federal government on the basis of the embedded commitments. In particular, states are required to have in place arrangements that promote the efficient use of water and thus align prices with costs. The range of costs recovered and the methodology for doing so is thus critical to the determination of prices faced by end users and this can vary between jurisdictions and between users within jurisdictions. The principles of cost recovery are nonetheless quite clear: there is an expectation that capital costs will be recovered, ultimately including a return to capital for many users; the user-pay principle applies; and the legitimate costs associated with water planning and management should also be met by end users. In this section, we draw heavily on the National Water Commission's (NWC) (2011) review of water pricing but supplement this with more contemporary detail where appropriate. We provide a brief description of each state to contextualize the price-setting processes and outcomes.

---

[4] Detailed descriptions of the policy activity within the Murray-Darling basin are available elsewhere (see, e.g., Crase 2012).

[5] This is not to say that the environmental ills of the basin have been "cured." Rather, the political solution for the time being involves a plan to deliver additional water and to achieve environmental restoration. This will undoubtedly be the source of additional political maneuvering as the basin plan is progressively implemented.

2.3.1 *New South Wales*

New South Wales (NSW) is Australia's most populous state (7.4 million) but, as with most jurisdictions, the population is concentrated in the coastal region close to the capital city (Sydney) (ABS 2013). Accordingly, the population is settled mostly east of the Great Dividing Range while west of the divide lies the Murray-Darling basin. NSW is a relatively large state by land area, comprising around 800,000 km^2, and the western portion of the state is generally arid and sparsely settled (Geoscience Australia 2010).

2.3.1.1 Urban Water Pricing

Urban water and sewerage services in metropolitan areas are provided by three state-owned metropolitan water utilities. The Sydney Catchment Authority manages bulk supplies to the Greater Sydney region with Sydney Water then fulfilling retail functions for residents within this area. Hunter Water operates to the north of Sydney, undertaking both bulk and retail functions for residents of Newcastle and proximate towns and cities. Water and sewerage prices are subject to economic regulation for these entities on the basis that they are monopolies, and rents and gold plating need to be kept in check. Economic regulation is undertaken by the Independent Pricing and Regulatory Tribunal (IPART), which also ensures license compliance.[6] Importantly, price determinations are binding, although the scope of the tribunal is determined by legislators.

The price-setting arrangements in these instances are built around entities producing sets of detailed plans for future infrastructure along with estimates of operating and maintenance expenditures and demand. IPART employs the long-run marginal cost (LRMC, sometimes called building-block) methodology. LRMC aims to estimate the cost of providing an extra unit of consumption, based on bringing forward the future capital program to efficiently balance supply and demand. On the basis of these costs, an entity's revenue requirement is set and then matched against anticipated demand. Because the NWI includes a commitment to two-part tariffs, with a volumetric charge signaling the impact of use, the estimate of demand forms a key part in determining the adequacy of the actual revenue received. Initially, Sydney Water opted for an inclining block tariff accompanied by a fixed fee, but this has now been simplified to a two-part tariff with a single volumetric charge, which also has economic efficiency advantages (for an explanation of alternative water tariff arrangements, see Crase et al. 2007).

Most residential customers face a fixed sewerage charge, based on the nature of their dwelling (i.e., stand-alone house versus unit or flat) largely because sewage is

[6] The tribunal also has the power to determine the prices charged by Gosford and Wyong Shire Councils, north of Sydney, and the water-related services of Country Energy in Broken Hill, located in the far west of the state.

not metered. Commercial customers face trade-waste charges set in line with the volume and level of contaminates in waste. In this instance, the volumes and constituents of waste are metered and monitored. The volumetric charge for water use is similar for most commercial purposes, although the fixed charge varies with the size of the water inlet to properties. Differential charges apply in some residential areas where water-recycling infrastructure has been put in place to provide non-potable water for gardens and other fit-for-purpose uses. Charges for using this alternative supply are set below potable water, primarily justified on the basis of the avoided costs associated with deferred potable supply augmentation (IPART 2011a). The average water prices paid by residential customers in selected locations in NSW appear in Table 2.1, along with details of pricing structures for a sample of large utilities in other jurisdictions.

A key driver of prices charged by regulated water utilities is the asset base associated with water and sewerage services and the cost of capital. In NSW, this requires an estimate of the regulatory asset base (RAB) for each business.[7] The initial RAB set for each business followed the "line-in-the-sand" process common when existing entities first enter a regulatory regime. Additions to the RAB should seemingly be straightforward thereafter, with only efficient capital expenditures approved and added to the base, but this is not always the case.

During the extended drought at the beginning of the twenty-first century, the NSW government intervened in several large infrastructure investments, thereby overriding the regulatory process. A desalination plant was constructed at Kurnell, and the Sydney Desalination Plant was established as a wholly owned subsidiary of Sydney Water. The costs of the desalination plant were thus initially reflected in the asset base of Sydney Water; however, a range of operating scenarios were subsequently investigated by IPART as dam inflows reduced the need for the plant to operate continuously (IPART 2011b). Subsequently, a change of government saw the desalination plant leased to private interests for 50 years, such that the leasing payments now form part of Sydney Water's operating costs (Malone 2013). The point is that arm's length economic regulation in NSW does not completely isolate the regulator from the preferences of legislators.

Government influence over water prices is arguably more overt in the entities not directly subjected to economic regulation by IPART. Water and sewerage services are provided by local governments outside the "regulated" metropolitan areas of NSW. There are around 100 of these local water utilities (LWUs) with each being "regulated" by a best-practice management framework administered by the NSW Office of Water. While entities regulated by IPART have been required to achieve what is known as "upper-bound" pricing, which involves a return to capital in addition to depreciation, the evidence on returns achieved by LWUs is mixed. For example, the NWC (2011, p. 27) noted that the proportion of LWUs in NSW generating a positive rate of return actually fell between 2005–2006 and 2008–2009

[7] The weighted average cost of capital and the chosen depreciation methodology are also major influences on costs. We highlight instances of the importance of these in our discussion of other jurisdictions.

Table 2.1 Example 2012/2013 tariff structures and charges

<table>
<tr><th>Area</th><th>Utility</th><th>Tariff structure</th><th>Fixed charge</th><th>Step usage charge/s ($/kl)</th><th>Annual bill[a]</th></tr>
<tr><td>ACT</td><td>ACTEW</td><td>Two-part tariff with 2-step inclining block</td><td>99.83</td><td>2.43|4.86</td><td>585.83</td></tr>
<tr><td>NSW</td><td>Hunter Water Corporation</td><td>Two-part tariff</td><td>18.92</td><td>2.08</td><td>434.92</td></tr>
<tr><td></td><td>Sydney Water Corporation</td><td>Two-part tariff</td><td>135.12</td><td>2.13</td><td>561.12</td></tr>
<tr><td></td><td>Wyong Shire Council</td><td>Two-part tariff</td><td>167.40</td><td>2.12</td><td>606.00</td></tr>
<tr><td>NT</td><td>Power and Water – Darwin</td><td>Two-part tariff</td><td>263.71</td><td>1.73</td><td>609.71</td></tr>
<tr><td>QLD</td><td>Gold Coast City Council</td><td>Two-part tariff</td><td>201.50</td><td>3.27</td><td>855.34</td></tr>
<tr><td></td><td>Townsville Water</td><td>Standard plan[b]</td><td>681.00</td><td></td><td>681.00</td></tr>
<tr><td></td><td>Toowoomba Regional Council</td><td>Two-part tariff with 2-step inclining block</td><td>590.00</td><td>2.10|3.30</td><td>1010.00</td></tr>
<tr><td></td><td>Unitywater</td><td>Two-part tariff with 3-step inclining block</td><td>292.97</td><td>2.37|3.04|3.50</td><td>746.09</td></tr>
<tr><td></td><td>Queensland Urban Utilities</td><td>Two-part tariff with 3-step inclining block</td><td>167.16</td><td>2.72|2.76|3.32</td><td>733.29</td></tr>
<tr><td>SA</td><td>SA Water – Adelaide</td><td>Two-part tariff with 3-step inclining block</td><td>293.00</td><td>2.42|3.45|3.73</td><td>897.40</td></tr>
<tr><td>VIC</td><td>Barwon Water</td><td>Two-part tariff</td><td>168.32</td><td>2.21</td><td>611.00</td></tr>
<tr><td></td><td>City West Water</td><td>Two-part tariff with 3-step inclining block</td><td>170.40</td><td>1.79|2.10|3.10</td><td>543.41</td></tr>
<tr><td></td><td>Coliban Water</td><td>Two-part tariff with 3-step inclining block</td><td>97.84</td><td>1.95|2.36|3.90</td><td>500.00</td></tr>
<tr><td></td><td>South East Water Ltd</td><td>Two-part tariff with 3-step inclining block</td><td>82.44</td><td>1.75|2.13|3.44</td><td>452.00</td></tr>
<tr><td></td><td>Western Water</td><td>Two-part tariff with 3-step inclining block</td><td>215.26</td><td>1.38|1.84|3.67</td><td>514.00</td></tr>
<tr><td></td><td>Yarra Valley Water</td><td>Two-part tariff with 3-step inclining block</td><td>120.26</td><td>1.78|2.08|3.08</td><td>488.00</td></tr>
<tr><td>WA</td><td>Water Corporation – Perth</td><td>Two-part tariff with 3-step inclining block</td><td>188.10</td><td>1.34|1.75|2.40</td><td>475.85</td></tr>
<tr><td>TAS</td><td>Cradle Mountain Water</td><td>Two-part tariff</td><td>384.49</td><td>0.90</td><td>564.49</td></tr>
</table>

Modified from National Water Commission (2014a, b)
[a]Based on 200 kL of residential water supplied
[b]Refer to section 2.3.3.1 for further detail

and the average return stood at only 0.6 %. In contrast, the rates of return set by IPART are usually around 6–7 %, with the state government being the beneficiary of these returns.

2.3.1.2 Rural Water Pricing

In Australia, the term "rural" water pricing relates to the charges imposed on irrigators, although in some jurisdictions, rural water agencies also provide bulk water to urban retailers. Many of the larger irrigation supply organizations in NSW are located in the Murray-Darling basin, and ownership and management of these entities was devolved to farmers during reforms of the 1990s. A key component for water prices paid by irrigators is the charges set for the delivery of water to the irrigation district, where it is then controlled by the irrigation infrastructure operator (IIO). In NSW, responsibility for the delivery of bulk water for irrigators resides with State Water Corporation, which initially had its charges regulated by IPART (see IPART 2010) but is currently regulated by the Australian Competition and Consumer Commission (ACCC). The methodology for estimating efficient costs is broadly similar to that applied for urban utilities, although the movement toward "upper-bound" pricing has been slower for this sector. Charges comprise a fixed fee, based on entitlement type, the valley where the entitlement is held, and the size of entitlement plus a variable charge.

A key difference in rural water pricing in NSW has been the significant progress made to isolate costs related to water planning and water management. These charges are also subject to independent review and, unlike neighboring jurisdictions, are specifically recovered from end users. Arrangements for cost recovery precede the ACCC's assumption of responsibilities for economic regulation in the Murray-Darling basin. For an environment in which water can be traded between jurisdictions, differences in charging regimes have been a source of contention between states, although the NWC (2011, p. 38) noted that the opportunity cost of water is the main determinant of farmer behavior rather than differences in bulk water charges.

Charges levied by State Water Corporation form the foundation of prices paid by farmers, but additional costs also derive from IIOs. As part of recent reforms in the Murray-Darling basin, the ACCC assumed additional responsibilities for monitoring and regulating prices charged by IIOs.[8] An important task of the ACCC was to establish prices and rules that related to irrigators selling their water entitlements to others outside the area controlled by an IIO. Initially, IIOs had imposed exit fees on these farmers, but the basis of those fees was considered to unfairly act against trade. The upshot was that water entitlements were further unbundled and delivery entitlements identified.

[8] This is limited to larger ISOs and those not subject to regulation by accredited regulators. The form of regulation is arguably more "light-handed" than that applied by IPART.

Delivery entitlements constitute a right to access irrigation infrastructure with a specified delivery capacity. Irrigators now have the option of selling water access entitlements and maintaining delivery rights or "shares," which in turn attract an annual charge. Alternatively, farmers can terminate their delivery shares and IIOs are constrained to charging no more than ten times the annual delivery share charges. These funds aim to compensate remaining irrigators for the increased cost of maintaining a network.

The annual prices paid by individual irrigators in NSW vary greatly in their complexity. For those irrigators not part of a communal scheme (i.e., pumping directly from rivers or aquifers), the charges levied by State Water Corporation and private pumping expenses represent the only pertinent costs. Farmers serviced by an IIO can expect to face account administration charges, delivery entitlement fees, fees related to outlets, drainage fees, standard water-use fees, and casual water-use fees (in which a premium is paid for exceeding an allocated entitlement).

Given that ownership of IIO assets was principally vested in private hands in NSW, the scope for gaining ongoing rents from government might be expected to be limited. However, the impacts of drought coincided with expanded government enthusiasm to reduce extractions in the Murray-Darling basin in the early 2000s, and this has placed at risk the cost recovery principles agreed in the NWI. The Commonwealth and, to a lesser extent, the NSW government, have undertaken so-called irrigation infrastructure renewal as part of a wider program to deliver more water for environmental purposes. The accounting that relates to these investments and the deflating impacts on prices paid by irrigators is important but potentially less problematic in NSW than for other jurisdictions in which IIOs remain in public hands.

2.3.1.3 Environmental Water Pricing

To understand the prices paid for environmental water, it is important to distinguish the different forms. First, some water is held by state agencies for environmental purposes and is based on the operating rules for regulated streams. This is often called "rule-based" water. For example, a volume of water might be held in storage for servicing a wetland and released, subject to downstream flow parameters being reached. Generally, this type of environmental water is not subject to management fees. Second, separate volumes of water have been acquired by environmental agencies that were previously assigned to consumptive uses. This is often called "held" water. The rules that govern water trade in the Murray-Darling basin mean that this second form of "e-water" carries similar costs and constraints that attended the rights when held in private hands. Accordingly, the environmental agency that owns e-water must also meet the statutory charges imposed by State Water Corporation in the storage and release of that water. The agency that "owns" most e-water on behalf of NSW is known as RiverBank, although the holdings of the Commonwealth Environmental Water Holder vastly outstrip those secured by RiverBank, meaning the

Commonwealth is obliged to pay fees and charges to NSW State Water Corporation for its e-water holdings.
RiverBank has historically reduced some of its water management costs by trading the allocations that accrue to e-water when it is deemed surplus to environmental need (DECCW 2010), and this approach is now being tentatively pursued by the Commonwealth.

2.3.2 *Victoria*

Victoria is Australia's second most populous state (5.8 million) but covers a much smaller land area than NSW (around 230,000 km^2) (ABS 2013; Geoscience Australia 2010). Like NSW, the capital city (Melbourne) is densely settled—by Australian standards—and enjoys a coastal location. The metropolitan area is also separated from the Murray-Darling basin and lies south of the Great Dividing Range, although some hydrological connectivity exists since the construction of pipeline linking Melbourne to the irrigation water supplies in the north of the state.

2.3.2.1 Urban Water Pricing

The institutional arrangements for urban water pricing share some similarities with those described for NSW, but there are also important differences. First, all of Victoria's urban water suppliers are in the form of water corporations owned by the state government, not simply those in metropolitan areas. In the metropolitan area, Melbourne Water undertakes responsibilities for bulk water supply and bulk sewerage services and also manages the drainage systems in the regions. Retail water and sewerage services in the metropolitan area reside with three entities—South East Water, Yarra Valley Water, and City West Water. An additional 13 regional water utilities operate outside the metropolitan area, many controlling their own bulk water supplies. All urban entities are subject to economic regulation and must have water plans approved by the Essential Services Commission (ESC), which manifests in price determinations lasting 5 years.

Second, like IPART, the ESC favors the building-block approach when reviewing water and sewerage prices, but the establishment of the RAB and related parameters differ in some instances. To illustrate the importance of these differences, we briefly examine the alternative principles that circumscribe asset valuations for water utilities within Victoria. The minister for water initially set the RABs of urban water businesses in 2004. Consistent with their designation as "for-profit businesses," the RABs of metropolitan businesses were initially set above the corresponding statutory values. By comparison, the RABs of the 13 "not-for-profit" regional urban water businesses were initially set below statutory values (Pawsey and Crase 2014).

Unsurprisingly, given the different approach taken in establishing opening RABs, the financial performance of metropolitan and regional urban businesses has been contrasting. As reviewed by Pawsey (2014), over the period 2005/2006–2012/2013, the average annual before tax profits of all metropolitan water businesses exceeded $A50 million.[9] These reported profits permitted metropolitan water businesses to return a total of $A1.3 billion in dividend payments and $A774 million in income tax payments to the state. By comparison, over the same period, only two regional urban businesses had an average annual before tax profit of more than $A5 million, and many reported average before tax losses. Furthermore, only three regional urban water businesses made any dividend and/or income tax payments to the state.

Notwithstanding these differences, all Victorian urban water utilities are reported as being "substantially compliant" with the notion of "upper-bound" pricing (NWC 2011, p. 25). But it is difficult to reconcile the stark differences between the treatment of metropolitan consumers and regional/rural urban water users. One of the basic tenets of LRMC pricing is that infrastructure augmentation should occur on the basis of economic merit. Put simply, economic regulation should ensure that the most cost-effective augmentation works are supported first. This has been broadly true for regional utilities in Victoria but is not the case for Melbourne. For example, the pipeline that links Melbourne with irrigation water north of the divide was constructed at the height of the drought and now represents low-cost water for Melbournians. However, in November 2012, it was announced that the pipeline could only be used to boost Melbourne's water supply during times of "critical human need," and this was defined as a period when water storage is below 30 % on 30 November (Office of Living Victoria 2013, p. 14).[10] It is worth noting that the commissioning of the desalination plant in Wonthaggi in 2012 means that the minister for water is at liberty to order up to 150 Gl of water in April of any year, implying that the "30 % at 30 November" dam threshold will likely be met in all but the most extraordinary years (Crase et al. 2014b).

Interventions like these bring into question the extent to which arm's length economic regulation can lead to efficient pricing outcomes in metropolitan Victoria. Similarly, in January 2014, the minister for water announced that the government would undertake a review of water prices even though the economic regulator completed its price determinations the previous year. The review titled "Fairer Water Bills" was launched leading up to the state election and included a commitment to "lower water bills in future despite labor's [i.e., the previous government's] legacy of waste and mismanagement" (Walsh 2014, p. 1). An integral component of the "Fairer Water Bills" initiative is a strong commitment to foster integrated urban water management, the costs and benefits of which have not been publicly tested. It is not clear the extent to which the ESC will be given responsibility for regulating price increases should they be deemed necessary after the election.

One of the major achievements of the ESC in its recent round of determinations has been to engender greater innovation among water retailers in tariff design. The fact that there are three retailers in Melbourne has encouraged a sense of competi-

[9] $A1 = $US0.91 in September 2014. Throughout most of the last decade, the Australian dollar varies from close to parity with the $US to about $US0.9.

[10] There is also a minor provision to draw water from the pipeline for firefighting purposes.

tion by comparison and this has recently shifted focus onto customer satisfaction and value. Thus, while retailers continue to offer water charges that comprise a fixed service fee and a volumetric tariff based on use,[11] there is considerable experimentation on this front. For example, Yarra Valley Water announced in April 2014 that it would pilot a "volumetric-only" tariff for customers wishing to engage in the trial (YVW 2012). The motivation for the pilot appears to be discontent among some customers that radical reductions in their household water use in response to conservation messages from government had yielded only modest financial savings.

2.3.2.2 Rural Water Prices

Unlike NSW, Victoria's irrigation infrastructure largely remains in public hands. Bulk water is supplied by several government-owned corporations, some of which also act as IIOs. Goulburn-Murray Water is the largest of these entities and provides bulk water and irrigation services in the north of the state. Prices set by Goulburn-Murray Water (and other state-owned IIOs) are subject to economic regulation by the ESC and, as with most water businesses, are heavily influenced by the asset base, as measured by the RAB. The opening RABs of rural water businesses were, however, initially set in 2004 at zero by the minister for water (VAGO 2013). Similarly, the so-called gifting of assets (i.e., government-subsidizing infrastructure provision) creates an additional long-term conundrum for generating even lower-bound prices.

We noted that in NSW, the Commonwealth and state governments had embarked on programs that subsidized irrigation infrastructure upgrades, in part to secure water access for environmental ends. This approach has been particularly prevalent in Victoria, where Goulburn-Murray Water has been the beneficiary of around $A2 billion of public investment in recent years (Crase et al. 2013a). Setting aside the cost of this policy approach and the potential for miscalculating water savings[12] (see Perry 2009), the impact on current prices paid for water services and the long-term consequences of underfunding are of concern. Since infrastructure that is "gifted" by government does not add to the RAB, it follows that insufficient monies are currently being collected to fund the depreciation of those gifted assets, let alone generate a positive rate of return.

---

[11] Two of the retailers offer an inclining block tariff, while the other employs a single-step usage rate.

[12] One of the major challenges with this policy approach is that it potentially double-counts water savings. In Australia, this is further complicated by the way irrigation entitlements are specified as "gross" entitlements that take little account of the impacts of return flows on downstream users. Thus, when a farmer "saves" water by increasing localized water-use efficiency, there is a real risk that other existing beneficiaries are deprived of water. Ironically, this stands to undermine efforts to improve environmental outcomes insomuch as environmental water uses are often third-party recipients of "inefficient" irrigation practices.

The management of Goulburn-Murray Water has been cognizant of the emerging challenges on this front and, in 2013, set about to reform its complex tariff regime and establish prices that better captured the benefits of the new infrastructure.[13] However, there is much ground to be made up, and Pawsey and Crase (2013) estimate that prices would need to increase by about 300 % to achieve upper-bound pricing.

2.3.2.3 Environmental Water Pricing

Victoria has access to both rule-based and "held" water, with the latter vested in the Victorian Environmental Water Holder (VEWH). Operations of the VEWH in the Murray-Darling basin are subject to the same conditions as those described for NSW.

As part of the irrigation infrastructure renewal projects in northern Victoria, some marginal irrigation networks were closed. For example, the Campaspe system was decommissioned and, together with the Commonwealth, the VEWH now holds most entitlements on that system. Peculiarly, both environmental water holders now find themselves paying fees for dam managers to release water in a manner that replicates the absence of the dam.

At a broader level, Victoria has been criticized by the NWC for failing to adequately establish the costs of water planning and management and attributing these to users (NWC 2011). Rather, the Victorian water utilities impose a so-called environmental contribution, which is set at 5 % of revenue for all urban water utilities and about half that for rural water utilities. Funds are appropriated as general revenue for the state. In metropolitan areas, a "park-and-garden" charge is also directly levied on water users and distributed to Melbourne Water for the management of waterways, the Botanic Gardens and Parks Victoria, which manages environmental and recreation sites near Melbourne.

2.3.3 Queensland

Queensland (Qld) is a large state with a land area in excess of 1.7 km^2. It has considerable climatic variation with tropical climates in the north and subtropical climates in the south. The inland is much dryer than coastal areas, and the southwestern portion of the state lies within the Murray-Darling basin. The population (4.7 million) is heavily concentrated in the southeast corner, near Brisbane and the Gold Coast, which continues to grow rapidly, in part from migration from other states (ABS 2013; Geoscience Australia 2010).

---

[13] A detailed assessment of farmer responses to tariff reform in this context is available at Crase et al. (2014a).

2.3.3.1 Urban Water Pricing

Like NSW, the pricing of urban water varies with the institutional backdrop, which is, in turn, determined by proximity to the metropolitan region. In the metropolitan areas that occupy Brisbane and the Gold Coast, an entity known as SEQ Water provides bulk water, while retail services are provided by local governments in the region. During the drought in the early 2000s, the Queensland and Commonwealth governments constructed a "water grid," which included a desalination plant and connectivity between remote storages. In 2008, the state government committed to price increases in bulk water to reflect these costs, but they were to be phased in over 10 years. Bulk water is charged on a volumetric basis only and appears separately on water users' accounts. Subsequently, the state government adjusted the price path such that different councils will meet full-cost recovery for bulk water at different times (DEWS 2014, p. 1). The shortfall in revenue is funded by SEQ Water debt, and while these arrangements are transparent, the NWC notes that the outcome is "inconsistent with Queensland's commitment to implement upper-bound pricing in metropolitan areas" (NWC 2011, p. 24).

Similar inconsistencies with the intent of the NWI are evident in the tariff regimes deployed by local governments with different rates applied to "business" customers and residential users. Sewerage charges are based on fixed access fees, and water charges comprise a fixed fee and a three-tier inclining block tariff.

The bulk water charges are subject to economic oversight via the Queensland Competition Authority (QCA), although its rulings are not binding and stand as recommendations to government. Retail charges are simply "monitored" by the Authority "to assess whether households and businesses are paying a price that is comparable with the costs of providing the relevant services" (QCA 2014, p. 1).

Beyond southeastern Qld, a further 71 "water service providers" deliver water and sewerage services to urban areas. Most of these (62) are local governments (NWC 2013, p. 149) and most administer a two-part tariff, including a component that reflects consumption. Townsville, in northern Qld, was noted by the NWC (2011, p. 29) as one of the remaining water providers that had not moved to consumption-based pricing and retained a water "allowance" per property. The NWC further noted that attempts to introduce a uniform two-part tariff resulted in some community opposition and the Qld government then advised the local government that it was not required to adjust its tariff regime. In 2014–2015, households in Townsville paid a standard fee of around $A700 per annum and were permitted to use up to 772 kl[14] before attracting a volumetric charge (Townsville City Council 2014, p. 1). The NWC (2011, p. 29) noted that "this example highlights a lack of commitment to the principle of pricing reform and a lack of enforcement powers at the state or national level."

[14] A kiloliter is one cubic meter.

2.3.3.2 Rural Water Pricing

Qld has two government-owned entities involved in the distribution of irrigation water and irrigation services. Both are subjected to economic oversight by the QCA although, again, recommendations are not binding on government. QCA's pricing reviews cover a 5-year period, and irrigation prices are presently set for 2012–2017. SunWater provides water to regional interests (including mining), while SEQ Water also services nonurban customers, notably irrigators, in southeastern Qld.

In the most recent price reviews undertaken by QCA, the government directed that prices be established that "reflect efficient operational, maintenance and administrative costs, and prudent and efficient expenditure on renewing and rehabilitating existing assets through a renewals annuity. Prices are to exclude dam safety and metering upgrade costs related to changes in national standards, and any rate of return on existing assets" (QCA 2013, p. xxi). The government also directed that while irrigation prices would likely break even with lower-bound pricing, any shortfall would be "expected to be paid by government in the form of a community service obligation (CSO)" (QCA 2013, p. xxi).

During the drought, the Qld government introduced fixed-charge drought-relief measures, but the most recent recommendations from QCA have been accepted by government and allow for a transition to a two-part tariff. Collectively, these revenues are expected to generally match lower-bound costs. The fixed component of charges approximates about 90 % of the revenue, although CSOs also make up a substantive contribution to revenue.

2.3.3.3 Environmental Water Pricing

Most provisions for environmental water in Qld are in the form of rule-based allocations. These are detailed in water resource management plans. The costs of water planning and management are purportedly embedded in the prices paid by end users. Nevertheless, the NWC (2011, p. 42) noted that on a statewide basis, only about 5 % of the costs associated with water planning and management activities are recovered via end users.

2.3.4 Western Australia

Western Australia (WA) is a vast land area of around 2.5 million square kilometers and has a population of around 2.5 million, mostly located in the southwest near the capital city, Perth (ABS 2013; Geoscience Australia 2010). The southwest has experienced marked declines in rainfall over the past three decades (CSIRO 2005). The metropolitan region has historically been heavily reliant of groundwater supplies, and Perth now has two desalination plants. Inland areas of the state are generally arid, the north is tropical and subtropical, and the southwest is temperate.

2.3.4.1 Urban Water Pricing

Almost all urban water supplies, sewerage services, bulk water, and irrigation are administered through a single government-owned entity, the WA Water Corporation. Two additional entities, the Bunbury and Busselton Water Board, are self-funded statutory authorities that provide water and sewerage services to their district populations, south of Perth. Historically, WA had applied postage stamp pricing for water services across the state. More recently, an attempt has been made to divide nonmetropolitan towns into classes, based on the cost of extracting, treating, and distributing water. There are five classes of town across the state and a single-tariff regime for the metropolitan area (Water Corporation 2014).

Water tariffs comprise a fixed service fee and a variable, three-tier inclining block component based on usage. Sewerage tariffs are based on the gross rental value of properties, and a minimum amount is set for metropolitan users, while an upper and lower bound applies in country areas.

Tariffs are subject to economic oversight by the Economic Regulation Authority (ERA), although its recommendations must be approved by the government.

2.3.4.2 Rural Water Pricing

Irrigation activity in WA is restricted to the southwest (Harvey Water Irrigation Area; Preston Valley Irrigation Cooperative) and the far northwest (Ord Irrigation Cooperative; Gascoyne Water Cooperative). The older of these entities were vested in farmers as part of the reforms in the late 1990s, and fees and charges are levied on members/owners. The tariff structure in the Ord comprises a flat fee, partially based on land area, and a single volumetric fee, although a surcharge applies if pumping is required (Ord Irrigation 2014). Harvey Water tariffs are more complex, in part because it must recoup funds from users to pay the Water Corporation for storage and dam safety services. Payment for these services is included within the fixed component of fees, as is a surcharge for access to pressurized supply via a pipeline. A variable charge also applies and is based on volumetric use.

The ERA has oversight of prices and receives a written submission on water charges from each IIO, subsequently embedding these in the operating licenses of each entity.

2.3.4.3 Environmental Water Pricing

WA does not currently specify or recoup from end users the costs associated with water planning and water management for environmental purposes (NWC 2011, p. 43). Nonetheless, in 2011, the ERA undertook a review and identified the efficient costs related to such activities and recommended that they be phased in over a 3-year period (ERA 2011). The interface between mining activities and water resources is also contentious in this jurisdiction and remains largely a work in progress (see, e.g., Department of Water 2013).

2.3.5 *South Australia*

The South Australian (SA) land area exceeds that of NSW at around 980,000 km^2, although the population is only about 20 % of that of NSW, at 1.7 million (ABS 2013; Geoscience Australia 2010). Again, most residents are clustered in a zone close to the capital city, in this case Adelaide. The capital relies to some extent on water pumped from the River Murray, and the southeast corner of the state also sits within the Murray-Darling basin. The inland and western zones are generally arid.

2.3.5.1 Urban Water Pricing

Water and sewerage services are provided by a single state entity, known as SA Water. Urban water prices in SA are largely uniform across the state (i.e., so-called postage stamp pricing). The gap between cost recovery from users in regional areas and the upper-bound revenue requirement is funded by the state government as a CSO. Economic regulation of prices is vested in the Essential Services Commission of South Australia (ESCOSA), but as with Qld and WA, the determinations of the commission are not strictly binding and prices are set "with government." Progress to distance the economic regulator from government was made in May 2013, when ESCOSA released its first "independent" determination of the maximum allowable revenues that could be collected to cover upper-bound costs[15] (ESCOSA 2013). However, illustrative of the pervasive influence of government in this arena, the premier and minister for water announced in the same month that rebates from water bills were to be increased for low-income earners and pensioners[16] (see Weatherill and Hunter 2013).

The water prices for residential customers in SA comprise a fixed charge and a usage charge that is made up of three tiers—prices increase as usage exceeds the relevant threshold (i.e., an inclining block tariff). As in NSW, some residential areas are serviced by a dual pipe system that supplies recycled water (e.g., Mawson Lakes). Consumption of this water is priced below the lowest tier for potable water (see SA Water 2014b).

Business customers pay a single-usage tariff that is almost equal to the highest block tariff for residential users. The fixed component of charges for nonresidential

---

[15]The first determination covers a three-year period. The upper revenue bound in SA is made up of operating costs, depreciation, and a return on assets. The latter was set at a pretax WACC of 6 % in 2008–2009 (NWC 2011, p. 26) but revised to 4.5 % in the most recent determination (ESCOSA 2013).

[16]A rebate was similarly announced a year earlier.

users is based on either a flat rate (of about $70 per quarter) or a fraction of the land valuation, whichever is higher. Some commercial premises are also subject to trade-waste charges. The sewerage charge for residential customers in SA is based on the highest of either $80 per quarter or a portion of the property value.

As with other jurisdictions impacted by drought in the 2000s, the SA government opted to construct a desalination plant to shore up potable supplies, with financial assistance from the Commonwealth. The Commonwealth contribution of $328 million (Department of Environment—Commonwealth 2013) sits outside the RAB that drives water prices, and as with other desalination plants in the eastern states, the appropriate operating costs for now "month-balled" assets remain contentious (see, e.g., SA Water 2014a).

2.3.5.2 Rural Water Pricing

The majority of SA's irrigation sector is managed through privately owned irrigation trusts. The largest of these is the Central Irrigation Trust (CIT), which manages delivery of water to ten irrigation districts via pumping infrastructure on the River Murray (DPIRSA 2013). Charges comprise a fixed service fee, based on the size of the water delivery right, and a volumetric usage fee. The usage fee varies according to time of use (i.e., peak/off-peak) and the pressure associated with delivery (i.e., low, medium, high, high lift high pressure) (CIT 2014).

As with other jurisdictions with irrigation interests in the Murray-Darling basin, SA's irrigators have accessed public funds to upgrade infrastructure. The Private Irrigation Infrastructure Program for South Australia was nominated by the SA government as a priority project to draw monies from the Commonwealth's Sustainable Rural Water Use and Infrastructure Program. As with irrigation entities in NSW, the gifting of capital necessarily deflates current prices, though the legacy issues are only problematic for the state to the extent that future governments offer to refurbish run-down but private assets with more public monies.

2.3.5.3 Environmental Water Pricing

SA applies a natural resource management water levy on all water license holders in the Murray-Darling basin region of the state. These charges are based on the size of water access entitlements or the allocation or use, depending on district and type of activity. In addition, the SA minister for water imposes a "Save the River Murray" levy on all customers of SA water. The levy currently sits at about $10 per quarter for residential customers and $40 per quarter for nonresidential customers (Department of Environment Water and Natural Resources 2013, p. 7). In updating these charges, the SA government ostensibly fulfilled its obligation under the NWI to make water planning and management costs transparent, although it is not clear that the charges have yet been subject to independent review.

2.3.6 *Tasmania*

Tasmania is an island state south of the mainland with a modest population of about half a million and land area of 68,000 km^2 (Geoscience Australia 2010). The state enjoys a cool/temperate climate and relatively high rainfall. Hydroelectricity is a major user of water.

2.3.6.1 Urban Water Pricing

Major reforms in urban water and sewerage services occurred in Tasmania in 2009, with services being delivered by three regional corporations owned by constituent local governments. Each corporation also became subject to economic regulation with the Office of Tasmanian Economic Regulation making its first independent price determination in 2012 (OOTER 2014). The initial regulatory period was for 3 years. In 2013, corporations agreed to form a single entity, TasWater, which bills customers in line with the initial price determinations in 2012.

Water tariffs aim to comprise a fixed and variable component, although metering has not yet been universally installed throughout the state. Where meters exist, users face a single tariff per kiloliter, although the charge varies with water quality. Non-potable supplies are set at about 70 % of the potable rate and when water quality declines and boil-water notices are issues, the lower rate applies. For customers with unmetered properties, prices are based on the size of the water inlet to the property. Sewerage charges are fixed and based on an estimate of equivalent tenements (TasWater 2014).

Overall, rates of return to the water utilities remain significantly below full-cost recovery. In addition, political decisions about the accounting treatment of assets undermine efforts to put the sector on a firmer footing. For example, OOTER (2014, pp. vi–vii) notes that:

> In terms of the corporations' long-term financial stability, the fact that all three water and sewerage corporations have been required to adopt 'impaired' asset values means that current levels of revenue are insufficient to fund the repair and replacement of existing assets. Without increases in revenue the corporations are not financially sustainable in the long-run based on their existing assets, let alone being able to fund the significant capital expenditure required to meet environmental and public health regulatory requirements.

2.3.6.2 Rural Water Pricing

Tasmanian Irrigation Pty Ltd (TI) was established in 2011 as a state-owned enterprise with the aim of developing and managing irrigation schemes across the state. Irrigation is relatively undeveloped in this jurisdiction and state and Commonwealth governments have set aside $220 million to progress irrigation projects. Such projects are viewed as public-private partnerships, with the private contribution coming in the form of the purchase of tradable water entitlements within schemes. It is envisaged that lower-bound pricing will be achieved with TI (2014) noting that "[o]ngoing operating costs, including provision for asset renewal, will not be subsidized and will be met by annual charges levied on water entitlement holders."

Prices vary between established schemes with most opting for a two-part tariff, with the fixed component based on entitlements held at the commencement of the irrigation season and the variable charge related to water delivered during the season (TI 2012a). In some instance (e.g., Lower South Esk Irrigation Scheme), a fixed charge is levied, based on entitlements and unused entitlements, and then attracts a rebate, set at about half the fixed charge, at the end of the season (TI 2012b). OOTER is restricted to urban water regulation and does not regulate irrigation prices.

2.3.6.3 Environmental Water Pricing

The NWC (2011, p. 43) noted that Tasmania specifies charges that relate to environmental water considerations, and these are paid by license holders. It is not clear if these charges have been reviewed by OOTER.

2.3.7 Australian Capital Territory

The Australian Capital Territory (ACT) is a separate jurisdiction that is circumscribed by NSW and houses the national capital, Canberra. The population, of around 380,000, is primarily urban and there are few substantive irrigation interests (ABS 2013).

2.3.7.1 Urban Water Pricing

Water and sewerage services are provided in the ACT by ACTEW, an unlisted public company owned by the ACT government. The company also operates gas and electricity distribution facilities through a joint partnership with commercial interests. The prices set for water and sewerage services are subject to economic regulation via the Independent Competition and Regulatory Commission (see ICRC 2013a). Illustrative of the challenge of regulators meeting competing goals, the commission recently modified its traditional approach to water pricing following ACTEW's extensive capital works in the wake of the 2000s drought. Ideally, water assets should be paid for by the generation of beneficiaries, implying long-lived assets would be paid for over a long period of time. However, the price direction issued in 2013 noted that "the Commission's analysis found that it was not possible to transfer the burden of ACTEW's costs to future water users without an unacceptable risk to ACTEW's financial viability." Similarly, in order to balance the impacts of higher prices with equity concerns, the commission opted to reduce the rate of return to the ACT government and put in place a price path toward higher rates of return in future (ICRC 2013b).

Water prices are levied as a fixed fee with a two-tier inclining block tariff applicable to water use. The water use is based on average daily consumption over the billing cycle. Sewerage charges are levied at a flat rate, based on the nature of dwellings (ACTEW 2014a).

2.3.7.2 Environmental Water Pricing

The ACT government imposes a water abstraction charge set at about $A0.50 per kiloliter for urban use and around half this rate for nonurban uses. The charge purports to cover costs related to catchment maintenance and related government expenditure, reflect the scarcity value of water, and capture environmental effects (ACTEW 2014b).

2.3.8 Northern Territory

The Northern Territory (NT) is Australia's smallest jurisdiction by population (around 240,000) but nonetheless has an extensive land mass of 1.3 million square kilometers (ABS 2013; Geoscience Australia 2010). The NT is also home to a large portion of Australia's indigenous population, some of whom live in isolated settlements located at considerable distances from the capital, Darwin.

2.3.8.1 Urban Water Pricing

Prices for urban water are set directly by the NT government, which owns and operates the combined Power and Water Corporation. The water tariff comprises a fixed fee, based on the size of the connection or meter, and a single volumetric charge. Sewerage services are levied at a flat rate on properties with access, regardless of connection. Prices are similar for residential and commercial users. The NWC (2011, p. 25) contends that revenues were sufficient in the metropolitan area to meet lower-bound cost recovery only, although prices were increased substantially in 2012. The subsequent election of a new government saw seemingly arbitrary reductions in water and sewerage charges (see Giles 2013). The provision of water and sewerage services in remote communities is largely supported by funding from state and Commonwealth governments. The NT Power and Water Corporation operates a not-for-profit subsidiary known as Indigenous Essential Services for these purposes (PWC 2014).

2.3.8.2 Rural Water Pricing

Large-scale irrigation is uncommon in the NT, although the planned expansion of the Ord irrigation scheme in WA would see its extension into NT. Contributions to this project from Commonwealth and WA governments were announced in late 2012 (DPIF 2014).

2.3.8.3 Environmental Water Pricing

No discernible prices for environmental water have been developed in NT (see NWC 2011), and "water resources are generally considered to be under relatively little pressure due to a comparatively small population base and low intensity of land use" (DLRM 2014).

2.4 Summary and Concluding Remarks

This chapter offers a concise overview of water pricing arrangements in Australia. The analysis reveals considerable variation by jurisdiction, notwithstanding shared commitments to important principles, like full-cost recovery. Some of these differences are a manifestation of the demographic, geographic, and hydrological contexts and the institutional apparatus that matches those settings. It is simply not possible to administer identical pricing arrangements in a country with such stark variations.

Nonetheless, there are also common and sometimes worrying similarities. The proclivity for political will to wane under pressure of drought is a common theme across jurisdictions, especially evident in those states with agrarian interest in the Murray-Darling basin. The enthusiasm for political intervention to shore up urban water supplies and to circumvent planning that would lead to increased scrutiny of costs is also evident in the Australian experience of drought. In addition, it is clear that legislative arrangements that seek to establish arm's length economic regulation cannot completely insulate against the vagaries of legislative intervention.

These trends are problematic on several fronts. First, Australia's water reforms of the last three decades have focused heavily on ensuring economic incentives are in place that support judicious management. Water markets were introduced with the view that water would not be held in less-efficient uses and would move, over time, to deliver the greatest net benefit to society. This should have resulted in greater coherence between water allocation among the competing demands of urban users, agriculture, and environmental interests. In practice, political intervention with water pricing limits the capacity of these wider policy instruments to take hold. In simple terms, irrigators in existing communal irrigation districts are advantaged by subsidies that drive down the charges faced for water use while metropolitan water users, in particular, often pay higher-than-cost prices. The extent to which water is bid toward environmental interests is also influenced by the manner in which "held" water attracts charges, namely, paying water prices that reflect the usefulness of the water for agriculture and not the environment. In sum, the nuances of water pricing place a constraint on the way water markets are supposed to operate.

The Australian experience also highlights areas where economic regulation of water needs additional research, especially in locations where water resource availability is so variable. The widespread use of LRMC as the basis for setting revenue requirements for utilities resulted in under-recovery of costs during drought and over-recovery during wetter years.

Nonetheless, overall Australian jurisdictions have made substantial progress to price reform in water, and the innovations now emerging in competitive environments offer at least some promise and lessons for others.

References

ACTEW. (2014a). *Customer accounts and services*. Retrieved March 12, 2014, from http://www.actew.com.au/Customer%20Accounts%20and%20Services/Pricing%20and%20your%20Customer%20Contract/Water%20and%20sewerage%20FAQs.aspx

ACTEW. (2014b). *Water abstraction charge*. Retrieved March 12, 2014, from http://www.actew.com.au/Customer%20Accounts%20and%20Services/Pricing%20and%20your%20Customer%20Contract/ACT%20network%20water%20prices/Water%20Abstraction%20Charge.aspx

Australian Bureau of Statistics. (2013). *Australian demographic statistics*. Retrieved January 10, 2014, from http://www.abs.gov.au/ausstats/abs@.nsf/mf/3101.0

Central Irrigation Trust. (2014). *CIT water prices 2013–14*. Retrieved January 15, 2014, from http://cit.org.au/

Crase, L. (2009). Water policy in Australia: The impact of change and uncertainty. In A. Dinar & J. Albiac (Eds.), *Policy and strategic behavior in water resource management* (pp. 91–108). London: EarthScan Publishing.

Crase, L. (2012). The Murray-Darling basin plan: An adaptive response to ongoing challenges. *Economic Papers, 31*(3), 318–326.

Crase, L., O'Keefe, S., et al. (2007). Inclining block tariffs in Australia's urban water sector: Theoretical, political and practical considerations. *Agenda, 14*(1), 69–80.

Crase, L., O'Keefe, S., et al. (2013a). Talk is cheap, or is it? The cost of consulting about uncertain reallocation of water in the Murray-Darling basin, Australia. *Ecological Economics, 88*(3), 206–213.

Crase, L., O'Keefe, S., & Pawsey, N. (2013b). A note on contradictions in Australian water policy. *Economic Papers, 32*(3), 353–359.

Crase, L., Cooper, B., et al. (2014a). *Cost reflective pricing as a market-based approach to water management: Empirical insights to overcome farmer resistance to price reform*. European Association of environmental and resource economists annual conference, Istanbul.

Crase, L., Pawsey, N., et al. (2014b). *The north-south pipeline: A case of hydraulic autarky*. Australian agricultural and resource economics annual conference, Port Macquarie.

CSIRO. (2005). *Climate change is real* (CSIRO Fact Sheet). Canberra: CSIRO.

Department of Agriculture Fisheries and Forestry and Department of the Environment Water Heritage and the Arts. (2010). *Natural Resource Management Ministerial Council—seventeenth meeting. Record and resolutions*. Darwin: Commonwealth of Australia.

Department of Energy and Water Supply (Qld). (2014). *Bulk water prices*. Retrieved January 12, 2014, from http://www.dews.qld.gov.au/policies-initiatives/water-sector-reform/water-pricing/bulk-water-prices

Department of Environment (Commonwealth). (2013). *Adelaide desalination plant*. Retrieved February 20, 2014, from http://www.environment.gov.au/node/25461

Department of Environment Climate Change and Water (NSW). (2010). *NSW RiverBank Business Plan 2006–2011*. Sydney: DECCW.

Department of Environment Water and Natural Resources (SA). (2013). *Murray-Darling basin regulated water charges*. Adelaide: DEWNR.

Department of Land Resource Management (NT). (2014). *Water resources*. Retrieved March 13, 2014, from http://www.lrm.nt.gov.au/water#.U1hZbFWSx8F

Department of Primary Industries and Fisheries (NT). (2014). *Ord development project*. Retrieved March 10, 2014, from http://www.nt.gov.au/d/Primary_Industry/index.cfm?header=Ord%20Development%20Project

Department of Primary Industries and Regions (SA). (2013). *Water intensive development*. Retrieved January 11, 2014, from http://www.pir.sa.gov.au/wid/regions/river_murray/water_considerations/irrigation_trusts

Department of Water (WA). (2013). *Western Australian water in mining guideline* (Water licensing delivery series, 12). Perth: Department of Water.

Economic Regulation Authority (WA). (2011). *Inquiry into water resource management and planning charges*. Perth: ERA.

Essential Services Commission of South Australia. (2013). *Final determination of SA water's water and sewerage revenues: 2013/14–2015/16*. Adelaide: ESCOSA.

Geoscience Australia. (2010). *Areas of Australia – States and Territories*. Retrieved January 10, 2014, from http://www.ga.gov.au/education/geoscience-basics/dimensions/area-of-australia-states-and-territories.html

Giles, M. (2013). *Giles acts on power and water concerns – media release*. Retrieved January 14, 2014, from http://newsroom.nt.gov.au/adminmedia/mailouts/10494/attachments/130314%20-%20Giles%20Acts%20on%20power%20and%20Water%20Concerns.pdf#search=%22water%22

Independent Competition and Regulatory Commission (ACT). (2013a). *Regulated water and sewerage services – Final report*. Canberra: ICRC.

Independent Competition and Regulatory Commission (ACT). (2013b). *Releases of final report and price direction – Regulated water and sewerage services*. Media release. Retrieved January 4, 2014, from http://www.icrc.act.gov.au/wp-content/uploads/2013/02/Final-report-media-release.pdf

Independent Pricing and Regulatory Tribunal (NSW). (2010). *Review of water bulk water charges for State Water Corporation – July 2010 to June 2014*. Sydney: IPART.

Independent Pricing and Regulatory Tribunal (NSW). (2011a). *Assessment process for recycled water scheme avoided cost*. Sydney: IPART.

Independent Pricing and Regulatory Tribunal (NSW). (2011b). *Review of water prices for Sydney Desalination Plant Pty Ltd*. Sydney: IPART.

Malone, U. (2013). *New South Wales desalination plant deal to cost consumers $10 billion over 50 years*. Retrieved September 27, 2013, from http://www.abc.net.au/news/2013-09-27/nsw-desalination-plant-deal-costing-customers-10-billion/4985168

National Water Commission. (2011). *The National Water Initiative – Securing Australia's water future: 2011 assessment*. Canberra: NWC.

National Water Commission. (2013). *National performance report 2011–12 – Urban water service providers*. Canberra: NWC.

National Water Commission. (2014a). *National performance report 2012–13 – Rural water service providers*. Canberra: NWC.

National Water Commission. (2014b). *National performance report 2012–13: Urban water utilities*. Available at: http://www.nwc.gov.au/publications/topic/nprs/npr-2013-urban

Office of Living Victoria. (2013). *Melbourne's water future*. Melbourne: OLV.
Office of the Tasmanian Economic Regulator. (2014). *Water*. Retrieved March 1, 2014, from http://www.economicregulator.tas.gov.au/domino/otter.nsf/water-v/000
Ord Irrigation Cooperative Limited. (2014). *Fees and charges*. Retrieved February 12, 2014, from http://www.ordirrigation.com.au/grower-information/fees-and-charges/
Pawsey, N. (2014). *Victorian water businesses: Profits, taxes and dividends* (Commissioned report for essential services commission). Melbourne: Centre for Water Policy and Management.
Pawsey, N., & Crase, L. (2013). The mystique of water pricing and accounting. *Economic Papers, 32*(2), 328–339.
Pawsey, N., & Crase, L. (2014). *Review of the statutory asset values of Victorian water businesses* (Commissioned report for Essential Services Commission). Melbourne: Centre for Water Policy and Management.
Perry, C. (2009). Pricing savings, valuing losses and measuring costs: Do we really know how to talk about improved water management? In J. Albiac & A. Dinar (Eds.), *The management of water quality and irrigation technologies*. London: Earthscan.
Power and Water (NT). (2014). *What we do?* Retrieved March 6, 2014, from http://www.powerwater.com.au/about_power_and_water/what_we_do
QCA. (2013). *Final report: Seawater irrigation price review 2013–17* (Vol. 1). Brisbane: QCA.
Queensland Competition Authority. (2014). *Retail water*. Retrieved February 12, 2014, from http://www.qca.org.au/Water/Urban-retail-water/Retail
Tasmanian Irrigation. (2012a). *Lower south esk irrigation district 2013–14 water charges*. Retrieved March 1, 2014, from http://www.tasmanianirrigation.com.au/uploads/docs/LS_water_charges_2013-14_(comparison).pdf
Tasmanian Irrigation. (2012b). *Meander Valley irrigation district 2013–14 water charges*. Retrieved March 1, 2014, from http://www.tasmanianirrigation.com.au/uploads/docs/MV_Water_charges_2013-14_(caveside_update).pdf
Tasmanian Irrigation. (2014). *About Tasmanian Irrigation Pty Ltd*. Retrieved March 1, 2014, from http://www.tasmanianirrigation.com.au/index.php/about
TasWater. (2014). *Your account*. Retrieved March 1, 2014, from http://www.taswater.com.au/Your-Account/Variable-Water-Charges/Variable-Water-Charges
Townsville City Council. (2014). *Utility charges: Water charges 2014/15*. Retrieved March 31, 2015, from http://www.townsville.qld.gov.au/resident/fees/rates/Documents/Water%20Utility%20Charges%202014-15.pdf
Victorian Auditor General Office. (2013). *Water entities: results of the 2012–13 audits*. Retrieved December 12, 2013, from http://www.audit.vic.gov.au/reports_and_publications/latest_reports/2013-14/20131212-water-entities.aspx
Walsh, P. (2014). *Media release: Coalition introduces Fairer Water Bills*. Retrieved February 15, 2014, from http://www.livingvictoria.vic.gov.au/announcements/media-release-coalition-introduces-fairer-water-bills
SA Water. (2014a). *Guide to SA water's regulatory business proposal 2013 – Background information fact sheet*. Retrieved March 1, 2014, from http://www.sawater.com.au/NR/rdonlyres/4E8FAEAF-85D0-4B9F-89BE-36AC2DB0C1F1/0/RBP_Guide.pdf
SA Water. (2014b). *Pricing information*. Retrieved March 1, 2014, from http://www.sawater.com.au/SAWater/YourHome/YourAccountBillPaymentCharges/Pricing+Information.htm
Water Corporation (WA). (2014). *Rates and charges*. Retrieved February 10, 2014, from http://www.watercorporation.com.au/my-account/rates-and-charges
WaterFind. (2014). *The next evolution in Australia's water markets*. Retrieved January 15, 2014, from http://www.waterfind.com.au/forward-market-info.html
Weatherill, J., & Hunter, I. (2013). *Increased water bill concessions for low-income earners and pensioners – Media release*. Retrieved January 22, 2014, from http://www.premier.sa.gov.au/images/news_releases/13_05May/waterprice.pdf
Yarra Valley Water. (2012). *Yarra Valley future water: Water plan for 2013/14 to 2017/18*. Melbourne: YVW.

Chapter 3
Water Pricing in Brazil: Successes, Failures, and New Approaches

Moises de Andrade Resende Filho, Javier Santiago Ortiz Correa, and Marcelo de Oliveira Torres

Abstract Brazil is marked by its vastness and contrast in terms of availability and access to water. We select and provide a description of the water pricing experiences in place during the past 15 years at the Doce, Verde Grande, Paraíba do Sul, Piracicaba-Capivari-Jundiaí (PCJ), and São Francisco River basins, which are under federal jurisdiction and, thus, under the responsibility of the Brazilian National Water Agency (Agência Nacional de Águas-ANA). The pioneer pricing system of the Paraíba do Sul River basin has been a reference for others throughout Brazil. Generally, water users are charged for water withdrawal and consumption and for effluent discharge in terms of quantity and concentration of Oxygen Biologic Demand per m^3. While ANA is responsible as the federal agency in charge, local basin committees were empowered and make the ultimate decision on setting basic unit prices for water, adjustment coefficients, and granting water permits.

Keywords Brazil • Water law • Water basic unit prices • Adjustment coefficients • Granting water permits

3.1 Introduction

Brazil is a federal republic divided into five geographical regions (North, South, West Central, Southeast, and Northeast) composed of 26 states and a federal district, where the capital city of Brasilia is located. Brazil's population of nearly 200 million inhabitants lives mostly (84 %) in urban areas of Brazil's current 5,565 municipalities. For the management of water resources, the 8.5 million km^2 of

M.A. Resende Filho (✉) • J.S. Ortiz Correa • M.O. Torres
Departamento de Economia, Universidade de Brasília, Campus Darcy Ribeiro., Prédio da FACE., 70910-900 Brasília, DF, Brazil
e-mail: marf0013@gmail.com; javiersantiago.ortiz@gmail.com: motorres@hotmail.com.br

A. Dinar et al. (eds.), *Water Pricing Experiences and Innovations*, Global Issues in Water Policy 9, DOI 10.1007/978-3-319-16465-6_3

Fig. 3.1 Hydrological regions of Brazil (Source: National Water Agency – ANA 2013)

Brazil's territory has been divided into 12 hydrological regions, as shown in Fig. 3.1 (ANA 2013).

Brazil is a country of spatial contrasts in terms of climate, ecosystems, distribution of population, and socioeconomic indices. Despite the high average annual flow of rivers in Brazil—179,000 m^3/s or 12 % of the world's available freshwater resources—the spatial heterogeneity is also a characteristic of water availability across Brazil's 12 hydrological regions (Ministry of Environment/Secretariat for Water Resource 2006).

The climatic variability that characterizes the Brazilian hydrological regions is responsible for the unequal spatial distribution of water availability. For instance, in the region of the highest water scarcity, the Atlântico Nordeste Oriental hydrological region, water availability in the rivers is less than 100 m^3/s, while in the Amazônica hydrological region it is almost 74,000 m^3/s (see Table 3.1). Also, the

Table 3.1 Brazil's hydrological regions' population, water availability, and water demand for year 2010

Hydrological region	Population (Number of persons)	Surface water availability (m^3/s)	Water demand as water withdrawal (m^3/s)
Amazônica	9,694,728	73,748	78.8
Tocantins-Araguaia	8,572,716	5,447	135.6
Atlântico Nordeste Ocidental	6,244,419	320	23.7
Parnaíba	4,152,865	379	50.9
Atlântico Nordeste Oriental	24,077,328	91	262.0
São Francisco	14,289,953	1,886	278.8
Atlântico Leste	15,066,543	305	112.3
Atlântico Sudeste	28,236,436	1,145	213.7
Atlântico Sul	13,396,180	647	295.4
Paraná	61,290,272	5,956	736.0
Uruguai	3,922,873	565	155.4
Paraguai	2,165,938	782	30.0
Brazil	191,110,251	91,271	2,372.6

Source: ANA (2013)

sparsely populated Amazônica region (5 % of Brazilian population) accounts for 45 % of the area and 81 % of Brazil's surface water availability, which makes the other 55 % of the country's area responsible for less than 20 % of all surface water resources available.

In fact, most of the Brazilian population lives in its coastal area. For instance, the Atlantic hydrological regions (Leste, Nordeste Ocidental, Nordeste Oriental, Sudeste, and Sul) account for 45.5 % of the population but own only 2.7 % of the country's surface water availability (see Table 3.1). The Paraná River basin concentrates 32 % of the population of the country but owns only 6.5 % of the country's available surface water resources. Thus, the challenge from the standpoint of water supply in Brazil is that its population is concentrated in those areas in which water supply is unfavorable (ANA 2010c).

In these areas, urban water pollution related to the discharge of untreated sewage in rivers, lakes, and beaches is in some cases severe. In 2012, according to IBGE (2011), 85 % of the Brazilian population was urban, and 73 % of the urban households were connected to the sewage network (56 %) or linked to a septic tank (17 %), with the remaining households discharging their sewage volumes directly into rivers, lakes, and beaches. The situation is more critical in the northern and northeastern cities in which only 50 % and 55 %, respectively, of the households had some form of sewage collection. In this same year, according to the National System of Sanitation Information (SNIS 2013), only 38.7 % of the total sewage produced in Brazil faced some form of treatment. In the northern and northeastern regions, the proportion of sewage treated is lower than the national average (14 % and 31 %, respectively).

Under the current scenario in which water resources are scarcer and public resources are insufficient for increasing water supply, operating and maintaining the existing infrastructure and introducing incentive charges for water use may be a way to generate funds and induce more efficient water allocation. In line with this proposition, the Brazilian Congress approved in 1997 the Federal Law 9433, the so-called Water Law, which establishes charging for water use as one of the five instruments to manage water resources in Brazil. But a charge for water use is not a tax; it is a fee paid for the use of a public good whose revenues belong to the federal union or to the state in which the water resource is situated (ANA 2013).

More than 15 years after the Water Law was enacted, it has not yet made an impact. The low number of red dots and blue areas in Fig. 3.2 shows that only a few river basins are actually charging or moving toward charging for water use. We hypothesize that this is a consequence of the length and complexity of the legal process necessary to make possible charging for water use in Brazil. Charging is intertwined with a process of water management decentralization by which each river basin committee (water parliament) takes on the duties of managing water resources and promoting their sustainable use.

Roughly speaking, the proposal for charging for water use must be formulated and agreed upon by the river basin committee, and it should be approved by the regional and national water councils, the state legislatures, and ANA. An important prerequisite for water pricing and charging is the existence of a state law approving and setting the guidelines for this mechanism. Once the law is enacted, the first step is the creation of a river basin committee, which has to express the willingness to start the discussions leading to setting the prices to be charged for the different water uses. When the values are defined, the second step consists of presenting the proposal to the National Council of Water Resources, the highest authority in the country. If the pricing and water scheme is approved by the National Council, the third step consists of the creation of a river basin agency, which has as its main functions to undertake all the investments considered in the River Basin Investment Plan. One of the key features of the Brazilian Water Act is that it separates and differentiates the functions of the committees (set up for deliberation of users) from the functions of the river basin agency (the operational arm). The collection of fees is performed by the National Water Agency. This agency has to return the collected fees to the river basins' agencies.

As it will be shown, only a few river basins falling under federal jurisdiction (Doce, Paraíba do Sul, Piracicaba-Capivari-Jundiaí, Verde Grande, and São Francisco) have completed the process. But charges are negligible if compared with international standards, having generated only 209 million Brazilian *reais* by 2011, which is US$ 63.3 million dollars (e.g., US$1 = R$2.3; Monteiro 2012). The majority of basins are still in the stages of consulting and fulfilling legal requirements.

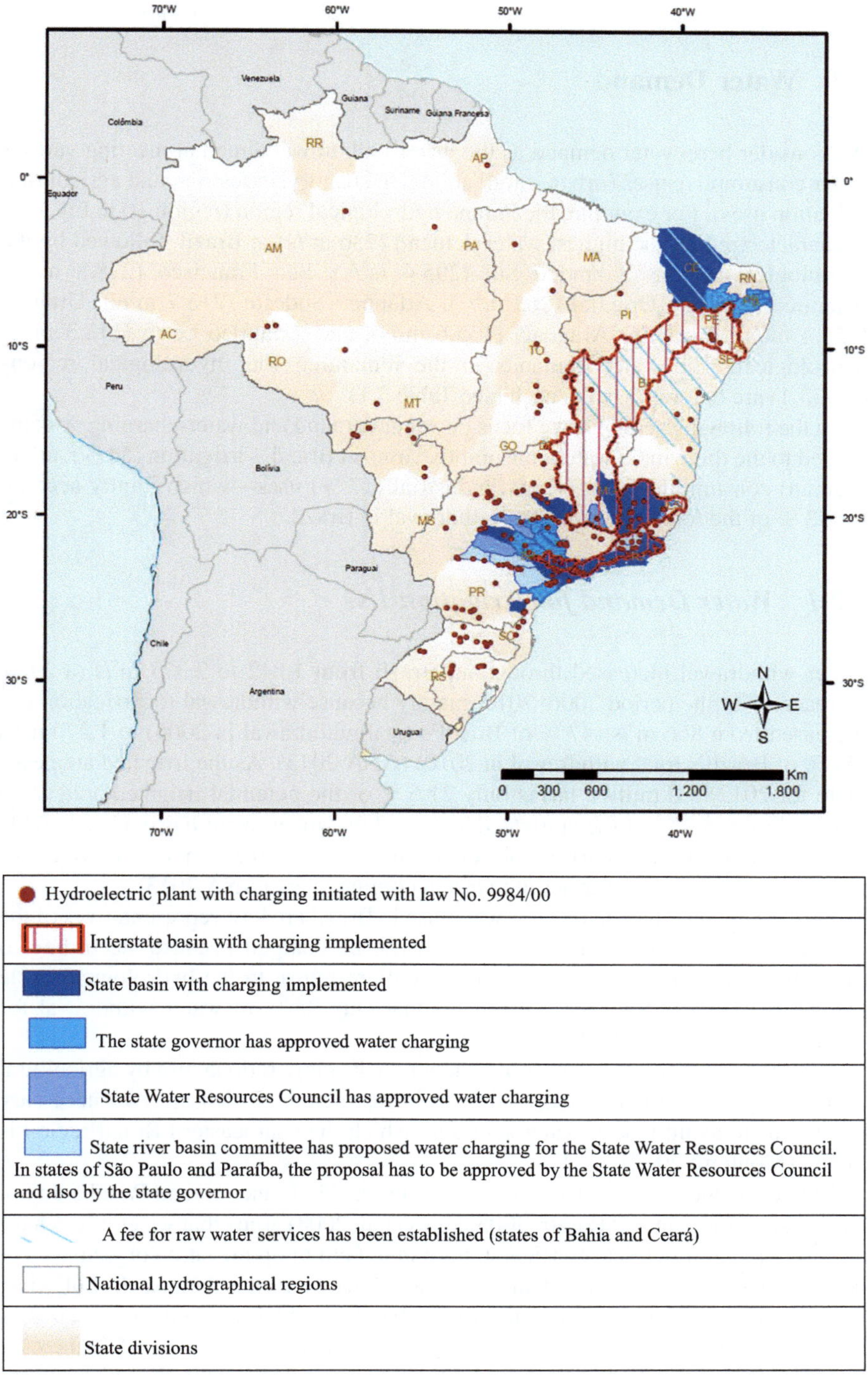

Fig. 3.2 River basins approving or applying water-charging schemes (Source: National Water Agency – ANA 2013)

3.2 Water Demand

We consider here water demand as the water withdrawal aimed at meeting various water consumptive uses (urban, rural, animal husbandry, industrial, and agricultural irrigation uses). For example, the Paraná hydrological region (region 10 in Fig. 3.1) is characterized by the highest water demand (736 m^3/s) in Brazil, followed by the hydrological regions Atlântico Sul (295.4 m^3/s), São Francisco (278.8 m^3/s), Atlântico Nordeste Oriental (262 m^3/s), Atlântico Sudeste (213.7 m^3/s), Uruguai (155.4 m^3/s), Tocantins-Araguaia (135.6 m^3/s), and Atlântico-Leste (112.3 m^3/s) hydrological. The water demands in the remaining four hydrological regions (Fig. 3.1) are below 100 m^3/s each (see Table 3.1).

In the following sections, we focus on water demand and water-charging systems related to the three main water consumptive uses in Brazil—irrigation (54 %), urban (human) consumption (22 %), and industrial (17 %) uses—which jointly account for 93 % of the total current water withdrawal in Brazil.

3.2.1 *Water Demand for Irrigation Use*

Water withdrawal increased throughout Brazil from 1.842 to 2.373 m^3/s (a 29 % increase) over the period 2006–2010, mainly because withdrawal for irrigation use increased from 866 m^3/s (47 % of Brazil's total withdrawal in 2006) to 1.270 m^3/s (54 % of Brazil's total withdrawal in 2010) (ANA 2013). As the irrigated area estimate for 2012 (5.8 million ha) is only 21.6 % of the potential irrigated area (26.9 million ha) or 8.3 % of the cultivated area, and the government has made available around US$5 billion for irrigation-related investments, the demand of water for irrigation is expected to further increase in coming years (ANA 2013).

Despite the fact that charges for water use in Brazil are low, representatives of the agricultural sector argue that they should not be charged because, according to them, 90 % of water withdrawals for agriculture return to the hydrological cycle (Monteiro 2012). ANA (2013) estimates that only 28 % of water withdrawal for irrigation returns to the hydrological cycle.

Higher prices for water use would encourage its more rational use by agricultural users and provide funds for investing in infrastructure, which are two necessary conditions to avoid risk of water shortage, which the southeastern Brazil region is facing (Monteiro 2012).

Studies on the impacts of water charging in the São Francisco and Paraiba do Sul river basins (Kelman and Ramos 2005; Féres et al. 2008) argue that water prices have been set up out of political considerations and not out of optimization objectives. For instance, the representatives of the agricultural sector in the Paraíba do Sul River Basin Committee demanded that the water charging, which was initiated in 2003 as the pioneer experience in Brazil, was set up to respect an upper limit of 0.5 % increase in agricultural production costs. In order to respect this upper limit, the unitary water price was set by the Paraíba do Sul River Basin Committee 40 times lower to the agricultural users than to sanitation and industrial users. But, in fact, water charging in the basin resulted in agricultural cost increases below 0.5 %, except for rice production costs, which went up around 1 % (Kelman and Ramos 2005).

3.2.2 Water Demand for (Human) Urban Use

Water supply for urban use, including water withdrawal and distribution, is provided by state-owned sanitation companies in 69 % of municipalities (e.g., COPASA in Minas Gerais and SABESP in São Paulo), municipality-owned sanitation companies in 27 % of municipalities, and private companies in 4 % of municipalities (ANA 2010c).

In terms of the charging schemes, the Law n. 6.528/78, endorsed in 1978, established that water tariff setting for residential users should take into account economic and social aspects, pursuing the economic and financial health of the water utility companies, without neglecting the social nature of the system. In this regard, tariffs should be differentiated among users and consumption levels with richer users cross-subsidizing poorer ones. By the late 1980s and early 1990s, however, a deregulation process took place and currently tariff setting is decentralized and performed, without a specific federal law, by the state, regional, and local or municipal companies (Faria et al. 2005).

In fact, each state-owned sanitation company has its own water-charging policy to the municipalities in which it operates, without any linkage with other federal, state, or municipality, to directly provide the services. Also, each municipality that directly provides water services has its own policy for charging water. There are municipalities that do not charge for water services and others that apply water service fees instead of water charges. But as a public service, the urban water supply is maintained through the collection of fees (i.e., the user pays for the service the provider has to withdraw, pump, treat, store, and distribute water through distribution networks until domicile). The costs involved are high and, for this reason, most consumers are subsidized, usually by increasing tariffs according to their consumption and through tariffs equalized between the areas served (Pereira Jr. 2007).

The vast majority of the water companies use increasing block tariffs. Up to 10 m^3, users pay either a flat rate per month or a linear tariff. Above this threshold, users pay a volumetric tariff that increases with consumption. In general, water companies also make a distinction of tariff values according to the users' income levels and living standards. For example, if the household demands, on average, up to 10 m^3 per month and lives in poor conditions (based on access to infrastructure and income), it may pay a *residential social* flat rate per month that is a fraction of the tariffs paid by the more privileged users. The households that end up consuming more than 10 m^3 per month but are considered as low income are subject to a *residential social* block tariff. The other types of residential users pay a *residential normal* block tariff.

Table 3.2 presents a sample of water companies in major capitals and their most updated charging schemes and tariffs in 2014. In the northeast, CAGECE,

Table 3.2 Sample of water companies in major capitals and their charging schemes and tariffs in 2014

Consumption levels (m^3)	CAGECE[a] (Fortaleza)	COMPESA[b] (Recife)	EMBASA[c] (Salvador)	CASAN[d] (Florianópolis)	SABESP[e] (São Paulo)
Residential social					
10]	0.34 or 0.70	2.79/month	4.09/month	2.61/month	2.48/month
]10–15]	1.19		1.79		
]10–20]					0.43
]10–25]				0.73	
]15–20]	1.27		1.95		
]20–25]			2.91		
]20–30]					1.51
]20–50]	2.18				
]25–30]			3.24		
]25–50]				3.52	
]30–40]			3.59		
]30–50]					2.16
]40–50]			4.11		
50]	3.84		4.94	4.30	2.39
Residential normal					
10]	0.93	13.04/month	9.09/month	13.93/month	7.31/month
]10–15]	1.20		2.54		
]10–25]				2.56	
]10–20]		1.50			1.14
]15–20]	1.28		2.72		
]20–50]	2.18				
]20–25]			3.05		
]20–30]		1.78			2.86
]25–30]			3.41		
]25–50]				3.58	
]30–40]			3.75		
]30–50]		2.45			2.86
]40–50]			4.11		
]50	3.84		4.94	4.30	3.15
]50–90]		2.90			
]90		5.49			

Source:
[a]http://www.cagece.com.br/atendimentovirtual/faces/publico/home.xhtml?page=estrutura_tarifaria
[b]https://lojavirtual.compesa.com.br:8443/gsan/exibirConsultarEstruturaTarifariaPortalAction.do
[c]http://www.embasa.ba.gov.br/centralservicos/index.php/tarifas
[d]http://www.casan.com.br/menu-conteudo/index/url/tarifas#0
[e]https://www9.sabesp.com.br/agenciavirtual/pages/tarifas/tarifas.iface
Note: (1) In the table, when there is no indication of the time extent, tariff values are per cubic meter, otherwise per month. (2) Exchange rate: 2.3 Brazilian reais per 1 US$. This exchange refers to the monthly average during the first semester of 2014 according to the Brazilian Central Bank at http://www4.bcb.gov.br/pec/taxas/port/ptaxnpesq.asp?id=txcotacao

COMPESA, and EMBASA provide water for the metropolitan regions of Fortaleza, Recife, and Salvador, respectively. In the south, CASAN provides water for the city of Florianópolis, and in the southeast, SABESP supplies water to the city of São Paulo.

Despite the tariff differentiation among users and consumption levels, the structure of residential water prices still amplifies income inequalities. For example, in the metropolitan region of São Paulo, the largest urban area in the country, wealthier households spend 0.37–0.52 % of their income on water, while the poorest households end up spending 4.2–4.7 % of their income (Ruijsa et al. 2008).

3.2.3 *Water Demand for Industrial Use*

Industrial water use concentrates in geographic regions of the south (hydrological regions 10 and 9 in Fig. 3.1) and southeast (regions 8 and 11 in Fig. 3.1) and in the São Francisco hydrological region (region 6 in Fig. 3.1). Those regions, which are rich in capital, labor, and infrastructure, concentrate on industrial activities and have received 80 % of permits issued for their water use throughout the country (ANA 2013).

Using data from 500 industrial plants in São Paulo state, Féres and Reynaud (2005) estimate that the water demand's price elasticity is −1.0, which suggests that increasing the price of water is an effective tool to reduce industrial water demand and, consequently, industrial effluent discharges. They also found that a 1 % water price increase has an almost negligible effect on cost structure (a nearly 0.07 % increase only in total costs). Those results suggest that increasing the price of water may be a strategy to reduce water use and protect the environment from industrial effluent discharges.

The reuse of water is an option that industrial firms and water utility companies may pursue to reduce water cost and water withdrawal. Simulation results for 2,311 industrial firms in the state of Sao Paulo have shown that by performing 60 % of industrial water reuse, an amount perfectly feasible for the largest majority of firms results in a significant reduction in firms' water costs (Hespanhol 2010). In addition, municipal water utility companies plan to purchase reused water and sell it for prices below the drinking water prices, since the drinking water prices are, according to the block tariff, from US\$2.2/m^3 to US\$ 4.8/m^3 and reused water prices range from US\$0.7/m^3 to US\$0.95/m^3 (Hespanhol 2010).

3.2.4 *Price Elasticities*

Estimates of price elasticities for residential water demand are less than 1 (Ribeiro et al. 2000; Worthington and Hoffman 2008; Casey et al. (2006); Ruijsa et al. 2008). Despite the fact that water demand for industrial use tends to vary substantially by sub-sector, in general, it is less inelastic than the water demand for residential use (Olmstead and Stavins 2007). For instance, Féres and Reynaud (2005) found an elasticity of −1.0 % for water demand for industrial use. On the other hand, water demand for agricultural use tends to be more price elastic, especially when it is used for the production of cheap crops, which tend to be irrigated with less advanced and efficient irrigation systems. For instance, Resende Filho et al. (2011) estimate for growers in São Francisco river basin that water price elasticities are −1.357 for sprinkler/micro-sprinkler irrigated horticulture (banana, guava, coconut, mango, cherry, and grape) and −1.8649 for horticulture (coconut, guava, mango, banana, and grape) irrigated by gravity.

Therefore, authorities acting as price setters, taking the pattern of water price elasticities of demand for different uses, would set higher prices for residential users than industrial users and higher prices for industrial users than agricultural users.

3.3 The Legal Foundations of Water Charging

The 1997 Federal Law 9433 (Water Law) relies on the principles of the 1992 Dublin Statement on Water and Sustainable Development. The principles of Water Law are: (1) water is a collective good; (2) water is a limited natural resource endowed with economic value; (3) the management of water resources should always provide for their multiple uses; (4) but human consumption shall have priority on all other water uses; (5) a river basin is the territorial unit of implementation of the National Policy of Water Resources (PNRH) and of action of the National System of Water Resources Management; and (6) the management of water resources should be decentralized, relying on the participation of the government, users, and communities (ANA 2013).

The Water Law aims at regulating the Brazilian Federal Constitution of 1988, establishing the National Water Resources Policy (PNRH,) and the following managerial instruments to support its implementation: (1) plans for water resources, (2) the classification of water bodies according to their main use, (3) the granting of water use rights by issuing water permit, (4) the charge for water use, and (5) the national system of water resources management (Singreh) (ANA 2013). The 1997 Water Law also envisioned the creation of state water agencies and the National Water Agency (ANA) to enforce the PNRH, grant permits for water use, prevent flood and drought, and stimulate the creation of river basin management committees (Tucci 2004). ANA was created in 2000 by Federal Law 9984.

Charging for water use is intended to create incentives for rational use of water, raising funds to implement river basins management plans (Benjamin et al. 2005). Indeed, the Water Law guarantees users the right to retain control over the revenue generated by stipulating that not more than 7.5 % of financial resources collected in a basin can be transferred out of the basin (Kraemer et al. 2003).

Charging for water use also provides a better water demand management across users and regions to induce the redistribution of social costs associated with water use, by establishing compensation mechanisms to reduce effluent emissions. Reduction of emissions is achieved by improving the quality of the effluents discharged onto water courses and to incorporate social and environmental considerations in the management of the water resources (Confederação Nacional da Indústria—CNI 2002).

But the implementation of the Water Law has been marked by legal voids. First, the special characteristics of groundwater, whether or not it falls within federal or state dominion, are not clearly defined by the law, which is troublesome since groundwater supplies a large amount of the surface water courses (e.g., rivers, lakes, and lagoons) especially in periods of drought. Groundwater supplies around 51 % of potable water (Benjamin et al. 2005) and serves fully or partially 61 % of the municipalities (ANA 2010c). The definition of the river basin as the planning unit often ends up in institutional conflicts and legal voids, since the 1988 Brazilian Constitution defines a river as falling within federal domain when it flows through two or more states or it has an international reach (Tucci 2004).

As will be discussed in the following section, only few river basins in the country are moving into the path set by the Water Law. The implementation of charges for water use has faced many problems at state or federal levels and has been delayed because a specific law establishing the legal standards for river basin organizations is still needed. Additionally, quantity and quality permits are issued by different government agencies, which harm suitable controls, and funds from water charges are usually not tied to investments in water infrastructure and management in the basin (Porto and Kelman 2000).

3.4 Water-Charging Experiences in Brazil

The application of water use charges is one of the management tools provided by the Water Law, which also establishes that the jurisdiction over water resources is the river basin committees (water parliaments). A river basin committee is composed of water users, civil society, and public authorities. It is responsible for internally discussing and proposing to its respective board of water resources the mechanisms and values to be adopted for charging for water resources in areas under its jurisdiction. Indeed, charging for the use of water resources is an indicator of the stage of implementation of the National Water Resources Policy as it follows the implementation of other policy instruments (ANA 2013).

3.4.1 Charging for Water Use in the River Basin System

Despite the fact that a charging scheme is basin specific, in general terms, water users of all types and in all basins are charged for three uses: water withdrawal, water consumption, and effluent discharge according to its quantity and concentration of Oxygen Biologic Demand (OBD) per m³. Each item is separately priced, according to the charging scheme approved for the basin committee. In general terms, the total value of water charges per year is calculated according to the general Eq. 3.1:

$$\text{Total Charge} = \text{withdrawal charges} + \text{consumption charges} + \text{effluent discharge charges} \quad (3.1)$$

where the annual charges for each item $i \in \{withdrawal, consumption, discharge\}$ are calculated according to

$$\text{Charge}_i = Q_i BUP_i K_i \quad (3.2)$$

where Q_i is *water quantity for item i*, calculated on the basis of water rights granted or permits issued, self-reported use, or, in a small scale, by remote sensing. For instance, in the PCJ river basin, a user may request to the agency granting water rights to be allowed to install his/her own metering equipment to be accredited by the agency. The difference between the volume of water rights granted and the quantity of water actually used may lead to the revision of the amount of rights grated to avoid unnecessary reservations of water and, thus, provide more water for other users in the basin. *BUPi is the Basic Unit Price for item i* as *reais* per m³ and is established within the river basin committee (see Table 3.3 for some examples); and *Ki* is an *adjustment coefficients for item i* and is set, usually, less than one so to give price discounts on the basis of the type of water use (e.g., irrigation), the class (defined by the quality of water of the river such that the better the water quality is the higher is the value of *K*, which can be set greater than one) at the point where water is withdrawn, and the sector (e.g., rural). For instance, $K_{consumption}$ may reflect a price discount on the basis of the percentage of the withdrawal that is estimated to return to the water source for a type of water use, as the PCJ basin committee has set $K_{consumption}$ for irrigation equal to 0.5 (ANA 2007).

Table 3.3 Basic unit price for item i (BUP_i) in Federal Dominion River Basins

River basin	Catchment (US$/m³)	Consumption (US$/m³)	Disposal (US$/kg of OBD)	River basin transposition (US$/m³)
Paraíba do Sul	0.0043	0.0087	0.0304	
Piracicaba, Capivari and Jundiaí	0.0043	0.0087	0.0435	0.0065
São Francisco	0.0043	0.0087	0.0304	
Rio Doce	0.0104	0.0135	0.0652	
Verde Grande	0.0043	0.0087	0.0304	

Source: ANA (2007, 2008a, b, 2010a, b, 2013), AGEVAP (2012), Bronzatto and Amorim (2012)
All figures denote the Basic Unit Price for water transposed from one river basin to other, and are in dollars (US$1=R$2.30)

Some examples of uses for Eq. (3.2) are charges for water withdrawal = $Q_{\text{withdrawal}} \times BUP_{\text{withdrawal}} \times K_{\text{witdrawal}}$; charges for consumption = $Q_{\text{withdrawal}} \times BUP_{\text{consumption}} \times K_{\text{consumption}}$; and charges for effluent discharge = $Q_{\text{withdrawal}} \times C_{\text{OBD}} \times BUP_{\text{discharge}} \times K_{\text{OBD}}$, where Ki denotes the adjustment coefficient for item i = {withdrawal, consumption, OBD}. It is worth noticing that charges for effluent discharge are calculated using C_{OBD}, that is, the average concentration of organic load in effluent as Oxygen Biologic Demand (OBD) discharged per m^3, and K_{OBD}, that is, an adjustment coefficient to make it possible discounting $BUP_{\text{discharge}}$ on the basis of the class the river is classified in the point where effluent is discharged.

Table 3.3 presents the values of *Basic Unit Price for* water withdrawal, water consumption, and water discharge in five federal river basins that have implemented water-charging schemes in Brazil.

In general, the river basin committees also use specific mechanisms for charging the rural sector for water withdrawal and consumption, the irrigation sector for the water usage, and the companies that use water for the generation of electricity, which is transposed to other basins. For instance, according to ANA (2007), the PCJ river basin committee established that the rural sector should pay only 10 % of the sum of *charges for water withdrawal* and *charges for consumption*; the irrigation sector is charged according to the scheme, charges for consumption = $Q_{\text{withdrawal}} \times BUP_{\text{consumption}} \times 0.5$, which makes the irrigation users pay only 50 % of the calculated charges for consumption. Finally, companies that use water for the generation of electricity are charged in the amount of 1 % of their annual revenues with electricity.

The general remarks about the charging for water use are:

- For the uses of water for agriculture, irrigation, and electricity generation, adjustment coefficients are set less than one, which gives these water uses price discounts.
- There is a trend to penalize users for the negative externality they generate on others when they pollute water (e.g., effluent discharges reduce the quality of water for downstream users in the basin). In general terms, the higher the average concentration of organic load in effluent, such as OBD per m^3 for a given quantity of water withdrawal, the higher the *charges for effluent discharge* will be.
- *Basic unit prices* vary little across river basins, because river basin committees take Paraiba do Sul River basin as a benchmark, as it was the first to implement a water-charging system.

Overall, quantitative (i.e., withdrawal + consumption + transposition) water use comprises the biggest majority of collected water charges. In fact, the quantitative use represents 83 % of the collected charges in Paraiba do Sul basin, 93 % in PCJ basin, 98 % in São Francisco basin, and 87 % in Rio Doce basin (ANA 2013). Among the types of quantitative uses, charges due to water transposition are really the main source of revenue for PCJ (58 %), São Francisco (65 %), and Rio Doce (59 %) basins (ANA 2013).

In year 2012, 1,563 users paid around $30 million for water use in federal dominion rivers, with 78 (or 5 %) of them paying 90 % of total water charges and 981 (or 63 %) of them paying only 0,04 % of the total charges. Also, the amount of collected charges has increased in all basins over the 2008–2013 period, since water prices have been updated upward, and more users have been registered in the system (ANA 2013).

Table 3.4 shows that, for most cases, the amount of collected charges is lower than the expected amount of charges, which is calculated as the amount that would have been collected if all users in the basin had paid for everything they consumed. A reason for this is that in most river basins, payment is still voluntary with no enforcement mechanism in place, which may indicate that instead of actually inducing changes in water use, river basin committees are more interested in creating a culture of payment for water use at this stage of the institutional development (ANA 2013).

3.5 Main Basins at a Glance

Since 2004, water charging was approved in all Rio de Janeiro state's river basins. In the state of São Paulo, the process started in 2007 and is still going on as more basins adopt the water pricing and water-charging schemes. For some basins in this state, only an approving decree is missing to start charging for water. In Minas Gerais state, the process began in 2010, but the pricing and charging mechanisms are still being discussed, and it will be a long time before it is finally approved by the state water council. Espírito Santo state is waiting for the regional legislature to approve the charging mechanism and to lay down the rules to implement it. In the state of Paraiba, the process started in 2008 and the legal process was completed in 2012, when a decree was issued approving the charging mechanism in all rivers of the state dominion. Nevertheless, the process has not started.

Two states deviate from the pattern previously mentioned. Since 1996, the state of Ceará, through a regional- and state-owned company for water management, has charged for the use of water resources and water infrastructure. But the system has the same problems it has with the price of a public good (e.g., it is difficult to determine the actual willingness to pay those providing the public good). Something similar happened in the state of Bahia, where another state-owned company has been charging for the use of water supplied to the reservoirs.

3.5.1 Paraíba do Sul River Basin

Paraíba do Sul River basin is located in the southeastern region between the states of São Paulo (13,605 km^2), Minas Gerais (20,500 km^2), and Rio de Janeiro (22,600 km^2, that is, nearly half of this state's area). It covers 184 municipalities—88 in Minas Gerais, 57 in Rio, and 39 in the state of São Paulo—draining one of the most developed regions of the country (ANA 2008a). The total urban population is close to 96 % of the total population in the basin which, according to IBGE's

Table 3.4 Water charging (expected and collected) in river basins of Federal Dominion ($1,000)

	River basin							
	Paraíba do Sul		Piracicaba, Capivari and Jundiaí		São Francisco		Doce	
Year	Expected	Collected	Expected	Collected	Expected	Collected	Expected	Collected
2008	4,768.73	3,519.65	7,794.78	7,388.28				
2009	5,170.99	4,247.77	7,444.91	7,368.06				
2010	5,683.82	5,396.59	7,913.84	7,633.38	4,972.84	3,752.63		
2011	5,378.74	11,115.40	7,398.28	7,180.12	9,512.54	9,095.52		
2012	4,707.44	4,482.68	8,019.17	7,876.55	10,040.38	9,348.37	4,497.96	1,500.37
2013	4,994.40	4,737.69	7,833.20	7,621.48	10,368.32	9,460.44	3,388.12	2,819.64

Source: ANA (2013). All figures in dollars (US$1 = R$2.3)

census of 2010, is 6,425,301 inhabitants. The Paraíba do Sul River basin was the first in the country to establish a charge for water resources use.

The basin accounts for 1.75 % of the country's hydroelectric potential and 13 % of the country's GDP and supplies 85 % of the water of the metropolitan region of Rio de Janeiro city. In 2001, charging for water use was approved, but the industrial users tried to block the agreement. As their strategy failed, they decided to influence the negotiation in their favor. But it was not only the industrial sector, because agricultural users and dams used to supply power had also been resistant. Other challenges are that the amount of collected charges has been too low to fund the basin development plan and the federal government is slow in returning the funds to the basin (Nelson 2008).

The number of users being charged went up from 186 in 2003 to 296 in 2011. As more users adhere to the culture of water payment for water use, the difference between what is being charged and what is being collected is only around 2.9 %. Water utilities and sanitary service companies pay for 56 % of what is collected in water charges, while the industrial sector pays for 43 %. Paradoxically, agricultural and residential users, as well as power dams, only pay for 1 % of the total collection. The quantitative uses (withdrawal, consumption, and transposition) represent 78 % of the total charged, and the qualitative uses (effluent and pollutant discharges) account for the remaining 22 % (ANA 2013).

It is worth mentioning that in the Paraíba do Sul River basin, 70 % of the value of the difference between water granted and actually withdrawn if water is metered is added to *charges for consumption*; otherwise the quantity of water charged equals the water volume granted. The use of this type of mechanism aims at discouraging the creation of "water reserves" (ANA 2008a). Agricultural users pay for the water they consume according to average measures of consumption. The charging for effluent discharge is based on the quantity and quality of pollutants disposed into the river basin. Although all users face the same price per unit of water, the pricing mechanism in this basin uses an adjustment coefficient by the type of use.

3.5.2 Piracicaba, Capivari, and Jundiaí Rivers Basin

The Piracicaba, Capivari, and Jundiai (PCJ) rivers basin covers an area of 15,303 km^2, with 92.6 % of its area in the state of São Paulo and 7.4 % in the state of Minas Gerais, with a population of 5.5 million. The PCJ basin is located between the meridians 46° and 49° W and latitudes 22° and 23.5° S, with an approximate length of 300 km in the east-west and 100 km north-south direction (ANA 2008b). The three main water consumptive uses (nearly 36.34 m^3/s) in the basin are urban (human) consumption (52.5 %), industrial uses (29.1 %), and irrigation (18.4 %) uses (Comitês das Bacias Hidrográ cas dos Rios Piracicaba, Capivari e Jundiaí 2012).

The committee for these river basins decided that water withdrawals below 5 m^3/day and effluent dilution below 0,058 l/s are insignificant and, thus, should not be charged. The charging formula also includes a coefficient to adjust for the irrigation system efficiency, which encourages the adoption of more efficient technologies. Irrigation users face incentives to invest in better irrigation technology as the use of adjustment coefficients rewards the investment they already undertook, by reducing the price irrigation users pay per unit of water. On top of this, the ANA indicates that the impact of water pricing on agricultural production is, at most, a 3.11 % increase (ANA 2008b).

3.5.3 São Francisco River Basin

The São Francisco River basin covers a drainage area of 638,576 km^2 (nearly 8 % of the country's area), accounts for about 7.5 % of the Brazilian population in 2010 (14.2 million people), has an average flow of 2,846 m^3/s, and crosses 521 municipalities (9.4 % of the 5,565 municipalities in the country) along six states (Bahia (48.2 %), Minas Gerais (36.8 %), Pernambuco (10.9 %), Alagoas (2.2 %), Sergipe (1.2 %), Goiás (0.5 %)), and the federal district (0.2 %). The three main water consumptive uses in the basin are irrigation (68 %), urban (human) consumption (18 %), and industrial (9 %) uses (ANA 2010b).

In 2008, the São Francisco River Basin Committee (CBHSF in Portuguese) defined the values and mechanisms for water prices in the part of the basin that falls within federal dominion. The process began with the accounting of old and new water users through an online system in which they declared their quantities and types of water use. Irrigation represents 68 % of the water consumption in the basin. Despite this large share, the revenues coming from agricultural users are low, since water prices for agriculture are 40 times lower than for other types of use. Effluents from sanitary services are 89.9 % of the total water disposals in the basin (ANA 2010b).

The water-pricing scheme adopted in this river basin has two aspects that make it unique. First, it has led to a 25.9 % decrease in organic matter discharged into the basin (ANA 2010b). The reduction of organic discharges has been accomplished by charging water and utility companies on the basis of the concentration of organic matter in effluents. Second, the pricing scheme involves an aridity, based adjustment coefficient that increases water prices for all uses (ANA 2008b). This aridity coefficient is important, because the basin crosses the very arid region of the Brazilian Northeast, where water availability suffers from very sharp fluctuations.

3.5.4 Doce River Basin

The Doce River basin is located in the southeastern states of Minas Gerais (86 % of the basin's area) and Espirito Santo (14 % of the basin's area), with 3.3 million people living in its area of 86,711 Km^2. The basin covers in full or partially 229 municipalities, 203 in Minas Gerais and 26 in Espírito Santo state. Within its boundaries are the largest metallurgical complex in Latin America and mining and forestry industries.

The Doce River basin suffers from deforestation and poor soil use that leads to silting of water courses, erosion, and flooding, as well as from the untreated discharge of sanitary services. Water pricing was adopted by the Rio Doce Basin Committee (CBH) in 2011. Different from the other charging mechanisms in the country, water consumption (the difference between water withdrawal and water disposal) is not charged due to technical difficulties in calculating how much of the water used in irrigation actually infiltrates back to the hydrological system. Also, water prices will be progressively growing from 2011 to 2015, a progression that depends on collection targets and the improvement of the charging and collection mechanisms (Amorim et al. 2011).

When comparing the water charges collection and the investment plan needed to improve water quality and quantity available in the basin, due to efficiency and technological improvements, current collection levels are insufficient and, at most, can cover 73 % of total investments. This is low when considering that 25 % of the investments up to 2020 are related to sanitary services and water quality and to the reduction of water losses in urban and rural water systems (ANA 2010a).

3.5.5 Verde Grande River Basin

As part of the São Francisco River basin, the Verde Grande River basin has an area of 31,410 Km^2, houses 5 % of the population of the São Francisco River basin, and flows through the states of Minas Gerais and Espírito Santo (southeastern Brazil). To implement the charging mechanism, basin authorities decided to stop issuing water permits as a way of reducing conflicts among users and to protect the critical groundwater resources. Groundwater is critical in this basin. Created in 2003, the Verde Grande River Basin Committee (CBH-VG) considers the following aspects in carrying out its water pricing scheme: water consumption equals 80 % of water withdrawals; prices for agriculture should be 40 times lower than for other users (sanitary services, mining, and industry); good irrigation practices should be rewarded by lower prices (but this reward is in fact indiscriminate and is rather an adjustment for assumed payment capacity that is based on the total area of the rural property. Certainly, more has to be done to promote better practices and to adjust for irrigated area size (Bronzatto and Amorim 2012).

3.6 Current Debates and Future Directions

3.6.1 Legal and Political Issues

Despite the fact that the 1997 Water Law relies on the principle of decentralized water management, this principle is under threat. For instance, in the state of Para (Lemos and Oliveira 2004), the Water Resources Management Company (COEGRH, its acronym in Portuguese) develops the management of the Jaguaribe River basin on the creation and involvement of users' commissions. But the effectiveness of this scheme depends on the role played by technicians and users' policy networks.

The political economy of water shows the user, with studies and research of the hydrological cycle of the basins, that charging for water is necessary to invest in better water-related infrastructure and to promote a more rational use of water. This is particularly important in regions that are suffering from a disorganized urban sprawl (like in the southeastern part of the country), and where extensive agriculture is rapidly expanding (like the in the east), or where climate change may induce more drastic variations of water availability (like in the northeast) Also the success of the Ceará's water-charging experiment depends on how the policy networks influence the implementation of the regulatory framework, their ability to gain support to protect the reforms, and how they resist opposition from technocrats and politicians opposed to the participatory schemes.

There is a need to increase the involvement of water users in the river basins' management, since the involvement of more water users is likely to lead to higher levels of willingness to pay for water. In this context, authorities and river basin organizations should explore how social marketing, aimed at creating social change, can be employed to tailor targeted campaigns that appeal to water users and lead to social change (Nelson 2008).

3.7 Conclusions

More than 15 years after the implementation of the Water Law, only a few river basins are actually moving toward charging for water use. The length and complexity of the legal process may be one of the reasons for the delay. Indeed, only a few river basins under federal jurisdiction (Doce, Paraíba do Sul, Piracicaba-Capivari-Jundiaí, São Francisco, and Verde Grande) have completed the process, with the majority of other basins still in stages of consulting and fulfilling legal requirements.

The amount collected for water use in Brazil is still negligible as compared with international standards. For most river basins, payment is still voluntary with no enforcement mechanisms in place, indicating that, instead of inducing changes in water use, river basin committees are more interested in creating a culture of payment for water use. Indeed, increasing the involvement of water users in river basin management is necessary to convince them to pay water charges, and this may be accomplished if authorities and river basin organizations explore social marketing strategies, such as tailoring targeted campaigns that appeal to water users and lead to social change (Nelson 2008).

Also the low values set for the *basic unit price*s may explain why charged and collected amounts are so little across river basins. The permits issued for water withdrawal by ANA and state water agencies are nontradable, which makes the existence of water markets nearly impossible. Missing markets for water make it easier to set *basic unit prices* and adjustment coefficients in water-charging schemes on the basis of water basin investment plans and budget needs, without any consideration of water scarcity (in terms of quantity and quality) or water use efficiency. In other words, water prices are set out of political considerations and not for the reason of inducing efficient use.

In terms of water pollution, the main issues are related to the discharge of effluents and untreated sewage into rivers, lakes, and beaches near or within urban centers, where there is a concentration of population and industrial activities. The practice of reusing water may be an option for industrial firms and water utility companies to reduce their cost of withdrawal and consumption of water and costs of effluent discharges.

Last, but not least, the supply of residential water and sanitation services face challenges as they are moving toward a scenario with lack of incentives for private sector participation and consolidation of public sector ownership. Thus, water and sanitation companies and municipalities, which are still constitutionally responsible for providing those services, will have to negotiate over the concession terms to create incentives for short- and medium-run investments, and a better regulation should be created to ensure the quality of those services.

References

Agência Nacional de Águas—ANA. (2007). *A implementação da cobrança pelo uso de recursos hídricos e Agência de Água das bacias dos rios Piracicaba , Capivari e Jundiaí* (112 pp). Brasília: ANA, SAG.

Agência Nacional de Águas—ANA. (2008a). *Relatório 2008: cobrança pelo uso de recursos hídricos: Bacia Hidrográfica do Rio Paraíba do Sul* (97 pp). Brasília: SAG-ANA.

Agência Nacional de Águas—ANA. (2008b). *Subsídios para as discussões visando o estabelecimento dos valores dos coeficientes que comporão a metodologia de cobrança pelo uso de recursos hídricos da Bacia Hidrográfica do Rio São Francisco* (9 pp). Brasília: SAG-ANA.

Agência Nacional de Águas—ANA. (2010a). *Cobrança pelo uso de recursos hídricos na Bacia Hidrográfica do Rio Doce* (91 pp). Brasília: SAG-ANA.

Agência Nacional de Águas—ANA. (2010b). *Relatório 2010: processo de regularização de usos e operacionalização da cobrança na bacia do Rio São Francisco* (61 pp). Brasília: SAG-ANA.

Agência Nacional de Águas—ANA. (2010c). *Atlas Brasil: abastecimento urbano de água – panorama nacional* (Vol. 1, 72 pp). Brasília: Engecorps/Cobrape.

Agência Nacional de Águas—ANA. (2013). *Conjuntura dos recursos hídricos no Brasil 2013* (432 pp). Brasília: SPRH-ANA.

AGEVAP. (2012). *Avaliação da implementação da cobrança pelo uso de recursos hídricos de domínio da União na bacia hidrográfica do rio Paraíba do Sul – Período 2003/2011* (71 pp). Resende: Agência da Bacia do Rio Paraíba do Sul.

Amorim, M. A. M., Carvalho, G. B. B., deThomas, P. T., de Freitas, N. N., & Alves, R. F. F. (2011). A cobrança pelo uso de recursos hídricos na bacia hidrográfica do Rio Doce.In *Proceedings of XIX Simpósio Brasileiro de Recursos Hidrícos, 2011* (14 pp).

Benjamín, A. H., Marques, C. L., & Tinker, C. (2005). The water giant awakes: An overview of water law in Brazil. *Texas Law Review, 83*, 2185–2244.

Bronzatto, L.A., & Amorim, M.A.M. (2012). A Cobrança pelo Uso de Recursos Hídricos e seu impacto no setor agropecuário da bacia hidrográfica do rio Verde Grande.In *Proceedings Sociedade Brasileira de Economia, Administração e Sociologia Rural*, Vitória, 22 a 25 de julho de 2012 (20 pp).

Casey, J. F., Kahn, J. R., & Rivas, A. (2006). Willingness to pay for improved water service in Manaus, Amazonas, Brazil. *Ecological Economics, 58*(2), 365–372.

Comitês das Bacias Hidrográ—cas dos Rios Piracicaba, Capivari e Jundiaí. (2012). *Plano das bacias hidrográ—cas dos Rios Piracicaba, Capivari e Jundiaí 2010 a 2020 – relatório síntese* (3rd ed., 120 pp). São Paulo: ArtPrinter Grácos.

Confederação Nacional da Indústria—CNI. (2002). *O Valor econômico da água: impactos sobre o setor industrial nacional* (20 pp). ANAIS-CNI.
de Faria, R. C., Nogueira, J. M., & Mueller, B. (2005). Políticas de Precificação do Setor de Saneamento Urbano no Brasil: As Evidências do Equilíbrio de Baixo Nível. *Estudos Economicos, 35*(3), 481–518.
Féres, J. G., & Reynaud, A. (2005). Assessing the impact of environmental regulation on industrial water use: Evidence from Brazil. *Land Economics, 81*(3), 396–411.
Féres, J. G., Reynaud, A., Seroa da Motta, R., & Thomas, A. (2008). Competitiveness and effectiveness concerns in water charge implementation: A case study of the Paraiba do Sul River basin, Brazil. *Water Policy, 10*(6), 595–612.
Hespanhol, I. (2010). Conservação e reuso como instrumentos de gestão para atenuar os custos da cobrança pelo uso da água no setor industrial. In C. E. M. Bicudo, J. G. Tundisi, & M. C. B. Scheuenstuhl (Eds.), *Águas do Brasil: Análises Estratégicas* (pp. 59–76). São Paulo: Instituto de Botânica – Imprensa Oficial do Estado de São Paulo.
Instituto Brasileiro de Geografia e Estatística – IBGE. (2011). *Demographic census 2010.* Brasília, DF, Brasil. Available at http://www.censo2010.ibge.gov.br/
Kelman, J., & Ramos, M. (2005). Custo valor e preço da água utilizada na agricultura. *Rega, 2*(2), 39–48.
Kraemer, R. A., Castro, Z. G., SeroadaMota, R., & Russell, C. (2003). *Economic instruments for water management: Experiences from Europe and implications for Latin America and the Caribbean* (92 pp). Washington, DC: Inter-American Development Bank, Integration and Regional Programs Department.
Lemos, M. C., & Oliveira, J. L. F. (2004). Can water reform survive politics? Institutional change and river basin management in Ceará, Northeast Brazil. *World Development, 32*(12), 2121–2137.
Ministry of the Environment., & Secretariat of Water Resources. (2006). *National water resources plan. Executive synthesis – english* (135 pp). Brasilia: MMA.
Monteiro, S. (2012). Cloudy waters: After 15 years there's not much to show for Brazil's national water policy. *The Brazilian Economy, 4*(5), 11–17.
Nelson, B. M. (2008). *Water reform in Brazil: An analysis of its implementation in the Paraíba do Sul basin and a consideration of social marketing as a tool for its optimal success.* A practicum submitted in partial fulfillment of the requirements for the degree of Master of Science at the University of Michigan (48 pp).
Olmstead, S. M., & Stavins, R. N. (2007). *Managing water demand orice vs. non-price conservation programs* (47 pp). Pioneer Institute for Public Policy Research.
Pereira Jr., J. S. (2007). *Tarifas dos serviços públicos de água e esgotos no Brasil, Consultoria Legislativa, Câmara dos Deputados* (14 pp).
Porto, M., & Kelman, J. (2000). Water resources policy in Brazil. In R. A. Kraemer, Z. G. Castro, R. Seroa da Motta, & C. Russell (Eds.), *Rivers: Studies in the science, environmental policy and law of instream flow* (Policy Dialogue Study Series), *7*(3). Washington, DC: Inter-American Development Bank.
Resende Filho, M. A., Araújo, F. A., Silva, A. S., & Barros, E. S. (2011). Precificação da água e eficiência técnica em perímetros irrigados: uma aplicação da função insumo distância paramétrica. *Estudos Economicos, 41*(1), 143–172.
Ribeiro, M.M.R., Lanna, A.E., & Pereira, J.S. (2000). *Elasticidade-preço da demanda e a cobrança pelo uso da água* (16 pp), Electronic version.
Ruijsa, A., Zimmermann, A., & van den Berg, M. (2008). Demand and distributional effects of water pricing policies. *Ecological Economics, 66*(2–3), 506–516.
SNIS. (2013). *Diagnósticos dos serviços de água e esgoto—2012* (18th ed., 164 p). Ministério das Cidades, Secretaria Nacional de Saneamento Ambiental.
Tucci, C. (2004). *Development of water resources in Brazil, Global Water Partnership South America* (27 pp).
Worthington, A. C., & Hoffman, M. (2008). An empirical survey of residential water demand modelling. *Journal of Economic Surveys, 22*(5), 842–871.

Chapter 4
Water Pricing in Canada: Recent Developments

Steven Renzetti and Diane P. Dupont

Abstract The purpose of this chapter is to provide a critical review of past and current practices related to water pricing in Canada's irrigation, residential, and industrial sectors, as well as water pricing related to the provision of environmental services. The chapter demonstrates that water prices in most sectors have historically been quite low, relative to the costs of supply and relative to international standards. Both residential water users and irrigators have had subsidized access to water distribution networks, and self-supplied water users (such as large manufacturing facilities) have gained access to water supplies at little cost. More recently, some provinces, irrigation districts, and municipalities have raised rates to promote conservation and increase the supply network's financial sustainability. The chapter concludes by pointing to a number of important emerging issues related to water pricing.

Keywords Canada • Industrial water demand • Residential water demand • Agricultural water demand • Efficiency

4.1 Past Pricing Systems

4.1.1 Water Resources, Population, and Issues in Water Supply

Canada is blessed with abundant freshwater supplies, relatively low population densities, and levels of income and technological knowledge seemingly adequate to meet the challenges associated with the provision of potable water and treatment of wastewater.

Another feature of the context for Canadian water resource management, allocation, and pricing is the legal framework that governs these activities. Under the Canadian Constitution, all water resources are owned by the Crown (as represented by

S. Renzetti (✉) • D.P. Dupont
Department of Economics, Brock University,
500 Glenridge Avenue, St. Catharines, ON L2s 3A1, Canada
e-mail: srenzetti@brocku.ca; ddupont@brocku.ca

A. Dinar et al. (eds.), *Water Pricing Experiences and Innovations*,
Global Issues in Water Policy 9, DOI 10.1007/978-3-319-16465-6_4

the government), and the primary responsibility for allocating water resources rests with the provincial governments (Saunders and Wenig 2007; Brandes et al. 2008). All provinces exercise this responsibility by requiring that major water users hold a government license that specifies the water source and permitted volume, which may be withdrawn. Because of differences in the relative abundance of water across Canada and differences in circumstances under which individual provinces joined Canada, there are differences in the way in which provincial water licensing regulations have evolved. Those provinces in which water is relatively scarce and which joined Canada relatively late have water allocation schemes largely based on first-in-time, first-in-use (FITFIR) principles (these are British Columbia and the western prairie provinces). The older provinces in the east of Canada have water allocation frameworks that evolved from English common law and riparian rights doctrines.

Despite the presence of relatively abundant supplies and a relatively well-functioning system of water regulations, a number of challenges related to water allocation have been identified. For example, almost two decades ago, a prominent think tank provided the following assessment regarding the state of Canada's municipal water systems (National Roundtable on the Environment and Economy 1996, p. 3):

> Canada's water and wastewater system is under pressure: the infrastructure—water and wastewater treatment facilities, sewers, supply lines—is severely deteriorating, primarily due to shortages of public funding. If the decline continues, the health of the country's water resources will suffer. At the same time, due to subsidized and below-cost pricing for water and wastewater services, innovative environmental technologies that conserve water resources are failing to find a market.

More recently, the Conference Board—another prominent Canadian think tank—issued its annual "report card" on social and economic issues in Canada and assigned a low score for Canada's efforts on water management. The authors explained that "two major reasons for Canada's excessive use of water are inadequate water conservation practices and prices that are too low to encourage efficiency" (Conference Board of Canada 2014).

4.1.2 Past Experiences with Irrigation Water Pricing

While official statistics on non-water-related farm operations have been collected by Statistics Canada via census over many years, it was not until the 2006 Census of Agriculture that detailed questions were asked about the nature of irrigation on farms (Statistics Canada 2008). Since climate conditions vary widely across Canada, most irrigation takes place in the relatively drier, western part of the country. Census data shows that more than half of the land irrigated in Canada is located in Alberta (60 %), and Alberta farms use more water per unit on irrigated land than other provinces. A follow-up pilot study to examine agricultural water use in 2007 reveals that 73 % of the total volume of water used for irrigation occurred in Alberta, with British Columbia as the second-largest irrigation user (Statistics Canada 2009).

An integral part of the issue around the pricing of irrigation water has to do with the source of the water. While water is owned by the Crown (federal or province), the majority of water for Alberta farms is obtained from 1 of 13 irrigation districts in the province. Irrigation districts operate under the authority of the Alberta Irrigation Districts Act (Alberta Agriculture and Rural Development 2013b). They are administered by elected boards of directors that set policies. Each district owns licenses that identify the volume of water that may be withdrawn and the priority of access. This is based upon FITFIR or first in time, first in right and is according to the date of the license. The districts collectively control 37 % of the licensed volume of water (Alberta Irrigation Projects Association 2002). Each irrigation district provides and maintains the infrastructure required for the transport and delivery of water from surface sources to farms. Irrigation districts serve over 1.3 million acres of irrigated land (Alberta Agriculture and Rural Development 2013b). Farmers pay an annual water charge to irrigation districts, and this is intended to support operations and annual maintenance costs. The irrigation rate is per acre assessed per year. For example, in 2000, one district charged $8.50 (year 2000 dollars) per acre, while another charged $12 per acre. These prices do not keep pace with inflation and, in fact, can even fall in nominal terms, e.g., the per acre price in 2004 for the latter fell to $11 (year 2004 dollars) (Alberta Agriculture and Rural Development 2013a). In the case of the few private farmers in Alberta who own their water licenses, representing about 300,000 acres of land (Alberta Irrigation Projects Association 2002), the province does not charge an annual water rate per acre. However, owners may be subject to a one-time license fee, depending on the volume of water diverted (this is zero if the volume is less than 62,500 m^3 per year and rises to $150 for double the volume.)

Water used by British Columbia farms comes both from on-farm sources (wells or surface water) and from irrigation districts. Self-supplied farmers pay an annual license fee and a storage fee, if applicable. For example, in 2003, a 40-acre farm using 76 acre feet (approximately 94,000 m^3) would pay $41.80 as its water license fee. By way of contrast, the estimated pumping costs are estimated at $1,128 (Tam et al. 2005). Some farmers may purchase water from one of the five major irrigation or improvement districts. Improvement districts are local authorities responsible for providing local services and operate according to the Local Government Act (British Columbia 1996). Actions are governed by elected trustees. Costs are either charged by the acre irrigated or by the total volume of water used. For comparison purposes, the 2004 water rates estimated on a per cubic meter basis were between $0.02 and $0.037, depending upon the district. Districts also set limits on the amount of water allowed per acre. For 2004, these ranged between 0.5 and 1 ft^3 per second per acre (Tam et al. 2005).

Farms in Saskatchewan draw water from both on-farm surface water supplies and off-farm water irrigation districts or private irrigation projects. In Saskatchewan, the largest irrigation project is located in the Lake Diefenbaker development area and made possible by the construction of the Gardiner Dam (Diaz et al. 2009). There are four irrigation development areas, and within each, there are a number of district areas (Saskatchewan Irrigation Projects Association Inc. 2014). While farmers are nominally supposed to pay for delivery and maintenance, they have been given both federal and provincial assistance to convert from dry land to irrigated crops. They have been charged less than the full development cost for the water conveyance infrastructure and face no explicit charge for water (Klein and Kulshreshtha 1991).

As for the rest of Canada, farms in Manitoba use on-farm sources almost entirely, while farms in the Atlantic provinces, Ontario and Quebec, use mostly on-farm surface sources and/or do not irrigate since precipitation is more regular. For them, the marginal cost of irrigation water is essentially the pumping/transportation costs. There is no external charge for the use of water.

In conclusion, water for irrigation purposes has essentially a zero marginal price for most farmers in Canada. Although some farmers pay on the basis of assessed acres irrigated, the charges are so small as to effectively amount approximately $0.01 US $ per cubic meter (Vander Ploeg 2011). Without the incentive to conserve on irrigation water that would arise from facing an increasing (or even positive) marginal cost, water use in the agriculture sector has been inefficient. Crops of low value (forage and field) are irrigated with irrigation methods that are least efficient at delivering the water (sprinkler irrigation) (Statistics Canada 2012).

4.1.3 Past Experiences with Residential Water Pricing

Under the Canadian Constitution, jurisdiction for regulation of matters related to municipalities and to natural resource exploitation resides almost exclusively with the provincial governments. The provinces have a long history of providing regulatory oversight to many features of the operations of municipal water suppliers. This oversight, however, did not extend in the past to a concern for municipal water pricing. The result of this regulatory environment was that municipal water pricing evolved in a very decentralized and largely unregulated fashion in Canada.[1]

For a number of years, detailed information on municipal water prices was collected regularly by Environment Canada in the Municipal Water and Wastewater Survey. Although recently discontinued, past reports from the survey provided snapshots of the state of Canadian residential water pricing. For example, Burke et al. (2004) summarized the evolution of municipal water pricing over the period 1991–1999. Over that time period, there was a remarkable diversity of forms of water prices across Canadian municipalities. The report indicated that 37 % of households paid a fixed monthly fee, independent of the volume of water used, while 39 % paid water prices that were invariant to the amount consumed. In 1999, 13 % and 10 % of Canadian households faced decreasing and increasing block rate structures for water, respectively. Those authors illustrated the implications of the prevalence of flat-rate (i.e., non-volumetric) water charges by demonstrating that

[1]Water supply systems in First Nations communities face their own serious challenges relating to governance issues, source water protection, institutional capacity, and adequate funding. Space does not allow us to discuss these challenges—the interested reader should consult Phare (2012).

water use was 70 % higher on average when consumers faced these types of water charges rather than volume-based rates. One particularly telling implication of these pricing practices is that Canada has developed an unenviable reputation as a jurisdiction with very low water prices. Repeated international comparisons show Canada as having among the lowest residential water prices in the world (OECD 1999, 2004, 2010).

There has been some analysis in the past of the economic characteristics of Canada's residential water prices (Renzetti 1999, 2007, 2009). Statistical analyses of the cost structure for Canadian water agencies indicated that prices were significantly less than the marginal cost of supply. For example, it was found that "Prices charged to residential and commercial customers are found to be only a third and a sixth of the estimated marginal cost for water supply and sewage treatment, respectively" (Renzetti 1999, p. 688). Furthermore, the cost analysis indicated that marginal costs varied by level of output, distance from source, and season. However, past Canadian residential water prices were invariant to these factors, further contributing to inefficient water use. The impacts of inefficient prices included inefficiently high levels of consumption, overbuilt systems, inadequate investment in system maintenance, and impaired aquatic ecosystems. These negative consequences of inefficient water prices were compounded by greater degrees of inefficiency regarding the pricing of sewerage services. In the past, pricing of residential sewerage services was rarely done on a volumetric basis. Instead, sewerage costs were recovered through lot levies or fixed charges independent of wastewater volumes.

Additional evidence of the past state of Canadian residential water prices comes from a survey conducted with Canadian water researchers in which their judgments related to a number of water issues, including pricing (Canada West Foundation 2011a, b; Vander Ploeg and Sommerfeld 2011). The survey results demonstrated that there was a strong degree of consensus surrounding the failings of past water pricing practices. The specific criticisms included the following:

- Municipalities' emphasis on water pricing as a means of generating revenue rather than signaling scarcity and promoting efficient water use
- Municipal water suppliers' failure to account for the complete life-cycle costs of capital and the external costs arising from sewage treatment plants' discharge
- The artificial separation of water supply and sewage treatment in cost accounting and pricing
- Municipal water suppliers' reliance on average rather than marginal costs and failure to design rates to signal differences in marginal cost of supply across users, distance, time of year, and time of day

4.1.4 Past Experiences with Industrial Water Pricing

Industrial water users comprise a wide range of firm types, including manufacturers, resource extraction operations, and mining companies. Smaller manufacturing firms are typically connected to public water and sewage systems, but other larger industrial firms are usually "self-supplied." This means that they withdraw water directly from a surface or groundwater source and may also dispose of their wastewater directly to a receiving body of water. In 2009, 78 % of manufacturers' water supply was self-supplied from surface water sources, while 13 % came from public utilities. The remainder of water intake was self-supplied from groundwater or brackish source waters (Statistics Canada 2012).

In all Canadian provinces, large self-supplied industrial facilities must hold a government-issued permit that specifies the volume and source of intake water. If these facilities also return their wastewater directly to the environment, then they must also hold a separate permit that usually sets limits on volumes of wastewater flows, as well as concentrations of any pollutants in the wastewater. While the volumes of intake water and discharge water are closely linked for a given facility, an important feature of the regulation of industrial water use in Canada is that the permit to take water and the permit to deposit wastewater are usually issued by two different provincial government agencies without coordination.

The provincial water use permit system has been criticized in the past for failing to promote efficient water use (Renzetti 2007; Vander Ploeg and Sommerfeld 2011). In addition to inefficient features such as a lack of transferability of licenses, a lack of coordination with the regulation of wastewater discharges, and incomplete coverage (in some provinces, groundwater withdrawals were not regulated historically), a major part of the criticism has related to the fees charged for permits. Past criticisms of prices have been based on the fact that permit fees were largely administrative and were not designed to promote efficient water use by reflecting the value of water or the opportunity cost of alternative uses. Permit fees varied widely by province with some provinces only charging a one-time application fee (e.g., Ontario, Quebec, and Alberta) while others added a nominal volumetric annual charge (e.g., Saskatchewan and Newfoundland). In those provinces where a fee was charged, the base for the charge was the permitted volume of water rather than actual recorded use. Furthermore, water permit fees were unrelated to the volume and nature of wastewater discharges.

4.1.5 Past Experiences with Pricing of Environmental Services

Like many other countries, Canada's experience with the pricing of environmental services has been largely focused on encouraging agriculture to undertake activities that are less detrimental to the environment. Unfortunately, these efforts have been hampered by other aspects of Canadian agricultural policy that has often operated in opposition (van Kooten 1991). As van Kooten notes, the combination of the financial safety net for farmers and subsidies to encourage capital investment and the draining of wetlands to bring marginal land into productive use have all contributed to a deterioration of environmental quality. Although the National Agri-Environmental Health Analysis and Reporting Program has now expired, its final report noted that water quality had declined since 1981, largely due to the increased use of nutrients (Agriculture and AgriFood Canada 2010). Natural Resources

Canada (NRCAN 2014) notes that much conversion of wetlands to agricultural or urban uses has taken place over the last few decades, with an estimate of 80 % of wetlands in proximity to these locations being subject to such conversion. Since wetlands provide a number of valuable ecosystem services and goods, this loss implies a reduction in environmental goods and services.

Limited efforts to "price" environmental goods/services have involved all three levels of government (federal, provincial, and municipal), as well as Conservation Ontario and its local conservation authorities, and private and nongovernmental organizations. Conservation authorities, in particular, have mandates to promote actions aimed at improving water quality (O'Grady 2011). The majority of programs have been either conservation agreements that involve the retirement of fragile land or payments/tax credits to encourage best management practices/beneficial land use choices. In the case of the former, conservation agreements such as ALUS (Alternative Land Use Services) began with farmers and a private organization interested in preserving waterfowl conservation. Subsequently, the program was expanded to include input from a number of provinces (Manitoba, Prince Edward Island, Ontario, and Saskatchewan). ALUS is based on the concept that farmers should be compensated for the "public goods" that they jointly produce when they make environmentally friendly land use decisions, such as crop residue management, grazing rotation, creation of forage areas, conversion of land to conservation cover, riparian area management, conservation of wetlands, and wildlife management zones. Both provincial and federal governments have provided support money to a number of ALUS programs. Funding has also come from private organizations aimed at improving habitat for waterfowl.

On its own, the federal government began Greencover Canada in 2003. But, this ended in 2009 (Agriculture and Agrifood Canada 2014a). This program was aimed at providing financial support to farmers who removed environmentally sensitive land from farming activities. Successful applicants were given both a one-time payment per acre to cover the costs of seeding or planting forage and trees and a second one-time payment per acre dependent upon a follow-up visit to ensure establishment of perennial cover. The maximum amount of money that a given landowner could receive was $50,000. Almost all of the land enrolled was concentrated in Manitoba, Saskatchewan, and Alberta (Knight 2010).

A separate effort directly focused on water quality improvements was that undertaken by the Grand River Conservation Authority, with the support of the regional government of Waterloo (Ontario). Under the Rural Water Quality Program, farmers had to submit an environmental farm plan to be eligible for either an outright grant or a co-pay subsidy, depending upon the chosen best management practice they intended to implement. In 1998—the first year of the program—for example, the maximum grant was $15,000 for adoption of a manure storage unit. Using data on the first 7 years of the program, Dupont (2010) jointly estimated the yes/no decision to participate in the different components of the program, along with overall participation rates (the extent of participation by farmers within a number of different geographical regions). Not surprisingly, the size of the maximum grant payable and percentage of costs reimbursed were significant determinants of the "yes" decision to participate. The extent of participation was also significantly related to the size of the maximum grant but with diminishing returns. This is consistent with the data that show that the largest share of projects are those that have removed fragile land and put in fencing to prevent livestock from entering stream and river courses (Dupont 2010).

In addition to these co-pay subsidy efforts or conservation agreement programs, a number of provincial governments have provided tax credits to farmers to encourage preservation of riparian strips and prevent livestock fouling of watercourses (Gagnon et al. 2005). In all cases, the amounts of money received by farmers have been either tied to the anticipated costs of undertaking projects and/or land rental rates. In no cases, however, were the amounts linked to the provision of specific environmental services. Moreover, there were no competitive bidding processes associated with the decisions regarding successful applications. Finally, there was no effort to ensure additionality, that is, to ensure that the program actually encouraged new acreage to be seeded, as opposed to providing payments to farmers who had already made the decision to retire lands.

4.2 Present Water Pricing Practices

4.2.1 Present Experiences with Irrigation Water Pricing

Over the last few years, some jurisdictions have made efforts to alter the manner by which irrigation water is priced in an effort to encourage conservation. While a number of irrigation districts in the province of British Columbia charge per acre rates for delivered water, one irrigation district is notable for its efforts to raise prices beyond the basic rate in order to determine the extent to which agricultural water demands can be reduced. The South East Kelowna Irrigation District (SEKID), located in a semiarid part of the interior of the province, has been proactively seeking to enhance understanding of water scarcity and encourage more efficient water use. In 1994, the Irrigation District had water meters installed for the purposes of being able to monitor water use and better inform the decision-making of the irrigators it served (largely, orchard owners). Each metered property was also given tensiometers to assist in irrigation planning. The board of trustees used historical water consumption information gained over a number of years from the use of these meters to identify basic seasonal allocation decisions for each licensed irrigator. Beginning in 2000, irrigators were charged a flat fee for their basic allocation and a volumetric fee for water usage over and above the basic allocation. In 2003, the volumetric fee became a very steep increasing block rate structure that aimed to penalize heavy water users who went over their allotments (Pike 2003). In 2004, the per acre basis charge was \$60 per acre, estimated to be approximately \$0.02 per cubic meter (Tam et al. 2005). However, for example, an irrigator who went 10 % over the allotment faced a volumetric charge of \$0.10 per 1,000 US gallons (since 1,000 US gallons is equivalent to 3.78 m^3, the marginal cost of 1 m^3 is \$0.026). An irrigator who went 50 % over would face a higher block volumetric charge of $0.082 per cubic meter, and someone who went more than 90 % over would face a volumetric price of $0.17 per cubic meter (SEKID 2014). Irrigators assessed excess water use penalties were required to pay their previous year's bills before their water would be turned on for the next season. After controlling for seasonal precipitation, Pike et al. (2007) estimated a 40 % reduction in irrigation water demand associated with the increasing block rate structure.

This same irrigation district recently examined options for water supply improvements (Econics 2013). The actual rate for delivered water for agricultural purposes for 2013 is $72.50 per acre per year (this represents a 21 % increase since 2004 where inflation has increased approximately 17 % over the same period). Among the options considered were annual increases in the per acre water rates for delivered water. Depending upon the assumed level of funding to be received from the provincial government, the anticipated annual increases in the per acre delivered cost of water could be as low 4 % and as high as 13 % (on top of the existing 2013 water rates). These increases, however, are intended to support the infrastructure and do not reflect the scarcity cost of the water itself.

A few irrigation districts in the province of Alberta charge irrigators additional fees for volumes above the annual allocations; however, these are flat-rate fees based upon additional acre inches (Alberta Agriculture and Rural Development 2013a). In 2012, the annual per assessed (cubic meter) charges for farm allocations ranged from $0 to $0.19, with the additional fee per cubic meter ranging from $0 (for those jurisdictions that do not use this) to $0.97 (Alberta Agriculture and Rural Development 2013b). As He and Horbulyk (2010) note, since the marginal price of water is close to zero for most of the irrigators in the province of Alberta, farms do not have an incentive to conserve water, decrease irrigation water demand, or switch to higher-valued crops.

The relatively lower prices for irrigated water in the province of Alberta are not coincidentally matched with the data on irrigation use provided by Statistics Canada (2013). According to the most recent data for 2012, Alberta farmers reported using about three-fourths of the total volume of irrigation water for Canada (1.7 billion cubic meters). This total was more than double the volume reported in 2010; however, reported irrigation volumes in the province of British Columbia were unchanged over the period. For 2012, the crops irrigated were mostly field (61 %) and forage (34 %). There is clearly room for improvement to encourage more efficient use of water for irrigation purposes. For example, in the southern part of the province, the government of Alberta no longer issues new water licenses to growing municipalities, since water is completely allocated. This makes very clear the opportunity costs associated with water and emphasizes the negative aspects associated with inefficient use. Since the province of Alberta is experiencing the fastest population growth in Canada—some 3.5 % per year—some communities are unable to accommodate this growth because water has become the limiting factor (Querengesser 2014). In order to accommodate multiple demands, the province has been actively looking at the potential for water markets to improve the situation (Bjornlund 2010).

4.2.2 Present Experiences with Residential Water Pricing

In recent years, there has been an increasing recognition of the inefficient and unsustainable nature of past municipal pricing practices (Canada West Foundation 2011a, b). This growing awareness has been evidenced by a number of reports drawing attention to the issue and, in some cases, attempting to document the size of the water and wastewater infrastructure "deficit"—that is, the gap between the current state of municipal water and wastewater systems compared to some desired state. For example, according to estimates made by the Federation of Canadian Municipalities in 2007, the deficit exceeds \$30 billion (Federation of Canadian Municipalities 2012).

Some provinces have responded by increasing regulatory requirements for municipal water providers (e.g., Ontario has mandated that municipalities must develop asset management plans and employ full-cost accounting as the basis for rate setting). One troubling trend, however, is the recent decrease in participation by the federal government in the water policy field. This change in position is demonstrated by budget and staff cuts at Environment Canada and the cancellation of the federal government-sponsored survey of municipal water pricing.

The increased level of provincial government scrutiny has had implications for residential water prices. In its report on the 2009 municipal water pricing survey, Environment Canada highlights several recent developments (Environment Canada 2012a, b). First, over the last 30 years, the proportion of households facing volumetric water prices nationally has risen from approximately 55 % to over 80 %. Second, for Canada as a whole, average and marginal water prices have risen recently. At an assumed household consumption level of 25 m^3 per month, for example, marginal prices for water and sewerage combined (averaged across all cities with volumetric rates) rose from \$1.45 in 2006 to \$1.84 in 2009.

As Table 4.1 demonstrates, however, this national average hides a wide range of experiences, as some municipalities have aggressively increased prices while others have increased water rates by less than the general rate of inflation. Toronto, for example, embarked in 2006 on an aggressive program in which water rates were to be increased by 9 % for each of 9 years (City of Toronto 2014) Third, both prices and price structures continue to vary considerably by province (due largely to the large-scale absence of metering in Quebec and the Maritime provinces). Prices also vary by city size, but the impact is much more muted with larger cities having slightly lower prices. Fourth, cities that employ volumetric water prices have much lower residential water use rates. Without correcting for any other factors that might influence demand, Environment Canada reports that the average daily household water use rate is 275 and 437 l in cities with and without volumetric rates, respectively (Environment Canada 2013). Finally, a noteworthy trend relates to the pricing of sewerage services. Part of this trend involves the growing use of volumetric pricing of sewerage services by municipalities. There is some evidence that a growing

Table 4.1 Distribution of metering, rate types, and marginal prices across provinces (2009)

Province/territory	% residential metered[a]	% pop served by treated water[a]	% pop served flat rate[b]	% pop served constant or decreasing block rate[b]	Average of marginal prices[c] (2006$)
Newfoundland and Labrador	0.02	75.4	84.2	7.0	NA
Prince Edward Island	1.5	68.9	31.6	68.4	0.75
Nova Scotia	92.6	68.7	0.2	99.8	1.56
New Brunswick	49.1	52.4	17.9	81.0	1.79
Quebec	16.5	88.5	67.9	14.4	0.91
Ontario	91.2	88.4	2.1	87.3	1.95
Manitoba	97.2	87.8	0.1	99.9	3.07
Saskatchewan	98.2	94.4	0.8	92.2	1.75
Alberta	84.8	90.7	0.2	62.6	1.78
British Columbia	32.6	83.4	20.3	70.1	1.28
Yukon	7.9	4.7	2.6	97.4	1.48
Northwest Territories	97.3	88.0	2.6	97.4	1.48
Nunavut	76.1	28.1	2.6	97.4	1.48

[a]Data are for 2009 (Source: Environment Canada 2010)
[b]Data are for 2006 (Source: Environment Canada n.d.). Source only provides single estimate for all territories (Yukon, Northwest Territories, and Nunavut). Provinces for which figures do not sum to 100 % use increasing block rates for the remaining population
[c]Marginal price is calculated at a 25 m^3/month consumption level. Rate is for water and sewer. Averaged over those municipalities reporting volumetric rate structures

number of cities and towns are shifting toward pricing sewerage directly (on a volumetric basis) or charging for sewerage on a volumetric basis but setting the unit price as a percentage of the price of water. Another part of the trend is that the relative costs of water supply and sewerage services have changed in recent years. Across Canada, many municipalities' sewerage charges have been rising more rapidly than water prices. A result of this is that, for Canada as a whole, sewerage charges now constitute 48 % of Canadian households' combined water and sewerage bills.

Despite these positive developments, Canadian residential water prices still lag in some important respects. There continues to be little evidence of municipalities being willing to adopt innovations, such as seasonal prices, time-of-use prices, or zonal pricing. An exception is the city of Vancouver, which recently introduced seasonally differentiated water prices for those residential customers who are metered. The water rate is $2.385 per 100 ft^3 (2.83 m^3) during October to May and $2.988 for the remainder of the year.[2]

[2] http://vancouver.ca/home-property-development/metered-rates.aspx

4.2.3 Present Experiences with Industrial Water Pricing

Most major industrial water users remain self-supplied, and total industrial water use has risen with increasing levels of output recently. Interestingly, however, Bruneau and Renzetti (2010) demonstrate that the intensity of manufacturing water use (i.e., volume of water intake relative to the value of output) fell by 4 % over the period 1981–1996. The major source of increases recently in industrial water use has occurred due to increases in the rate of extraction from Canada's oil sands (Rivers and Groves 2013). At the same time, a number of provincial governments have responded to the growing pressures on their water resources by adopting new water management frameworks. In some cases, these reforms have included changes to permitting systems for self-supplied water users. Some provinces have raised or are planning to raise water license fees where they were already in place (Newfoundland and Labrador), and others have added an annual rental fee to supplement the one-time application fee (Vander Ploeg and Sommerfeld 2011). Ontario, Canada's most populous province, is a case in point. Historically, large self-supplied water users were able to gain access to water through the payment of a one-time application fee. In recent years, however, Ontario has announced its plan to introduce modest annual rental fees for withdrawals by certain categories of commercial and industrial water users. In the first step of this plan, firms in water bottling; food and beverage processing; concrete manufacturing; pesticide, fertilizer, and other agricultural chemical manufacturing; and inorganic chemical manufacturing must pay a fee of \$3.71 per 1,000 m^3 annually (Ontario 2014). Other provinces have moved to bring groundwater use under management (British Columbia).

Alberta stands out among provinces because, rather than maintain control over water withdrawal permits and set administrative fees for those permits, it has introduced a limited market for water permits—allowing holders of permits to lease or sell water rights. Despite research employing numerical simulation methods showing potential welfare gains from water trades (Horbulyk and Lo 1998), participation by permit holders has been low. This situation has led a number of researchers to point to a number of features of the Alberta water market, which appear to be constraining participation (Bjornlund 2010; Adamowicz et al. 2011). These features include uncertainty about how instream flow needs would be dealt with during droughts and how the existing FITFIR seniority system can be integrated into Alberta's water market.

It appears that recent policy reforms have created the regulatory environment to potentially improve water management. Fewer sectors and sources of water are exempt from regulations. Record-keeping and decision-making processes have improved. Despite these improvements, however, there still remain major deficiencies in provincial permitting systems and, in those provinces that charge for self-supplied water use, fees remain very low (Horbulyk 2010). The major shortcomings continue to be that there is no mandated or practiced linkage between the value of water and how application fees or annual rentals (when they exist) are calculated. Furthermore, in most provinces, licenses are not transferable. Finally, there continues to be a fundamental disconnect between the way in which water withdrawals and the deposition of wastewater flows are regulated.

4.2.4 Present Pricing Experiences of Environmental Services

The federal government has recently replaced Greencover Canada with Growing Forward 2, a 5-year policy framework (Agriculture and Agrifood Canada 2014b). This initiative is shared with the provinces and has a component specifically aimed at improving water quality (with an overall budget of $2 billion). The framework involves adoption of cost-sharing programs similar to those employed by the Rural Water Quality Program in the Grand River (and discussed in Sect. 4.1.5). Under Growing Forward 2, while following the federal framework, each province is encouraged to manage its own set of programs in order to meet local conditions, e.g., Alberta farms must create a long-term water management plan prior to applying for program support (Government of Alberta 2014). Like the Rural Water Quality program, payments to farmers are based on the anticipated costs of various farm improvements but unrelated to the potential types of environmental services provided by better quality water.

Aside from the federal government continuing with programs that are similar to those adopted in the past, there are a few novel programs in place. As a result, Canada lags many other countries in terms of efforts to use economic frameworks and tools. First, the South Nation River Total Phosphorus Management Program allows point-to-nonpoint emissions trading (O'Grady 2011). This was begun as a pilot project in 2000 and involved a number of provincial ministries, municipal governments, farmers, and the local conservation authority (South Nation Conservation). It allowed industrial and municipal wastewater dischargers to purchase phosphorus offsets from farmers. For 2010, the "price" per kilogram of phosphorus that new dischargers put into the river is $390 (Knight 2010). This money goes into a fund administered by the South Nation Conservation and used to provide support funding for a number of best management practices. Despite the documented success of this small program, it has not been repeated elsewhere in Canada.

Second, there has been an effort to introduce a competitive aspect to the determination of remuneration for the conversion of land from production into land capable of providing environmental services. This is the use of a reverse auction (Hill et al. 2011). Using two different rounds of bidding, the participants provided bids related to changes in either the extent of cultivated cropland or perennial forage. In addition, bids were evaluated against potential environmental benefits associated with predictions in the number of hatched waterfowl nests. Successful bids ranged between $20.83 and $391.22 per acre per year (with an average bid of $118.52) under conditions of a 12-year agreement. Not surprisingly, higher bids were associated with cropland than with land for forage, due to the marginal value to farmers associated with cropland. However, the only successful bids were ones on forage land, because they combined lower bid values with greater environmental benefits in terms of the number of hatched waterfowl nests.

Third, using stated preference methods, Pattison et al. (2011) estimated the willingness to pay (WTP) of residents of the province of Manitoba for wetland retention and restoration. Survey respondents were shown possible benefits from improved wetlands, namely, water filtration, flood control, reductions in soil erosion, and carbon capture. Estimates of the total WTP for specific levels of improvements ranged between \$296 and \$326 per household, per year for a 5-year program. The upper-end values were associated with full restoration of wetlands back to a 1968 level. Unfortunately, these values are essentially the only Canadian values available for researchers and policy analysts to employ when examining policies that have implications for water quality and, hence, ecosystem services. This is problematic, given the findings of Johnston and Thomassin (2010) in a meta-analysis of water quality values using American and the few Canadian studies that exist. Namely, they find a systematic downward pattern in Canadian WTP, relative to American values, even when examining otherwise identical policies that provide water quality improvements.

4.3 Current Debates and Future Directions

4.3.1 Cross-Cutting Issues and Challenges

We will briefly highlight a number of issues that are relevant to many Canadian water suppliers and receiving a significant amount of attention at all levels of government. Firstly, there is a growing concern regarding the potential impacts of extreme weather events on municipal infrastructure. Major recent floods in Calgary and Toronto caused billions of dollars in damages to private property and public infrastructure, and heightened fears that municipal water systems built with combined storm sewers and based on possibly outdated climatic assumptions are vulnerable (Freek and Sanford 2014). The impacts of required changes to municipal infrastructure arising from changes in hydrologic and climatic conditions, however, have not received sufficient research to know what the implications will be for water prices.

Another challenge arises from the previously mentioned failure on the part of Canadian governments to adequately price externalities associated with diminished water quality. The rapid increase in the number and severity of large-scale algal blooms being observed in Canada is illustrative of how the impacts of this policy failure are being compounded by climate change. These blooms are caused by excessive nutrient loadings (phosphorus, nitrogen) that stem primarily from agricultural operations and sewage treatment systems (International Joint Commission 2014). Furthermore, these blooms are known to pose serious threats to human and ecosystem health and to cause significant economic damages in reduced recreation opportunities, decreased property values, and increased water treatment costs. These massive blooms are occurring on some of Canada's largest lakes, including Lake Winnipeg, Lake Erie, and Lake Simcoe.

4.3.2 New Approaches for Water Pricing

The preceding sections illustrate how Canadian water prices have historically been quite low, and only recently has there been some movement on pricing reform. There are still substantial sources of inefficiency for Canadian water prices: the failure to price externalities arising from diminished water quality and the lack of connection between municipal water prices and the factors determining marginal costs of supply are only two important examples. Nonetheless, there are some limited instances of new approaches for water pricing in Canada at the provincial and municipal levels, which we discuss here.

First, at the provincial level, British Columbia's recently announced Water Sustainability Act represents a significant increase in the scope of one provincial government's efforts to manage water and regulate water use. For the first time, groundwater withdrawals will be regulated in the same fashion as surface water withdrawals. In addition, the government has signaled an intention to raise water use fees. The complexity of the political and administrative challenges associated with doing this are reflected in the following principles, which the government says will guide its choices for new water withdrawal charges: simplicity, fairness and equity, implications for water users, impacts on water resources, cost recovery, efficiency, food security, and public health (British Columbia 2014). A number of other provinces are also moving slowly in the direction of higher fees for self-supplied water users. However, fees remain remarkably low and remain disconnected from water discharges, and there appears to be continuing institutional hesitancy to employ economic instruments more widely or more aggressively other than where they already exist (Adamowicz 2007; Kenny et al. 2011). The debate concerning the appropriate role for economic instruments, including water pricing, is particularly strong in the case of water used in Canada's oil sands (Griffiths et al. 2006; Horbulyk 2010; Adamowicz et al. 2011).

Second, there are also a number of initiatives that are in the pilot or testing stage at a number of Canadian cities (Brandes et al. 2010). First, there is a growing interest among provincial governments to have water agencies move toward full-cost accounting. While typically not including external environmental costs, these rules would require complete life-cycle accounting for capital costs. A second initiative concerns developing pricing structures for water and sewer agencies' collection and treatment of storm runoff water. The fact that most Canadian cities have combined sewer-stormwater systems, combined with the apparent increasing frequency and severity of storm events, has motivated a number of Canadian cities to investigate the introduction of a separate charge for the provision of stormwater services (Saxe 2009). Since the service provided by the municipal agency is not directly related to a customer's water use, cities have been considering using the size of property as a basis for the charge. Richmond Hill (a rapidly growing municipality north of Toronto) is moving in this direction with a phased-in stormwater management charge (Richmond Hill 2014). Finally, sewerage pricing is likely to become more important and perhaps surpass water supply costs for most households. Provinces and municipalities will have to invest in sewage treatment infrastructure upgrades in order to be in compliance with recently revised federal regulations that raised the required level of treatment for many sewage systems in Canada (Environment Canada 2012b). While undoubtedly promising improved water quality, these regulations will be costly to implement, and the impacts will likely be greatest for those smaller and rural systems that are furthest from compliance.

References

Adamowicz, W. (2007). Reflections on environmental policy in Canada: Presidential address to the Canadian Agricultural Economics Society. *Canadian Journal of Agricultural Economics, 55*, 1–13.

Adamowicz, W., Percy, D., & Weber, M. (2011). *Alberta's water resource allocation and management system: A review of the current water resource allocation system in Alberta*. Report prepared for the Alberta Water Research Institute's project on Financial Market-Based Instruments for Sustainable Water Management. Retrieved July 8, 2010, from website: http://www.seawa.ca

Agriculture and AgriFood Canada. (2010). *Environmental sustainability of Canadian agriculture: Agri-Environmental Indicator*. Report Series—Report No. 3. 2010. Retrieved July 29, 2014, from website: http://publications.gc.ca/collections/collection_2011/agr/A22-201-2011-eng.pdf

Agriculture and AgriFood Canada. (2014a). *Greencover Canada*. Retrieved July 17, 2014, from website:http://www.agr.gc.ca/eng/programs-and-services/list-of-programs-and-services/expired-programs-and-services/greencover-canada/?id=1181580137261

Agriculture and AgriFood Canada. (2014b). *Growing forward 2*. Retrieved July 17, 2014, from website: http://www.agr.gc.ca/eng/about-us/key-departmental-initiatives/growing-forward-2/?id=1294780620963

Alberta Agriculture and Rural Development. (2013a). *Irrigation in Alberta: A statistical overview 2012*. Basin Water Management Branch. Retrieved June 3, 2014, from website: http://www1.agric.gov.ab.ca/$Department/deptdocs.nsf/all/irr7401/$FILE/altairriginfo2012.pdf

Alberta Agriculture and Rural Development. (2013b). *Alberta irrigation information, 2012*. Basin Water Management Branch, Irrigation and Farm Water Division.

Alberta Irrigation Projects Association. (2002), *Irrigation in the 21st century* (Summary Report, Vol. 1). Lethbridge: AIPA.

Bjornlund, H. (2010). *The competition for water: Striking a balance among social, environmental, and economic needs* (C.D. Howe Institute Commentary No. 302). Retrieved April 10, 2014, from website www.cdhowe.org/pdf/commentary_302.pdf

Brandes, O., Nowlan, L., & Paris, K. (2008). *Going with the flow? Evolving water allocations and the potential and limits of water markets in Canada*. Report prepared for the Conference Board of Canada. Retrieved August 2, 2014, from website: http://www.conferenceboard.ca

Brandes, O., Renzetti, S., & Stinchcombe, K. (2010). *Worth the charge: A primer on conservation-oriented water pricing*. POLIS Project on Ecological Governance, University of Victoria.

British Columbia. (1996). *Local government act, revised statues of British Columbia chapter 323*. Retrieved August 18, 2014, from website: www.bclaws.ca/

British Columbia. (2014). *Pricing B.C.'s water*. Report retrieved March 18, 2014, from website http://engage.gov.bc.ca/watersustainabilityact/

Brock University. (1999). *Canada [computer file]*. St. Catharines: Brock University Map, Data & GIS Library. Available: Brock University Map, Data & GIS Library Controlled Access. Retrieved August 20, 2014, from website: http://www.brocku.ca/maplibrary/maps/outline/local/Canada.jpg

Bruneau, J., & Renzetti, S. (2010). Water use and the composition of production in Canada. *Journal of Water Resources Planning and Management, 136*, 72–79.

Burke, D., Leigh, L., & Sexton, V. (2004). *Municipal water pricing 1991–1999*. Ottawa: Department of the Environment. Report retrieved December 5, 2005, from website http://www.ec.gc.ca/eau-water/default.asp?lang=En&n=ED0E12D7-1

Canada West Foundation. (2011a). *Summary of recent Canadian literature on water pricing*. Canada Water Policy Backgrounder report 5. Retrieved April 8, 2014, from website http://cwf.ca/projects/water-pricing

Canada West Foundation. (2011b). *Water, water use & water pricing around the world* (Canada Water Policy Backgrounder report 8). Retrieved April 8, 2014, from website http://cwf.ca/projects/water-pricing

City of Toronto. (2014). *Financial condition and performance: City of Toronto 2014*. Retrieved August 1, 2014, from website: http://www1.toronto.ca/City%20Of%20Toronto/Strategic%20Communications/City%20Budget/2014/PDFs/Public%20Book/Financial%20Condition%20and%20Performance.pdf

Conference Board of Canada. (2014). *How Canada performs: Water withdrawals*. Retrieved April 3, 2014, from website: http://www.conferenceboard.ca/hcp/details/environment/water-consumption.aspx

Diaz, H., Hurlbert, M., Warren, J., & Corkal, D. (2009). *Saskatchewan water governance assessment final report unit 1E institutional adaptation to climate change*. Project. Report retrieved July 20, 2014 from website: http://www.parc.ca/mcri/pdfs/papers/gov01.pdf

Dupont, D. P. (2010). Cost-sharing incentive programs for source water protection: The Grand River's rural water quality program. *Canadian Journal of Agricultural Economics, 58*, 481–496.

Econics. (2013). *South East Kelowna irrigation district water rate review 2013*. Report retrieved July 31, 2013, from website: http://www.sekid.ca/documents/SEKIDWATERRateReviewReportDRAFTV5.pdf

Environment Canada. (2010). *2010 Municipal water use report: Municipal water use, 2006 statistics*. Ottawa: Environment Canada.

Environment Canada. (2012a). *2011 Municipal water pricing report: Municipal water pricing 2009 statistics*. Ottawa: Environment Canada.

Environment Canada. (2012b). *Factsheet: Environment Canada's wastewater systems effluent regulations*. Report retrieved March 3, 2014, from website http://www.ec.gc.ca/eu-ww/

Environment Canada. (2013). *Water-how we use it*. Report retrieved April 11, 2014, from website http://www.ec.gc.ca/eau-water/default.asp?lang=En&n=0BBD794B-1#a10

Environment Canada. (n.d.). *Municipal water and wastewater survey: Municipal water pricing, 2009 summary tables*. Report retrieved April 20, 2014, from website: http://ec.gc.ca/eau-water/default.asp?lang=En&n=ED0E12D7-1

Federation of Canadian Municipalities. (2012). *Canadian Infrastructure Report Card volume 1: Municipal roads and water systems*. Report retrieved March 15, 2014, from website: http://www.fcm.ca/home/resources/reports.htm

Freek, K., & Sanford, R. (2014). *Flood forecast: Climate risk and resiliency in Canada* (An RMB Manifesto). Calgary: Rocky Mountain Books.

Gagnon, B., St-Onge, F., & Gagnon, M-F. (2005). *Remuneration for ecological goods and services produced by agriculture: Elements for a Quebec analysis*. Agri-environmental Policy Branch. Ministère de l'Agriculture, des Pêcheries et de l'Alimentation du Québec. Library and Archives Canada.

Government of Alberta. (2014). *Growing forward 2*. Report retrieved August 19, 2014, from website: http://www.growingforward.alberta.ca/Programs/index.htm?contentId=ONFARM_WTRMGT_PRG&useSecondary=true&active=yes

Griffiths, M., Taylor, A., & Woynillowicz, D. (2006). *Troubled waters, troubling trends: Technology and policy options to reduce water use in oil and oil sands development in Alberta*. Pembina Institute. Report retrieved August 9, 2010, from website www.pembina.org/reports/TroubledW_Full.pdf

He, L., & Horbulyk, T. (2010). Market-based policy instruments, irrigation water demand, and crop diversification in the Bow River basin of southern Alberta. *Canadian Journal of Agricultural Economics, 58*, 191–213.

Hill, M., McMaster, D., Harrison, T., Hershmilller, A., & Plews, T. (2011). A reverse auction for wetland restoration in the Assiniboine River Watershed, Saskatchewan. *Canadian Journal of Agricultural Economics, 59*, 245–258.

Horbulyk, T. (2010). *Water pricing: An option for improving water management in Alberta. Report prepared for the Alberta Water Research Institute's project on Financial Market-Based Instruments for Sustainable Water Management*. Retrieved July 8, 2010, from website: http://www.seawa.ca

Horbulyk, T., & Lo, L. (1998). Welfare gains from potential water markets in Alberta, Canada. In K. W. Easter, M. Rosegrant, & A. Dinar (Eds.), *Markets for water: Potential and performance* (pp. 241–257). Boston: Kluwer Academic Press.

International Joint Commission. (2014). *A balanced diet for Lake Erie: Reducing phosphorus loadings and harmful algal blooms*. Report of the Lake Erie Ecosystem Priority. Retrieved August 3, 2014, from website: www.ijc.org

Johnston, R., & Thomassin, P. (2010). Willingness-to-pay for water quality improvements in the United States and Canada: Considering possibilities for international meta-analysis and benefit transfer. *Agricultural and Resource Economics Review, 39*, 114–131.

Kenny, A., Elgie, S., Sawyer, D., & Wichtendahl, C. (2011). *Advancing the economics of ecosystems and biodiversity in Canada: A survey of economic instruments for the conservation & protection of biodiversity*. Report prepared for Sustainable Prosperity Institute, Ottawa. Retrieved April 8, 2014, from website http://www.sustainableprosperity.ca/article1431

Klein, K., & Kulshreshtha, S. N. (1991). Irrigation development and cost of general agricultural programs: A case study of the South Saskatchewan River Irrigation District. *Canadian Water Resources Journal, 16*, 143–160.

Knight, T. (2010). *Enhancing the flow of ecological goods and services to society. Key principles for the design of marginal and ecologically significant agricultural land retirement programs in Canada*. Report prepared for the Canadian Institute for Environmental Law and Policy. Retrieved February 17, 2010, from website: http://www.cielap.org/pdf/EnhancingTheFlow.pdf

National Round Table on the Environment and the Economy. (1996). *State of the debate: Water and wastewater services in Canada*. Ottawa: National Round Table on the Environment and the Economy.

Natural Resources Canada. (2014). *Areas of wetlands by ecozone*. Retrieved July 14, 2014, from website: http://www.nrcan.gc.ca/forests/canada/sustainable-forest-management/criteria-indicators/13299

O'Grady, D. (2011). Sociopolitical conditions for successful water quality trading in the South Nation River Watershed, Ontario. *Canada Journal of the American Water Resources Association, 47*, 39–51.

Ontario. (2014). *Water taking report and charges*. Retrieved May 28, 2014, from website: https://www.ontario.ca/environment-and-energy/water-taking-report-and-charges

Organisation for Economic Co-operation and Development. (1999). *The price of water: Trends in OECD countries*. Paris: Organization for Economics Co-operation and Development.

Organisation for Economic Co-operation and Development. (2004). *OECD environmental data: Compendium 2004*. Paris: Organization for Economics Co-operation and Development.

Organisation for Economic Co-operation and Development. (2010). *Pricing water resources and water and sanitation services. OECD Environment Directorate*. Paris: Organization for Economics Co-operation and Development.

Pattison, J., Boxall, P., & Adamowicz, W. (2011). The economic benefits of wetland retention and restoration in Manitoba. *Canadian Journal of Agricultural Economics, 59*, 223–244.

Phare, M.-A. (2012). *Denying the source: The crisis of first nations water rights* (An RMB Manifesto). Calgary: Rocky Mountain Books.

Pike, T. (2003) *Agricultural metering program review and metered rate options internal report South East Kelowna Irrigation District*. http://www.sekid.ca/pdf/reports/Agricultural%20Metering%20Program%20Review.pdf. Accessed 10 Jun 2014.

Pike, T., Campbell, I., Neilsen, D., & Aït-Ouyahia, M. (2007). *Does pricing water reduce agricultural demand? An example from British Columbia. Government of Canada* (Policy Research Initiative. Sustainable Development Briefing). Report retrieved June 5, 2014, from website: http://www.obwb.ca/fileadmin/docs/sekid_pricing_impacts.pdf

Querengesser, T. (2014). *Alberta's future water wars.* Alberta Venture. January 13, 2014. Report retrieved June 5, 2014, from website: http://albertaventure.com/2014/01/future-water-wars-alberta.

Renzetti, S. (1999). Municipal water supply and sewage treatment: Costs, prices and distortions. *Canadian Journal of Economics, 32*, 688–704.

Renzetti, S. (2007). Are the prices right? Balancing efficiency, equity, and sustainability in water pricing. In K. Bakker (Ed.), *Eau Canada: The future of Canada's water* (pp. 263–279). Vancouver: UBC Press.

Renzetti, S. (2009). *Wave of the future: The case for smarter water pricing.* C.D. Howe Institute Commentary 281, Toronto.

Richmond Hill. (2014). *Water and wastewater billing and collection.* Retrieved March 20, 2014, from website: http://www.richmondhill.ca/subpage.asp?pageid=finance_water

Rivers, N., & Groves, S. (2013). The welfare impact of self-supplied water pricing in Canada: A computable general equilibrium assessment. *Environmental and Resource Economics, 55*, 419–445.

Saskatchewan Irrigation Projects Association Inc. (2014). *Districts and areas.* Retrieved August 18, 2014, from website: http://www.irrigationsaskatchewan.com/SIPA/sipa_013.htm

Saunders, O., & Wenig, M. (2007). Whose water? Canadian water management and the challenge of jurisdictional fragmentation. In K. Bakker (Ed.), *Eau Canada* (pp. 119–142). Vancouver: University of British Columbia Press.

Saxe, D. (2009, July/August). *Rain check: Getting a charge out of stormwater.* Canadian Water Treatment, 40–42. Retrieved May 5, 2010, from website: http://www.watertreatment.ca/

South East Kelowna Irrigation District. (2014). *SEKID agricultural metering program.* Retrieved June 4, 2014, from website. http://sekid.ca/agri_metering_program.html

Statistics Canada. (2008). *Snapshot of Canadian agriculture. 2006 census of agriculture.* Retrieved June 5, 2014, from website: http://www.statcan.gc.ca/ca-ra2006/articles/snapshot-portrait-eng.htm

Statistics Canada. (2009). *EnviroStats 2007* (Catalogue no. 16-002-X, Vol. 3 no 3). Ottawa: Minster of Industry.

Statistics Canada. (2012). *Industrial Water Use 2009* (Catalogue no. 16-401-X). Ottawa: Ministry of Industry.

Statistics Canada. (2013). *Agricultural water use in Canada 2012* (Catalogue no 16-402-X). Ottawa: Ministry of Industry.

Tam, S., Nyvall, T. J., & Brown, L. (2005). *British Columbia irrigation management guide.* Report retrieved August 20, 2014, from website: www.agf.gov

Van Kooten, G. C. (1991). Improving policy instruments for sustainable agriculture. *Canadian Journal of Agricultural Economics, 39*, 655–663.

Vander Ploeg, C. (2011). *Water pricing. Seizing a Public Policy Dilemma by the Horns. Canadian water policy backgrounder.* Prepared for the Canada West Foundation. Report retrieved September 30, 2011, from website: http://cwf.ca/pdf-docs/publications/Water_Backgrounder_8_Sept_2011.pdf

Vander Ploeg, C., & Sommerfeld, L. (2011). *Charging for water use in Canada: A workbook of the central principles, key questions, and initial steps.* Report prepared for the Canada West Foundation. Retrieved April 8, 2014, from website: http://cwf.ca/projects/water-pricing

Chapter 5
Water Pricing in Chile: Decentralization and Market Reforms

Guillermo Donoso

Abstract The water sector in Chile underwent major changes as a result of decentralization and market reforms. This chapter focuses on recent pricing experiences in the urban residential and rural sectors. Over the last 30 years, the Chilean government has successfully incorporated private participation in the water and sanitation sector and implemented a regulatory framework that has contributed to cost recovery and affordability of the reform. The service offered has greatly improved in quality and coverage reaching, in 2013, 99.9 % of urban population. National coverage of sewage treatment has significantly increased from 17 % in 1999 to 99.8 % in 2013. However, the privatization and decentralization of water utilities is facing new challenges, such as increasing extreme climatic events and a more informed and organized consumer base. In addition, there are concerns with respect to sustainability of groundwater extraction and deterioration of water-dependent ecosystems due to over allocation of water rights. This chapter also presents an overview of Chile's national Rural Potable Water (APR) program, which has reached almost 100 % coverage in semiconcentrated rural areas. Unlike urban service providers, the rural water-supply and sanitation sector has not been subject to regulation like urban services.

Keywords Chile • Water affordability • Urban water and sanitation sector reform • Rural water subsidy • Water markets

5.1 Introduction

A long narrow strip of land, Chile's unique geography provides a variety of climatic conditions and a number of short river valleys running from the Andes to the Pacific Ocean. Two primary mountain ranges, the Andes and the Coastal Mountains, span the length of central Chile and provide the limits to the coastal plain and the central

G. Donoso (✉)
Department of Agricultural Economics, Pontificia Universidad Católica de Chile, Vicuña Mackenna 4860, Macul 7820436, Chile
e-mail: gdonosoh@uc.cl

A. Dinar et al. (eds.), *Water Pricing Experiences and Innovations*,
Global Issues in Water Policy 9, DOI 10.1007/978-3-319-16465-6_5

valley. Chile's total land area is 743,800 km^2, of which 21.2 % is agricultural land (157,687 km^2) and 21.8 % is forest (162,148 km^2). Arable agricultural land is 1,294,000 hectares (ha), which is 1.7 % of the total land surface. Chile has just over 1 million ha of irrigated agricultural land. Urban area covers approximately 0.06 % of total surface. Currently, the area of wetlands in Chile is about 5 million ha, which is equivalent to 5.9 % of Chile's total land area.

In the last 30 years (1980–2010), Chile's real GDP has grown at an annual rate of 6.2 %. The economy is based mainly on exports concentrated on natural resource production processes that are highly dependent on water, such as mining and agriculture. Chile had a per capita GDP measured in purchasing power parity of US$15.331 in 2013.

Precipitation ranges from near zero in the north to an annual 2,000 mm in the south. Additionally, the spatial distribution of water flows follows the same pattern as rainfall, generating three hydrologic systems: the dry Pacific, central Chile, and southern humid Pacific systems (Fig. 5.1). The characteristics of these systems are the following:

Dry Pacific: In this system water flows reach their peak during the summer months (November–February), which coincide with rainy season of the Bolivian Altiplano. Thus, water flows in these basins are mainly rain driven. Average water flows in this sector are 45 m^3/s.

Central Chile: This system has large snow pack reserves, and water flows are highest during the summer months due to snowmelt. Water flows are significantly greater than those of the dry Pacific system, reaching an average 2.800 m^3/s.

Southern Humid Pacific: Higher rainfall and lower temperatures increase the annual water flow/annual rainfall to values close to 0.9. Rivers in the north of this hydrological system present a mixed regime, snowmelt and rainfall. Toward the south, water flows become more driven by rainfall. This system presents an average water flow of 27.600 m^3/s, the highest water flows of all three systems.

Water withdrawals in Chile average approximately 4,000 m^3/s/year (World Bank 2011). Of this, almost 85 % is used in nonconsumptive hydroelectric generation. Consumptive water use in Chile is dominated by irrigation, with 73 % of consumptive water use. Industrial use of water is 12 % of consumptive withdrawals, and mining and potable water supply account for 9 % and 6 % of total water consumptive water use, respectively. It is interesting to note that all consumptive water uses have increased since 1990; total consumptive water use has increased 13 % since 1990. Industry is the sector with the highest consumptive water-use increase (79 %), followed by potable water and mining (48 % and 46 %, respectively).

The five classes of water-consuming activity with the largest share of GDP were manufacturing (12 %); retail, restaurants, and hotels (10 %); mining (8 %); agricul-

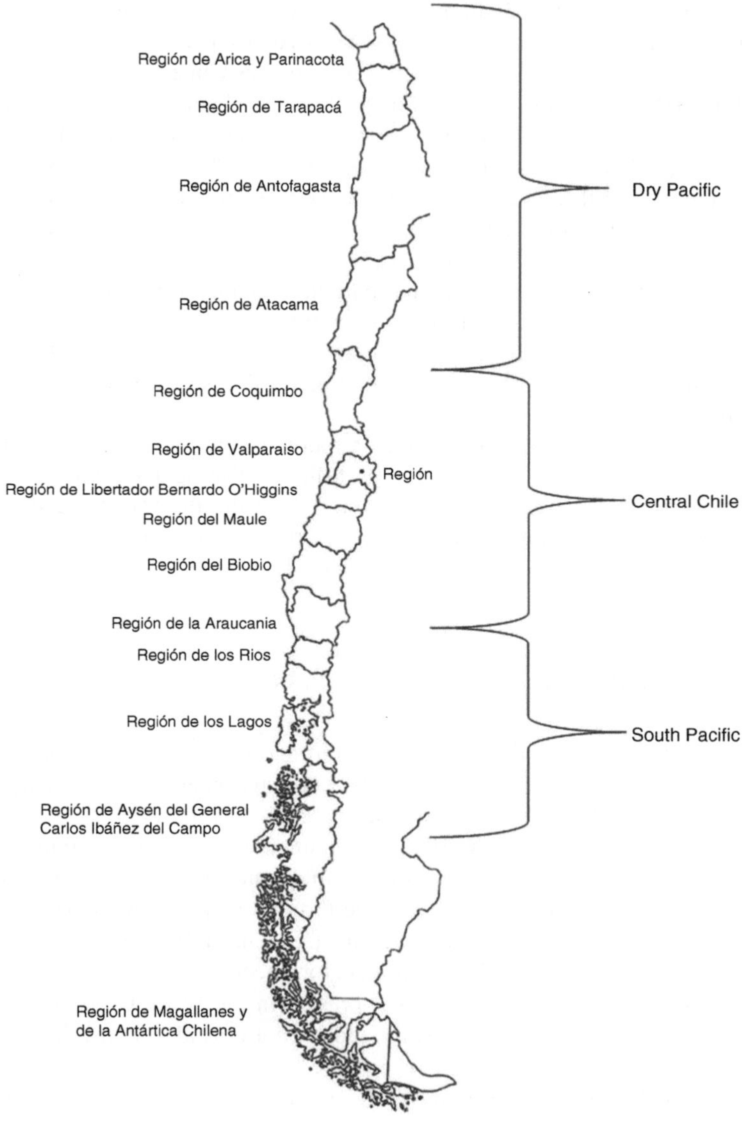

Fig. 5.1 Map of Chile

ture and forestry (4 %); and electricity, gas, and water (3 %). In 2005, the contribution to merchandise exports were mining (57 %); agriculture, forestry, and fishing (7 %); and industrial (31 %) (World Bank 2011).

In the dry Pacific hydrological system, the scarce water resources are divided among Chile's principle mining operations, agriculture toward the south of this system, and a sparse population. The northern portion of the central Chile hydrological system, irrigated agriculture is important and concentrated on fruit crops grown for international markets. Additionally, nearly one-third of the nation's population is located in this system. The southern humid Pacific concentrates Chile's forests, fisheries, and aquaculture industries. A low density of population also characterizes this system.

Chile has a high level of coverage of water, sewerage, and wastewater treatment systems. In 2013, water and sewer coverage reached 99.9 % and 96.3 % of total urban population (SISS 2013), while wastewater treatment coverage was 99.8 % (SISS 2013). On the other hand, only 72 % of the rural population had access to improved potable water (WRI 2003).

During the late 1970s, the economic paradigm changed from one in which the state must protect and oversee optimal allocation of resources to one in which the market is responsible for allocating resources in an efficient manner. The government thus introduced neoliberal economic policies that supported private property rights and free markets.

This chapter focuses on recent pricing experiences in the urban residential and rural sectors. The chapter is structured as follows. The next section presents the market reforms in the water sector. Section 5.3 covers the transformation of the urban water and sanitation sector, while the rural water sector is described in Sect. 5.4. Finally, Sect. 5.5 concludes the paper.

5.2 Market Reforms in the Water Sector

The case of Chile is illustrative of a transition from command and control to market-based water management policy, in which economic incentives play a significant role in water-use rights (WUR) allocations.

The Water Code of 1981 (WC 1981) maintained water as "national property for public use," but granted permanent, transferable water-use rights to individuals to reach an efficient allocation of the resource through market transactions of water-use rights (WUR). The WC 1981 allowed for freedom in the use of water to which an agent has WUR; thus, WUR are not sector specific. Similarly, the WC 1981 abolishes the water-use preferential lists, present in the Water Codes of 1951 and 1967. Additionally, WUR do not expire and do not consider a "use it, or lose it" clause.

The WC 1981 established that WUR are transferable in order to facilitate WUR markets as an allocation mechanism. Although private water-use rights existed in Chile prior to 1981, the previous water codes restricted the creation and operation of efficient water markets. The framers of the 1981 Water Code sought to achieve the efficiencies of market reallocation of water; the objective of the governmental action in this field was to create solid water-use rights in order to facilitate the proper operation of the market as an allocation mechanism. Thus, the WC 1981 was designed to protect traditional and customary WUR and to foster economically beneficial reallocation through market transfers.

WUR markets have received wide attention, both in Chile and internationally. Although market reallocation of water has not been common throughout most of Chile, the existence of water markets has been documented. Studies have shown active trading for WUR in the Limarí Valley, where water is scarce with a high economic value, especially for the emerging agricultural sector. Inter-sectoral trading has transferred water to growing urban areas in the Elqui Valley and the upper Mapocho watershed, where water companies and real estate developers are continuously buying water and account for 76 % of the rights traded during the 1993–1999 period. Other studies have shown limited trading in the Bío Bío, Aconcagua, and Cachapoal valleys. In all of these studies, some permanent transactions of water-use rights have occurred.

A key conclusion of these studies is that water markets are more prevalent in areas of water scarcity. They are driven by demand from relatively high-valued water uses and facilitated by low transaction costs in those valleys where water user associations (WUAs) and infrastructure present assist the transfer of water. In the absence of these conditions, trading has been rare and water markets have not become institutionalized. It should be noted that during the 2000s, the market was more active than in the previous two decades, 1980s and 1990s. This is largely due to a slow maturation in the public's knowledge concerning the new legislation.

The average permanent WUR price is US$215,623 per WUR (Hearne and Donoso 2014). Permanent WUR prices in the north of the country are greater than in the South, which indicates that the market at least in part reflects the relative scarcity of water. WUR prices present a standard deviation of US$100,460,800 per WUR; price dispersion is lower in the more active WUR markets. Thus, Chilean WUR markets are characterized by a large price dispersion for homogeneous WUR.

This large price dispersion is due, in great part, to the lack of reliable public information on WUR prices and transactions. Given the lack of reliable information, each WUR transaction is the result of a bilateral negotiation between an interested buyer and seller of WUR in which each agent's information, market experience, and negotiating capacity is important in determining the final result.

As a result of the WC 1981 2005 reform, combined with the performance of the Antitrust Commission, the monopolistic distortion due to speculation and nonconsumptive WUR hoarding has been reduced. In turn, WUR that still are not used are generally no longer a major obstacle to the development of the water basin, and it is likely that nonuse of WUR will continue to reduce in the future, due to the projected increase in the nonuse tariff.

A major challenge of the WUR markets in Chile is how to ensure optimal water use without compromising the sustainability of rivers and aquifers. The sustainability of northern rivers and aquifers is compromised due to the over-provision of WUR related to the practice of allocating WUR based on foreseeable use.

The WC 1981 did not pay much attention to the sustainable management of groundwater because, at that time, groundwater extraction was marginal during the early 1980s. Recognizing the need to improve groundwater management regulation due to increased groundwater pumping, the 2005 amendment of the WC 1981 introduced procedures to reach a sustainable management of underground water resources. World Bank (2011) concludes that these groundwater regulations have not been fully implemented over time, and, thus, there exists various problems associated with groundwater management. An additional challenge for a sustainable groundwater management is the fact that presently ground and surface waters are managed independently, despite their recognized interrelations. The 2005 amendment of the WC 1981 established that the Juntas de Vigilancia (surface water user associations) must, in the future, integrate groundwater user associations. However, there are only two groundwater user associations, and, thus, there is no conjunctive management of surface and groundwater, which has proven to be an effective adaptation mechanism for climate change.

The literature on WUR markets in Chile indicates that these markets have helped (1) facilitate the reallocation of water use from lower to higher value users (e.g., from traditional agriculture to export-oriented agriculture and other sectors, such as water supply and mining); (2) mitigate the impact of droughts by allowing for temporal transfers from lower-value annual crops to higher-valued perennial fruit and other tree crops; and (3) provide lower cost access to water resources than alternative sources, such as desalination.

The analysis of the problems that have been resolved through water-use rights indicates that the use of this allocation mechanism (1) has allowed users to consider water as an economic good, internalizing its scarcity value; (2) constitutes an efficient mechanism that has facilitated the reallocation of granted rights; (3) has permitted the development of mining in areas in the semiarid northern region of Chile where this resource is scarce by buying water rights from agriculture; (4) has solved problems associated with water deficits derived from a significant increase in water demand, caused by significant population growth in the central region of Chile; and (5) solved water-scarcity problems when a quick response was required.

In the Paloma system, for example, a semiarid water basin located in the dry Pacific hydrological system of the country, water is a scarce resource with a high economic value (especially for the export-oriented agricultural sector). This scarcity generates strong competition for water between users which, in turn, causes the temporary and permanent water market to be very active (e.g., during the 1993–1999 period, 6,000 water-use rights were traded). In the Maipo system, in the central region of the country, water supply is greater and demands from the agricultural sector lower. In the first section of this river basin, only 793 WUR were traded in the period 1993–1999 (Donoso et al. 2014).

There is an incentive for the adoption of water-saving technologies by farmers (Law N°. 18,450). This program subsidizes small-scale, private irrigation investments. It has supported much of the installation of drip irrigation systems in the dry north and spray systems in the humid south. Present estimates indicate that 30 % of agricultural operations concentrated in the northern water-scarce regions use water-conservation technologies. However, there has been no assessment of the impacts of this incentive instrument on groundwater recharge and sustainability. Hence, it is essential to strengthen the coordination between sectoral policies and water management policies. Other sectors present a significant increase in water-use efficiency, as a response to the scarcity value signal through WUR prices. The mining sector, for example, has reduced its water footprint from 1.7 m^3/ton of copper ore in 1980s to 0.5 m^3/ton of copper ore in the 2000s.

The problems that water-use rights markets have not been able to resolve are water-use inefficiency in all sectors, not only in the agricultural sector, environmental problems, and maintenance of ecological water flows. Additionally, integrated water resource management has not been implemented, although it has been established as a priority in the 1999 and 2013 National Water Strategy.

5.3 Urban Water and Sanitation Sector

During the 1980s, the sector was dominated by governmental water-supply utilities, which supplied water and sanitation (WSS) services to most of Chile. The inclusion of private operators began in 1988. The urban water-supply coverage in 1993 was 97.6 %, and the service was provided mainly by state-owned water-supply operators (see Fig. 5.2). Furthermore, only 85.9 % of urban population had access to sewer collection, and only 13 % of wastewater was treated. The driver of this situation was the low investment in infrastructure; the estimated investment cost deficit for the 1993–2000 period was $2.4 billion, and 63 % of the deficit was in wastewater treatment. Before 1993, the average annual investment of the state-owned operators was $150 million.

During 1994, several of the 13 state-owned WSS operators presented losses; for example, Essat presented −4.1 % and Emssa −3.2 %. This was, in part, due to a 30 % increase in average costs between 1990 and 1994. Administrative costs increased during this period; ESVAL increased its administrative costs by 140 %, while EMOS increased by 40 %. Furthermore, nonrevenue water varied between 24 % and 43 %.

The actual legal framework of the WSS sector established in 1988 presented the following objectives that water and sanitation tariffs must satisfy:

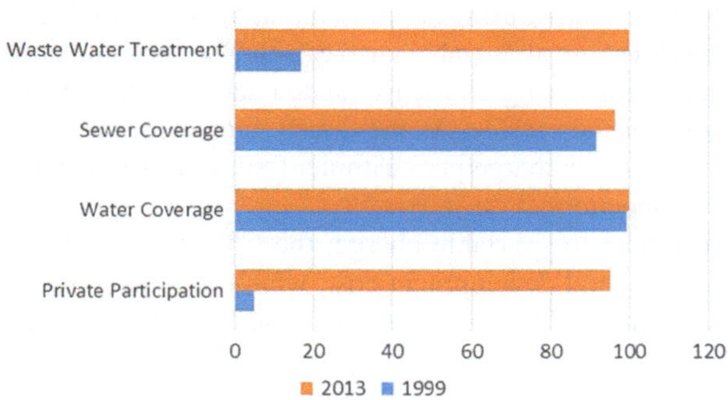

Fig. 5.2 Growth and evolution of regulated water and sanitation sector (Own elaboration based on SISS 2013)

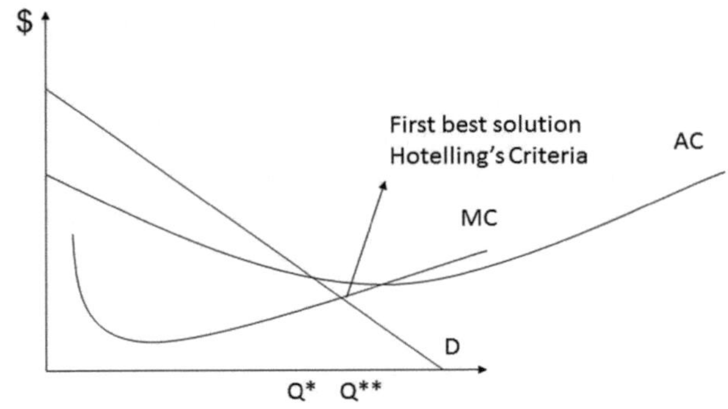

Fig. 5.3 Tariff setting principles

(a) Full recovery of operation and maintenance costs
(b) Funding of necessary infrastructure reposition and development plan investment
(c) Tariff reductions when operators increase efficiency
(d) Operational margins that are consistent with the opportunity cost of capital

The legal framework of the Chilean water and sanitation tariff system establishes that tariffs must satisfy the principles of (1) economic efficiency, (2) water-conservation incentives, (3) equity, and (4) affordability (Chavez 2002).

In order to comply with economic efficiency, the WSS tariffs are based on a two-part tariff, following Coase's solution: a variable and fixed tariff. The variable tariff is set following Hotelling's principle; thus, variable water tariff is consistent with the first best solution where marginal benefits are equal to long-run marginal costs[1] (MC) and social welfare is maximized (Fig. 5.3). However, this variable tariff set at MC does not cover the operator's average costs; that is, the WSS providers operate with losses. In order to satisfy the full cost recovery principle, a fixed tariff is included to cover the natural monopoly's losses at the first best solution.

The Executive Decree 453 of the 1988 Law N° 70, of the Ministry of Public Works (*Ministerio de Obras Públicas*, MOP) establishes a variable tariff that is set for periods of high demand, during summer months (peak variable tariff $/m^3), and for nonpeak periods (nonpeak variable tariff $/m^3). The peak and nonpeak tariffs are considered to internalize changes in seasonal demand and, thus, cover differences in the provision costs of the service. As previously indicated, the current tariff structure also considers a fixed charge per customer (connection), which depends on the diameter of the connection.

In order to estimate the variable charge, the Chilean tariff law introduced the concept of an incremental development cost, which is defined as the value applied to the incremental forecasted demand that generates the necessary revenues to cover

[1] Long-term infrastructure investment costs are included in the water and sanitation services tariff rates.

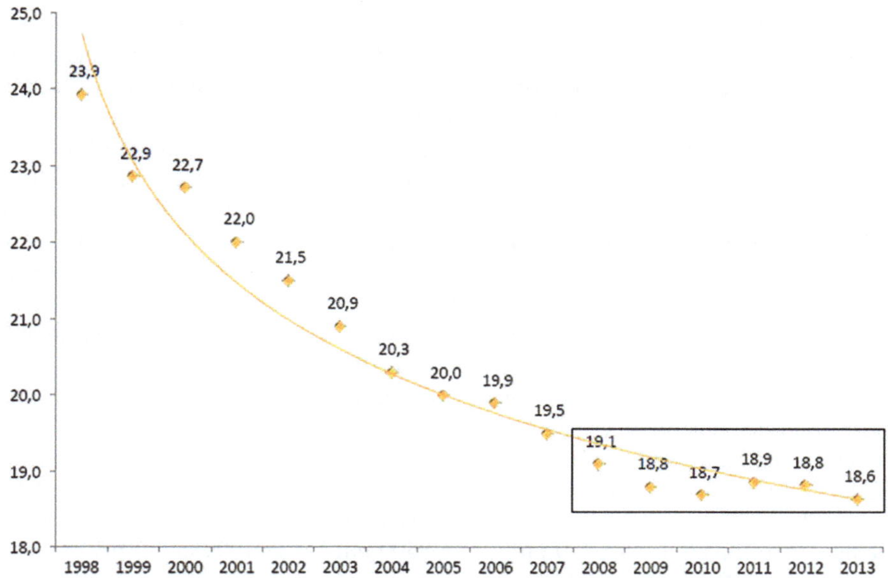

Fig. 5.4 Average monthly household water consumption (m^3/household/month) (SISS 2013)

incremental operation efficient costs, and the required investment for an optimized expansion project of the WSS firm. The incremental development cost is determined such that the net present value of the optimized expansion project is equal to zero (D.F.L. No 70/1988).

The variable tariff also considers the value of water so that consumers consider the scarcity of water in their water usage decisions.[2] This generates correct incentives to conserve water in resource-scarce areas. For example, average variable non-peak and peak tariffs in the dry Pacific arid system are \$1.3/m^3 and \$2.2/m^3. In the southern humid Pacific system, on the other hand, they are \$0.88/m^3 and \$1.3/m^3, respectively. Fixed tariffs also vary according to water scarcity, representing \$1.9/ m^3 and \$0.8/ m^3 in the dry and southern humid Pacific systems, respectively (SISS 2013). Additionally, evidence that tariffs send the right signals to consumers is that average monthly household consumption has significantly fallen since 1998, from approximately 25 to 18.6 m^3/household/month in 2013 (Fig. 5.4).

The affordability criteria is met by the provision of subsidies directly to the most vulnerable households. Households are classified based on an annual survey (*Encuesta Casen*), which estimates household per capita income. In order to qualify for the subsidy, households must not have payment arrears with the service provider. The central government transfers the block subsidy to the municipalities; the latter use this to pay a share of each of the eligible household's water bill; the payment share ranges from 15 to 85 % of the water bill, with the poorest families get-

[2] The value of water for each WSS provider is determined by market prices of traded WUR.

ting the highest share. The subsidy covers a consumption of up to 20 m^3. The Social Development Ministry (*Ministerio de Desarrollo Social*, MDS) uses the household survey information for each region of Chile to determine the size of the block subsidy that needs to be transferred to the municipalities. The WSS providers bill the benefiting households for the net of the subsidy amount, indicating the full consumption cost, and then charge the municipality for the subsidies granted.[3] The municipality will be charged interest for late payment, and the WSS provider can discontinue service to benefiting households if there is nonpayment by the municipality. In 2011, 15 % of WSS provider customers benefited (6 % of total sales), at a cost of $80 million and an average monthly subsidy per household of $10.

In order to obtain the necessary investment funds to improve its performance, the WSS sector instituted during the period 1989–1999, a model in which the regulatory and supervisory functions were separated from the investment, production, and sale of service functions. The new regulatory regime, which considered concessions to establish, build, and operate water and sanitation services by private providers, led to an increase in private participation in the provision of WSS services from 5 % in 1999 to 95.5 % in 2013. This process also led to a significant increase in average annual investments from $200 million to $500 million in 1999 and 2013, respectively (Fig. 5.2). This is mainly due to the increased rate of return on capital, due to increases in tariff rates. Tariff rates are determined so that investors receive a low-risk return of at least 7 % on capital expenditures, and therefore, private WSS providers have the incentive to invest in water provision, wastewater collection, and treatment (Hearne and Donoso 2005). For example, sewage treatment coverage increased from 17 % in 1999 to 99.8 % in 2013 (Fig. 5.2).

This reform period coincided with the era of high economic growth (6.2 % per year) with real incomes rising significantly. Williams and Carriger (2006) proposed that the transformation of the WSS sector would not have been so successful without these high rates of growth. The level of investment needed to attain this coverage could not have been reached if the Chilean government were responsible for investment. With tariffs set centrally for water and sanitation, efficiency incentives exist for the companies to increase returns on investment. This has happened and these companies perform well on the Chilean stock exchange (Bitran and Arellano 2005).

Currently there are 53 water and sanitation service providers operating in the urban areas of Chile. They function as private companies, although the state investment company, ECONSSA, still owns a considerable number of shares in most companies (Hearne and Donoso 2005). Five of Chile's 13 regional water companies were fully privatized with partial sale to multinationals in 1998.

The WSS providers service more than 4.5 million clients[4]; 94.4 % of clients are domestic, 4.7 % commercial, 0.2 % industrial, and 0.7 % other. Additionally, 95 %

[3] This practice does not distort the price signals.

[4] A client is determined by the property, rather than the individual, that receives services and is billed for these (more than one person may live in the same property, benefiting from the services).

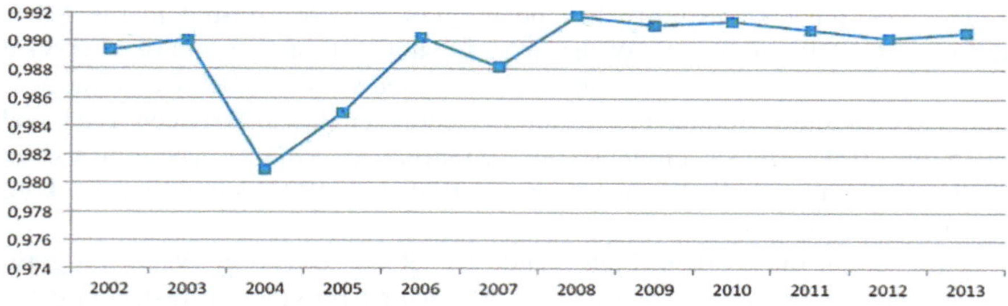

Fig. 5.5 Water service quality (SISS 2013)

of all clients have both drinking water and wastewater connections. The other 5 % have either one or the other, with most having only drinking water connections.

The large and medium service providers (8 of the 53) serve 84.2 % of all clients. It is interesting to note that a municipality (SRMPA of the Maipu municipality) owns one of them. Private providers service 95.5 % of all clients.

With respect to the service quality, Fig. 5.5 shows that customer satisfaction levels since 2008 are over 99 %. WSS clients were quite satisfied with the service, rating it with a 5.3 on a scale of 1–7 (GWI 2013).

Therefore, the new regulatory scheme in the Chilean WSS sector has provided the right economic signals for an efficient allocation of resources. It has also led to meeting the set goals for service coverage. Additionally, the transformation of the WSS sector has led to an:

(a) Improvement in quality of service
(b) Increase in WSS provision coverage, despite rapidly increasing urban populations
(c) Increase in water conservation by customers

In summary, Chile's policy of providing water-supply and sanitation (WSS) services through privatized regional and local water companies has been a notable success.

5.4 Rural Water Sector

In 1960, only 6 % of the rural population had an adequate supply system of water. During this period, there was no public agency in Chile responsible for supplying drinking water in rural communities. As of 1964, the government adopted the Rural Sanitation Master Plan, which appointed the National Health Service, as the executing agency of the first stage of the National Rural Drinking Water Program. The Inter-American Development Bank (IDB) funded the program to supply drinking water to 199 concentrated rural localities.

In 1975, the responsibility for the program was transferred to the MOP, through the Directorate of Water Works, later National Sanitation Service (SENDOS). In 1977, a second contract was signed with the IDB, called the second stage of the program, benefiting 142 rural localities. Between 1981 and 1985, the third stage was implemented, benefiting 233 villages. The fourth stage of the program was established between 1986 and 1991, supplying drinking water to 240 villages.

Since 1991 the Chilean state funds the Rural Drinking Water Program (*Programa de Agua Potable Rural, APR*), which provides infrastructure for water provision in rural areas. The program is directed to rural communities living in concentrated towns[5] and semiconcentrated towns[6] that lack WSS or have a WSS service that needs to be expanded or improved. The state subsidizes the installation of infrastructure, and rural water user committees (RWC) manage water provision in their areas, supervised by Chile's Department of Health. Additionally, tariff setting is the responsibility of the RWC. Evidence shows that these committees have not set tariffs at the correct levels in order to fully recover costs. The evaluation of the APR Program conducted by the Budget Directorate of the Ministry of Finance (Dirección de Presupuestos, Ministerio de Hacienda Dirección de Presupuestos, Ministerio de Hacienda 2007) shows that only 57 % of the total rural water-supply installations have been maintained or improved. This is mainly due to tariffs that, in general, only cover operating costs. The majority of the RWCs have not set tariffs that allow for recovery of maintenance costs. Moreover, the RWCs have not been able to finance the required investments to attend growing demands. This explains the deterioration of the systems over time, requiring further subsidies to recuperate the systems.

More importantly, these RWCs set the tariffs without a supervising regulatory agency. During Bachelet's first presidential administration (2006–2010), the government submitted a bill to the Chilean congress to give this sector a new institutional framework in the form of a specialized agency. However, to date no changes have been implemented, and this agency still has not been created.

Thus, the APR program subsidy has been effective in installing water-supply infrastructure in concentrated and semiconcentrated rural towns. However, due to funding problems, these water-supply installations are precarious and vulnerable. For example, the dry Pacific and northern portion of the central Chile hydrological systems have suffered a severe drought during the past 7 years. Due to this drought, the majority of the rural water-supply installations have not been able to supply water to their clients, and the state has had to supply water with cistern trucks. On the other hand, all of the regulated urban water service providers have been able to satisfy water demands.

---

[5] Towns with over 300 inhabitants and a minimum density of 15 households per km.

[6] Towns have at least 80 people and a minimum density of 8 homes per km.

5.5 Conclusions

The water sector in Chile underwent major changes as a result of decentralization and market reforms. Over the last 30 years, the Chilean government has successfully incorporated private participation in the urban water and sanitation sector and implemented a regulatory framework that has contributed to cost recovery and affordability of the reform. The service offered has greatly improved in quality and coverage, reaching in 2013, 99.9 % of urban population. National coverage of sewage treatment has significantly increased from 17 % in 1999 to 99.8 % in 2013. Thus, Chile's policy of providing water-supply and sanitation (WSS) services through privatized regional and local water companies has been a notable success.

Unlike urban service providers, the rural water-supply and sanitation sector has not been subject to regulation like urban services. This has led to tariffs that do not allow for full cost recovery. More importantly, tariffs have not allowed for adequate funding and maintenance to satisfy growing demand. Thus, rural WSS systems are precarious and vulnerable.

There is no irrigation water pricing in Chile. However, agricultural producers face the opportunity cost of water through markets for WUR. This has led to significant increases in the adoption of water-conservation technologies. Average irrigation efficiencies have increased to levels above 50 % in all three hydrological systems (Comisión Nacional de Riego 2010).

References

Bitran, G. & Arellano, P. (2005). *Regulating water services, sending the right signals to utilities in Chile.* Public Policy for the Private Sector, World Bank, Note number 286, March 2005. Available at Background Document http://rru.worldbank.org/Documents/PublicPolicyJournal/286Bitran_Arellano.pdf

Chavez, C. (2002). Public-private partnership and tariff setting: The case of Chile, OECD global forum on sustainable development conference on financing the environmental dimension of sustainable development, 24–26th April. Paris: OECD. http://www.oecd.org/sweden/2083795.pdf

Comisión Nacional de Riego. (2010). *Mejoramiento de la eficiencia hídrica sector agrícola.* Gobierno de Chile.

D.F.L. No 70/1988. http://www.leychile.cl/Navegar?idNorma=4427

Dirección de Presupuesto, Ministerio de Hacienda. (2007). *Informe Final de Evaluación Programa de Agua Potable Rural.* http://www.dipres.gob.cl/595/articles-31606_doc_pdf.pdf

Donoso, G., Melo, O., & Jordán, C. (2014). Estimating water rights demand and supply: Are non-market factors important? *Water Resources Management, 28*(12), 4201–4218. (ISI), doi:10.1007/s11269-014-0739-3.

Executive Decree 453 of the 1988 Law N° 70, of the Ministry of Public Works. http://www.leychile.cl/Navegar?idNorma=98618

GWI. (2013). *Global water intelligence market insight: Chile.* UK: Media Analytics Ltda.

Hearne, R., & Donoso, G. (2005). Water institutional reforms in Chile. *Water Policy, 7*, 53–69.

Hearne, R., & Donoso, G. (2014). Water markets in Chile: Are they meeting needs? In W. Easter & Q. Qiuqiong (Eds.), *Water markets for the 21st century: What have we learned? Global issues in water policy* (Vol. 11, pp. 103–126). Dordrecht: Springer.

Ministerio de Justicia, Chile. (1981). Código de Aguas de De Chile, 1981 (WC1981). http://www.leychile.cl/Navegar?idNorma=5605

SISS. (2013). *Informe de Gestión 2013*. http://www.siss.gob.cl/577/articles-10684_recurso_1.pdf

Williams, S. & Carriger, S. (2006). *Water and sustainable development: Lessons from Chile* (Global Water Partnership, Policy Brief 2, Technical Committee (TEC)). Available at www.gwpforum.org/gwp/library/policybrief2chile.pdf

World Bank. (2011). *Chile: Diagnóstico de la gestión de los recursos hídricos* (No. 63392). Available at http://water.worldbank.org/node/83999

WRI (World Resources Institute). (2003). *Earth trends country profiles*. http://earthtrends.wri.org/country_profiles/index.cfm?theme=2

Chapter 6
Water Pricing in China: Impact of Socioeconomic Development

Yue Che and Zhaoyi Shang

Abstract This chapter briefly introduces the basic information about water resources in China and discusses the price changing trends of irrigation, residential, and industrial water in representative regions or cities. After continuous water price reform, water-pricing mechanisms have become more scientific and rational. Water supply pricing has completed a transformation from public welfare to commercialization, and resource value and waste treatment costs are now included in pricing mechanisms. The prices for irrigation, domestic, and industrial water have increased significantly during the past two decades. During the reform process, China launched multiple compulsive laws and regulations, economic incentives, and rewards to promote water-pricing reform and water-saving measures. At present, the water volume quota system is enforcing industrial water consumption in the country, and a block rate structure mechanism has been established in most cities for regulation of industrial and residential water usage.

Keywords China • Socioeconomic development • Industrialization • Urbanization • Price reform

6.1 Introduction

6.1.1 Water Resources, Population, and Issues of Water Supply

With a land mass of 9.6 million square kilometers and 45,203 rivers, each having catchment areas larger than 50 km^2 (Ministry of Water Recourses P.R. China and National Bureau of Statistics P.R. China 2013), China has abundant water resources with minor interannual disturbances. According to the *China Water Resources Bulletin 1997–2012* (Ministry of Water Recourses P.R. China 1997–2012),

Y. Che (✉) • Z. Shang
Department of Environmental Science, East China Normal University,
Room 421, Building ZiHuan, No. 500 Road DongChuan, MinHang, Shanghai, China
e-mail: yche@des.ecnu.edu.cn

A. Dinar et al. (eds.), *Water Pricing Experiences and Innovations*,
Global Issues in Water Policy 9, DOI 10.1007/978-3-319-16465-6_6

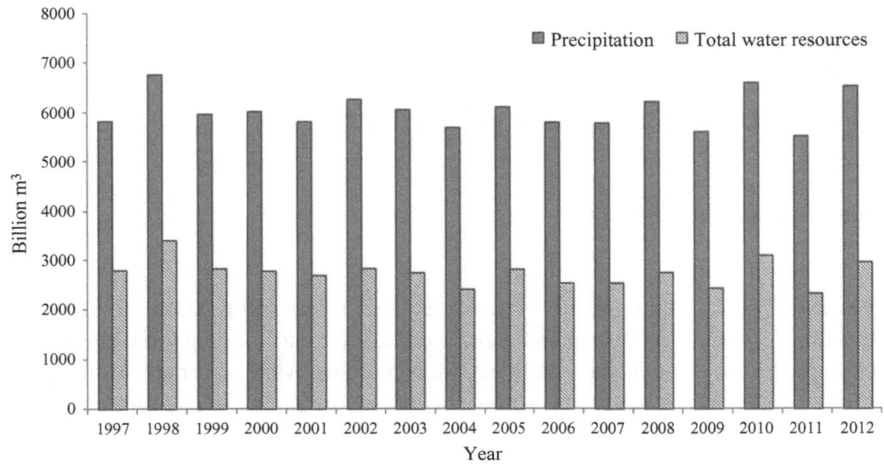

Fig. 6.1 Precipitation and total water resources of China in the past 15 years (Data Source: The Ministry of Water Recourses P.R. China 1997–2012. Bulletin of China Water Resource. http://www.mwr.gov.cn/zwzc/hygb/szygb/)

precipitation volumes on the mainland[1] were around 5,500–6,500 billion cubic meters, and total available water resources were about 2,500–3,500 billion cubic meters, varying with climatic conditions. These absolute figures are among the largest in the world (Fig. 6.1). However, the rapidly growing population results in small per capita resource volumes. From 1953 to 2010, for example, the population in China doubled, yielding a per capita water resource of only one-fourth of the global average (Table 6.1).

Moreover, spatiotemporal distributions of water resources, population, farmland, and economic development were not well balanced. This created greater pressure on water resources, especially in northern and central China, including Beijing, Tianjin, Hebei, Henan, Shandong, and other provinces. In the near future, the growing population and its migration toward the eastern coast of China will compound the water shortage problem there. The population growth and the economic activities that go with it result in an increase in wastewater discharge and water pollution, which compounds water shortages in that region.

The growing population and rapid economic development have increased water demand. From 1997 to 2012, total water consumption in China grew about 10 %, with slight fluctuation. Although agricultural water use decreased by 8 %, the total amount remained at about 65 % of the total water consumption, with agriculture still the largest water consumer. The usage amounts of industrial and domestic water grew by small percentages, and ecological and environmental water use constituted less than 2 % of total consumption (Fig. 6.2). Water-use efficiency rose considerably because of technological development. Water consumption per 10,000 RMB

[1] It includes Hainan Province but does not include Hong Kong, Macao, and Taiwan.

Table 6.1 Population of China according to national census

Census series	Year	Population of China (million)
1st national census	1953	601.912371
2nd national census	1964	694.580000
3rd national census	1982	1008.180000
4th national census	1989	1133.680000
5th national census	2000	1242.600000
6th national census	2010	1339.724852

Data resource: http://baike.baidu.com/view/46884.htm?fr=aladdin#3

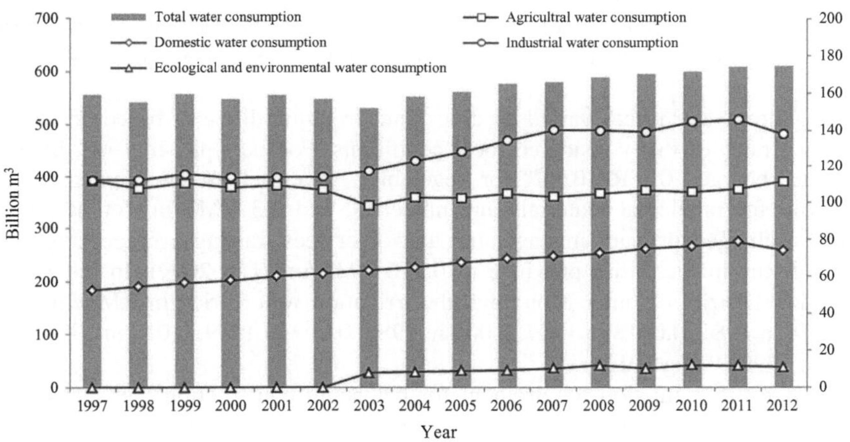

Fig. 6.2 Water consumption in China from 1997 to 2012 (Data Source: The Ministry of Water Recourses P.R. China 1997–2012. Bulletin of China Water Resource. http://www.mwr.gov.cn/zwzc/hygb/szygb/) (left vertical axis is for total water consumption and agricultural water and right axis the remainder) (Ecological and environmental water was not measured before 2002. Livestock water consumption was calculated as domestic water before 2012 but as agricultural water in 2012)

(Chinese currency)[2] of China's gross domestic Product (GDP) decreased from 726 to 118 m^3, and irrigation water per mu (Chinese measure of area: 1 ha = 15 mu) decreased from 492 to 404 m^3 from 1997 to 2012.

6.2 Past Water-Pricing Practices

6.2.1 Past Experiences with Irrigation Water Pricing

As a traditional agricultural country, China has a long history of using an irrigation water-pricing system. Early in the second century BC, feudal officials charged 5 kg rice per measurement unit (mu) for this water. Before 1965 but after the founding of

[2] See Appendix Table 6.8 for the information about exchange rate.

the People's Republic of China, the state practiced irrigation without charging, and construction of farmland irrigation and drainage mainly relied on farmers' labor. In 1965, the act *Regulation of Hydraulic Engineering Water Price Collection, Usage, and Management* was enhanced. Although the regulation demanded management institutions charge water consumers, it was not vigorously enforced. In 1985, the State Council published *Regulation of Hydraulic Engineering Water Price Assessment, Collection, and Management*, which established water pricing according to various industries, construction, and districts. In 1997, the *Water Conservancy Industry Policy* stated that irrigation water prices should cover construction, operation, and maintenance costs, plus profit, and taxes of waterworks. But later, in 2002 and 2003, considering the acceptance and payment ability of farmers, the new act *Regulation of Hydraulic Engineering Pricing Management* asserted that irrigation water prices should compensate production costs but not profits and taxes (Wang 2012).

Irrigation water prices varied by cities and irrigation districts, based on water-pricing strategies that considered local conditions. For example, in 1993, Shanxi Province charged 0.1 RMB/m^3[3] for vegetables, watermelons, and peanuts; 0.15 RMB/m^3 for medicinal materials and nurseries; and 0.3 RMB/m^3 for orchards. Later, it charged primary management and other fees, raising average irrigation water prices in the entire province to 0.216 RMB/m^3 (Lin 2009). In the Hetao Irrigation District of Inner Mongolia, the irrigation water price in RMB/m^3 was 0.00114 in 1981, 0.0018 in 1987, 0.006 in 1988, 0.009 in 1989, 0.017 in 1995, and 0.04 in 1999 (Wang 2012).

Despite the regulation requirement for balancing construction fees and costs, irrigation water prices in various districts in general only covered national construction fees but not local investment or labor costs and were very low. Figure 6.3 shows the average water supply price of 100 large water management agencies from 1994 to 2001. This shows that agricultural water prices had increased, but the growth rate was small compared with industrial and domestic water prices (Zhang et al. 2003). According to available statistics, the average irrigation water price in central and northern China was 0.0198 RMB/m^3, and in northeastern China, it was 0.02–0.05 RMB/m^3 in 2002–2003 (Zhang et al. 2003). The average comprehensive irrigation water price in the country was 0.03 RMB/m^3 in 2002 (Tang 2006). It was estimated that the price of irrigation water only represented 50–60 % of the water supply cost (Zhang et al. 2003).

Furthermore, the practical collection rate of irrigation water prices was relatively low. The average collection ratio for 100 large water management agencies from 1999 to 2001 was 74 %, and some provinces had low ratios; for example, Hainan and Hebei Provinces had ratios of only 14 % and 47 % (Zhang et al. 2003).

Beginning in 2003, the Ministry of Water Resources began to regulate the irrigation water consumption quota mechanism and evaluated water consumption norms

---

[3] See Appendix Table 6.9 for the information about exchange rate of RMB to USD from 1990 to 2013.

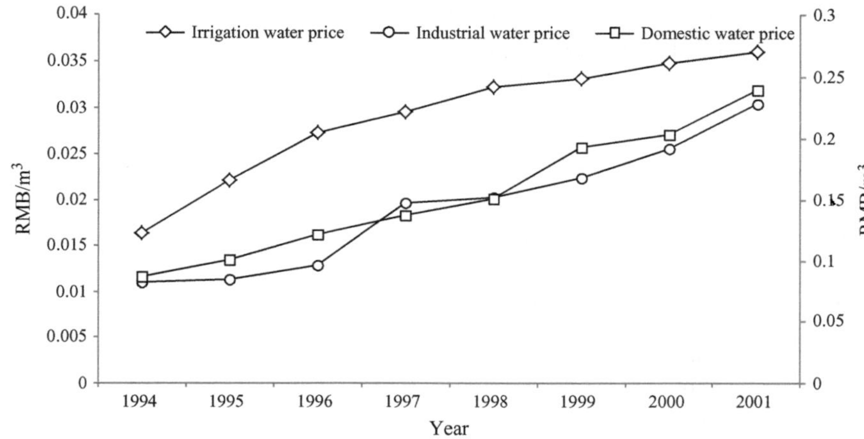

Fig. 6.3 Average water supply price of 100 large water management institutions from 1994 to 2001 (left vertical axis is for irrigation water price and right axis for the remainder)

of 193 crops in 2007 (Liu and Sang 2007). However, due to the landscape and public ownership of land, farmland in China is too scattered to manage. The lagging economy and lack of equipment for measuring water usage further obstructed the macro-control. Thus, the regulation was apparently not effective because of inadequate enforcement.

6.2.2 *Past Experiences with Residential Water Pricing*

Because of the planned economy system in China, water was regarded as a public welfare service, and water prices were not adjusted for more than 30 years after the foundation of the People's Republic of China. In the late 1980s, increasing water consumption due to demand and supply costs brought upon by the economic reform made price adjustment inevitable. In contrast to irrigation water prices, those of residential water increased rapidly beginning in 1986 (Yan and Shao 2006).

Beginning in the 1990s, cities such as Shanghai began to charge discharge fees in addition to the tap water fee. Later, the price structure in most cities included both these fees. In other cities, such as Chongqing and Tianjin, the price also included water resource and public service management fees that were based on local water management conditions.

Table 6.2 shows price change in four representative municipalities of China from 1990 to 2010, based on Yan and Shao (2006), Liu (2010), and Sun and He (2010). There may be some omission in certain years because of lost historical records. Over the past 20 years, the price changed more than ten times, and Chongqing raised the price twice in 2004. The growth rate was more rapid from 1995 to 2005 than in other periods. In 1990, the residential water price was less than 0.2 RMB/m^3.

Table 6.2 Residential water price change in four Chinese cities from 1990 to 2010

	Residential water price (unit: RMB/m^3)			
Year	Beijing	Shanghai	Tianjin	Chongqing[a]
1990	0.12	0.18	/	
1991	0.30	/	/	
1992	/	0.28	/	
1993	/	0.40	/	
1994	/	0.50	/	
1995	/	0.48	0.40	
1996	0.50	/	0.65	
1997	/	/	0.78	/
1998	0.80	0.68	0.98	/
1999	1.10	/	1.40	/
2000	1.60	/	1.80	1.27
2001	2.00	1.84	2.20	1.67
2002	2.50	/	2.60	2.42
2003	2.90	/	2.90	/
2004	/	/	/	2.46 and 2.66
2005	3.70	/	3.40	/
2006	/	/	/	2.80
2007	/	/	/	/
2008	/	/	/	/
2009	4.00	2.30	3.90	/
2010	/	2.80	/	/

Data Source: See Footnote 5
Only change is shown and slashes mean no price change in a given year
[a]Chongqing Municipality was established in 1997

And until 2010, prices rose by 3,233.3 % in Beijing and 1,455.5 % in Shanghai, compared with the initial stages.

The price trend in Tianjin was nearly synchronous with that of Beijing, while pricing in Chongqing was similar to that of Shanghai. This was mainly attributable to local water resource conditions. Beijing and Tianjin are in water-deficient areas of northern China: consequently, water prices were approximately 1 RMB/m^3 higher.

In 1998, the *Regulation of Water Supply Tariff Management* stated that urban water tariffs should gradually execute a multipart or block rate structure mechanism. It also defined the fundamental proportion of three water volume tiers as 1:1.5:2. Later in 2004, the State Council published circulars that required block rate structure mechanisms and extended the range between different tiers after guaranteeing basic domestic water volume (Dong 2007). Shenzhen, Xiamen, Yinchuan, and other cities were the first to start pilot price reform projects. Table 6.3 indicates the price tiers according to variable water consumption volumes of four spearhead cities. Volume levels and prices varied spatially and temporally. The reformation process was less advanced in many cities. It was reported that by 2010, only 80 cities in the

Table 6.3 Residential price tiers of four cities in early stage of the reform

City	Year. Month	Item	Unit	First tier	Second tier	Third tier
Xiamen	1997.1	Water volume	m^3/house/month	[0, 15]	(15, 20]	(20, +∞)
		Water price	RMB	0.6	1.4	2.1
Yinchuan	2004.1	Water volume	m^3/house/month	[0, 12]	(12, 18]	(18, +∞)
		Water price	RMB	1.15	1.75	2.3
Zhanjiang	2006.7	Water volume	m^3/house/month	[0, 20]	[21, 25]	[26, +∞)
		Water price	RMB	1.8	2.43	3.05
Taiyuan	2008.9	Water volume	m^3/house/month	[0, 9]	(9, 13.5]	(13.5, +∞)
		Water price	RMB	2.3	4.6	6.9

Data Source: See Footnote 5
The symbol of "+∞" means no upper limit

country had implemented residential block rate structure systems. Among 36 large and midsized cities, the reformation rate did not exceed 50 %.

6.2.3 Past Industrial Water Pricing

Table 6.4 presents change of industrial water prices in four representative municipalities in China from 1990 to 2010, based on available references (Jia and Zhang 2003; Yan and Shao 2006; Liu 2010). The prices in these four municipalities increased greatly over the analyzed period, and the frequency of adjustment was high during the past two decades. Industrial water prices in Beijing and Shanghai were fixed at 0.25 and 0.26 RMB/m^3 in 1990. Prices increased more than ten times, growing to 6.21 and 3.62 RMB/m^3, respectively, in 2009. The price in Tianjin increased from 0.7 to 7.5 RMB/m^3 between 1995 and 2010, and the adjustment frequency was even greater. From 2000 to 2010, the price in Chongqing doubled to 3.35 RMB/m^3 in 2006. The decade from 1995 to 2005 witnessed the most rapid increase in price.

The price change of industrial water was similar to residential water in the four cities. The water resource deficiency in Beijing and Tianjin resulted in prices higher by 2–3 RMB/m^3, compared to those in abundant water areas.

The *Water Law of the People's Republic of China* was revised in 2002 and required that water use in China should follow total volume control, planning management, and water consumption quota management. Among various fields, industrial water use was the most strictly regulated. Some cities began quota management in the late 1970s and early 1980s. In 1984, the Ministry of Construction and State Economic Committee confirmed a water-use quota for more than 200 industrial products from 14 businesses. And by 2007, more than 30 provinces and municipalities had published regulations and compulsorily implemented provincial water quota management systems (Liu and Sang 2007). On the basis of quota accounting,

Table 6.4 Industrial water price change in four cities from 1990 to 2010

	Industrial water price (unit: RMB/m^3)			
Year	Beijing	Shanghai	Tianjin	Chongqing
1990	0.25	0.26	/	
1991	0.45	/	/	
1992	/	0.36	/	
1993	/	0.51	/	
1994	/	0.70	/	
1995	/	0.68	0.70	
1996	0.80	/	0.82	
1997	1.00	0.90	1.30	/
1998	1.60	/	1.70	/
1999	2.10	/	2.00	/
2000	2.40	/	2.40	1.52
2001	/	1.73	3.00	1.93
2002	2.90	2.38	3.80	2.68
2003	4.40	/	4.60	/
2004	5.60	/	/	2.71 and 3.21
2005	/	/	5.60	
2006	/	/	/	3.35
2007	/	/	6.20	
2008	/	2.94	/	
2009	6.21	3.62	6.70	
2010	/	/	7.50	

Data Source: See Footnote 5

the water management administration set planned water consumption volumes for each enterprise consumer, based on product water-use quotas and production values of the previous year, and water supply institutions charged higher prices for volumes that exceeded planning.

Table 6.5 lists four examples of the block rate structure mechanism for industrial water consumption in excess of the quota volumes. There was a wide variation of access ranges and added charge ratios above prices between various cities. Fuzhou and Jinan established four tiers, and Beijing and Hangzhou developed three pricing tiers. The price tiers of Fuzhou were the most tolerant, while that of Jinan was the strictest, with the largest ratio reaching eight times the base price.

6.2.4 Past Experiences with Pricing of Environmental Services

The concept of environmental water usage was not well defined in China. In the *China Water Resources Bulletin*, ecological and environmental water usage included water supply to urban environments, such as greenbelt irrigation, as well as

Table 6.5 Price tiers of industrial water use in four cities

District	Year	Item	First tier	Second tier	Third tier	Fourth tier
Fuzhou	2004	Amount above quota	(0, 10 %]	(10 %, 20 %]	(20 %, 30 %]	(30 %, +∞)
		Price charge ratio	1.15	1.35	1.8	2.5
Beijing	2005	Amount above quota	(0, 20 %]	(20 %, 40 %]	(40 %, +∞)	
		Price charge ratio	1	2	3	
Hangzhou	2006	Amount above quota	(0, 20 %]	(20 %, 30 %]	(30 %, +∞)	
		Price charge ratio	1	2	3	
Jinan	2007	Amount above quota	(0, 20 %]	(20 %, 30 %]	(30 %, 40 %]	(40 %, +∞)
		Price charge ratio	2	4	6	8

Data Source: See Footnote 5

groundwater, rivers, lakes, and wetlands. In 2012, the total ecological and environmental water consumption exceeded 11 billion m^3. However, except for water use by residents, industries, and public and commercial service and special uses such as car washing and drink production, prices of ecological and environmental water had not been taken seriously in most cities.

A few cities set the price of municipal water use, including urban green space, environmental sanitation, and fire control. Even fewer cities established prices for water recharge to groundwater, rivers, lakes, and wetlands. For example, the price of urban green space irrigation was 1.35 RMB/m^3 in Taiyuan in 2001, 1.04 RMB/m^3 in Wulumuqi in 2005, 1.95 RMB/m^3 in Dongying in 2006, and 2 RMB/m^3 in Wanning in 2012. In contrast to irrigation, industrial, and residential water, the price of environmental services did not include sewage treatment fees. Beijing was an advanced city in environmental services and water pricing. The city set prices of tap water for urban green space, surface water for parks and lakes, and reclaimed water at 3.9, 1.3, and 1 RMB/m^3, respectively. Even so, maintenance of public green space was supervised by the government, and costs were covered by public finance. Moreover, in many other cities, maintenance workers stole water for green spaces from fire hydrants. Imperfect measurement and supervision of environmental service water made pricing even more inefficient.

6.3 Present Water-Pricing Practices

6.3.1 Present Experiences with Irrigation Water Pricing

At present, there are multiple pricing patterns for irrigation water. For example, based on surveys, there are four types of pricing systems in Beijing for two irrigation areas and 18 villages. The Tongzhou Xinhe irrigation area charges by water volume, and the price is 0.04 RMB/m^3. The Tongzhou Dongxiaying irrigation area is irrigated by conjunctive use of surface and groundwater and is charged by area, with a price of 300–450 RMB/(hm^2.year[4]). Some villages that are irrigated by groundwater wells charge by electricity consumption, and others like Shangmo do not charge for irrigation water (Gu et al. 2008). In other locations, such as Qinghai Province, the government charges for irrigation water by a certain amount of agricultural good, and the farmers pay the water fee based on the current year's price of such agricultural good. For example, in the Hehuang Irrigation District, the price is 90–180 kg wheat/ (hm^2.year), equaling 75–225 RMB/ (hm^2.year) (Li and Lei 2002).

Some irrigation areas have practiced block rate structure mechanisms. In the Zhanghe irrigation area of Hubei Province, the basic price is 0.033 RMB/m^3, and added price ratios are 0.5 and 1.0 for water usage above 0–30 % and above 35 %, respectively (Ma et al. 2009). Beijing requires an additional charge of 0.08 RMB/m^3 for grains and 0.16 RMB/m^3 for other crops when its limits are exceeded (Gu et al. 2007).

Taocheng District in Hubei Province, Pingyuan County in Shandong Province, and other districts experiment with the pricing system of "price increase with government subsidy." For example, the irrigation price for farmers to pay in Taocheng District of Hengshui City is increased by 0.15 RMB/m^3 (from 0.35 to 0.5 RMB/m^3), and, at the same time, the government grants a subsidy of 0.05 RMB/m^3. The increment of water fees collected from consumers and the subsidy from the government are returned to the local community to establish a water saving of 0.2 RMB/m^3, which is distributed annually to each farmer in the community, based on the proportion of farmland (Zhang 2013). The result is that farmers that paid water fees less than 582 RMB/(hm^2 per year) would receive a financial reward. This pricing mechanism can reward water-saving behavior and punish for water waste and raises awareness among locals of the need to save water.

Although irrigation water pricing has been reformed, the price is still too low to cover irrigation water supply costs. Additionally, the lack of an effective water measuring system, difficulty in gathering statistics, and lack of supervision continue to make irrigation water pricing unreasonable and difficult. Plot survey results indicated that the farmers preferred the cost of irrigation water to be no more than 10–12 % of the total cost, equaling 5–7 % of the production value (Zhou and Wu 2005). Policy requests from the government to lighten the burden of farmers and the opposition from peasants to price increases both retard the pace of price reform.

6.3.2 *Present Experiences with Residential Water Pricing*

Table 6.6 illustrates the present water price and its structure in 11 large cities in various parts of China. Among these cities, Yinchuan, Lanzhou, and Harbin have the lowest price at around 2 RMB/m^3, while Beijing and Tianjin have the highest at around 5 RMB/m^3. For other cities, those in northern and eastern China have higher

[4] See Appendix Table 6.8 for the information about exchange rate.

Table 6.6 Present water price of cities

Regions	Item	Water-use volume	The sum of fees	Tap water fee	Water resource fee	Sewage treatment fee
	Unit	m^3/household/ year[a] or month[b]	RMB/m^3			
North China	Beijing[a]	(0, 180]	5.00	2.07	1.57	1.36
		(181, 260]	7.00	4.07		
		(260, +∞)	9.00	6.07		
	Tianjin	/	4.90	/	/	/
	Jinan	/	3.15	1.85	0.40	0.90
East China	Shanghai[a]	(0, 220]	3.45	1.92		1.70
		(220, 300]	4.83	3.30		
		(300, +∞)	5.83	4.30		
Southwest	Chongqing	/	3.50	2.50		1.00
	Chengdu	/	2.94	1.98	0.06	0.90
South China	Guangzhou[b]	(0, 27)	2.88	1.98		0.90
		[27, 35)	4.17	2.97		1.20
		(35, +∞)	5.46	3.96		1.50
Northwest	Lanzhou	/	2.25	1.75		0.50
	Xi'an	/	2.90	1.95	0.30	0.65
	Yinchuan[b]	(0, 12]	1.70	/	/	/
		(12, 18]	2.80	/	/	/
		(18, +∞)	4.00	/	/	/
Northeast	Harbin	/	2.40	/	/	/

Data Source: See Footnote 5

The slashes mean that details could not be found in available references. The symbol of "+∞" means no upper limit

[a]Yearly consumption volume

[b]Monthly consumption volume

prices, while those in water-deficient northern districts and water-abundant southern districts enjoy lower prices.

Only five cities in Table 6.6 practice block rate structure pricing systems, and regional differences are significant. According to national research, among 484 surveyed cities, only 133 practiced block rate structure pricing systems for residential water use at the end of 2013, representing less than one-third of the total. The general trend is that southern regions of China have higher water amount quotas, whereas western ones have lower quotas, which is in accordance with the uneven distribution of water resources in China and lifestyles resulting from such conditions.

The present block rate structure system in China is being diversified because of various natural endowments and socioeconomic development, for example, (1) pricing in some cities, such as Baotou in Mongolia, has two tiers, while those in other cities, such as Hong Kong and Fuxin, have four. (2) The price ratio of range difference is as high as 1:2:10 in Langfang or as low as 1:1.4:1.8 in Shanghai and Beijing. (3) Most cities, such as Beijing and Shanghai, charge block rate structure prices only for tap water fees, while others, such as Guangzhou, do so for sewage treatment fees. (4) Units of water amount are m^3 per household per year, per month, or per 2 or 4 months. One household might have three or four people, and the added amount for another person could be 3 or 6 m^3. (5) Hefei, in the Anhui Province, broadens the boundaries in the block rate structure systems during water-abundant months from June through September and reduces the amount during the other months.[5]

There were reasons for the lag in block rate structure pricing. First, the basic water amount was large and the range difference for most cities was small. Second, reconstruction of water meters for each household called for a substantial money and labor investment, and managers and workers had to learn and adapt to new pricing systems. Third, block rate structure pricing heightened the awareness of water saving among residents, which reduced the operating income of water supply enterprises. Thus, those enterprises faced new threats to their profits and objected to reforms.

To solve this problem, on December 31, 2013, the State Development and Reform Commission and Ministry of Housing and Urban-Rural Development jointly released *The Guideline to Promote Block Rate Structure for Residential Water* to accelerate block rate structure pricing reform through administrative and legal measures. The guideline mandated that all district cities execute block rate structures for residential water citywide before the end of 2015. The guideline also stated that the price system should have more than three tiers and that the first and second tiers should guarantee 80 and 95 % of residential water demand. In principle, the price ratio of the three tiers should be larger than 1:1.5:3; for areas of deficient water quantity and quality, the ratio could be even larger.

The guideline suggested water amount tiers in six different districts for reference and emphasized cost disclosure and supervision, which would guarantee more effective practice. Under the guideline requirements, more effort will be expended and the block rate structure system will be widely implemented in China.

6.3.3 Present Experiences with Industrial Water Pricing

Table 6.7 lists industrial water prices in the 11 large cities of China. All these prices were higher than those of basic residential water given in Table 6.6, with increases ranging from 30.0 % in Chongqing to 79.2 % in Harbin. Although water resource fees of industrial water in Jinan and Chengdu were equal to or lower than those of residential water, all three types of fees for industrial water were higher than for residential water.

All 11 cities in Table 6.7 had block rate structure mechanisms of industrial water consumption for planning volume exceeding the quota. The number of tiers varied

[5] Data Source: See Notice Present Experiences with Residential Water Pricing.

Table 6.7 Present industrial water price and price tiers for amount exceeding the quota

City	Total price	Tap water fee	Water resource fee	Sewage treatment fee	No. of tiers	Amount exceeding the quota and corresponding price charge ratio
	RMB					
Beijing	7.15	3.52	1.63	2.00	3	(0,20 %]: 1; (20 %, 40 %]:2
						(40 %, +∞):3
Tianjin	7.85	/	/	/	5	(0,10 %]:1; (10 %, 20 %]:2
						(20 %, 30 %]:3 (30 %, 40 %]:5
						(40 %, +∞):10
Jinan	4.40	2.90	0.40	1.10	3	(0, 10 %]:1; (10 %, 30 %]:2
						(30 %, +∞):3
Shanghai	5.00	2.89		2.34	1	(0, +∞):2
Chongqing	4.55	3.25		1.30	3	(0, 10 %]:1; (10 %, 30 %]:2
						(30 %, +∞):3
Chengdu	4.39	2.93	0.06	1.40	3	(0, 10 %]:1; (10 %, 30 %]:2
						(30 %, +∞):3
Guangzhou	4.86	3.46		1.40	5	(0, 10 %]:1; (10 %, 20 %]:2
						(20 %, 30 %]:3; (30 %, 40 %] 4
						(40 %, +∞):5
Lanzhou	3.33	2.53		0.80	5	(0,5 %]:0.5; (6 %,10 %]:1
						(10 %, 20 %]:2; (20 %, 30 %]:4
						(30 %, 40 %]:6
						(40 %, +∞):8
Xi'an	4.90	3.08	0.72	1.10	3	(0,50 %):1; (50 %,100 %]:2
						(100 %,+∞):4
Yinchuan	2.60	/	/	/	5	(0, 10 %]:1; (10 %, 20 %]:2
						(20 %, 30 %]:3; (30 %, 40 %]:4
						(40 %, +∞):5
Harbin	4.30	/	/	/	3	(0, 10 %]:1; (10 %, 20 %]:3
						(20 %, +∞):4

Data Source: See Footnote 5
The slashes mean that details could not be found in available references. The symbol of "+∞" means no upper limit

from one to five and the added price charge ratio from one- to tenfold. Although with variable pricing, water planning and water consumption quota management has been widely implemented in China and is effective for water saving and improvement of industrial water-use efficiency.

6.4 Current Debates and Future Directions

6.4.1 Current Problems of Chinese Water Pricing Under Rapid Urbanization

China is experiencing rapid industrialization and urbanization. The urbanization rate of China grew from 10.64 % in 1949 to 53.73 % in 2013, and annual GDP growth remained steady at a rate of 8 %. The increasing and centralizing economic output and population will exert an even greater demand on the water supply.

In 2012, the amount of water consumption per person was 454 m^3/year, water consumption per 10,000 RMB of GDP was 118 m^3, water consumption per 10,000 RMB of industrial added value was 69 m^3, water consumption per mu area of farmland was 404 m^3, and the effective coefficient of irrigative water use was 0.516 at the national level, based on sampling plot surveys. Even after long-term development of water-saving technology, legislation, and policy management, compared with advanced foreign countries, water-use efficiency remains low, especially in agriculture. Additionally, China is struggling with a severe water pollution problem. In 2012, among 4,847 assessed water districts, only 47.4 % reached water quality goals. Organic pollution, eutrophication, persistent organic pollutants, and heavy metal pollution are threatening water quality in rivers, lakes, reservoirs, and even urban water supplies. Growing water consumption will result in more point and nonpoint sewage discharge and substantially increase water pollution.

At present, water price reform in China lacks systematic planning and has difficulty in executing and encouraging consumer water saving. The reasons that the price lever loses effectiveness in water use are listed below.

First, water prices are too low. In cities like Jinan, Chengdu, and Xi'an, the water resource fee makes up less than 10 % of the total price and, in almost every city, the sewage treatment fee is about half the tap water fee. Water pricing in the country only emphasizes supply engineering fees but not water resource or sewage treatment fees, which are high and several times the tap water fee in developed countries. This unreasonable pricing mechanism ignores resource and environmental costs of water utilities, so the market mechanism does not work to balance the price and supply-demand relationship.

Second, the collection rate of water fees is too low because of long periods of low water prices and lack of efficient supervision, especially for irrigation water. This results in financial deficits of water supply enterprises and funding gaps for repair, maintenance, and management of the government. This triggers a domino effect and delays construction of advanced measuring, monitoring, and statistical systems, which in turn blocks promotion of water price reform.

Third, public consciousness of water saving has not been established. Owing to long periods of preferential price subsidization by the government, residents became used to paying little for water consumption and therefore resist accepting higher but reasonable water prices. Particularly in rural areas, peanuts are exempted from various fees and taxes because of the government's agricultural policy. Growers take low water prices as a matter of course and expect water for free. Such deficient understanding of water prices and water saving in society retards the promotion of water price reform.

Fourth, market competition has not been implemented. For a long time, public service has been regarded as a government responsibility. Water service in China is provided and supervised by the government and is a monopoly market. Private enterprises are limited to market access, and existing water supply and treatment enterprises have not become true market entities under government restriction. Under such conditions, healthy competition can scarcely develop.

In addition, public disclosure of cost mechanisms is not well developed. The monopoly of water service and lack of such disclosure makes supervision difficult. In most cities in China, water supply and treatment services are provided by few or even single enterprises. The public cannot receive information on whether enterprises are in the red and if the reason is low water prices and can hardly supervise the cost of water service enterprises.

Furthermore, the government has a dilemma with raising water prices and guaranteeing residents' living standards. Cheap water prices can lead to water waste and low efficiency. Conversely, raising water prices will increase the cost of living (especially for low-income families) because per capita income of the country ranks only 89th in the world. The immature water service market also generates a low bearing capacity of the public.

Finally, the management framework is huge but complex. Water supply and pricing in China is under multiple management of different ranks of institutions, from the state to local enterprises. The State Development and Reform Commission, Ministry of Housing and Urban-Rural Development, Ministry of Environmental Protection, Ministry of Water Resources, and all levels of administrative organs are involved in management. Ownership, executive power, and operational rights of water resource management coexist and have not been completely separated, and the current management system has become an obstacle to water-pricing progress.

6.4.2 New Approaches to Water Pricing

China faces new challenges in water price reform. Based on existing problems and foreign experiences and faced with new challenges of social and economic development, the country should improve water price reform in the following ways.

First, water pricing should follow the principle of sustainable development and cost recovery. Current water pricing includes engineering cost, resource value, and waste treatment cost but has not yet covered all costs. The pricing process should give weight to external costs of water use that includes economic loss from water pollution and the cost of restoring the water environment, although it is quite a challenge to assess. Additionally, water pricing should take equality and bearing ability into consideration, balancing differences between developing and developed districts and rich and poor communities. A block rate structure mechanism based on variable quantities, qualities, and timescales will greatly assist the achievement of this goal.

Second, the government should gradually establish and improve a water service market with healthy competition. First, government administration should be separated from enterprise management, and water supply and treatment enterprises should bear responsibility for their own profit and loss. Then, the governor should accelerate the marketization of water services, quicken system reform of water supply and treatment enterprises, open operating rights, attract nongovernmental investment, and build water rights trading markets. To ensure national security and public benefit, higher thresholds, stricter supervision, broader level of public regulation, and macroeconomic control remain equally important.

Third, effective motivation mechanisms should be involved in any water-saving policy system. Previous water-saving policies were mainly compulsory and dominated by the government and did not link water saving with consumer behavior. These administrative or legal measures will not be suitable for diversification and decentralization of decision-makers and stakeholders or decentralization under a market economy. The governor should set up motivation mechanisms driven by economic interest gradually from domestic and industrial water pricing to agricultural pricing and encourage the passive behaviors of "I have to save water" to "I want to save water."

Fourth, administrations should continuously amend relevant laws and regulations and reform management systems. The current legal system of water resource management is fragmented, which causes managers to lack a legal basis in the performance of their duties. After years of water resource management reform, overlap between different laws and regulations persists, and there is duplication of executive functions between different agencies. In the future, management responsibilities of various institutions should be clarified in detail through legislation and regulation. Communications channels should be connected, and relationships between managers should be harmonized through cooperation.

6.5 Conclusion

This chapter examines the price changing processes of China for agricultural, domestic, industrial, and environmental water usage. From the past to present, irrigation water prices have been raised by a hundred percent, from a certain thousandth of 1 RMB to about half of 1 RMB. During the past two decades, domestic water pricing in major cities increased from less than 0.2 RMB/m^3 to about 3–4 RMB/m^3, and industrial water pricing increased from about 0.25 RMB/m^3 to even higher at about 6–7 RMB/m^3.

The reform resulted not only in price increases but also in multiple control measurements. China has launched multiple compulsive laws and regulations, economic incentives, and honor and reward systems to promote water-pricing reform and water savings. At present, the water volume quota system is ruling industrial water consumption in the entire country, and a block rate structure has been established in most cities to regulate industrial and residential water usage. In the agricultural sector, some irrigation districts conducted pilot experiments using government subsidies to encourage water savings.

Looking back over more than 20 years of water reform in China, the pricing mechanism is becoming more scientific and rational, and the effect of this reform is significant. Water supply pricing has transformed from public welfare to commercialization, and resource value and waste treatment costs are now included in the pricing mechanisms. The average price of different sources of water supply has increased significantly and can now nearly cover its cost.

However, water price reform has not completely covered all sources of water supply in China. For example, only a few cities have established prices for environmental water supply, and very few cities have regulated the price for ecological water resupply and underground water extraction recharge.

Some cities have conducted water market practices in China. For example, in 1998, four businessmen paid 5.18 million RMB to purchase water usage rights of the Nanxi River in Zhejiang Province for 12 years. In 2000, Yiwu City ordered 50 million cubic meters per year from Dongyang City by a one-time payment of 200 million RMB. In 2001, Zhangyan City granted water rights market among farmers in Hongshuihe Irrigation District. However, results showed that because of lack of a legal guarantee, mechanism support, moral education, and technical instruments, such as monitoring systems and information sharing, the price management of the water market was not as effective during this development period in China.

On the whole, China is still on the road to water price reform. China has a population of over 1.3 billion, 34 provincial administrative divisions, different natural endowments, and economic development and water price reforms, and the country must gradually explore methods that are better suited to national and local conditions. There remain many problems with pricing mechanisms, water-use measures, management, and supervision in various industries. Marketization is the most effective way to optimize distribution of the resource. Through more policy measures, research, and practices and by reforming the water-pricing system and constructing water markets, water resources in China can be optimally allocated and efficiently managed.

Appendix

Table 6.8 Chinese unit converter rate to international unit

	Chinese unit	To international unit
1	mu	1 mu = 666.67 m^2
2	hm^2	1 hm^2 = 0.01 km^2

Table 6.9 Annual average exchange rate of RMB against USD from 1990 to 2013

Year	RMB per 100USD	Year	RMB per 100USD
1990	445.15	2002	827.70
1991	532.22	2003	827.70
1992	551.46	2004	827.68
1993	576.20	2005	819.17
1994	861.87	2006	797.18
1995	835.09	2007	760.40
1996	831.42	2008	694.51
1997	828.98	2009	683.10
1998	827.91	2010	676.95
1999	827.83	2011	645.88
2000	827.84	2012	631.26
2001	827.70	2013	619.56

Data resource:
1. Editorial Office of Almanac of China's Finance and Banking, Almanac of China's Finance and Banking 1991–2013
2. The People's Bank of China Statistics and Analysis Department, Accessible on 2014/9/19 http://www.pbc.gov.cn/publish/diaochatongjisi/4032/index.html

References

Dong, L. (2007). The study of ladder's measure water price of urban household water consumption. Xi'an: Xi'an University of Technology.

Gu, Y. G., Zhang, F. G., Hao, Z. Y., & Wu, W. Y. (2008). Survey and analysis of agricultural water price in Beijing. *Beijing Water Resource, 2*, 26–28.

Jia, S. F., & Zhang, S. F. (2003). Response of industrial water use to water price rising in Beijing. *Shuili Xuebao, 4*, 108–113.

Li, X., & Lei, H. Z. (2002). Current status and strategy for agricultural water price reform in Northwest China. *Water Resources Development Research, 2*(3), 23–25,41.

Lin, X. M. (2009). Analysis and thinking about current situation of irrigation water price. *China Flood and Drought Management, 1*, 35–38.

Liu, W. Z. (2010). Discussion on Tianjin water tariff reform. *Journal of Economics of Water Resources, 28*(3), 36–38.

Liu, Q., & Sang, L. H. (2007). Problems and counter measures on water quota management mechanism in China. *Journal of Yangtze River Scientific Research Institute, 24*(1), 16–19.

Ma, J. Q., Xia, J., Liu, X. J., & Liu, J. (2009). Comparison between irrigation water prices in China and Australia and the policy reform of water price in China. *Resources Science, 31*(9), 1529–1534.

Sun, Y. N., & He, M. L. (2010). Rising extent analysis of Beijing residential water price: Based on the Markov chain and ARMA model. *China Price, 1*, 12–16.

Tang, L. (2006). *The study on prices accounting method of agricultural irrigation water*. Ulumuqi: Xinjiang Agricultural University.

The Ministry of Water Recourses P.R. China. (1997–2012). *Bulletin of China water resource*. http://www.mwr.gov.cn/zwzc/hygb/szygb/

The Ministry of Water Recourses P.R. China & National Bureau of Statistics, P.R. China. (2013). *Bulletin of First National Census for Water*, pp. 1–20. http://www.mwr.gov.cn/2013pcgb/

Wang, J. P. (2012). *Study on irrigation water price in Inner Mongolia Autonomous Region*. Beijing: Chinese Academy of Agricultural Sciences.

Yan, W., & Shao, E. S. (2006). *China urban water tariff. In the Ministry of Construction P. R. China edition*. Beijing: China Building Industry Press.

Zhang, Q. C. (2013). Example of water price reform: Mechanism of price increase with government subsidy. *Economy and Management, 11*, 68–70.

Zhang, C. L., Ruan, B. Q., & Luo, J. F. (2003). Analysis on water pricing in agriculture in China. *Journal of Shenyang Agricultural University, 34*(3), 207–210.

Zhou, Z. M., & Wu, H. (2005). Research on agricultural water price reform and bearing capacity of farmers. *Journal of Economics of Water Resources, 23*(3), 31–34.

Chapter 7
Water Pricing in Colombia: From Bankruptcy to Full Cost Recovery

Diego Fernández

Abstract Despite having one of the largest water resources of the world, in the early 1990s, the Colombian water and wastewater sector was in a deep crisis, with low coverage, low investment, and financial infeasibility of most of the companies responsible for the provision of these services. Rates were not consistent with the needs of maintenance, operation, management, and investment, and providing companies were almost totally dependent on state resources. The enactment of Law 142 of 1994 and regulations issued by the new (CRA acronym for Comisión de Regulación de Agua Potable y Saneamiento Básico) completely changed the landscape of the sector. In 1995, CRA set a clear methodology for the calculation of costs and charges for water and sewerage services, which aimed to cover the full costs of administration, operation, and investment. The law defined specific levels of subsidy only applicable to the basic consumption of poor families. Implementation of the new tariff scheme began in 2006 and lasted for several years, finally achieving financial viability of most businesses.

Keywords Colombia • Regulation • Social rates • Water services • Water companies

7.1 Introduction

Thanks to its geographical location and landscape, Colombia has one of the largest water resources on the planet, being the fourth in the world in volume of surface water, with $2.13 \cdot 10^9$ km^3, next to Russia, Canada, and Brazil—countries with surfaces three or more times that of Colombia's. It also has an average of 3,000 mm/year annual rainfall, which is twice the average rainfall of South America (1,600 mm/year) and three times the world average (900 mm/year).

D. Fernández (✉)
Econtec Consultores Ltda., Carrera 16 76-42 Oficina 702, Bogotá, Colombia
e-mail: aguayaseo@hotmail.com

A. Dinar et al. (eds.), *Water Pricing Experiences and Innovations*,
Global Issues in Water Policy 9, DOI 10.1007/978-3-319-16465-6_7

However, as shown by the Institute of Hydrology, Meteorology and Environmental Studies (Instituto de Hidrología, Meteorología y Estudios Ambientales, IDEAM), the natural water supply is not so evenly distributed over the different regions of the country, resulting in areas of high resource abundance, large areas of shortage, and areas completely deserted.

Analysis by the Vice Ministry of Environment (Viceministerio de Ambiente 2010) shows that the majority of Colombia's population and economic activities are located precisely in low water supply areas where the adverse effects on water resources from human activities are highest.

As a result, large cities, such as Bogotá, Medellín, Cali, and Cúcuta, among others, have a water demand that tends to exceed the available supply in each of their areas, forcing the use of water from surrounding areas (transfer between watersheds).

In contrast, in the southern (prominently jungle and mountain) and eastern areas of the country, with low population and low economic activity, water resources are abundant, and the demand for them is reduced.

The pressure on water resources in the most populated areas has increased in line with the deepening of the process of urbanization that the country has faced, from an urban population of only 39 % in 1951 to 66 % in 1985, 76 % in 1995, and around 80 % by 2014.

In that framework, it is not surprising that historically Colombia has always had a great concern for the prices of urban public services. Thus, since 1968,[1] there was a National Board of Public Utility Rates (Junta Nacional de Tarifas de Servicios Públicos, JNT), which was centrally (from Bogotá) regulated and even set prices of public services, such as water and wastewater (W&W), electricity, telephone, and mail for the entire country.

In 1994, Law 142 was issued. This law created a totally new institutional and regulatory framework for utility services. As part of this change, JNT was split into three specialized institutions: (1) the Regulatory Commission of Energy and Gas (CREG), (2) the Regulatory Commission of Telecommunications (Comisión de Regulación de Telecomunicaciones, CRT), and (3) the Regulatory Commission of Potable Water and Basic Sanitation (Comisión de Regulación de Agua Potable y Saneamiento Básico, CRA).

For the water sector, Law 142 created the free market entry for any service provider (by removing the ability of municipal governments to be the only providers), forcing public-owned companies to have corporate governance and achieve self-sustainability through rates plus consumption subsidies to poorer users.

The first part of this chapter explains the water pricing system for potable water and sanitation prior to the issuance of Law 142, while the second part focuses on the evolution of the price system implemented after that law.

[1] In fact, in 1936, a utilities section in the finance ministry was created, and, later, in 1960, an economic regulation superintendency was created, both of which controlled some utilities' prices until the creation of JNT.

7.2 Past Water Pricing System

During much of its existence (from 1968 to 1993), JNT was aimed at controlling the prices of the services under its scope in order to decrease the pressure of these prices on national inflation, which was a chronic problem suffered by the country for many years. It must be remembered that while Colombia never suffered from hyperinflationary processes like many other South American countries (Argentina, Brazil, Chile, and Perú), it was distinguished by its high and persistent annual inflation (between 25 and 35 %), which was a permanent concern of the governmental economic teams.

Public services regulated by JNT were provided (principally) by public-owned utilities. The national government owned almost all electricity companies and many water utilities in urban areas of small and medium municipalities. In contrast, municipalities owned the largest cities' water utilities.

The use of the tariffs of these services as an inflation anchor by the national government for many years led to almost all of these companies suffering huge financial deficits that had to be assumed by the public budgets.

The control of electricity rates played an important role in the national government's fiscal deficit. Under pressure of multilateral organizations (World Bank and International Monetary Fund) at the end of the 1980s, JNT started a process to review and increase energy rates to reflect the cost of services and reduce the level of subsidies given by the national government.

As a first step in the institutional reorganization of the water sector, in 1986 the National Institute of Municipal Promotion (Instituto Nacional de Fomento Municipal, INSFOPAL)[2] was closed. The responsibility of provision of W&W services was transferred to municipal and departmental[3] governments, claiming that decentralization would bring greater efficiency and political responsibility for the proper provision of public services.

The transfer of responsibilities of W&W provision services to local governments led JNT to be less concern about the water prices restriction on the public budgets. Controls of water prices by JNT were so acute that (combined with the administrative and operational inefficiency of these companies) the country's largest company, the Aqueduct of Bogotá (Acueducto de Bogotá), entered into default in the early 1990s, and the national government had to assume credits that the company had with the World Bank. The situation was quite similar to the vast majority of W&W utilities in the rest of the country.

[2] Until that moment, INSFOPAL was in charge of the delivery of W&W services in almost all medium and small municipalities by decentralized utilities (departmental ACUAS and Empresas de Obras Sanitarias, EMPOS) owned by the national government.

[3] Colombia has three political and administrative public levels: municipal, departmental, and national.

7.3 The Water Rate Regulatory Scheme

JNT had, at least on paper, two types of price regulations: "Regulated Freedom" and "Total Control":

- Using the first scheme (Regulated Freedom), JNT defined the methodology and tariff formulas that companies should use to calculate rates. Companies applied those methodologies and sent the results to JNT for its review, and JNT approved or modified the rates to be applied by utilities.
- In the Total Control scheme, there was no methodology, and JNT administratively defined rate increases to be applied by each utility in response to the respective utility's request.

The W&W sector was subject, throughout the period of existence of JNT, to the Total Control scheme so that, without any defined methodology, companies developed their studies to request rate increases, especially to recover the inflationary effect, and JNT arbitrarily decided which increases to authorize, often below past inflation. It is worth noting that in the early 1990s (a few years before its closure), JNT established automatic monthly tariff rate increases to recover past inflation, but existing rates, in almost all cases do not reflected the operational cost of the services and, of course, never the cost of investments (neither cost of capital nor depreciation).

7.4 Water Rate Differentiation

Since its inception, JNT established a tariff differentiation scheme (cross-subsidies) with higher-income families and nonresidential users paying much higher rates than lower-income families.

Because, in practice, the family income was unknown, for the classification of families, the houses in each municipality were divided into six groups, according to their cadastral (public record) value, calling each group "stratum" and defining stratum 1 as the poorest and stratum 6 as the richest. The nonresidential users were classified as commercial, industrial, or government.

The housing classification system based on the cadastral value lasted for 15 years (from 1968 to 1983). Until it was evident that high rates of Colombian inflation and the lack of an automatic adjustment scheme of cadastral values, led to poor families with houses of recent construction and registration in cadastral systems faced higher fees than homes of wealthy families registered some years before. Thus, in 1983 the classification system based on housing changed to one based on each house's characteristics, using a methodology defined by the National Administrative Department of Statistics (Departamento Administrativo Nacional de Estadística, DANE).

Initially, the methodology defined by DANE took into account both external (garden, front walls, and roof) and internal (floor, kitchen) housing characteristics and even families' holding of some electrical appliances (refrigerator, stove, or TV).

To define each house's stratum, an interviewer visited it and set it by direct observation of house features. This ad hoc system led to many problems, and it was common for houses with similar characteristics, even within the same town, to be classified in a different stratum because of the interviewer's judgment on the respective value. Later, DANE slightly modified its methodology, arguing that the classification had to be set based solely on the external housing conditions. However, some problems remained.

In 1992, the responsibility for defining the stratification methodology was passed to the National Planning Department (Departamento Nacional de Planeación, DNP), at that time (and for many years) the most recognized technical and official think tank entity within the national government. DNP's methodology kept the external features as the basis for classification of housing, but added environmental variables characteristic of the dwelling (such as quality of the streets, commercial activity, and existence of pollution sources). Additionally, each house's stratum ceased to be set by the interviewer and resulted in a statistical model that feeds on the information collected in the field.[4]

7.4.1 Charges and Blocks

In addition to user categories, the water rate structure for W&W services contemplated the existence of fixed charges, variable charges, and consumption blocks:

- The fixed charge was an amount payable monthly by the user ($/user/month), regardless of the level of consumption. In many Latin American countries, this fixed charge gives the user the right to a minimum consumption, but in Colombia this minimum was abolished in 1983.
- Variable charge is the value to be paid for each cubic meter of water consumed ($\$/m^3$) in each block of consumption.
- There was a block structure with increasing consumption rates. Until 1987, there were five blocks, but, since that year, it changed to three blocks: the first block, called *basic* (0–20 m^3/family/month), had a lower rate than the second block, called *complementary* (21–40 m^3/family/month), which had a lower rate than the third block, called *luxury* (41 or more m^3/family/month).

In summary, the rate structure contemplated categories of users, fixed charges, and consumption blocks, with increasing rates by user's category and blocks, as shown in Fig. 7.1.

The monthly water service bill was then equal to the sum of the fixed charge, plus the product of m^3 of water consumed in each block in the month, times the rate applicable to each block, with different fixed charges and different rates, per category of user and consumption blocks.

[4] For more details, see Cepal (2006).

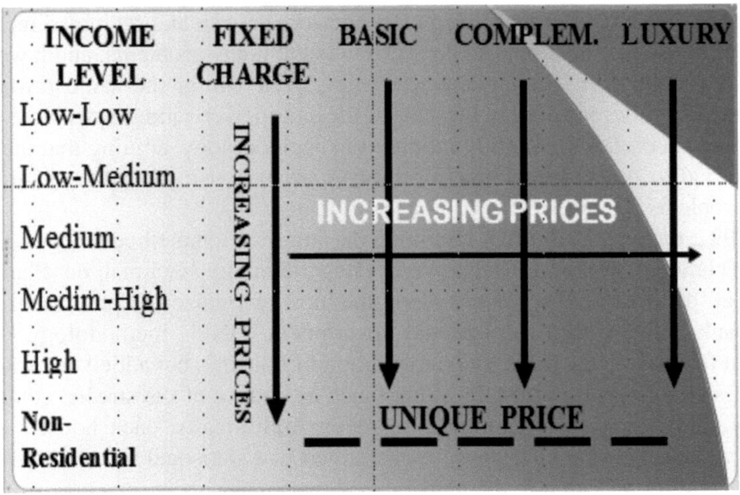

Fig. 7.1 Residential water rate structure in Colombia (Source: Diego Fernandez)

The pricing system applied to both residential and nonresidential users. As explained above, nonresidential users were classified as commercial, industrial, and government. In almost all cities, overprices applied to commercial users were similar to those applied to stratum 5 or 6; overprices applied to industrial users were higher (especially when the water was an important input in their production process), while the government users were rarely subject to surcharges.

For users without a consumption meter, a volume of consumption was assigned according to the category (based on user's consumption measured in the same category in each city) or their economy activity. Then, the bill was calculated as if they were a metered users.

7.4.2 Wastewater Rates

For wastewater service, a charge of 40 % of the invoice value of the water service was applied to users who had wastewater service.

7.4.3 Rates Differences

There were no rules about the difference in rates between the different categories of users or between blocks of consumption, although JNT established that luxury block's rate have to be equal for all types and user's categories within each town.

Significant differences between the rates applied in different cities, both in terms of the average tariff rates as distance between categories of users and blocks consumption arose. Those differences can't be attributed to differences in the cost of service delivery but to local policy decision.

An example that can be seen in Fig. 7.2 shows that in 1995 Bogotá families classified in stratum 6 paid a fixed charge almost 100 times higher than that paid by stratum 1. But this difference was reduced to 40 times in the city of Bucaramanga, 22 times in Cartagena, 20 times in Medellín, and only 10 times in Barranquilla.

Although not as large as in the fixed charge, the differences in variable charges according to the rates of consumption ($\$/m^3$) were also significant. As you can see from Fig. 7.3, in Medellín, the rate per m^3 paid by stratum 6 was almost 16 times

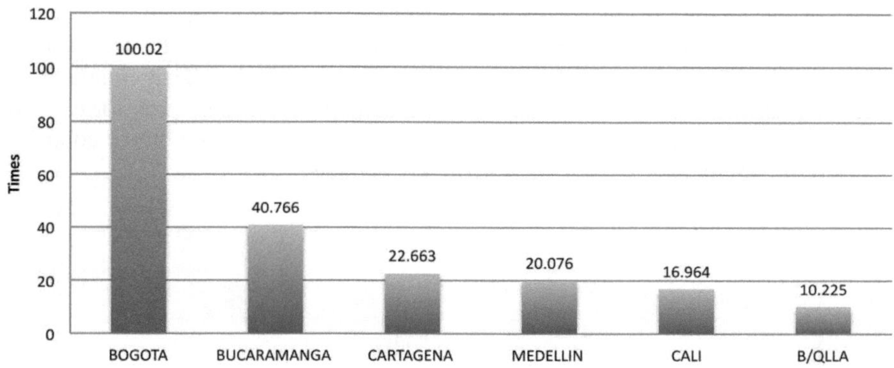

Fig. 7.2 Water rate distance between strata 1 and 6 in the fixed charge in some big cities in 1995. Source: CRA (2006) Comportamiento tarifario de los servicios públicos domiciliarios de acueducto, alcantarillado y aseo en Colombia

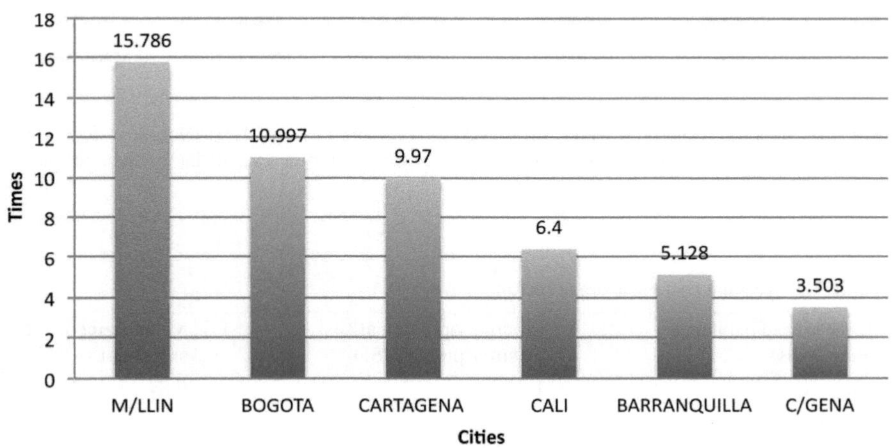

Fig. 7.3 Water rate distance between strata 1 and 6 in the consumption charge in some big cities in 1995. Source: CRA (2006) Comportamiento tarifario de los servicios públicos domiciliarios de acueducto, alcantarillado y aseo en Colombia

that paid by stratum 1; in Cali, that difference was 6.4 times, and in Cartagena it was around 3.5 times.

It is possible to identify some technical and conceptual problems in the tariff system used. However, neither the rate structure (categories of users, consumption blocks, fixed charges, and variable charges) nor the tariff differences (among categories of users or blocks) seemed to generate practical problems, so the real problem was the very low value of the average water bill, as can be seen in Fig. 7.4.

The Pan-American Health Organization (PAHO 2006) study evidenced the low tariff levels applied: in 1994, stratum 1 (poorest families) spent between 0.25 % and 2.5 % of legal minimum monthly salary (MMS) on their monthly water bill, and 60 % of this stratum paid less than 1 % of an MMS on their water bills.

Those low tariffs did not allow water utilities to recover at least the total operating costs (see Table 7.1) nor the depreciation or cost of capital. That entailed that, for the realization of any investments for expanding coverage or even for existing infrastructure renewal, public water utilities completely depended on the contributions of the municipal or national government budgets.

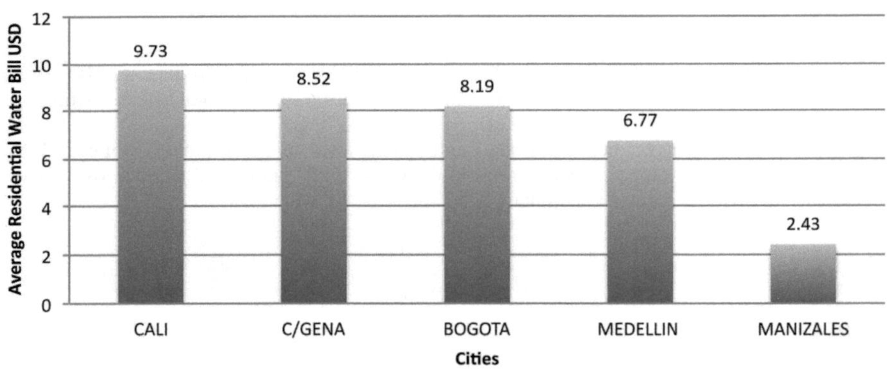

Fig. 7.4 Average residential water bill (USD) in some big cities in 1995 (Source: CRA (2006) Comportamiento tarifario de los servicios públicos domiciliarios de acueducto, alcantarillado y aseo en Colombia)

Table 7.1 Average cost vs. average income

Town size in habitants (thousands)	Monthly operational cost by suscriptor in USD	Monthly income by suscriptor in USD
<30	1.29	0.84
30–100	3.25	2.16
100–500	3.88	2.59
>500	7.48	6.43

Source: CRA (2001) Comportamiento tarifario de los servicios públicos domiciliarios de acueducto, alcantarillado y aseo en Colombia

The problems of very low rates and revenues were reflected directly into bad services. In its 1996 study (Organización Panamericana de la Salud 1996), PAHO found that although almost all water utilities have potable water treatment plants, the quality of water delivery was very low, water losses were very high, and the W&W coverage was low.

The political management of tariffs, along with citizen demands for better service, produced a crisis in 1980 in which the water and sewage companies were involved, as stated by Malinovitz (1998). This resulted in the need to propose a new model for providing potable water service.

7.5 Tariffs in 1995: Examples from Medellin and Cali

7.5.1 Differences in Fixed Costs

As can be seen, the families of the higher strata in both Cali and Medellin pay higher fixed charges. However, the absence of a clear pricing policy led to significant differences between these cities in tariff applied.

For example, the fixed charge applied to the families of stratum 6 in Medellin was 84 % higher than that applied to the same stratum in Cali. In total, Medellin families of higher strata had a bill significantly higher than the families of Cali (Fig. 7.5).

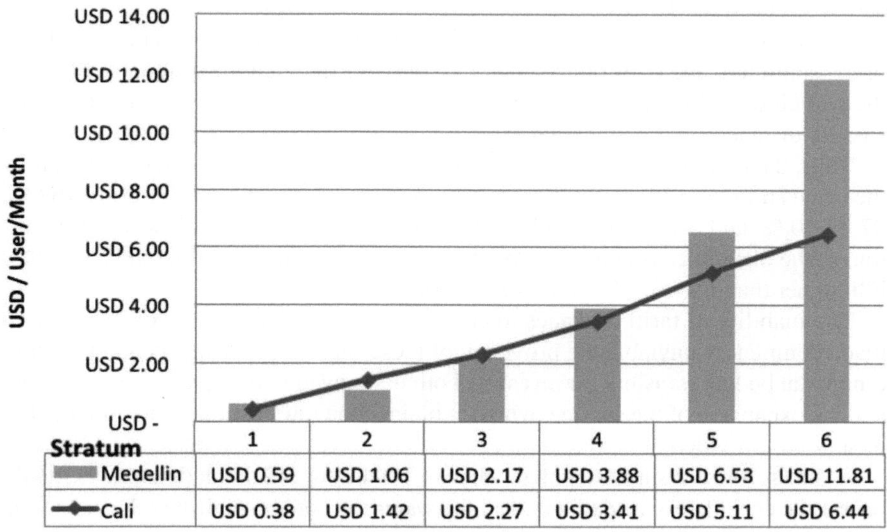

Stratum	1	2	3	4	5	6
Medellin	USD 0.59	USD 1.06	USD 2.17	USD 3.88	USD 6.53	USD 11.81
Cali	USD 0.38	USD 1.42	USD 2.27	USD 3.41	USD 5.11	USD 6.44

Fig. 7.5 Fixed charge in 1995 in Cali and Medellin (Source: Author with data of Comisión de Regulación de Agua Potable y Saneamiento Básico)

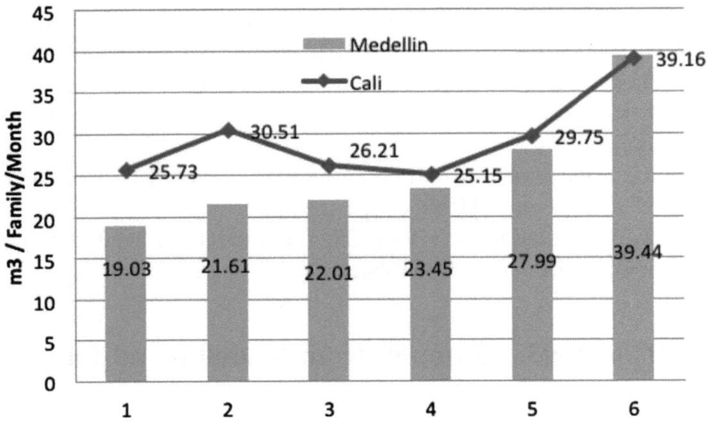

Fig. 7.6 Average monthly consumption in 1995 (Source: Ahutor with data of Comisión de Regulación de Agua Potable y Saneamiento Básico)

7.5.2 *Difference in Total Monthly Bill*

Before analyzing the difference in the monthly bill, it should remember that its total value depends largely on consumption. As shown in the graph, in 1995 Cali had average monthly consumption (per subscriber) higher than Medellin for all strata, probably explained by the difference in average temperature between these cities (25 % vs. 20 %). However, the difference in average consumption is quite variable across strata (see Fig. 7.6).

When calculating the total monthly bill (taking into account fixed costs, volume of consumption, and rates of each city in 1995), we find that the difference between the two cities follows a clear pattern, with the greater distances in lower strata and smaller distances in the higher strata.

Thus, the monthly bill paid by Cali stratum 1 was 216 % higher than that paid by the same stratum in Medellin. For strata 2, 3, 4, and 5, the differences were 168 %, 37 %, 10 %, and 6 %, respectively, always against users in Cali. However, for stratum 6, the difference is about 6 % but in this case users in Medellín paid a monthly bill higher than users of the same stratum of Cali.

The inability of tariff revenues to cover even the operating costs of the services made completely unviable the provision of these services, which depended on governmental budget transfers for investing both in renovation of existing infrastructure and the expansion of the service, which explained the low coverage and poor quality of services provided.

It is therefore not surprising that, when in 1994 Law 142 defined that rates should reflect the cost of service delivery, and the water regulator established the economic methodology for costing, the need for significant rate increases became evident, as we shall see in next section.

7.6 Current Water Pricing System

In 1991, seeking to make numerous political and institutional changes in Colombia, a new constitution was issued, which included as one of the most important issues the need to reorganize the management scheme of residential public services (water, wastewater, fixed landline telephone, energy, and gas).

The 1991 constitution created the Superintendence of Domiciliary Public Services (Superintendencia de Servicios Públicos Domiciliarios, SSPD) as the agency responsible for monitoring and controlling the adequate provision of residential public services. Additionally, the constitution states that the congress should issue a law regulating all aspects related to public services. In May 1994, the congress issued the Law 142—or Domiciliary Public Service Law.

Law 142 makes a real break in the scheme of providing these services, based on a clear definition of the roles of SSPD and other actors of the sector in regulation, control, definition of policy, and service delivery. It was eliminated JNT and created regulatory commissions. These sectors open to private participation in competition, defining the principles of tariff regime and the basic parameters of the subsidy system.

The new law defined the fundamentals of the public services' tariff system:

- Neutrality: Users of the same type that cause the same costs should have equal rates.
- Solidarity: Higher-income users must pay overprices, and lower-income users will receive subsidies.
- Transparency: Pricing formulas should be known by all players in the sector.
- Economic efficiency: Rates should reflect the economic cost of the service.
- Financial sufficiency: Tariffs must provide income for allowing efficient companies to provide the service.

Although that law gave priority to the principle of sufficiency, a further review of the Colombian Constitutional Court (Case C-150, 2003) determined that this principle could not have priority, but it should be of equal importance as other principles.

Combined with the rules defined by former Law 99 of 1993 about responsibilities with the water as natural resource, the authorities and functions related to W&W are those presented in Table 7.2.

Law 142 maintains the former rate structure that contemplated categories of users, fixed charges, and consumption blocks, with increasing rates by user's category and blocks (explained in detail above). As a transcendental point, the law states that the rates should be directly related to the economic cost of the service and that subsidies should be granted to the poorest families (strata 1 and 2 and eventually 3) and surcharges be applied to nonresidential users and less poor families (strata 5 and 6).

The subsidies apply only to the basic consumption and can't be more than 50 % of the economic cost of the service for stratum 1, 40 % for stratum 2, and 15 % for stratum 3, while maximum overprices would be 20 % of the economic cost of the

Table 7.2 Water and wastewater authorities and functions in Colombia

Sector	Authority	Functions
Water as a natural resource	Ministry of Environment and Sustainable Development	Definition of national policy
		Definition of price/rates of water consumption and discharge of contaminated water
	Regional Environmental Authorities (REA)	Grant licenses and use permits
		Definition of price/rates of water consumption and discharge of contaminated water where regional factors are affected
Drinking water and wastewater	Vice minister of drinking water and basic sanitation	Discharges quality control
		Definition of national policy
		Definition of public budget allocation
	Municipalities	Grant licenses and use permits
		Definition of ASE (with approval of the CRA)
		Service provision (in free competition)
		Responsible for the service
		Definition of local budget allocation
	Public and private companies	Service provision
	CRA (Comisión de Regulación de Agua Potable y Saneamiento Básico)	Definition of price methodology and rulings
		Definition of performance indicators
		Definition of overprice and subsidy %
		Allocation of subsidies in the budget
	Ministry of Health and Social Protection	Definition of water quality standards
	SSPD (Superintendencia de Servicios Públicos Domiciliarios)	User protection
		Utility performance control
	SIC (Superintendencia de Industria y Comercio)	Free competition control

Source: Diego Fernandez

service for strata 5 and 6 and nonresidential users. Both residential users from stratum 4 and official users were exempted from overprices and are not beneficiaries of subsidies.

As will be discussed below, subsequent laws took the maximum subsidy levels for stratum 1 to 70 % (of the economic cost of supply) and released the percentages of overprices. In almost all municipalities, complementary contributions from municipal budgets are necessary to attain balance between subsidies and overprices.

Both percentages of subsidies and overprices to apply, and the budgetary contribution to subsidies, must be defined by each municipal council at least once every 5 years, but can be modified each year. As part of this definition, each council has to guarantee the budget to cover the difference between contributions and subsidies.

Defining the first methodology for determining the economic cost of the service proved to be a great challenge for the CRA, because none of the companies of the country providing W&W had a consolidated accounting system. Practically, every company in the sector was governed by budget rules rather than for public business accounting standards; and almost none of the companies in the sector had an investment plan over medium term.

Resolutions 8, 9, 12, 14, 15, 16, and 17 of 1995[5] defined, for the first time in Colombia, a tariff methodology for W&W services. These norms were issued between August and November 1995, and the new tariffs were to be applied in June 1996. The tariff regime defined by CRA is based on economic cost of providing the service, taking into account administrative, maintenance, operating, and capital costs, both in terms of the amount of capital invested as the cost of capital (or profitability) expected from the inversion in a sector of similar risk. In the new methodology, fixed charges had to cover billing and management costs. The variable charge must cover operation, maintenance, and investment costs.

Included within operating costs are applicable environmental charges (charged for environmental authorities)—that is, a "water fee" for use of water (as a natural resource) in the case of drinking water and a "pollution fee" for wastewater service.[6]

As solution for the limited reliable financial information and investment plans, the first methodology defined by CRA stated that the cost to recover (by tariffs) were:

- Direct costs of administration and operation reported by each company, excluding only those who had no relation with the provision of the service
- Capital costs calculated based on the new value of renewal (VRA[7]) of existing assets and a cost of capital rate between 9 % and 14 % real annual, before taxes

7.6.1 Wastewater Rates

Unlike the existing rate schedule prior to Act 142, the new tariff methodologies defined water and wastewater costs and tariffs separately. To calculate the wastewater consumption charge, the volume of water used is taken into account.

[5] All these resolution were compied further in Resolution 151/2001.

[6] For details on the subject of environmental charges, see Rodriguez et al. (1996) or Rudas (2009).

[7] Valor de Renovación a Nuevo (VRA For its acronym in Spanish).

7.6.2 Implementation of the New Tariff Regimes

Between the last quarter of 1995 and the first half of 1996, the majority of the country's largest companies conducted their cost studies. A 1996 CRA report shows that the weighted average increase in the bill for eight of the major cities of the country would have to be 212 %. For strata 1 and 2 (the poorest), even after applying the maximum subsidies (50 % and 40 %), increases would be 569 % and 310 %, respectively (see Table 7.3). To soften the impact of increases, in 1996, through the Law No. 286, a term of 2 years (until 1998) was established for companies to make adjustments.

A study performed by the University of the Andes[8] showed that in December 1997 levels of subsidies and overprices in 16 of the major cities of the country were far from legal goals, but that the situation was worse in large cities, as shown in Table 7.4.

The analysis of the dynamics of prices between 1998 and 2001 submitted by the Superintendence of Public Services[9] showed that while the growth of the total national consumer price index was nearly 28 %, prices of the W&W sector increased at 82.5 %.

Although there was a significant increase in water rates in that period of time, they remained below the targets. So a new Law (632/2000) extended until 2005 the term to fulfill the rate adjustment plan. To ensure the achievement of the objective, CRA established that companies should decrease each year at least one-fifth of the difference between applied tariffs and target tariffs (resulting from the application of the methodology). Additionally, through Law 812, the subsidy level of stratum 1

Table 7.3 Expected increase in average bills of W&W of eight major cities due to the implementation of the tariff methodology

Stratum	Participation in total stratum (%)	Monthly consumption m^3/user	Bill in December 1995 USD/Month	Legal bill target USD/ Month	Required increase (%)
Stratum 1	5	24.72	1.56	7.54	569
Stratum 2	25	27.26	3.12	9.16	310
Stratum 3	40	24.95	4.87	10.46	190
Stratum 4	20	26.25	7.41	11.68	124
Stratum 5	7	28.37	11.73	15.20	62
Stratum 6	3	33.04	16.48	16.92	34
Average increasing required					212

Source: CRA (2001) Comportamiento tarifario de los servicios públicos domiciliarios de acueducto, alcantarillado y aseo en Colombia

[8] See CEDE (2004).

[9] SSPD (2002).

Table 7.4 Percentage of cost of reference covered with tariffs in December 1997 (%)

Stratum	Average 16 large cities	Medellín	Barranquilla	Bogotá	Cali	Legal target
Water						
1	25	12	17	7	14	50
2	38	26	23	18	31	60
3	58	49	34	30	43	85
4	80	80	82	44	68	100
5	103	120	109	65	95	120
6	124	159	135	79	118	120
Wastewater						
1	26	14	22	9	11	50
2	37	26	28	19	22	60
3	52	51	37	28	30	85
4	73	85	77	38	58	100
5	93	137	109	58	74	120
6	118	174	137	81	103	120

Source: CEDE-ANDESCO, Analysis of the evolution of the Public services, 2005, Tables 19 and 20

Table 7.5 Consumption reduction by stratum in five of the major cities of the country between 1997 and 2001 (%)

	Bogotá	Medellín	Cali	Barranquilla
Stratum 1	–30.9	–22.9	–16.1	11.8
Stratum 2	–32.2	–24.2	–20.6	–25.8
Stratum 3	–34.2	–17.4	–10.6	–35.0
Stratum 4	–21.5	–14.0	–8.2	–37.2
Stratum 5	–17.1	–17.3	–6.3	–39.4
Stratum 6	–13.1	–25.5	0.9	–31.1

Source: CRA (2001) Comportamiento tarifario de los servicios públicos domiciliarios de acueducto, alcantarillado y aseo en Colombia

expanded to 70 % and left without limits to overprice applicable to strata 5 and 6 and commercial and industrial users.

Price increases caused a substantial reduction in water consumption far beyond what was expected. As shown in Table 7.5, for example, between 1997 and 2001, the volume of monthly consumption (in m^3) of strata 1, 2, and 3 of Bogotá fell more than 30 %.

Not only residential user consumptions fell, but industrial and commercial user consumptions also fell. As shown in Fig. 7.7, water consumption as raw material significantly reduced in 1996 and 1997, after the implementation of the tariff methodology. However, the tariff increases were so significant that no matter the important reduction in consumption, the value of water used increased significantly as can be seen in Fig. 7.8.

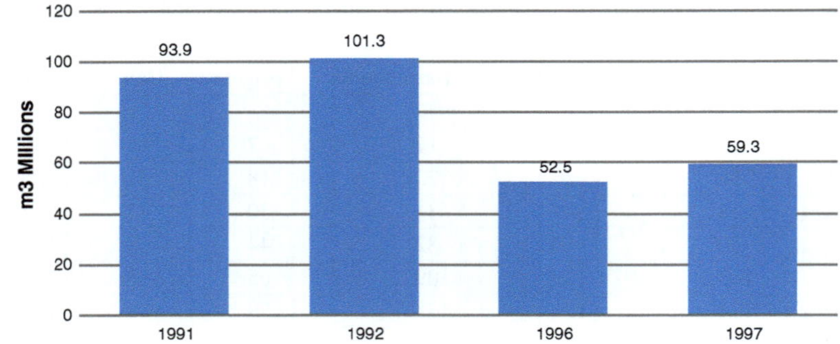

Fig. 7.7 Water consumption as raw material Source: Rudas (2009). Rates and tariffs for water usage. Impact over W&W services in residential areas and over industrial and agriculture profitability

Value of water consumed as raw material (millions of dollars in 2009)

	1991	1992	19996	1997
Millions of Dollars	31.3	51.9	40.1	51.4

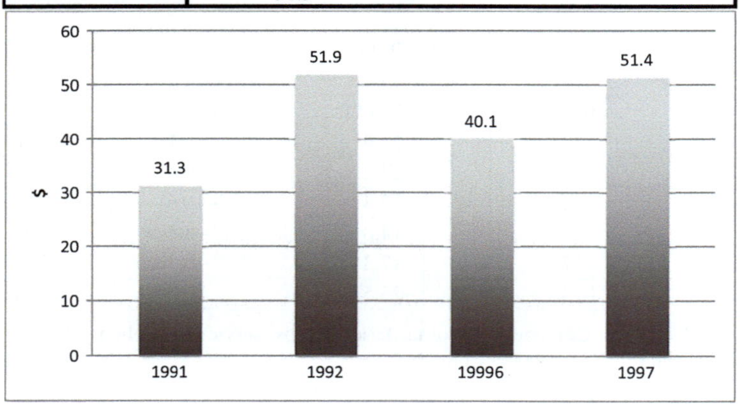

Fig. 7.8 Value of water consumed as raw material (millions of dollars in 2009) Source: Rudas (2009). Rates and tariffs for water usage. Impact over W&W services in residential areas and over industrial and agriculture profitability

The drastic reduction in consumption and the increase in the percentage of subsidy (stratum 1) led to increases in bills paid by users to be lower than initially expected, but still significant, as can be seen in Table 7.6.

The next big leap in the price regulation of the sector occurred in 2004 with adoption of CRA Resolution 287, which introduces the use of the Data Envelopment Analysis (DEA) model for the establishment of efficient management and operation

Table 7.6 Evolution of the average bill for each city between 1995 and 2005

		USD		
Water company	Principal city	1995	2000	2005
EAAB	Bogotá	15.02	15.08	27.83
EMP	Medellín	14.32	15.29	26.92
EMCALI	Cali	20.11	14.62	29.79
Triple A	Barranquilla	15.91	19.40	25.29
CAMB + CDMB	B/manga	13.23	10.28	22.22
ACUACAR	Cartagena	12.14	16.82	28.59
EPN	Neiva	10.41	7.08	11.84
ACUAVALLE	Regional	10.42	10.12	15.79
Aguas Kapital Cucuta	Cúcuta	13.19	6.65	16.73
Aguas y Aguas	Pereira	10.44	17.22	21.92
Average		14.93	14.60	26.26

Source: CRA (2001) Comportamiento tarifario de los servicios públicos domiciliarios de acueducto, alcantarillado y aseo en Colombia

costs. The new methodology has two application periods: in the first period, from July 2004 to December 2006, after computing the DEA scores, CRA ordered a temporary adjustment on the fixed charges of all companies to a maximum of USD 2.9/user/month for the two services (W&W). The second period began in January 2006, when companies started a full implementation of the results of the new methodology (regardless of the limit of the fixed charge) with the application of each DEA score.

By defining the ledger accounts that should be used to calculate each of the (fixed and variable) costs, the most important effect of this new methodology was to reduce the fixed charge ($/user/month). In contrast to the effect of decreasing the fixed charge, the new methodology increased the usage variable charge ($\$/m^3$).

The introduction of control efficiency through the DEA model had a fairly marginal effect, partly due to pressure from a political lobby by service providers, which led to revision of the initial models, eventually leading to the vast majority of companies finding themselves very near the efficiency borders of both their administration and operating costs. Additionally, the effect of methodology was eased up by the establishment of a period of tariff transition (gradual adjustment), which extended from January 2006 to May 2009.

Rates defined by Resolution 287 still apply until the date of preparation of this chapter (May 2014), with adjustments only for inflation, whenever the consumer price index (CPI) reaches a cumulative increase of 3 %.

This new methodology led to a marginal reduction of the average bill with respect to the level achieved in 2005. A long-term bill evolution is presented in Fig. 7.9, including the water bill (excluding sewerage) for four users' categories (poorest or stratum 1, middle or stratum 4, richest or stratum 6 and commercial), as an average for Bogotá and Barranquilla cities. Including sewerage, the bill will increase around 80 %.

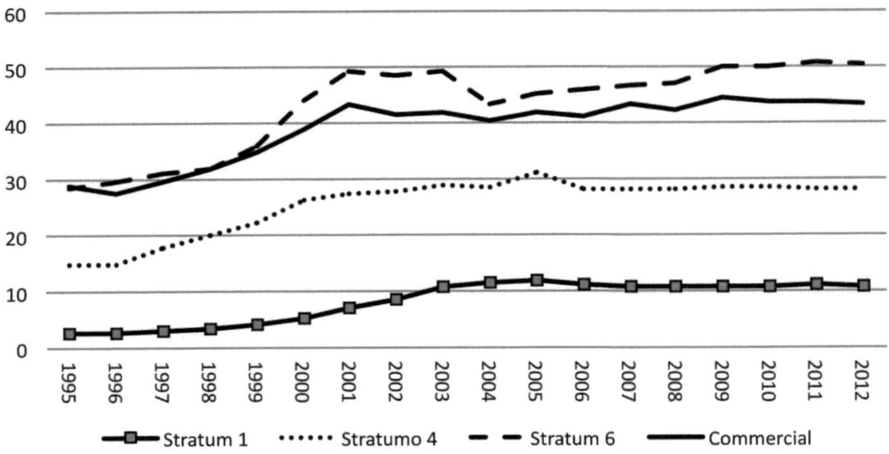

Fig. 7.9 Only water* bill for four users categories average for Bogotá and Barranquilla USD of December 2012 (Source: CRA (2001) Comportamiento tarifario de los servicios públicos domiciliarios de acueducto, alcantarillado y aseo en Colombia)

As can be seen in that figure, the poorest families' water bill increased from less than $3 in 1995 to almost $12 in 2004 and remains around $11 since then. That means bill of this user category increased almost 3.6 times. For the middle-class users (stratum 4), the increase between 1995 and 2012 was almost two times, and for the richest category, it was 1.5 times.

The only relevant changes (additional to the adjustment for inflation) that have occurred since 2006 are:

- The change in sewerage tariffs in some companies (with prior approval of CRA) to include the operational costs and/or investing in cost of their new systems of sewage treatment (such as Bogotá, Cartagena, Barranquilla, and Manizales).
- The introduction in Medellín (in 2010) and Bogotá (in 2012) of an additional subsidy for the poorest users to fully cover the consumption of up to 6 m^3 of potable water, as part of the introduction to the concept of the human right to water established by the United Nations (UN) in 2010. This additional subsidy is a local policy (initiative of its mayors) and not a national one.

Because the existing scheme provided a subsidy of up to 70 % for the first 20 m^3 per month, this local policy gives an additional allowance of 30 % on the first 6 m^3 water (not including wastewater) for some user groups in Medellín and all users in stratum 1 in Bogotá.

Since February 2013, CRA submitted for consideration and feedback from industry actors a new proposal to replace the tariff methodology defined in 2004 (Resolution 287). The regulations, with highly complex mathematical and conceptual terms, are still under discussion and are expected to start their final implementation in July 2015.

7.6.3 Irrigation Rates

Colombia has an area of 114.17 million hectares, of which 44.6 % (50.91 million hectares) are for agricultural use. Only some 900,000 ha of this agricultural land has an irrigation system by either gravity, sprinkler, or drip.

Ninety percent of the irrigated area is irrigated by gravity, recognized as the most inefficient system in water use, but most popular for its simplicity in infrastructure installation, easy maintenance, and low or no cost in electricity. According to Marin (1991), "The gravity method is used in the country mainly due to the traditional custom of considering water as an abundant and cheap resource, which can be spent without further rationality and probably also by the high investment initial demanding other systems."

Of the 900,000 irrigated hectares, two-thirds correspond to small-scale irrigation developed by the private sector, while the remaining area has public irrigation of medium or large scale, which currently is operated under the Colombian Institute for Rural Development's 24 irrigation districts.

Most of Colombia's climate is seasonal (with at least three dry months during the year) with driest months requiring irrigation for the production of permanent crops.

Since the 1930s, the national government has been promoting the adaptation of land with irrigation systems, which is reflected in the institutional evolution of the sector: the creation of Electraguas in 1936, the Colombian Institute of Agrarian Reform (INCORA) in 1961, the Institute of Hydrology, Meteorology and Land (HIMAT), and Instituto Colombiano de Hidrología, Meteorología y Adecuación de Tierras (HIMAT) in 1994. INAT was replaced in 2003 by the Colombian Institute for Rural Development (INCODER). Through these institutions, Colombia has tried to operate efficient irrigation districts that contribute to the modernization of agricultural production. However, the absence of a methodology for the definition or regulation of tariffs for irrigation services has led these rates to cover only a portion of the maintenance and operation cost of the systems, preventing investment in renovation, expansion, and modernization of irrigation systems.

The charges for irrigation are composed of a fixed charge per hectare (per year) and a variable metric or consumption volume charge. Currently, INCODER fixed proportion payment of these charges, which in the past were fixed by the HIMAT then by INAT. In the past, the government paid the difference between the proportion collected by INCODER and operation and maintenance needs.[10]

Fixed charges are payable in advance, prerequisite for irrigation services, while charges per unit of water consumed are paid at the end of each season. Due to the inefficiency of the paternalistic system of water charges in 1980, irrigation charges covered only reached 35 % of the costs of operation and maintenance. That percentage reduced in 1987 to 28.7 %, which precluded the state's investment in creating major areas covered with irrigation systems.

[10] Plusquellec (1989).

Despite being an agricultural country, Colombia does not have a proper irrigation system, and use of inefficient technologies for water management makes it vulnerable to dry spells. In addition, the subsidized tariff model has not been allowed to invest enough in public irrigation districts.

7.7 Future Measure Debate

Today, companies are in the process of discussing a new rate methodology proposed by the CRA for potable water and sewage, which would take effect from July 2015. The aim of the new methodology is simply defining the procedures and formulas to be used by companies to calculate their costs and rates for the next 5 years without changing any aspect (such as categories of users, block consumer subsidy levels, etc.) of the current rate schedule. The new methodology defines in detail how companies should make their projections for new users and billed volumes.

Similarly, projection methods and administrative and operating costs are detailed; maximum unit costs are defined by type of infrastructure and their useful lives (to calculate annual depreciation rates). Although the proposed methodology is certainly more complicated (or detailed) than what is expected from a regulator, it is not expected to have significant effects on the rates currently applied by companies.

In addition to the work of the CRA, the congress is discussing and preparing a legal definition for the theme "human rights to water," probably to accompany or follow the action already taken by Bogota and Medellin explained above.

Another discussion that is currently under way in Colombia is related to the provision of water and sewerage services in small communities. The definition of an assistance program and special regulations for small providers is anticipated.

7.8 Conclusion

The use of W&W service tariffs as an inflationary anchor for years in the past, led water companies into a severe financial crisis that became evident at the beginning of the 1980s. As part of the decentralization process in the mid-1980s, the national government transferred ownership of medium and small enterprises to local governments, which yet were owners of large cities' W&W utilities. The national government reserved for itself the power to define, through JNT, the tariffs that each of the W&W companies in the country could apply. But the low-level equilibrium of the sector did not change. By the early 1990s, almost all companies in the sector were in deplorable financial and technical conditions. Coverage of water and sanitation services, even in large cities, was low, and the continuity and water quality problems were common. The issuance of Law 142 in July 1994 was a break point that opened the sector to free private participation, defined the rules to determine rates, and established a clearly defined and transparent subsidy scheme granted to the lower stratum.

CRA issued the first methodology to define tariffs in the W&W sector in 1995, and Bogotá (the capital city) was the first one to apply it. The transition from the old level of rates to those resulting from this new methodology took 10 years. The regulations issued by CRA, establishing clear methodologies for calculating the economic cost of each service (water or wastewater) and the fact that Bogotá was the first to apply the new methodology and initiate rate increases, encouraged other large and medium companies in the country to implement the new regulation.

The tariff increases led some Colombian cities to have the most expensive services of W&W in Latin America. This fact, in turn, resulted in significant reductions in consumption in a way that some Colombian cities also have the lowest consumption per household in Latin America. The sector had significant strengthening to the point that almost all large and medium Colombian W&W utilities cover completely their costs (including depreciation and cost of capital), based on their tariffs.

In general, the companies of the sector have access to capital markets, and some of them have issued bonds (something very unusual in companies of this sector in Latin America), and all of them pay income taxes, like any other economic activity.

The regulation of the water sector in Colombia has had a much better performance than that achieved by any other Latin American country, predominantly served by publicly owned companies.

References

CEDE. (2004). Centro de Estudios sobre Desarrollo Económico Universidad de los Andes. *Análisis de la evolución de los servicios públicos domiciliarios durante la última década*. Bogotá D.C. 2004.

CEPAL. (2006). *Maria Cristina Alzate, La estratificación socioeconómica para el cobro de los servicios públicos domiciliarios en Colombia* (LC/L.2604-P). Septiembre 2006.

CRA. (2001). Comisión de Regulación de Agua Potable y Saneamiento Básico. *El estado del arte de la regulación en el sector de agua potable y saneamiento básico en Colombia*. Bogotá D.C: Colombia. Comisión de Regulación de Agua Potable y Saneamiento Básico. Bogotá D.C. 2001.

CRA. (2006). Comisión de Regulación de Agua Potable y Saneamiento Básico. *Comportamiento tarifario de los servicios públicos domiciliarios de acueducto, alcantarillado y aseo en Colombia*. Reporte Estadístico. Agosto de 2006.

Malinovitz, S. (1998). *Servicios públicos domiciliarios, vivienda y distribución del ingreso*. Bogotá: Mimeo.

Marín, R. (1991). *Estadísticas sobre el Recurso Agua en Colombia*. Instituto Colombiano de Hidrología, Meteorología y adeacuación de tierras (HIMAT). Bogotá. 3. 412.

Organización Panamericana de la Salud. (1996). *Análisis del sector de agua potable y saneamiento básico en Colombia*. Autor. Bogotá. Diciembre de 1996.

Plusquellec, H. (1989). *Two irrigation systems in Colombia. Their performance and transfer of management to user's associations*. The World Bank. 12–15. Retrieved in September 24, 2014, from http://www-wds.worldbank.org/servlet/WDSContentServer/WDSP/IB/1989/09/01/000009265_3960928071552/Rendered/PDF/multi_page.pdf

Rodriguez Becerra, M., Botero, U., & Carrizosa Umaña, E. J. (1996). *Instrumentos económicos para la gestión ambiental en Colombia*. Bogotá: FESCOL, CEREC.

Rudas, G. (2009). *Tarifas de las tasas por uso del agua. Impacto sobre los costos el servicio de acueducto residencial y sobre la rentabilidad industrial y agropecuaria*. Bogotá. Mayo de 2009.

SSPD. (2002). *Superintendencia de Servicios Públicos Domiciliarios*. SuperCifras, revista No. 6.

Viceministerio de Ambiente. (2010). Política Nacional para la Gestión Integral del Recurso Hídrico (124 p). Bogotá: Colombia, Ministerio de Ambiente, Vivienda y Desarrollo Territorial.

Chapter 8
Water Pricing in France: Toward More Incentives to Conserve Water

Marielle Montginoul, Sébastien Loubier, Bernard Barraqué, and Anne-Laurence Agenais

Abstract With an historical overview of the legislative and regulatory framework of water pricing in France, this chapter first describes how the focus of pricing policy progressively shifted from budget balancing to water conservation then to social protection. The next part focuses on pricing practices in the urban sector. Price levels and the evolution of tariff structures are analyzed using surveys and case studies results. The fourth section focuses on water pricing in the agricultural sector at different scales: large public irrigation schemes, smaller water user associations, and individual irrigation systems. The evolution of water abstraction fees collected by river-basin authorities is also analyzed, and we present how these fees can be modulated depending on the degree of collective management of agricultural water resources. To conclude, we discuss the efficiency of water pricing in urban and irrigation sectors and highlight some limits to take into account several uses.

Keywords France • Economic instrument • Economic incentives • Irrigation • Water tax

8.1 Introduction

Like in many regions of Europe, water is increasingly scarce in France, and as water demand goes up, environmental standards incite to let more in rivers' basins, and pollution reduces available resources. Simultaneously, the cost of producing water

M. Montginoul (✉) • S. Loubier
UMR G-Eau – Irstea, 361 rue JF Breton, BP5095, F34196 Montpellier cedex 5, France
e-mail: marielle.montginoul@irstea.fr; sebastien.loubier@irstea.fr

B. Barraqué
Agroparistech-Engref, 19 Avenue du Maine, F75015 Paris, France
e-mail: bernard.barraque@agroparistech.fr

A.-L. Agenais
ACTeon – Innovation, Policy, Environment, 5, Place Sainte Catherine, F68000 Colmar, France
e-mail: al.agenais@acteon-environment.eu

A. Dinar et al. (eds.), *Water Pricing Experiences and Innovations*,
Global Issues in Water Policy 9, DOI 10.1007/978-3-319-16465-6_8

rises, as water has to be transported over longer distances and/or treated at a cost that has been continuously rising over the last two decades—in particular for drinking water, due to the cost of removal of nitrates and pesticides, and the strengthening of quality standards. In response to these changes, water is now clearly perceived as an economic good that should be charged to users in order to provide economic incentives to save it (efficiency objective), to recover direct and indirect costs related to its production (cost recovery objective), taking into account equity considerations and constraints of administrative and political feasibility. In this chapter, we only cover France's territory in Europe (Metropolitan France), but we do not consider French overseas départements[1] and territories.

8.1.1 Climate Diversity

Metropolitan France is the largest country among European Union members, with 543 965 km^2, located in the northern temperate zone. The wide diversity of landscapes, from coastal plains in the north and west to a variety of mountain ranges in the southeast (Alps), center (Massif Central), and south (Pyrenees), results in four different climate areas:

- Oceanic climate (southwest, west, and north), with average rainfall all year long and a reduced range of temperatures
- Continental climate (inland and east), with a wider range of temperature from winter to summer and rainfall in spring and summer
- Mediterranean climate (southeast), with hot and dry summers, episodic but heavy rainfall (violent storms in autumn)
- Mountain climate, with important rainfall and a wide range of temperatures

8.1.2 Abundant but Unequal Distribution of Water Resources

France is endowed with abundant water resources and important natural water storage, due to the numerous mountain areas (south and east) and large littoral zones (west and north).

Yearly average rainfall volume is 486 km^3, of which 175 km^3 turns into effective rainfall. From these available water resources, 75 km^3 flows as surface water, while the remaining 100 km^3 percolates to aquifers. These latter volumes are then released over time to rivers to form the basic stream flow.

[1]The *département* is an administrative division created after the French Revolution. Territorial state services at this level are led by a prefect. There are also elected representatives who form the *Conseil Général*. This institution has gained in importance since decentralization laws of 1982–1983.

Surface water in France corresponds to more than 550,000 km of rivers (mainly small rivers and streams), with the five main rivers (Rhone, Rhine, Loire, Seine, and Garonne) draining most of France's surface water flows.

Still water bodies include more than 34,000 lakes, reservoirs, and ponds. The 9,800 largest ones cover 2,800 km^2 and have a cumulative capacity of 24 km^3, Lake Leman excluded. Groundwater resources are estimated to reach 2,000 km^3, with about 200 main aquifers and more than 6,300 small aquifers.

Despite abundant surface and groundwater resources on a national level, water resource availability is variable. In the southern and eastern regions, the weather is dry, while torrential rain episodes occur during short periods of time. The total volume of rainfall is thus equivalent to the national average. In most of the other climate areas, rainfall is common all year long, but only mountainous areas receive a higher volume of rainfall, compared to the national average.

In the end, due to a combination of climatic and human factors, drought-prone areas are located not under Mediterranean climate so much as in middle-range Garonne, Charente, western France, and Loire; North of France is in the same situation as south-eastern England: 600 mm of rain, no large rivers, and high population density.

8.1.3 Main Freshwater Uses in France

In 2007, about 31.6 km^3 of water was abstracted in France, mainly from surface water (82 %). On the total volume of water resources collected in 2007 (Table 8.1), 59 % was used in thermal power plant cooling (classical and nuclear power plants, excluding most hydroelectricity). The water was mainly pumped from rivers and almost completely returned after use.

- Eighteen percent was abstracted by public water supplies (drinking water), mostly for the needs of urban areas. The total volume collected for drinking water remains stable but undergoes a decline in downtown areas.
- Twelve percent was collected for irrigation, mainly from surface water catchments (rivers, ponds) in southwestern and southeastern France, where crops with high water consumption are grown (e.g., corn). The volume of water collected for irrigation slowly increased over previous years on a national level but faster in the south and west.
- Only 10 % was abstracted by industry. Industrial water use is chiefly taken from surface water (59 %) and is mainly located in northern and eastern France, mostly for paper production and metallurgy. Water volume collected for industry use continuously decreased over the past decades (−30 % since the 1970s).

Data for irrigation is usually underestimated due to difficulties in monitoring private wells. The equivalent figures 15 years before were (1) 6.0 for public water supply, (2) 3.9 for industry, (3) power plant cooling peaked at 24.2, and (4) irrigation was around 3.9 for collective schemes and 4.9 including private wells (our estimation in the Eurowater report). It is clear then that the only growing water

Table 8.1 Freshwater resources available in 2007 and their uses

In billion cubic meters and percents										
(Source)	Drinking water		Industry		Irrigation		Energy		Total	
Surface water	2.2	37 %	1.8	59 %	3.1	80 %	18.8	100 %	25.9	82 %
Groundwater	3.6	63 %	1.3	41 %	0.8	20 %	0.0	0 %	5.7	18 %
Sub total	5.8	100 %	3.1	100 %	3.9	100 %	18.8	100 %	31.6	100 %
% by use	18 %		10 %		12 %		59 %		100 %	

Bommelaer and Devaux (2012)

demand is from agriculture, which has an obvious consequence on water scarcity: agricultural water demand is concentrated in 20 % of national territory. Irrigated surfaces doubled between 1980 and 1990 in France and grew particularly in a large southwestern portion of the country. If abstracted volumes remain small compared to power plant cooling needs, water consumed (i.e., not returned to the ecosystem) by agriculture halves the total, and abstractions reach 80 % of the total in the summer. It is then clear that droughts reveal a man-made scarcity, which can be alleviated by water reallocation.

Water consumption depends on local conditions, on uses, and also on prices. Water pricing levels and structures can be explained in France by the French legislative and regulatory framework, which is presented in Sect. 8.2. The following sections will be dedicated to the presentation of the variety of water pricing implemented for the urban use (Sect. 8.3) and the agricultural use (Sect. 8.4) and to take into account environmental constraints (Sect. 8.5). Section 8.6 concludes giving an overview of current debates and future directions of water pricing.

8.2 Historical Overview of French Legislative and Regulatory Framework Concerning Water Prices

Historically, the focus of pricing policy progressively shifted from a budget balance mandate in the 1970s to water conservation (1992 water law) and more recently to a social protection objective (2006 water law and subsequent regulations).

Water prices in France are framed by a national history that seeks, since the creation of the Agences de l'eau (water agencies) in 1964, to price water at its economic value, including environmental cost. It is reinforced by the European legislative framework: the European Water Framework Directive—WFD (European Commission 2000)—published in December 2000 aims at recovering the quality of the aquatic environment and presents economic instruments as ways to reach it. To do so, the European Union (EU) member states have to estimate the full cost of water services (operational, capital, and environmental) and to try to recover it through water pricing. The European Water Framework Directive asks also to design water pricing policies in order to provide adequate incentives for an efficient water use.

8.2.1 Water Agencies on How to Target Full Cost Pricing

In the 1960s, the booming economy, the rapid urbanization process, and the catching up with sewerage infrastructure delays led to increased situations of scarcity and to massive pollution discharge in rivers. Under the Gaullist government, the planning system expanded to encompass more than initial industrial development, typically targeting global regional and urban planning. Concerning water, a special committee on water problems studies was set up to propose solutions, and it came up with the idea of controlling both pollution and scarcity at the river-basin level. They took members of parliament and of the senate to visit other countries, and they finally chose to adopt/adapt the Ruhrverband model: urban and industrial water users would be qualitatively represented in a Comité de bassin (basin committee), which would both decide priority investments on a 5-year planning basis and vote the levies, and each of them would have to pay to fund the resulting budget up to 35 %. Investments proposed by stakeholders would be subsidized at 10 % and granted a zero-interest loan for another 20–40 %. This system started to operate in 1970 and would lead to important water pricing increases. By the way, it can be compared with the United States Clean Water Act's revolving fund, with an important difference though: from the beginning the fund was made up with water users' contributions and not the government's.

In more decentralized countries like Germany and the Netherlands, typically water boards have this taxation power, plus some police powers, and also the possibility to build and operate infrastructure (dams, sewage works) by themselves. These two additional roles were not granted to the six French water agencies, which ended up being almost like mutual savings banks of water users, in which contributions would be mandatory. This system is described by Colin Green (personal communication to B. Barraqué) as "hypothecated levy," you must pay, but you can get your money back if you decide to go environmentally friendly. And as a matter of fact, this system allowed adding 16,000 sewage treatment plants to the 1,000 that existed in 1965. It also allowed to fund a few multipurpose reservoirs, initially for enhancing low summer flows (water supply of large cities and nuclear power plant cooling needed river regulation) and, eventually, also for flood control.

The taxation system was made up of two different levies: one is a (small) water abstraction levy, itself composed of a tiny levy for abstraction and of a larger levy for water consumed (i.e., not returned to the aquatic environment); another levy, about five times larger for urban water uses, was targeted on pollution discharge; the quantification was based on biological oxygen demand (BOD), chemical oxygen demand (COD), suspended particulates, heat, toxic substances, and, later, also phosphates and nitrates.

Why did this financial system impact water tariffs? Indeed, the taxation should have targeted the initial abstractors and the end dischargers (i.e., industrial premises non-connected to public sewers, large farms[2]) and, for cities, the water

[2] Initially farmers were protected and they only paid the abstraction levy when they pumped important amounts of surface water. They did not pay any tax for diffuse pollution discharge. Only later a taxation of battery cattle breeding would be introduced, but quite painfully.

supply and sanitation (WSS) public services, which are under the responsibility of local councils.[3]

However, everybody resisted the new green taxes. Industry, of course, claimed that it would reduce their competitiveness, but thanks to the ongoing national planning system, they obtained the signature of "branch contracts," in which they could pool the taxes they paid to the six water agencies at the national level, and received additional grants from the ministry of industry, while co-deciding with the industrial environment government services the phasing of pollution control works in the branch's premises. This practice was condemned by the European Commission, but by that moment, industry had understood it was in its own interest to play the depollution game.

The opposition from local authorities was more serious: the Association of French Mayors voiced against having to pay a tax to institutions that were not elected one man, one vote (i.e., which were not sovereign as they were). In order to escape the central-local conflict, which was already important, the government decided to charge for levies through the water supply bills. The government passed a decree in October 1967, which considered sewerage a service rendered the same as the water supply. Therefore, sewerage charges would be included in water bills and in proportion of drinking water purchased (metering is generalized in France) and no longer through local land and housing taxes. With that change, it became simpler to include the abstraction and pollution discharge levies in the water bills. This would lead to important price increases, since the long-term cost of sewage collection and treatment, without subsidies, was above water supply costs. Since the local water supply authorities and not the water customers received the financial support of the water agencies, the tariff system was criticized by some consumers and alter-globalist NGOs as being opaque and unfair to domestic users. It certainly still constitutes part of a water tariff crisis today.

However, if we recall that the pollution discharge levy is far higher than the abstraction levy, it can be readily understood that investments needed to improve the environmental performance of sewage collection and treatment has always been more important than those needed to improve drinking water reliability. Since money paid to water agencies acting as a savings bank will, over time, be returned to water users to support needed investments, adding the pollution levy to the wastewater portion of the bill represents the long-term (partly mutualized) average cost of sewage collection and treatment. Symmetrically, the sum of the abstraction levy and the drinking water portion of the bill will represent the long-term average cost of water supply (Fig. 8.1). Even though the wastewater charge paid to the local operator is only two-thirds of the water supply bill, once the long-term costs are considered, wastewater is above drinking water, just like in other countries where there is no such mutual funding system of the water agencies.

But there is another interpretation to be made of the two levies, in terms of full cost recovery: the pollution discharge levy can be considered as representing the environmental cost above the full internal cost (in France, it is mandatory for WSS

[3] Since the initial water agencies were lightweight institutions, in the beginning they did not target villages below 500 inhabitants, which additionally had no sewer systems.

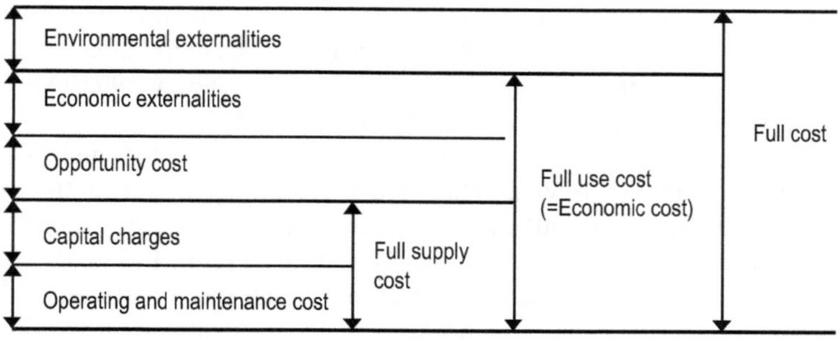

Fig. 8.1 The notion of full cost pricing (Agarwal et al. 2000)

services to cover operation costs, plus a reasonable figure for depreciation, i.e., full internal cost). And since the abstraction levy is there to fund investments to reduce situations of scarcity, one can consider it as the last part of full cost (i.e., users' costs). The only thing is that these two additions to reach full cost recovery are (a) mutualized and (b) not necessarily representing the real economic calculations of environmental and users' costs but rather a proxy obtained after the arbitration on the budget by the Comité de bassin (basin committee).

In the end, it implements the ideas of the "inventors" of the water agencies, which decided that France needed to introduce economic incentives both to reduce pollution and to reduce rivalries in quantity, as expressed in a book by Ivan Chéret, secretary of the committee on water problem studies, way back in 1967 (Chéret 1967).

8.2.2 A French Regulation as a Mix of European Legislative Framework and National Fluctuant Objectives

French water laws address only the case of urban water pricing, while other uses are being regulated by other instruments, such as quotas and levies at the river-basin level or at local levels. For instance, the 2006 French water law (Loi n°2006–1772 sur l'eau et les milieux aquatiques 2006) does not regulate raw water pricing (water used directly by farmers or industries) and prefers quantitative instruments to share water in scarcity areas, apart from the incentive put through water agency levies. It is at the local level, for instance, at the river-basin catchment, that we find regulation imposing to enhance incentive water pricing structures: an example is given by the SDAGE[4] Adour-Garonne (southwest part of France), implemented in 2010, which obliged water managers to generalize incentive pricing and then encouraged water conservation to guarantee water sustainability, particularly during low-flow seasons (Comité de bassin Adour-Garonne 2009).

[4] French acronym for master plan at hydrographic district level: *Schéma Directeur d'Aménagement et de Gestion des Eaux.*

For the case of urban water pricing, the first law addressing this aspect was voted in November 1992. It required all WSS services to balance their budget (except tiny villages), and not just the municipal global budget, by January 1993 (Montginoul 1997), as it was allowed previously for direct procurement.

The second law addressing water pricing is the 2006 French water law, which translated the 2000 European Water Directive to conditions in France. Its Article 57 is devoted to potable water and sewerage pricing and clearly aims at encouraging water conservation. In particular, Article 57 forbids (except for small cities and cases with plenty of water) flat rates and declining water rate structures. Forbidding declining block rates impacts large housing projects but also industry (i.e., large consumers who were granted this type of discount by many utilities).

Article 57 also limits the fixed part: it cannot represent more than 30 % (for urban districts) or 40 % (for rural services) of water bill (calculated for 120 m^3 annual consumption), except for utilities facing a high seasonal population. It is in fact more restrictive because this obligation is put separately on the two parts of water bills—potable and sewerage—not including taxes and fees.

8.3 Water Pricing Practices in Urban Sector (Including Industry)

The urban sector represents all users connected to the public potable water network, including households, hotels and commerce, public services, and industry. Urban water pricing is increasingly regulated in France: the 2006 law has induced drastic changes for some WSS units, in which water pricing structures did not fit the new rules.

The evolution of tariff structures is analyzed in this section using results from two national surveys conducted in 2003 and 2013 (Montginoul 2007). These surveys were carried out on the same 1,630 French districts selected following a stratified sampling procedure (taking into account three types of factors: geography, population size, and level of seasonal population). It was structured to collect information on the characteristics of water and wastewater management utilities, the detailed water bill, and the eventual existence of pricing specificities. The response rate was 29 % in 2003 and 40 % in 2013, with 429 (respectively, 393) answers totally exploitable. The results were adjusted to be fully representative of the French situation.

The average price (including VAT) in France in 2013 is 3.73 €/m^3 (Table 8.2). However, there is considerable variation in prices across municipalities, because water is priced at a local level, taking into account local conditions, and the fact that 25 % of water service units (small size, however) do not have collective sewerage and let households face the costs of decentralized solutions (not in bills).

On average, the fixed part is 44 euros for potable water (equivalent to a consumption of 29 m^3) and 23 euros for sewage collection and treatment (16 m^3 consumed). This low

Table 8.2 Average French water prices in 2003 and 2013

(2013 constant prices)	2003	2013
Water		
Proportional part	1.59 €/m³	1.61 €/m³
Fixed part	37 €/m³	44 €/m³
Fixed part in equivalent water consumed	29 m³	29 m³
Average price (for 120 m³)	1.91 €/m³	1.97 €/m³
Sewerage		
Proportional part	1.11 €/m³	1.63 €/m³
Fixed part	13 €/m³	23 €/m³
Fixed part in equivalent water consumed	14 m³	16 m³
Average price (for 120 m³)	1.21 €/m³	1.82 €/m³
Total		
Proportional part	2.69 €/m³	3.18 €/m³
Fixed part	51 €/m³	65 €/m³
Fixed part in equivalent water consumed	23 m³	22 m³
Average price (for 120 m³)	3.11 €/m³	3.73 €/m³

Montginoul (2007) and 2013 survey
2013 constant prices – *1 € 2013 = 1.3288 US$*

level is mainly explained by the fact that sewerage is chiefly priced with a volumetric rate, while the cost of metering and billing is usually attached to potable water.

In most cases (for 96 % of French utilities corresponding to 95 % of the French population), water is charged with a two-part structure. The simple volumetric rate is only found in 3 % of utilities (representing 5 % of the population). The flat-rate structure remains anecdotal, concerning only 1 % of French supply units (rural), which hardly represents a few per thousand of the population.

The proportional water part charged to users is constant in 61 % of the utilities, corresponding to more than 72 % of the population (Table 8.3). Thirty-six percent of the utilities used a declining block tariff structure in 2003 vs. only 4 % in 2013 following the new regulation. On the contrary, the proportion of utilities with increasing block structure has drastically increased, representing only 1 % (5 % of the population) in 2003 and 29 % (11 % of population) in 2013. An additional 4 % of French utilities have a more complex price structure, combining increasing and declining block rates.

We have described above the total bill (corresponding to both drinking water and sewerage services). This bill is highly influenced by the amount of drinking water consumed (Table 8.4). The sewerage part, for the 75 % of utilities that have a sewer system, is priced in a different way. This difference is particularly high in terms of population: when it exists, sewerage is priced with a volumetric rate for 51 % of French inhabitants (even if this weight has decreased since 2003).

Table 8.3 Distribution of the types of the volumetric part (for water and sewerage services)

	2003		2013	
	% of districts	% of population	% of districts	% of population
Simple	57 %	71 %	61 %	72 %
Declining	36 %	20 %	4 %	8 %
Complex	3 %	4 %	4 %	9 %
Increasing	1 %	5 %	29 %	11 %
Flat rate	3 %	–	1 %	0 %

Montginoul (2007) and 2013 survey

Table 8.4 Distribution of rate structure for drinking water and sewerage separately

		2003		2013	
		% of districts	% of population	% of districts	% of population
Drinking water	Volumetric rate	4	6	3	5
	Two-part rate	93	93	95	95
	Flat rate	3	–	2	–
Sewerage	Volumetric rate	22	63	21	51
	Two-part rate	34	27	52	44
	Flat rate	6	2	2	1
	No central sewer	39	8	25	4

Montginoul (2007) and 2013 survey

Moreover, the constant rate structure dominates for sewerage (61 % of districts, 72 % of inhabitants), and a block-rate structure is not common but rising and changing from a decreasing block-rate structure to an increasing one.

In addition to this structure, some specificities can be highlighted: 3 % of utilities (9 % of population) have implemented a "social access to water" principle. This is done through the definition of social water pricing or through subsidies directly given to poor households. Seven percent of utilities applied industrial water pricing in 2013 (one-third in 2003), mainly through a decreasing block rate. However, in order to follow the last French water law, a new water pricing structure has emerged: the optional one. This structure can be analyzed as a way to continue to propose a decreasing block price through the back door. Finally, the last 10 years has been the arena of multiple tests and implementation of innovative water pricing structures: optional water pricing and seasonal water pricing (sometimes combined with increasing block rates).

8.4 Water Pricing Practices in the Agricultural Sector

Irrigation water charges depend on water management types. We can distinguish roughly three types (Montginoul 1997): a farmer who individually extracts water without any intermediary, a farmers' association (small-scale water user association called ASA—Association Syndicale Autorisée) that extracts and distributes water to its members, and a regional development company (named SAR—Société d'Aménagement Régionale) that delivers water to farmers (or to farmers' associations) through a large collective network or a resupplied river.

8.4.1 Individual Extraction of Water Resource: Only Water Agency Fee

In that case, there is no water pricing, because there is no water service delivery. Investment and operation costs of the system are both fully supported by farmers. Irrigators can however be incited not to waste water through the water agency abstraction levy, the energy pumping cost, and the new water license due to a collective water management institution, called Organisme Unique de Gestion Collective (OUGC), when it exists in water scarcity river basins.

8.4.2 Farmers' Associations: A Water Price Built to Cover Financial Costs

Farmers' associations (mostly organized into a legal association format—ASA) deliver water through a collective network. They fix water prices to cover expenses (only rarely water pricing will also aim at managing water). The price is set to maintain the water delivery network, to cover exploitation costs and the part of investment costs not paid by subsidies (which usually represent 60–80 % of the capital cost).

Water pricing structures are highly diverse, reflecting a variety of situations. We illustrate this fact through the presentation of two former surveys, the main conclusions of which remain valid.

The first survey (Gleyses 1998) covers the situation in southern France (i.e., Adour-Garonne and Rhône-Méditerranée and Corse river basins). Seventeen water pricing structures were identified, with three main ones: all gravity-fed systems are billed with a flat-rate structure for 70 % of them, based on the subscribed surface; water pricing structures in pressure irrigation networks are more varied—81 % of them have binomial water pricing; the remaining 19 % apply a flat-rate structure based on subscribed discharge or surface or on a combination of subscribed and irrigated surfaces. For binomial pricing, the fixed part is priced for 41 % of cases on the subscribed surface and for 20 % on the subscribed discharge.

The second survey (Gleyses 2004) was done in northwestern France (Loire Bretagne river basin). In that region, traditional associations (ASA) represent only 23 % of collective irrigation structures, but 60 % of farmers are connected to a collective irrigation network. This region is also characterized by the absence of collective gravity networks. This survey confirms the predominance of binomial water pricing, widely implemented in pressure water networks, based on subscribed surface or discharge. It identifies three cases in which water is charged through a flat-rate

Table 8.5 Water pricing structures in 2003 in Loire Bretagne river basin

Water pricing structure	Proportion of water pricing structure			Average price	Average water pricing	
	Networks (%)	Farmers (%)	Volume (%)		Fixed part	Variable part
Flat rates						
Subscribed surface	19	17	23		198 €/ha	
Other 5 flat tariffs	5	2	5			
Total 6 flat tariffs	***24***	***19***	***28***	***0.09*** **€/m³**		
Binomial						
Subscribed surface	36	32	33		81 €/ha	0.06 €/m³
Subscribed discharge	4	13	8		38 €/m³/h	0.06 €/m³
Other 12 binomial tariffs	10	30	19			
Total 14 binomial tariffs	***50***	***75***	***60***			
Volumetric						
Strictly proportional	20	5	11			0.10 €/m³
Three other modalities	5	1	1			
Total 4 volumetric tariffs	*25*	*6*	*12*			
Total of 18 tariffs with a volumetric part	***75***	***81***	***72***	***0.12*** **€/m³**		
Total of 24 tariffs	**100**	**100**	**100**	**0.11 €/m³**		

Gleyses (2004)
1 € 2003 = 1.0622 US$

system, ten binomial water pricing systems, and four volumetric pricing structures, mainly implemented in very small irrigation systems that have not adopted the ASA legal structure. Table 8.5 describes levels of prices, water pricing structures implemented, and their weights in terms of number of networks (farmers' association), number of farmers, and of river-basin water volume.

The diversity of water pricing across farmers' association networks illustrates the autonomy of these associations in terms of water pricing. However, this heterogeneity does not always reflect service costs' heterogeneity: there is no statistical difference in terms of price level between the two main binomial water pricing structures (Table 8.6).

The tariff structure reflects in particular the age of the irrigation network (Fig. 8.2): flat rates are preferred in young networks, which have to repay loans. The proportional part increases with the age of the network. At the beginning, expenses are mainly fixed (annual loan charges can represent more than half of the total budget), and a flat-rate structure guarantees to cover charges even for a wet year. Moreover, implementing a binomial rate structure increases management costs

Table 8.6 Irrigation cost for four types of water pricing

River basin	Base of fixed part for binomial tariff	Average tariff (in € 2004)		Average cost for 2000 m³/ha/year
		Fixed part	Proportional part	
Adour-Garonne and RM&C (Gleyses 1998)	Subscribed surface	107 €/ha	0.062 €/m³	232 €/ha
	Subscribed discharge	45 €/m³/h	0.056 €/m³	208 €/ha
Loire Bretagne (Gleyses 2004)	Subscribed surface	81 €/ha	0.060 €/m³	201 €/ha
	Subscribed discharge	38 €/m³/h	0.060 €/m³	202 €/ha

Adapted from Gleyses (2004)
RM&C Rhône-Méditerranée and Corse. 1 € 2000 = 1.0137 US$, 1 € 2004 = 1.2613 US$

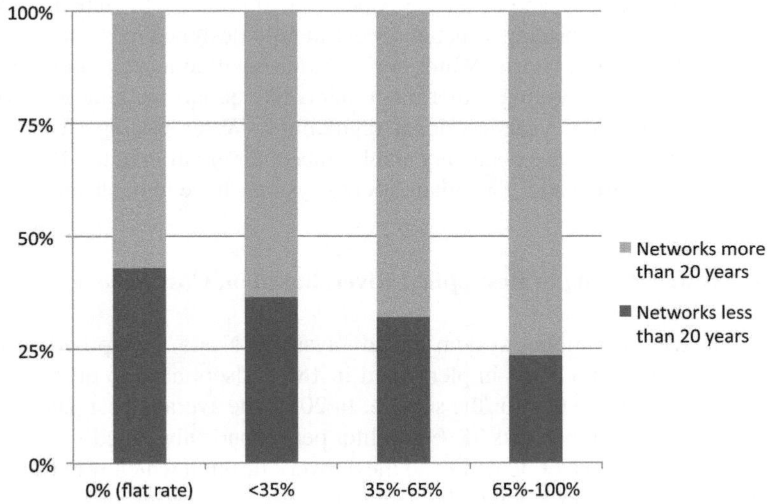

Fig. 8.2 Weight of the proportional part in farmers' association water price depending on the age of network in Loire Bretagne river basin (Gleyses 2006)

(reading water meters, preparing water bills, among other tasks) and water charges due to the obligation to buy water meters during a period in which water charges are already high due to the new investment.

Water tariff structure changes when there are no more loans to repay and when proportional expenses represent a large part of total expenses. Progressively, farmers' associations shift for binomial pricing. In parallel, average price levels decrease, following the decrease of expenses. This situation can also be explained for equity reasons, when some farmers who do not consume much water refuse to pay for the others (Garin and Loubier 2007). Another consequence of the shift from a flat-rate structure to a binomial one is the reduction of volume consumed (mainly wasted volumes) and then of the irrigation cost (Loubier and Garin 2008).

8.4.3 Regional Development Companies: Cost Recovery and Water Conservation

Regional development companies (SAR) are large public irrigation schemes, located in southern France. They were initially created in the 1960s to help economic development of the three regions (Adour-Garonne for the Compagnie d'Aménagement des Côteaux de Gascogne, CACG; Languedoc Roussillon for Bas-Rhône-Languedoc, BRL; and Provence-Alpes-Côte d'Azur for Société du Canal de Provence, SCP). Their pricing structures were mainly designed to incite farmers to irrigate but also to cover costs. Water conservation is often a secondary objective pursued through water pricing (other instruments like quotas were favored) or was only imposed in recent years by local regulations. Water pricing structure and underlying philosophy have been very stable since their creation in the 1960s. Only few adjustments were made. Two main pricing systems have to be differentiated.

8.4.3.1 Water Pricing in Resupplied River, Based on Cost Recovery

CACG manages a complex system called "système Neste" composed of dams, resupplied rivers, and a canal implemented in 1863. Users have to pump water in resupplied rivers and to pay for the service. In 2013, the average cost paid by users is 0.03 €/m^3, which represents 78 € per liter per second subscribed.

Because of a high water demand and the network through which water is distributed (rivers), CACG chose, since the beginning of its concession in 1991, to share water through a discharge quota: the user paid according to the 4,000 cubic meters per liter per second subscribed. If user exceeded the quota, he had to pay a considered-deterrent price, corresponding to eight times the average price level. This type of water pricing structure does not incite farmers to save water.

In order to respect the new local regulation (Comité de bassin Adour-Garonne 2009), CACG decided to introduce a volumetric portion of the water bill. This is however restricted following the assumption that irrigation is an "all-or-nothing" decision and the farmer has only the option to "take" or not take the last water turn and then can save at most 20 % of the allocated quota. Pricing consists therefore in a binomial structure with three increasing block rates: a fixed rate, representing 80 % of the previous bill associated with a first null volumetric part ($p_0 = 0$) for the associated volume; a volumetric rate (p_1) calculated to cover the remaining 20 %; and $p_2 = 8 * p_1$ to dissuade farmers from consuming more than the allocated quota.

8.4.3.2 Water Pricing in Collective Pressure Networks, Based on Equalization, Cost Recovery, and Incentive Principles

In their initial concession perimeter, SAR implemented water pricing that was based on the three principles of equalization, cost recovery, and incentives to save water. Each SAR has a specific manner to respect these three principles.

For SCP, equalization is designed at a use level: other uses (urban, industrial, etc.) compensate discount (corresponding to originally 40 % of the real cost and currently to 60 %) made for irrigation uses. For CACG and BRL, equalization is done at a territory level: CACG has defined three perimeters, including relatively identical farming practices in terms of income; water price level is higher in the perimeter corresponding to wealthy farmers and smaller in the one that regroups the poorest. BRL applies the same price to all farmers located in the same department, whatever the supported cost, which increases with the distance to the main canal. The underlying assumption is that it favors farmers located far from the canal, who are the poorest. In order to maximize water user surplus and also to improve the knowledge of the type of water utilization,[5] BRL has adopted an optional water tariff (Fig. 8.3). Finally, BRL applies reduced prices for young farmers or water bill discounts for farmers facing high water bills during dry years.

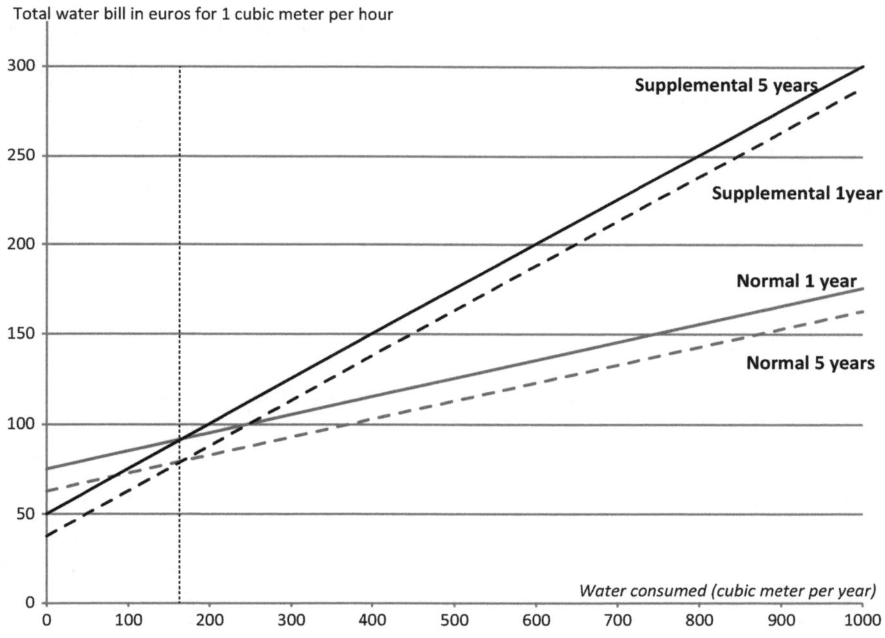

Fig. 8.3 Total water bill according to tariffs (subscribed flow: 1 cubic meter per hour) (Source: compiled by M. Montginoul from BRL 2013 water tariff. 1 € 2013 = 1.3288 US$)

[5] Farmers choose their contract (normal versus supplemental—five years versus one year) according to field characteristics, cropping pattern, and irrigation equipment.

Cost recovery is calculated for BRL and CACG, taking into account average cost, whereas SCP bases its water price on long-term development cost, pursuing the Boiteux pricing principles (Boiteux 1971). To follow the marginal pricing principle, SCP applies a tariff higher when users are far from the source or when it is needed to pressure water and in the peak season. SCP aims also to incite users to subscribe discharges at the needed level and not in excess. That is why, for irrigators, a flat rate is partially applied, corresponding to a consumption of 100 cubic meters per liter, per second subscribed.

The last principle that guides water pricing is linked to the incentive to save water, except, sometimes initially, when first developing an irrigation system. Water pricing structures are binomial everywhere, even if sometimes with a flat but limited rate. SCP applies seasonal water pricing: water tariffs during the summer season are higher than those applied in the winter season. Besides saving water, the underlying objective is to incite users to store water during the winter for use during the summer period to reduce peak demand and smooth water demand.

In the SAR perimeter, the water pricing structure takes a binomial form, with a fixed part priced per liter, per second subscribed. For instance, CACG prices water in its concession perimeter on average at 360 € per l/s + 0.065 €/m^3 (corresponding to the energy cost). Table 8.7 presents average prices for SCP and Table 8.8 tariff grid implemented by BRL. In this last case, BRL designed its tariff to benefit

Table 8.7 Average water prices in 2012 (€/m^3) in the SCP concession perimeter, detailed by subsectors

	Irrigation by farmers	Watering	Domestic raw water	Other uses	Water for industries	Urban (raw water)	Urban (potable water)
Area 3 SCP	0.11	0.49	2.10	0.58	0.60	0.28	–
Area 2 SCP	0.20	0.49	2.51	0.63	0.31	0.39	0.83
Area 1 SCP	0.21	0.58	2.34	0.72	0.47	0.56	0.58
Valensole	0.28	1.13	2.06	1.13	–	–	0.77
Montmeyran	0.20	0.48	–	–	–	–	–
Rieu Vancon Buech Durance	0.25	0.87	–	–	–	–	–
Manosque	0.18	0.80	–	–	–	–	–
Total	**0.19**	**0.54**	**2.46**	**0.66**	**0.43**	**0.48**	**0.61**

1 € 2012 = 1.2905 US$

Table 8.8 Main tariffs implemented by BRL in 2013 before taxes and environmental fees

Type of contract	Type of tariff	Subscription fee (per cubic meter subscribed)	Proportional fee (per cubic meter consumed)
Long term (5 years)	Normal	62.656	0.1003
	Supplemental	37.593	0.2505
Short term (1 year)	Normal	75.188	0.1003
	Supplemental	50.125	0.2505

1 € 2013 = 1.3288 US$

long-term contracts rather than short-term contracts. A supplemental irrigation contract is characterized by a cheaper subscription fee but a higher volume fee, which benefits supplemental needs, for example, a vineyard that does not need to be irrigated each year.

No fundamental changes have been made these last decades, except some adjustments to follow previously described principles. For instance, the index formula, which bases the adjustment of water price levels, is revised in the different SAR to take into account the evolution of weights of the different cost components. SCP is currently adjusting its pricing zones to homogenize the water price in similar and closed areas, which was not the case in the past due to the fact that different entities were in charge of water distribution. CACG engaged a reflection on the equalization principle as it was put in place in its concession perimeter, proposing to implement a uniform price or, on the contrary, to adapt prices to local costs. Farmers refused, arguing the equalization principle, and then preferred to maintain the in-place water price with three price areas.

8.5 Taking into Account Environmental Services: Water Agency Fees

As described in the historical part, water agencies levy taxes to follow the polluter/user-pays principle. Currently, there are ten taxes addressing the different water services (Table 8.9).

These taxes have increased in level and now represent a non-negligible part of water prices. Looking at the historical trend of urban water pricing, one can see that the water agency weight has increased, especially for the pollution levy after the adoption of the European Urban Waste Water Directive (EC 1991/271) (European Commission 1991). Indeed, since 1996, the addition of a pollution discharge levy with a wastewater fee is higher than the addition of the abstraction levy with a potable water price (Fig. 8.4).

In river basins where water demand largely exceeds water supply, the 2006 water law allows a modulation of the withdrawal tax. It can be divided by two, if a unique collective agricultural water management institution (generally managed by agricultural organizations) is implemented. These institutions are in charge of reallocating among farmers the global state allocation to the agricultural sector. The functioning cost of these institutions is partly supported by water agencies during the first years and progressively transferred to farmers through a service fee. This service pricing can be as variable as those of collective irrigation systems. However, it offers one more possibility, mostly chosen, that consists of pricing according to volume farmers choose to use. This system incites farmers to curb their water demands.

Table 8.9 Water agency taxes. Levels in 2013 in the Rhône-Méditerranée and Corse (RM&C) river basin

Water agency taxes	Uses	Calculation	RM&C level in 2013
Domestic pollution	Urban uses	Proportional to urban water consumption	0.23 €/m^3
Nondomestic pollution	Industrial or economic uses	Proportional to generated pollution	Rates depending on type of pollutants
Sewer systems' modernization	Users connected to a sewage public network	Proportional to volume discharged in sewer network	0.15 €/m^3
Water withdrawal	All users	Proportional to withdrawn water	Depends on the use, the level of water scarcity, and the collective or noncollective management
Hydroelectric production	Hydroelectric uses (>1 billion cubic meter per year diverted)	Proportional to diverted water	1.2 € per billion cubic meter and per meter of waterfall height
Non-point source pollution	Phyto-pharmaceutical uses	Proportional to toxicity	5.1 €/k when dangerous for wealth. 2 € when only dangerous for environment
Livestock pollution	Livestock (>90 livestock units)	Proportional to livestock unit	3 € per livestock unit from the 41e one
Barriers on rivers	Owners who modify natural river systems except hydroelectric uses	Proportional to meters' length of the barrier	150 € per meter
Water storage	Entities who store water	Proportional to water stored in peak period	0.01 €/m^3 stored
Aquatic protection	Recreational fishermen	Per recreational fisherman	8.8 € for 1 year and one adult + 20 € for specific species

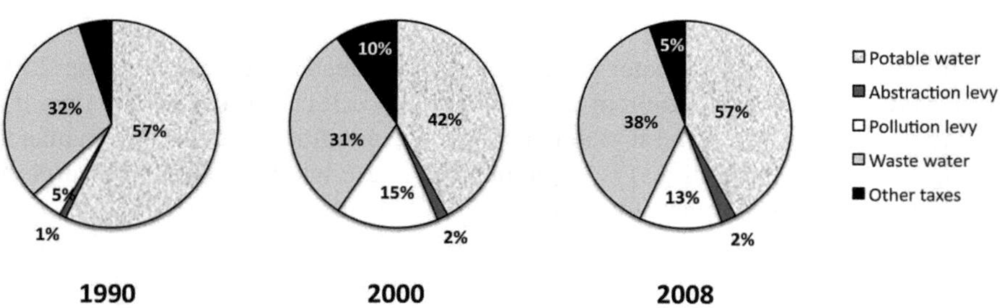

Fig. 8.4 Evolution of average urban water price and breakdown in France (Source: compiled by B. Barraqué from national environment statistics. Constant 2008 euros. 1 € 2008 = 1.4570 US$)

8.6 Conclusion: Current Debates and Future Directions

Water pricing is still currently subject to debate. However, it can be noted that conflicting signals between French policies have been reduced, increasing the weight of water price in consumption behavior. For instance, French agricultural policy diminishes incentives to irrigate: before the last European common agricultural policy reform, farmers received a subvention from 0.1 to 0.15 euro per cubic meter, a sum similar to or even higher than water price. Since 2013, France's irrigated area has been reduced, and we observe an intensification of irrigation. Indeed, direct subvention currently accounts for only 25 % of the previous one.

Water price is designed taking into account various (and sometimes conflicting) objectives, which have to be addressed directly or indirectly through other instruments (for instance, quotas). The three main objectives are balancing water budget, allowing water access (to poor households but also to maintain farmers, especially the smallest), and inciting users to save water. These objectives are more and more difficult to achieve, due to the current trends: a tightening of environmental constraints (European Water Framework Directive asks European states to achieve in 2015 good status of water bodies); a climate change, which enhances water demand; an economic crisis; and a sharp increase in energy prices, which impacts all incomes (households, farmers, industries).

In that way, for the next water agencies program, the SDAGE plans to condition aids to both climate change impacts and mitigation of the project, requiring subsidies and collective economic benefits of the project territory. In the SDAGE as well as in the WFD, special attention will be paid to water pricing (basic measure for reaching good ecological status) and cost recovery.

Water agencies will also recommend developing contrasted scenarios for the most important driving forces and uncertainties to insure their effectiveness regarding climate change and economic return.

Enhancing water agencies' fees does not seem a good solution to incite users to save water, because of the sharp increase it would require causing opposition by the users. For instance, for irrigation use, it would consist of an increase up to 20 times the actual water agency fee's level to provide a real incentive.

That is why France preferred quantitative instruments to share water between users, defining in each water-scarce area the maximum annual abstracted volume. For urban water uses, a combination of incentive instruments (see Article 57 of 2006 water law[6]) and water efficiency measures is preferred, and quotas remain the most efficient way to share water between farmers. However, costs incurred by the organization responsible to share water between farmers (OUGC) should take into

[6]Abstracts from Rule 57—Article L. 2224-12-4: I. Each water bill includes a sum depending on the consumed volume and can also include a fixed part taking into account water management fixed costs and the characteristics of the connection pipe (in particular the number of served flats. This fixed part cannot exceed a ceiling defined by a Ministerial Order. [...] In case of abundant resource water and of restricted connected users, a flat rate structure can be implemented. [...] III. From January 1, 2010, declining rates are forbidden, except in case of abundant resource water. IV. Seasonal water tariffs can be defined in districts facing seasonal water scarcity. [...].

account different billing bases, such as the requested volume (the real consumed volume is legally excluded as a billing base). This base may, however, have an incentive impact on water consumption: the requested volume defines the maximum volume that a farmer can withdraw. It is an ex ante incentive, whatever the real climate and every other factor influencing the irrigation season water consumption.

The question of social access to water has especially increased in France these last few years due to the economic crisis. To address this question, in March 2014, a new law (Loi Brottes) allows testing social water and sanitation tariffs. The underlined philosophy is to experiment tariffs, taking into account households' size, type, and/or income. Helping poor households to pay their water bill either ex ante or ex post (i.e., outside water price) is also allowed.

Current debates on water pricing are also in some ways linked with collective water distribution networks' sustainability. The first one concerns urban management and especially in rural areas. The legal limitation of the fixed part weight in the water bill may raise concerns for the water budget balance and then the sustainability of the delivery system, especially in rural areas facing a high seasonal population but not considered as touristic ones. This legal limitation also poses problems for sewage cost recovery, when secondary water inflows represent an important share of water to be treated. This is the case in particular when there are undeclared individual water supplies through tube wells, for instance.

The second one concerns industries. As soon as water agencies started to levy pollution discharge fees on industrial premises, companies started to change industrial processes to conserve water, and indeed water abstraction dropped from above 5 to around 3.5 km^3/year. But, as previously pointed out, districts with industries were used to price water with a decreasing block-rate structure. With the new water law and the ban of decreasing structure, industries are tempted to exit from the urban water network. Because they represent often a high part of district water consumption, it may question the sustainability of urban managers.

The third one refers to irrigation management and subsidies often given to finance infrastructures and their renewal. It is tempting to think that reducing subsidies dedicated to upgrade infrastructures will increase the "user-pays" principle and also expect users to reduce water consumption due to the induced water price increase. However, especially in the case of ASA farmers' associations, rights and obligations to pay water charges are attached to land. Therefore, a decrease in investment subsidies will augment loans and the fixed part of water pricing. In that case, the only impact is to reduce farmer income or to incite farmers to increase water consumption for irrigated crops to compensate for income decline resulting from the fixed part increase of the water bill. To sum up, once the irrigated system is created, this measure is inefficient or counterproductive (for instance, reducing maintenance expenses in a non-sustainable way). The only case in which it has a positive impact is at the time of an investment decision: reducing subsidies may give a signal of the nonnecessity to create new irrigation systems and then avoid intensifying pressures on water resources.

To go further, in many cases, collective networks have substitutes mobilized often individually. An increase in water prices induces users (households, farmers, industries, etc.) to decrease their consumption. But some unintended effects happen:

water prices rise, initially intended to generate environmental benefits through reduced water use, and may produce economic incentives for users to drill their own boreholes to satisfy their water needs (Montginoul and Rinaudo 2011). Once the investment has been made, the water cost is then really low, which discourages users to save water. Moreover, individual withdrawals are more difficult to control than collective ones, endangering water resource sustainability.

The challenge in France would be to implement water pricing structures that incite users to save water resources while reaching cost recovery and, last but not least, guaranteeing an access to water for all (households and farmers).

Acknowledgments The authors thank all the respondents to the urban water pricing survey conducted in 2013. Many thanks also to Daniel Lepercq (CACG), Céline Geoffroy (SCP), and Eric Belluau (BRL) for their help in understanding historical and current water pricing structures implemented in regional development companies.

References

Agarwal, A., de los Angeles, M. S., Bhatia, R., Chéret, I., Davila-Poblete, S., Falkenmark, M., Gonzalez Villarreal, F., Jønch-Clausen, T., Aït Kadi, M., Kindler, J., Rees, J., Roberts, P., Rogers, P., Solanes, M., & Wright, A. (2000). *Integrated water resources management*. Stockholm: Global Water Partnership and Technical Advisory Committee (TAC).

Boiteux, M. (1971). On the management of public monopolies subject to budgetary constraints. *Journal of Economic Theory, 3*(3), 219–242.

Bommelaer, O. & Devaux, J. (2012). *Le financement de la gestion des ressources en eau en France (actualisation de janvier 2012)*. Service de l'Économie, de l'Évaluation et de l'Intégration du Développement Durable (SEEIDD) du Commissariat Général au Développement Durable (CGDD), Paris.

Chéret, I. (1967) L'eau.

Comité de bassin Adour Garonne. (2009). *Dessinons l'avenir de l'eau dans notre bassin – le SDAGE 2010–2015*. Toulouse.

European Commission. (1991). Council Directive 91/271/EEC of 21 May 1991 concerning urban waste-water treatment.

European Commission. (2000). Directive 2000/60/EC of the European Parliament and of the Council of 23 October 2000 Establishing a framework for community action in the field of water policy. Official Journal 22 December 2000L 327/1. European Commission, Brussels.

Garin, P., & Loubier, S. (2007). Des associations d'irrigants se réforment en façonnant ce qu'elles trouvent juste et équitable. *Ingénieries – EAT, 49*, 27–38.

Gleyses, G. (1998). *La tarification de l'eau dans les réseaux collectifs d'irrigation en France – Résultats d'une enquête auprès d'associations d'irrigants et de syndicats de collectivités territoriales*. Montpellier: Cemagref.

Gleyses, G. (2004). Les structures tarifaires des réseaux collectifs d'irrigation – Méthodologie et test sur le Bassin Loire-Bretagne Annexe 4: D.C.E. et volet économique. Montpellier: Cemagref – Unité Mixte de Recherche G-EAU/Cemagref.

Gleyses, G. (2006). Caractère incitatif de la tarification dans les petits réseaux d'irrigation. Convention pluriannuelle Direction de l'Eau Cemagref Convention N° CV 03000102 – Annexe 4. oi n°2006–1772 sur l'eau et les milieux aquatiques (2006). Journal Officiel 303 (31/12/2006):20285.

Loubier, S., & Garin, P. (2008). Evaluation des fondements et impacts des réformes tarifaires au sein des ASA: Analyse détaillée d'une ASA du Lot. Convention 2007 MEDAD/DE – Cemagref; Rapport Cemagref – UMR G-EAU.

Montginoul, M. (1997). The case of France. In: A. Dinar & A. Subramanian (Eds.), *Water pricing experiences—An international perspective* (World Bank Technical Paper No. 386, pp. 46–53).
Montginoul, M. (2007). Analyzing the diversity of water pricing structures: The case of France. *Water Resources Management, 21*(5), 861–871.
Montginoul, M., & Rinaudo, J.-D. (2011). Controlling households' drilling fever in France: An economic modeling approach. *Ecological Economics, 71*, 140–150.

Chapter 9
Water Pricing Experiences in India: Emerging Issues

Kuppannan Palanisami, Krishna Reddy Kakumanu, and Ravinder P.S. Malik

Abstract The debate on growing water scarcity and the need to use the available water more efficiently among different sectors has once again brought in renewed focus in India. In this debate, a large part of the emphasis has been on the pricing of irrigation water, the sector which accounts for almost 80 % of the total water use but for which water is charged at a fraction of the supply cost. Low water rates, apart from encouraging the inefficient use of water, result in low revenue collections and contribute to the growing burden of government subsidies. Efforts to increase revenue collection through institutional reforms motivated mostly by international lending agencies have yielded mixed results. However, given the increasing demand for water and the resulting competition among sectors, there is scope to price water. In this context, the chapter aims to examine the issues relating to water pricing in India with a case study from Andhra Pradesh.

Keywords India • Volumetric pricing • Multiple uses • Cost recovery • Water privatization

---

K. Palanisami (✉)
International Water Management Institute, ICRISAT campus,
502324 Patancheru, Hyderabad, India
e-mail: Palanisami.iwmi@gmail.com

K.R. Kakumanu
International Water Management Institute (IWMI), C/o ICRISAT,
Patancheru, Andhra Pradesh, India
e-mail: k.krishnareddy@cgiar.org

R.P.S. Malik
International Water Management Institute, Office Block "C", NAS Complex DPS Marg, Pusa,
New Delhi 110012, India
e-mail: rps_malik@yahoo.com

A. Dinar et al. (eds.), *Water Pricing Experiences and Innovations*,
Global Issues in Water Policy 9, DOI 10.1007/978-3-319-16465-6_9

9.1 Introduction (Including Success and Failure of Past Pricing Systems)

Water pricing is one of the water-demand management tools advocated to improve water allocation, efficiency, equity, and sustainability (Saleth 2001; Rao-Garcia 2014). But at the same time, water pricing as a single instrument is limited in its effectiveness to control water use and demand (Berbel and Gomez-limon 2000). However, a wide range of water pricing methods exists, depending on the field conditions, as well as institutional and administrative capacities of government authorities of a given area.

In this connection, studies were also carried out to optimize the usage of water by estimating its value. Water extraction accounting systems, volumetric charging, direct charging at flat rates or indirect charging by taxes, optimal groundwater extractions, and supply-based pricings were introduced to optimize the extractions (Schuck and Green 2002; Natalia et al. 2003; Hellegers et al. 2001; Kumar 2005; Doppler 1977; Small and Carrauthers 1991; Sampath 1992). But the implementation of such methods such as volumetric pricing for agriculture is unfeasible in the developing and underdeveloped countries, because they need heavy capital inputs for the volumetric measure, as well as administrative inputs. Additionally, most of the irrigation systems are continuous and rotation distribution systems. Direct charging, such as area-based charges and flat rates per unit area irrigated, and output-based prices provide no incentive to conserve, as farmers are charged a fixed fee for a certain share of the water (Frederick 1992), or of free cost due to political influences, especially in the case of extracting groundwater for the agriculture sector. However, they are simple to administer, and revenue collection is fairly easy. Indirect water charges do not permit economically optimal methods of production, but they are relatively easy to administer (Katako 1990). Even though the scarcity value of water is increasing, the political incentive of subsidized water charges and poor recovery rates are undermining the efficient maintenance of existing water infrastructure, as well as additional investment in future water projects.

Water pricing for the domestic sector in most of the Indian cities and towns is deficit in several aspects. Water prices from the sale of water are less, at 22–25 % of the operational and maintenance cost. Cities like Chennai, Hyderabad, and Bangalore are charged at one-tenth of the operating and maintenance costs incurred (Mathur and Thakur 2003). The underpricing has resulted in poor and variable spatial coverage of services. Most of the water-supplying entities in India are operated at a loss. The state government political situations, and supplying safe drinking water to the rural poor at lower or no cost, are hindering the water prices in the country.

From a philosophical point of view, the introduction of water pricing would have had a positive impact on economic water allocation and use. The cost paid for a scarce resource leads to a more economic use of the resource. In addition, from society's point of view, water users should pay the social opportunity cost of water, especially if water is scarce and its use is critical. The opportunity cost of water could be used as an economic signal to improve water allocation decisions and encourage water conservation. The absence of a pricing mechanism for water may lead to market failure. To increase water use efficiency and conservation in water use, the economic instrument of pricing is needed. Economic incentives can encourage water users to conserve water (National Research Council 1997).

The fundamental role of prices is to help to distribute goods and services to consumers and to determine the allocation of limited resources among competing uses and users (Sampath 1992). Many concepts of water pricing were developed in the past, such as cost pricing, benefit pricing, and opportunity costs (Doppler 1977; Prasad and Rao 1991; Sampath 1992). However, pricing is often determined by the amount needed to recover at least the cost of maintenance and operation of an irrigation project. The concept of benefit pricing or income generation is considered as the return to irrigation, which will increase the agricultural income. This magnitude of water pricing will have a direct effect on farms and incomes. For efficient water allocation, irrigation water has to be priced based on the opportunity cost of water (World Bank 1993) although this may not seem practical.

Economic theory clearly shows that if perfectly competitive conditions are satisfied and externalities are absent, then market prices will reflect social values. And if long-run marginal-cost pricing is followed in the pricing of irrigation water, then the corresponding levels of investment in irrigation projects and the resulting social benefits will all be optimal. This is the principle of marginal-cost pricing. The concept of marginal-cost pricing of irrigation water acts as an incentive to efficient water use, yet this method has not been used. Sampath (1992) has summarized the reasons why marginal-cost pricing is not adopted. There are many indirect beneficiaries, such as consumers who benefit as much or even more than the direct beneficiaries (farmers), and it is unjust to expect the farmers to bear the full burden. If farmers bear every cost, then the food grain prices will be highly increased and passed on to the consumers. Among the consumers are low-income families who cannot afford such high prices. The water price also varies, depending on season, cropping, region, and climate. In India, the political and religious constraints affect water pricing, and increasing the water fees may reduce the farmers' motivation in farming activities to use less water for irrigation. On the other hand, where charges are significantly lower than the returns on irrigation, users tend toward wasteful consumption. Therefore, a complex pricing system is required for the charging for water usage.

This chapter includes a detailed discussion on the water resources in India and its challenges in water pricing for agricultural, domestic, and industrial uses. The chapter also gives the basis for fixing the water rates, charges in different sectors, periodicity of revision in water rates, cost allocation and subsidies in major irrigation projects, a case study of Andhra Pradesh, institutional arrangements for assessment, and collection of water charges. The future directions to improve the resources through institutional reforms are also discussed.

9.2 Water Resources, Population, and Issues in Water Supply

India has an estimated annual precipitation (including snowfall) of 4,000 billion cubic meters (BCM). The average annual potential in rivers is 1,869 BCM, and per capita water availability (2001) is 1,820 cubic meters (Cu. M). The estimated utilizable water is 1,122 BCM, of which surface water is 690 BCM, while groundwater is 432 BCM. India is among the foremost countries in the world in which water availability determines the land use and crop productivity under irrigated and nonirrigated areas. The country has a large-scale irrigated area, which consumes the higher amount of water for irrigation. To reach the required food grain levels and meet the needs of a growing population of 1.21 billion inhabitants (2011 census), the area under irrigation has increased significantly, more than double from 1961 to 2011 (i.e., from 24.6 million hectares (m.ha) to 63.6 m. ha, respectively). The increased irrigated agriculture alone is consuming 78 % of the total water resources, leaving the rest to domestic (6 %), industrial consumption (5 %), power (3 %), and evaporation losses (6 %) (Indiastat 2014). It was also predicted that the future water consumption would decrease for irrigation by increasing the domestic, industrial, and power consumption (Fig. 9.1).

The water withdrawal in India is from various sources, such as canals, tanks, groundwater, etc. The annual requirement of water for domestic purposes, including cattle, is increasing over the years: 31.84 (1991), 43.38 (2001), 47.49 (2004), and 50.23 BCM (2006). The expected demand is 72.81 BCM for 2025 (Indiastat 2011). The water release for irrigation, domestic uses, and industry is highly dependent on water levels in the reservoirs and precipitation rates. The drought years are putting high pressure on different sectors. One part of the country might experience floods

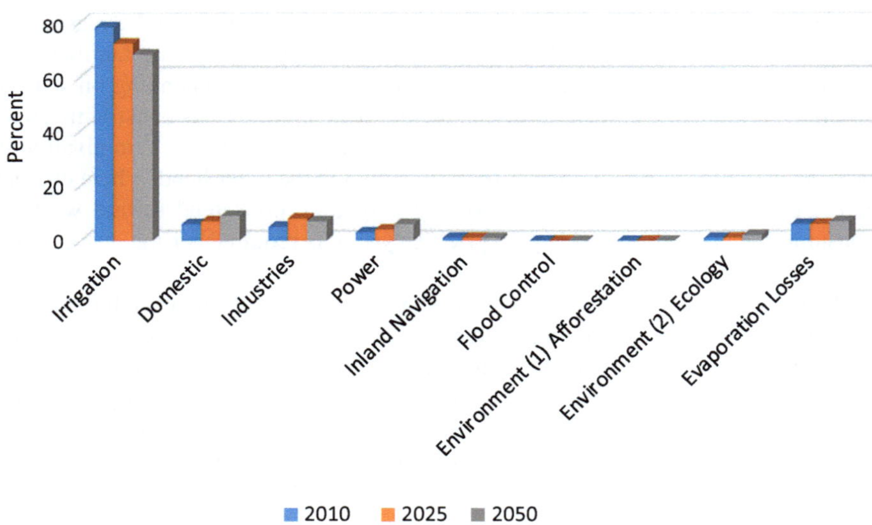

Fig. 9.1 Water requirements for different uses in India (Source: Indiastat 2014)

and waterlogging problems, while other parts might need to cope with droughts and scarcity at the same time. In periods when the need for water is larger than the supply, institutional setup for planning in the allocation of water between competing sectors can be useful.

Water management in India is primarily the responsibility of the states. The state water policies, along with the national water policy, are the instruments that spell out water management practices. Policymaking, water allocation, operation, and maintenance of irrigation canals are carried out by the respective state governments. The constitution at the national level provides certain powers to the government of India in the case of transboundary river issues, wherein tribunals are set up to resolve interstate issues in water allocation. The government of India also provides technical support to the states during extreme weather situations, including droughts as well as floods.

As per national as well as state water policy documents, the majority of the states have domestic water supplies as the top priority, followed by irrigation, hydropower generation, industry, fisheries, and environmental flows. The states have different mechanisms to allocate water among competing demands.

9.3 Experiences with Irrigation Water Pricing

India has an enviable irrigation system across the country providing support to the agriculture development process for centuries. During the British period, the rulers took some interest in promoting irrigation works of large magnitude, like the development of irrigation canals that could facilitate more revenue generation from agricultural lands. The British did not have much experience in developing the irrigated agricultural farms, as they were basically civil engineers with an excellent track record in the planning and maintenance of canals linking the rivers. It was only after India's independence in 1947, when significant irrigation development took place, particularly from the 1960s, through the construction of new dams and canals. From a situation of food scarcity and hunger, the country shifted to a situation of food adequacy and even exporting food grains due to irrigation development coupled with the Green Revolution (FAO hunger map 2007).

Figure 9.2 provides information on the increase of irrigated area with different sources and the increasing trend for the net and gross irrigated area in India over the decades. The increase is more in the case of groundwater sources than for the canals (i.e., major and medium irrigation). This is primarily due to the fact that groundwater offers greater control over the supply than other sources of irrigation, leading to higher yields. But the subsidized policies for groundwater extractions are increasing the overdraft problems. On the other hand, gravity-based tank-irrigated areas have declined over the years due to poor operation and maintenance and giving less preference to minor surface flows in the water resource budgets.

This shift in the usage of irrigation water sources is mainly attributed to the state interferences in the operation and maintenance of traditional irrigation structures

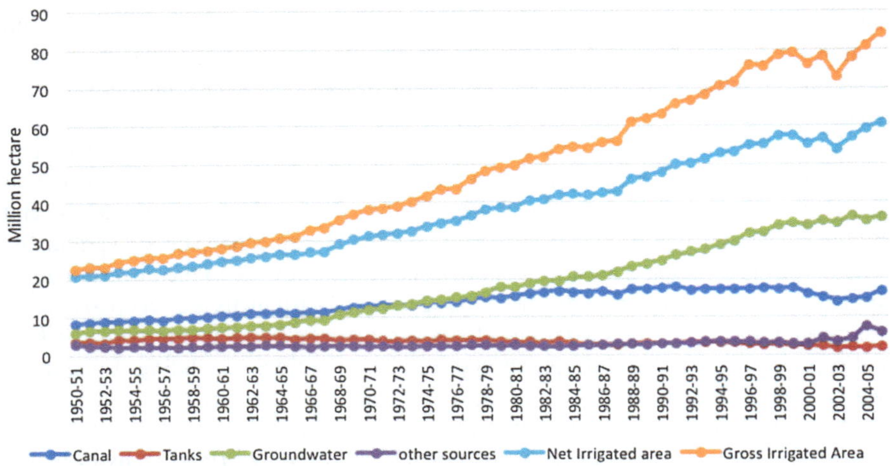

Fig. 9.2 Area irrigated by different sources in India from 1951 to 2006 (Source: DACNET 2014)

and biased provision of the share in state irrigation budgets to major and medium irrigation projects (more than 80 %) in the 5-year plans (PCI 2005).

Large-scale investments are made for major and medium irrigation projects, which have tied down considerable capital and operational costs. But prices are fixed on the basis of the mix of social, economic, and political factors (Narayanamoorthy 2011). The main criteria so far followed for fixing water prices are farmers' ability to pay, which is determined by output, area irrigated based on the volume of water used, quality of irrigation, and recovery cost of equipment. The current pricing system varies with season, crops, and between states, and, more importantly, volumetric pricing is not followed.

There is a considerable difference in the levying of water rates by the different states and union territories in India. No water rate is levied for agricultural purposes in most of the northeastern states, except for Manipur. In Orissa, a flat basic compulsory water rate is charged for staple crop paddy in lands coming under the major (if cultivated area is more than 10,000 ha) and minor (if cultivated area is between 2,000 and 10,000 ha) irrigation projects. The Central Water Commission of the government of India has tabulated the water rates for irrigation in major states and the year in which the revision was made (Table 9.1).

There has been a considerable time lag in the revision of water rates by the states. In many states, no revisions in the water rates have been carried out over several decades. The inordinate delay in the revision of water rates and the low water prices established by states were mainly due to the lack of linkages between fees collected and funds allocated to irrigation projects, lack of farmers' participation, poor communication, lack of transparency between farmers and irrigation departments, poor water delivery service, no users' penalties, low efficient water use, and system operation and management (Easter and Liu 2005).

In spite of the massive investments in the irrigation sector and the impressive growth achieved in agricultural production, a horde of problems appear to cast doubts

Table 9.1 Irrigation water prices in major states of India

States	Rate (Rs./ha)		Few crop specific rates			Last year in which rate was revised
	Maximum	Minimum	Paddy	Wheat	Sugarcane	
Andhra Pradesh	99	370	222	–	370	1986
Bihar	30	158	89	51	158	1983
Gujarat	40	830	110	110	830	1981
Haryana	20	99	74	62	99	1975
Karnataka	37	556	99	54	556	1985
Maharashtra	100	1,750	100	200	1,750	1990
Madhya Pradesh	99	741	198	24	741	1992
Orissa	6	185	40	32	100	1981
Punjab	14	82	49	29	82	1974
Rajasthan	20	180	99	74	143	1982
Tamil Nadu	6	64	49	–	49	1962
Uttar Pradesh	15	410	143	143	237	1983
West Bengal	37	134	37	49	124	1977

Source: Central Water Commission (2010)
The maximum and minimum rates were due to soil fertility and duration of water supplies in the canal systems (1US$ = Rs. 45 in 2010)

on future potentials and sustainability of the country's economic development. Increasing disparities and growing inefficiencies are clearly related to the patterns of investment and creation of networks in irrigation systems. In addition, failures of water management institutes (such as water user associations) created such problems. An increased irrigated area has made possible the increase in production and employment for the last several decades. But, the Indian irrigation system is known for its underperformance. The problems in major and medium irrigation sectors arise at two phases, the construction phase and operation and maintenance phases. At the construction phase, problems are due to inadequate project financing, increasing costs of new schemes, and faulty design. The problems are more pronounced in the second phase (i.e., in case of operation and maintenance), which leads to underutilization of the irrigation potential, inequity in irrigation, indifferent quality of irrigation, wastage of irrigation water, waterlogging, soil salinity and alkalinity, sustainability of irrigated farming, and financial losses due to low pricing of water. The rate of return on investments in the irrigation sector is too low, which is not proportionate with the costs incurred for the projects and their operation and maintenance (Vaidyanathan 2006). There is a wide gap between the actual and desirable performance, which threatens the sustainability of irrigated agriculture (Gulati et al. 2005).

Water taxes are estimated by different agencies and departments and applied to a diverse range of water users. In the case of irrigation, water charges are revised yearly by the irrigation and command area development and revenue department, based on the extent of irrigated area per crop—known as "Joint Ajamahish." The collection of water charges often falls short of the level requested, due to poor compliance by farmers despite the low charges. Various explanations, ranging from farmers' free-ride attitude to lack of willingness to pay and lack of trust in the system, are often put forth to explain the low recovery rates.

The groundwater extractions are also increasing over the years in the country (Fig. 9.2). Initially, groundwater is used as a supplementary source in command areas to augment the total water supply for irrigation and reduce the fluctuations of canal/surface water availability and, thus, of users' income (Ranganathan and Palanisami 2004). Later, groundwater was typically seen as a supplemental source to surface flows even in the years of normal surface flows and increased groundwater utilization. The increasing use of groundwater, due to low cost of extraction, and also technology development add woes to it. In addition to these, recently, some state governments are implementing the policy of free electricity for agriculture to retain farmers' vote bank, in which the majority of voters are farmers (e.g., Andhra Pradesh, Tamil Nadu, and Punjab). It is particularly more intensive in the areas where it should be restricted for maintaining groundwater levels for ecological and environmental consideration.

Electricity charges are the major cost components in operating the irrigation pumps during groundwater utilization. Lower electricity charges decrease the cost of pumping and increase the depth to which it is profitable for farmers to pump groundwater (Zilberman and Lipper 1999). Nonetheless, state governments are restricting the power supply from 7 to 9 h per day. Given that large populations depend for their livelihoods on agriculture and a large part of the food production depends on groundwater irrigation, pricing groundwater or removing the agricultural electricity subsidy is a politically sensitive issue (Kakumanu 2009). So, pricing, although it is an ideal solution to the water-energy problems (Mukherjiet al. 2009), would not always be feasible everywhere in the present political economy context.

9.4 Experiences with Domestic Water Pricing

At the central level, the Union Ministry of Water Resources (MoWR) is responsible for the development, conservation, and management of water as a national resource. MoWR also oversees the regulation and development of interstate rivers. The planning commission, MoWR, the Ministry of Environment and Forests, the Ministry of Urban Development, the Ministry of Health and Family Welfare, and the Housing and Urban Development Corporation are involved in taking care of various roles (planning and allocation, regulation, pollution watch, research and advocacy, and development funding). However, since water is a state subject, the state governments are primarily responsible for the use and control of the resources.

Access to drinking water is the first priority in India's national water policy (GOI 2002), and water for agriculture is the next. There is a variety of institutional arrangements in the provision of urban and rural water in India. Some have city-level water boards for water services and sanitation (like Bangalore, Chennai, and Hyderabad), and some have state-level water supply and sewerage boards (Delhi, Gujarat, Punjab, Tamil Nadu). The local village president, municipality, and metropolitan administrations fix the water taxes and collect them from the individual households. The water charges for many of the Indian rural and urban areas are flat. But in metropolitan cities, water charges are collected on a volumetric basis, which again depends on the residential category. For example, the

Table 9.2 Water tariff for different categories in Hyderabad Metropolitan

Slab (kiloliters)	Domestic (Rs.)	Commercial (Rs.)	Industrial (Rs.)
0–15 (slums)	7	–	–
0–15	10	40	50
16–30	12	70	80
31–50	22	70	80
51–100	27	70	80
100–200	35	100	120
>200	40	100	120
Outside the Greater Hyderabad Municipal Corporation	–	180	180

Source: Hyderabad Metropolitan (2014)

Hyderabad Metropolitan Water Supply and Sewage Board levies tariffs in order to provide sufficient revenues to cover its operating expenses, depreciation, debit serving, etc. The water tariffs for the different categories are presented in the Table 9.2. These water prices vary from state to state and between cities. For example, the flat rate for an individual domestic house in Hyderabad is Rs. 90, and in Chennai, it is Rs. 150 per month. Similarly, the volumetric pricing charge for multistoried residential apartments in Hyderabad is Rs. 6 for 0–15 m^3 and Rs. 2.5 for up to 10 m^3 for Chennai. These rates increase with the increase in consumption pattern (Hyderabad Metropolitan 2014; Chennai Metro Water 2004).

There are also various problems in the existing tariff systems in the country. For example, the present water charges are able to recover 22–25 % lower than the O&M costs (Sridhar and Mathur 2009) and do not recover the capital cost or cost for future expansion. The actual level of consumption is not known in many cities and states as metering is negligible and flat rates exist. Overcharging the commercial users and subsidizing the domestic consumption are also creating problems in the existing systems as there are no norms for fixing these varying rates. In addition to the inadequate pricing, the water sector is characterized by huge inefficiencies due to unaccounted-for water losses (up to 40 %), poor quality, low cost recovery, etc.

9.5 Experiences with Industrial Water Pricing

In India, industrial water use is similar to the domestic consumption (Indiastat 2014). The demand for industrial water is increasing with the pace of industrial development. For the purpose of economic growth, preferences are given to industries, which are further increasing the demand for water. Surface water from the rivers is the major source of water for the industries (41 %), followed by groundwater (35 %) and municipal water (24 %). It is limited in urban and peri-urban areas (FICCI 2011). There are no accurate estimates of water consumption by the industrial sector. However, it is expected that the water requirement for the industries by 2050 would be 103 BCM. This can be decreased to 81 MCM, if water-saving technologies are adopted on a large scale (Kumar et al. 2005). In the view of shortage and competition between different sectors, industries are expected to switch over to efficient technologies.

The industrial water price also varies from state to state in India (Table 9.3). The irrigation and command area development department calculates the tax, based on the individual industry's demand for bulk water supply from the reservoir, and cred-

Table 9.3 State-wise industrial water rates in India

Sl. no	State/union territory (UT)	Unit	Rate (Rs.)
1	Andhra Pradesh	'000 gallons	1.50–450.00
2	Arunachal Pradesh	–	No water rates
3	Bihar	'000 gallons	4.50
4	Chhattisgarh	Cu. M/month	0.06–3.60
5	Delhi	'000 liters	<10 KL pm = @10 + 400 SC
			10–25 KL pm = @ 20 + 600 SC
			25–30 KL pm = @ 50 + 700 SC
			50–100 KL pm = @80 + 800 SC
			>100 KL pm = @ 100 + 600 SC
			Sc = surcharge, KLPM = kiloliter per month
6	Goa	10,000 l	20
7	Gujarat	Kiloliter (KL)	10
8	Haryana	2,500 cubic feet (cft)	250–500
9	Himachal Pradesh	KL	Urban areas only = 11.7
10	Jammu and Kashmir	Per connection	NA
11	Jharkhand	'000 gallons	4.50
12	Karnataka	Million cft	1,800–3,200
13	Kerala	KL	25 and 250/month minimum charges
14	Madhya Pradesh	Cu. M	0.02–2
15	Maharashtra	10 KL	20–720
16	Manipur and Mizoram	–	No water rates
17	Orissa	Lakh gallon	60–250
18	Punjab		NA*
19	Rajasthan	'000 cft	20
20	Sikkim	Cu. M	NA
21	Tamil Nadu	KL	15–60
22	Tripura	Per month/connection	100–250
23	Uttar Pradesh	Annum/cusec	NA
24	West Bengal		NA
25	A and N Islands	NA	NA
26	Chandigarh	KL	6
27	Dadra and Nagar Haveli	Per month	150–430
28	Daman and Diu	Per month/tap	250–450
29	Lakshadweep		No water rates
30	Puducherry	NA	NA

Source: Central Water Commission (2010) (I US$ = Rs. 45 in 2010)
*Not available

its the water tax to the government account. The cost of water supply varies widely and can be in the range of Rs. 0.09 to 50 per cubic meter.

In India, the cost of water has three components: water excess paid to the pollution control boards, cost of buying water from the suppliers (municipalities), and cost of extracting water from rivers and groundwater (CSE 2004). Table 9.3 shows that there is no consensus on the range of industrial water demand, price elasticity, and sensitivity of the water demand to other factors, such as input prices and output levels. Kumar (2006) analyzed the economic value (shadow price) of water and found that the average shadow price of water is 7.21 Rs./KL. It varied between the firms, from 1.4 Rs./KL for petrochemicals to 30.54 Rs./KL for paper and paper products.

Industries are not only consuming water but also pollute the water resources. On an average, each liter discharged further pollutes about 5–8 l of water and increases the water demand by 35–40 % (CSE 2004). There are no clear environmental policies and fragmented responsibilities to control the industrial pollution. This will inevitably put pressure on the available freshwater resources, due to the future water demand.

9.6 Experiences with Pricing of Environmental Services

Most of the Indian rivers are monsoon-driven hydrological regimes with 70–80 % of annual flows in 3–4 months. The environmental flow water requirements for most of the Indian rivers range between 20 % and 27 % of the renewable water resources. The pricing of environmental services is poor in the country, as the major challenge is to maintain minimum flows in the rivers. Except for some of the perennial rivers, most of the rivers run dry during the summer due to variations in rainfall pattern and catchment characteristics.

In the Himalayas, to preserve a small dam, a downstream village decides to pay an upstream village to cease the grazing that causes soil erosion and the accumulation of silt. In economics, this is an example of "payment for environmental services" (PES). When payment compensates for the opportunity cost of lost income, PES is seen as a useful instrument for the preservation of nature through negotiations. However, this method of valuing nature or externalities can also have its pitfalls (Supriya Singh 2013).

9.7 Current Debates and Future Directions

9.7.1 Addressing Climate Change, Population Growth, Water Quality Problems, and Other Issues

Climate change is considered by many to be the greatest challenge to humanity (Alston 2013). According to Lynn et al. (2011) and Hackmann (2013), climate change is primarily a social problem and not an environmental one, and its primary causes and consequences are social. Thus, the solutions must be targeted toward the "transformation in a changing climate" that will establish an arena in which researchers, artists, entrepreneurs, and individuals will meet and discuss the transformational issues to build a common platform and language to address the key parameters of social transformation (i.e., natural, physical, human, and financial capital-related issues). Hence, there is a need to establish such a platform involving all stakeholders in addressing the climate change impacts on societal transformation.

The impact of climate change (CC) is likely to have a serious influence on the agricultural and water sectors and, eventually, on the food security and livelihoods of a large section of the rural population in developing countries (Lal et al. 2001; IPCC 2007; Aggarwal 2009; Jacoby et al. 2009). India is vulnerable and needs to address the threats, together with the respective states or regions, from climate change and extreme weather events. Vulnerability is defined as "the extent to which climate change may damage or harm a system; it depends not only on a system's sensitivity, but also its ability to adapt to new climatic conditions," characterized as a function of adaptive capacity, sensitivity, and exposure (IPCC 2001; McCarthy et al. 2001; O'Brien et al. 2004). Although a number of studies were conducted to analyze vulnerability and adaptability issues in the context of climate change, only a few of them have focused on the socioeconomic vulnerability and the adaptation of communities to extreme climate changes (Adger 2006; O'Brien et al. 2007; Moser 2008; Sekhar et al. 2010). In India, vulnerability assessment was conducted at the river basin level (Palanisami et al. 2014) by considering the climatic, demographic, agricultural, occupational, and geographical parameters. The authors have also considered the exposure, sensitivity, and adaptability for assessing the vulnerability of the river basins. O'Brien et al. (2004) have mapped the district-wise adaptive capacity, climate sensitivity index, and climate change vulnerability and sensitivity in India. The adaptive capacity is measured, based on biophysical, social, and technological indictors, but is unable to target the price-related indicators for resource management.

In the present scenario, the states are unable to provide adequate funds for O&M costs, in view of compelling necessities and requirements of other sectors. The cost recovery from water rate/charges has failed to compensate even the deficit. The low cost recovery is attributable to low water rates levied by the states and the ineffectiveness of the existing machinery to ensure full and timely collection of the assessed revenue. A revision in the level and structure of water rates, as well as strengthening the revenue collection machinery, is necessary both in the interest of efficiency and equity. A rational price structure, periodic review/revision of water rates, and due and timely recovery of water charges are essential to ensure the availability of more reliable services and also to promote savings, create disincentives for wastage, and expand services. There is an increase in cost every year on account of inflation, but there is no provision in the water rates in most of the states to take care of this important cost rider, which also needs to be built into the water rate structure. As the water rates presently being charged are at a high, subsidized rate, low revenue realization has created an adverse impact on ensuring satisfactory and adequate maintenance. Palanisami et al. (2011) and Easter and Liu (2005) have conducted detailed studies on cost allocation and subsidies by selecting multipurpose irrigation projects in Andhra Pradesh.

9.7.2 Cost Allocation and Subsidies in Major Irrigation Projects: A Case Study of Andhra Pradesh State

This irrigation project in Andhra Pradesh is an example of how the cost from a multipurpose water project can be allocated among different type of uses or purposes. But, before the reallocation, the case was different. Palanisami et al. (2011) measured irrigation subsidies in Andhra Pradesh and southern India, and the findings revealed that there was a systematic problem relating to the irrigation sector, due to the underutilization of the potential created through large financial investments. This leads to a 33 % gap between created and utilized irrigation potential in the state, due to defective water distribution systems, noncompliance of farmers in adopting cropping pattern for which the system was designed, and lack of operational plans. In addition to this underutilization gap, water distribution within the project's irrigated areas is often neither reliable nor equitable, with large differences in water availability between the head and tail end of irrigation canals.

The cost allocation among multiple water uses, based on the water delivery and benefits in Andhra Pradesh state, has indicated that in the multipurpose projects, irrigation is likely to be a major share of the costs allocated (Tables 9.4 and 9.5), but with the growing domestic and industrial demand for water, the irrigation's share of the cost is likely to drop significantly over time.

Palanisami et al. (2011) also estimated the cost allocation among multiple uses and found that cost allocation will vary greatly in old projects, where the irrigation share in the total O&M cost is only about 60 %, implying that under such cases, farmers may be paying the water charges to match the O&M expenses. In the

Table 9.4 Water allocation among multiple uses (%)

Water projects	Domestic water supply	Industrial	Irrigation
Nagarjuna Sagar (NSP)	2	0	98
Tungabhadra (TBP)	1	4	95
Sriram Sagar (SRSP)	2	3	95

Source: Easter and Liu (2005)

Table 9.5 Cost allocation based on benefits (%)

Purpose of the project	Multipurpose water project		
	NSP	TBP	SRSP
Irrigation	88.1	94.3	91.3
Hydropower	3.0	4.0	4.2
Domestic	3.0	1.6	2.1
Industry	4.3	0.1	2.3
Fisheries	1.6	0.1	0.1

Source: Easter and Liu (2005) and World Bank (2003)

absence of such cost allocation, the irrigation subsidy is looking overestimated. The authors have illustrated that the total subsidy for major irrigation projects in Andhra Pradesh was around Rs. 12, 627million (US$282 million).

9.7.3 Water Pricing and Privatization of Domestic and Industrial Water Supply: A Case Study of Tamil Nadu State

Tirupur is an industrial town with a high concentration of textile products meant for export. In order to meet the growing water demand in the textile industry, the New Tirupur Area Development Corporation Ltd (NTADCL) was set up in 1995 under a public-private partnership. Prior to the setting up of NTADCL, the water needs of the industry were being met by the supply of water through tankers, drawing water from open and borehole wells in the surrounding villages that ranged from 150 to 200 million liters per day (MLD). NTADCL was planned in this context to primarily meet the demands of the industry. The Tirupur Water Supply and Sewerage Project was operational under the NTADCL.

The Tirupur Water Supply and Sewerage Project has many firsts to its credit. It is the first project to be structured on a commercial format, the first project-specific public limited company for water and sewerage with equity participation of major beneficiaries, and the first concession by a state government to a public limited company to draw raw water for domestic and industrial uses and to collect revenues. The project reportedly has also a technical sophistication that is unmatched in the country.

Project capacity: The system will supply 185 MLD to about 900 textile firms and over 1.6 million residents in Tirupur, Tamil Nadu, and surrounding areas. About 135 MLD of water is to be supplied to the knitwear dyeing and bleaching industry; 25 MLD to residents of Tirupur, including 60,000 slum dwellers; and 35 MLD to the region's remaining rural towns, villages, and settlements.

Pricing: The Panchayats pay at the rate of Rs. 3.50 per 1,000 l, Rs. 5 per KL for domestic use in the Tirupur Municipality, and Rs. 45 per KL for industrial and commercial consumers. The price was later revised to Rs. 55/KL. The actual cost of supplying water by the NTADCL, including pumping and treating, is about Rs. 41.70/KL.

The industries are required to provide a bank guarantee to NTADCL, thus ensuring the offtake of water. The project was implemented with an assessed quantum of water of 108 MLD per day, but it is reported that the actual water draw by the Tirupur industries, even after 1 year, is estimated at 75 MLD on normal weekdays. The water rates have again been revised in February 2007 to Rs. 35 per 1,000 l (Table 9.6). NTADCL is offering a 10 % discounted rate for those industries whose monthly offtake of water is more than the agreed quantum.

The project does not envisage cost recovery to be effected through the water supplied to the municipality and the Panchayat areas. It seeks to cross-subsidize the water supply for domestic purposes with the water charges obtained from industry.

Table 9.6 Domestic and industrial water charges under privatization project

Month	Water charges – nondomestic per KLD (Rs.)	Water charges – domestic – Tirupur municipality – per KLD (Rs.)	Water charges – domestic – Panchayats – per KLD (Rs.)
July 2005–June 2006	45	5.00	3.50
July 2006–Jan 2007	23	5.00	3.50
Feb 2007–June 2007	35	5.00	3.50
July 2007–Mar 2008	45	5.00	3.50
Apr 2008–Mar 2009	50	5.00	3.50
From April 2009	55	5.00	3.50

I US$=Rs. 48 in 2009

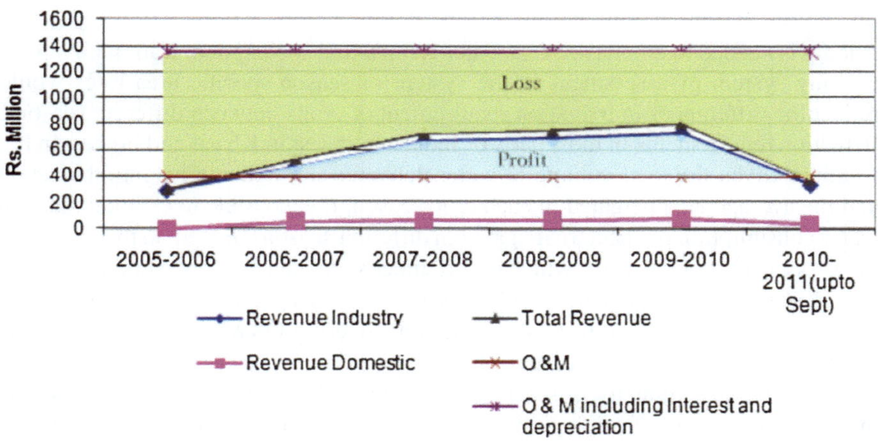

Fig. 9.3 Financial performance of the industrial and domestic water supply project

Current performance: The project cost is about Rs. 10,230 millions, with a project period of 30 years. It started operation in 2005, and now it is in the ninth year of operation. The O&M cost is Rs. 391 millions/year and, with interest and depreciation, it is about Rs. 1,354 millions/year. The accumulated loss is about Rs. 1,770 millions (Fig. 9.3). The financial loss has been managed through the use of a reserve fund created for servicing debt, which is only due to high interest and depreciation burden. The company is making operating profits.

The following are reasons for low capacity utilization: (a) overestimation of the demand in the planning period; (b) economic slowdown, which affected the exports of the industries; (c) the state government not enforcing the law to regulate the exploitation of groundwater directly by the industries and the many industries that use groundwater and, hence, less demand for NTADCL water; (d) recycled water use, which helps to reduce the demand by 20–25 %; and (e) technology advancement in the industry resulting in less water use.

It is suggested that the NTADCL can explore the options such as (a) rotating the water supplies among different users and increasing the charges marginally. The endogenous pressure will help for further price increases through institutional interventions and (b) expansion of the pipeline to nearby towns, so that the fixed cost will be reduced.

9.8 New Approaches for Water Pricing Under Consideration in the Country

Pricing is needed to emphasize that water is not a free social good. Rationalizing the existing system of assessment to the system of season-specific area rates would reflect the differences in irrigation requirement of crops between different seasons. Irrigation requirements in most parts of India are lowest in Kharif and highest in hot weather. Given the total volume of water delivered through any irrigation channel and the average area irrigated in each crop season, it is possible to estimate relative water consumption per hectare irrigated in different seasons. Based on this estimate, the variable part of the tariff, which is still under state management, should be fixed. Full-cost recovery cannot be sought without improving the quality of irrigation. Thus, the level of cost recovery should aim at covering O&M costs and reduce the financial burden on the Exchequer.

Group-based volumetric assessment can be a shift to a fully volumetric assessment system instead of pricing on crop and area base. This will call for additional investments to modify the distribution system for effectively regulating water supply volumes at the outlets and capital base for determining the cost recovery. If the efficiency of system and productivity is improved, the targets of cost recovery can be progressively increased. A volumetric system of assessment cannot be implemented if the managers are required to monitor deliveries and bill individual farmers. The system should take responsibility only for bulk deliveries to relatively large groups of farmers. In this manner, the burden on the system managers and therefore the costs to the government can be reduced substantially.

The system improvement will have an everlasting effect if the farmers' group management is consolidated and made to participate in programs involving the upgradation of systems at a higher level of efficiency in water use and productivity. At the technical level, improving the productivity of a surface system requires measures such as making the system capable of guaranteeing the delivery of a specified quantum of water of definite duration and raising overall irrigation efficiency.

The adoption of sophisticated techniques to coordinate the use of surface- and groundwater in conjunctive use, depending on the supply and demand, should be stressed in order to get the optimum production from available water. Besides, the conjunctive use of water and distribution networks should be made effective with the use of the latest technologies. Supply regulation should be fostered and perhaps linked with pricing (i.e., prices may vary as per supply). But, in any case, supplies need to be measured at the level closest to the actual users. In the Indian context, supplies should be metered at the water user association (WUA) level. The WUA should take the responsibility of distributing the water equitably among its members and collect the water charges. For this, the metering of supplies and strengthening of WUAs are mandatory. In few cases, such an allocation through WUA was observed (Palanisami et al. 2012). Pricing becomes affordable if the land and water productivities are enhanced. This, coupled with effective institutional arrangements, could pave the way for full-cost- or marginal-cost-based pricing in the long run. Water-saving technologies like sprinkler and drip systems need to be promoted through institutional arrangements, rather than through subsidies. WUAs need to be encouraged and capacitated to promote these technologies. Lately, labor-saving technologies, such as mechanical threshers and harvest combines, are promoted through WUAs in Andhra Pradesh, India (Deshpande 2008). Either way, institutional strengthening holds the key for effective demand management via the pricing of irrigation water.

References

Adger, W. N. (2006). Vulnerability. *Global Environmental Change, 16*(3), 268–281.

Aggarwal, P. K. (2009). *Vulnerability of Indian agriculture to climate change: Current state of knowledge*. Paper presented at the national workshop—Review of implementation of work programme towards Indian Network of Climate Change Assessment, Ministry of Environment and Forests, New Delhi. http://moef.nic.in/downloads/others/Vulnerability_PK%20Aggarwal.pdf. Last accessed on 14 Mar 2010.

Alston, M. (2013). Gender mainstreaming and climate change. *Women's Studies International Forum*. http://dx.doi.org/10.016/j.wsif.2013.01.016

Berbel, J., & Gomez-limon, J. A. (2000). The impact of water pricing policy in Spain: An analysis of three irrigated areas. *Agricultural Water Management, 4*, 219–238.

Central Water Commission. (2010). Pricing of water in public system in India, information systems organisation—Water planning and projects wing. http://www.cwc.gov.in/main/downloads/JS_Pricing%20of%20Water%20in%20Public%20System-%20Final-RN_.29.10.pdf. Accessed on 20 Sept 2014.

Centre for Science and Environment [CSE]. (2004, February). Not a non-issue. *Down to Earth, 12*(19). http://www.cseindia.org/dte-supplement/industry20040215/non-issue.htm. Accessed on 28 Mar 2015.

Chennai Metro Water. (2004). *Metered consumers' tariff*. http://www.chennaimetrowater.tn.nic.in/departments/finance/tariff.htm. Accessed on 22 May 2014.

DACNET. (2014). Department of Agriculture and Cooperation: E-governance project from Government of India. http://eands.dacnet.nic.in/LUS-2010-11/All-India-3.pdf. Accessed on 22 Sept 2014.

Deshpande, R. S. (2008). *Rating assessment of water users associations in KC canal modernisation project* (Project Report). Bangalore: Institute for Social Economic Change

Doppler, W. (1977). Towards a general of irrigation water charging policy. *Agricultural Administration, 4*(1977), 129–29.

Easter, W. K., & Liu, Y. (2005). *Cost recovery and water pricing for irrigation and drainage projects* (Agriculture and rural development discussion paper 26). Washington, DC: World Bank.

FAO. (2007). *Hunger map*. http://www.fao.org/es/ess/faostat/foodsecurity/FSMap/map14.htm. Accessed on 31 Oct 2007.

FICCI. (2011). *Water use in Indian industry survey*. FICCI Water Mission, New Delhi. http://www.ficci.com/SEDocument/20188/water-use-indian-industry-survey_results.pdf. Accessed on 29 May 2014.

Frederick, K. D. (1992). *The role of management in a world of increasing scarcity* (World Bank technical paper number 189, p. 72). Washington, DC: The World Bank.

Government of India (GOI). (2002). *National water policy*. Ministry of water resources, New Delhi. http://www.swaraup.gov.in/Downloads/nwp_2002.pdf. Accessed on 12 Apr 2014.

Gulati, A., Meinzen-Dick, R., & Raju, K. V. (2005). *Institutional reforms in Indian irrigation*. New Delhi: Sage Publications.

Hackmann, H. (2013, June 19–21). *Taking social transformation seriously*. Presented at the conference on Transformation in a changing climate. Oslo: University of Oslo.

Hellegers, P., Zilberman, D., & Ekko, V. I. (2001). Dynamics of agricultural groundwater extraction. *Ecological Economics, 37*, 303–311.

Hyderabad Metropolitan. (2014). http://hyderabadwater.gov.in/wwo/UI/TariffContent.aspxand; http://www.hyderabadwater.gov.in/wwo/ui/Revision%20of%20Water%20Tariff%20for%20Commercial%20and%20Industrial%20consumers%20of%20HMWSSB.pdf. Accessed on 22 May 2014.

Indiastat. (2011). *Indian statistical information*. www.indiastat.com.

Indiastat. (2014). *Indian statistical information*. www.indiastat.com.

IPCC (2001). *Climate change 2001: Impacts, adaptation vulnerability*. Contribution of WORKING group II to the third assessment report of the Intergovernmental Panel on Climate Change. Geneva: UNEP/WMO.

IPCC. (2007). *Climate change 2007: Synthesis report*. Contribution of Working Groups I, II and III to the fourth assessment report of the Intergovernmental Panel on Climate Change, Core Writing Team. Pachauri, R. K., Reisinger, A. (Eds.) (104 pp). Geneva: IPCC.

Jacoby, H., Skoufias, E., Rabassa, M. (2009). *On the distributional implications of climate change: A methodological framework and application to rural India*. http://go.worldbank.org/XDJN920PA0. Last accessed on 14 Mar 2010.

Kakumanu, K. R. (2009). *Economic analysis and management of irrigation systems under subsidized electricity regime in the semi-arid region ofAndhra Pradesh, India* (Vol. 99). Weikersheim: Margraf Publishers.

Katako, T. S. (1990). Cost recovery in water supply in developing countries. *Water Resource Development, 6*(2), 86–94.

Kumar, M. D. (2005). Impact of electricity prices and volumetric water allocation on energy and groundwater demand management: Analysis from western India. *Energy Policy, 33*, 39–51.

Kumar, S. (2006). Analyzing industrial water demand in India: An input distance function approach. *Water Policy, 8*, 15–29. Reprinted in Kumar, S., & Managi, S. (2009). *Economics of sustainable development: the case of India*. New York: Springer.

Kumar, R., Singh, R. D., & Sharma, K. D. (2005). Water resources of India, special section: Water. *Current Science, 89*(5), 794–811.

Lal, M., Harasawaand, H., & Murdiyarso, D. (2001). Asia. In J. J. McCarthy et al. (Eds.), *Climate change 2001: Impacts, adaptation and vulnerability* (pp. 533–590). Cambridge: Cambridge University Press.

Lynn, K., MacKendrick, K., Donoghue, E. M. (2011). *Social vulnerability and climate change: Synthesis of literature* (Gen. Tech. Rep. PNW-GTR-838, 70 p). Portland: U.S. Department of Agriculture, Forest Service, Pacific Northwest Research Station.

Mathur, O. P., & Thakur, S. (2003). *Urban water pricing: Setting the stage for reforms*. New Delhi: National Institute of Public finance and Policy.

McCarthy, J. J., Canziani, O. F., Leary, N. A., Dokken, D. J., & White, K. S. (Eds.). (2001). *Climate change 2001: Impacts, adaptation, and vulnerability*. Cambridge: Cambridge University Press.

Moser, S. C. (2008). *Adaptation and vulnerability* (Draft CEC PIER-EA Discussion Paper). Sacramento: California Energy Commission.

Mukherji, A., Das, B., Majumdar, N., Nayak, N. C., Sethi, R. R., Sharma, B. R., & Banerjee, P. S. (2009). Metering of agricultural power supply in West Bengal: Who gains and who loses? *Energy Policy, 37*(12), 5530–5539.

Narayanamoorthy, A. (2011). Development and composition of irrigation in India. Temporal trends and regional patterns. *Irrigation and Drainage, 60*, 431–445.

Natalia, P., Ramaswamy, S., Norma, R. S., Eloise, K., & Tammo, S. S. (2003). Water accounting for conjunctive groundwater/surface water management: Case of the Singkarak – Ombilin River basin. *Journal of Hydrology, 292*, 1–22.

National Research Council. (1997). *Valuing of groundwater: Economic concepts and approaches*. Washington, DC: National Academy Press.

O'Brien, K., Leichenkob, R., Kelkarc, U., Venemad, H., Aandahla, G., Tompkinsa, H., Javedc, A., Bhadwalc, S., Bargd, S., Nygaarda, L., & West, J. (2004). Mapping vulnerability to multiple stressors: Climate change and globalization in India. *Global Environmental Change, 14*, 303–313.

O'Brien, K., et al. (2007). Why different interpretations of vulnerability matter in climate change discourses. *Climate Policy, 7*(1), 73–88.

Palanisami, K., Mohan, K., Giordano, M., Charles, C. (2011). *Measuring irrigation subsidies in Andhra Pradesh and south India: An application of the GSI method for quantifying subsidies*. Report. Geneva: International Institute for Sustainable Development.

Palanisami, K., Doraiswamy, R., Khaishagi, M. K. (2012).*Water resources group: Expert input on preparing detailed project report for water-enabled growth in Karnataka*. Water management interventions for Narayanpur right bank canal and Ghataprabha project in Krishna Basin in Karnataka. Hyderabad: International Water Management Institute and ICRISAT.

Palanisami, K., Ranganathan, C. R., Nagothu, U. S., & Kakumanu, K. (2014). *Climate change and agricultural situation in India: Studies from selected river basins*. New Delhi: Routledge-Taylor & Francis group. ISBN 978-0-415-73599-5.

PCI. (2005). Planning commission of India, New Delhi. (www.planningcommission.nic.in).

Prasad, K., & Rao, P. K. (1991). Irrigation water pricing in India. *Water Resource Development, 7*(4), 274–280.

Ranganathan, C. R., & Palanisami, K. (2004). Modelling economics of conjunctive surface and groundwater irrigation systems. *Irrigation Drainage System, 8*(2), 127–143.

Rao-Garcia, M. C. (2014). Equity, efficiency and sustainability in water allocation in Andes: Trade-offs in a full world. *Water Alternatives, 7*(2), 298–319.

Saleth, R. M. (2001). *Water pricing: Potential and problems: Overcoming water scarcity and quality constraints* (Policy Brief No. 10, 2020 focus 9). Washington, DC: International Food Policy Research Institute. http://www.ifpri.org/sites/default/files/publications/focus09.pdf. Accessed on 2 Jun 2014.

Sampath, R. (1992). Issues in irrigation pricing in developing countries. *World Development, 20*(7), 967–977.

Schuck, E. C., & Green, G. P. (2002). Supply-based water pricing in a conjunctive use system: Implications for resource and energy use. *Resource and Energy Economics, 24*, 175–192.

Sekhar, Udaya Nagothu, Geethalakshmi, V., Annamalai, H., & Lakshmanan, A. (Eds.). (2010). *Sustainable rice production on a warmer planet: Linking science, stakeholders and policy*. New Delhi: Macmillan Publishers. ISBN 978-0-23033-168-6.

Small, L. E., & Carrauthers, I. (1991). *Farmers financed irrigation: The economics of reform* (p. 233). Cambridge: Cambridge University Press.

Sridhar, K. S., & Mathur, O. P. (2009). *Costs and challenges of local urban services: Evidence from India's cities*. New Delhi: Oxford University Press.

Supriya Singh. (2013). Payments for ecosystem services in India from the bottom-up. In H. Healy, M. Martinez-Alier, L. Temper, M. Walter, & J. Gerber (Eds.), *Ecological economics from the ground up*. Abingdon/Oxon: Routledge/Taylor & Francis Group.

Vaidyanathan, A. (2006). *India's water resources, contemporary issues on irrigation* (p. 2006). New Delhi: Oxford University Press.

World Bank. (1993). *Water resources management* (A World Bank policy paper, p. 140). Washington, DC: The World Bank.

World Bank. (2003). *The incident of canal irrigation subsidies in India* (Policy Note). Washington, DC: South Asia Development Unit.

Zilberman, D., & Lipper, L. (1999). The economics of water use. In J. C. J. M. van den Bergh (Ed.), *Handbook of environmental and resource economics* (pp. 141–158). Cheltenham: Edward Elgar.

Chapter 10
Water Pricing in Israel: Various Waters, Various Neighbors

Nir Becker

Abstract Israel manages its water scarcity by a relatively unique combination of quantitative and pricing tools. As a semiarid climate country, efficient water pricing might prove to have much more potential welfare implications. The chapter contains a summary of the theoretical background of the different water pricing policies and reforms that have been recently implemented. The summary will then be accompanied by an effort to explain the rationale of the reforms. The chapter covers water pricing schemes in the various sectors and links them into one consistent policy vision. Currently, water pricing in Israel is more closely connected to the true scarcity value of this natural resource. Yet the goals and targets faced by water planners in Israel do not allow water prices to be the only allocation mechanism, and as such, a mixture of quantities and prices will be explored. The challenges faced now by the water regulators are new and contain pricing of different water sources (treated wastewater, desalinated water, etc.) for a variety of uses, including those that are characterized as nonmarket in nature (e.g., in-stream value) and those that should be based on basin cooperation among different countries (e.g., the Palestinian Authority, Jordan, and, potentially, Syria and Lebanon in the future).

Keywords Israel • Desalination • Marginal pricing • Integrated management • Water corporations

10.1 Introduction

Following the establishment of the state of Israel in 1948, the new nation's government invested heavily in developing water infrastructure and institutions to support a safe, secure, and affordable water supply. Facing the need for domestic food self-sufficiency in its early years, combined with the important role of domestic farms in encouraging settlement while protecting borders, priority was assigned to irrigated

---

N. Becker (✉)
Department of Economics and Management, Faculty of Social and Humanities science, Tel-Hai College, Upper Galilee 12210, Israel
e-mail: nbecker@telhai.ac.il

A. Dinar et al. (eds.), *Water Pricing Experiences and Innovations*,
Global Issues in Water Policy 9, DOI 10.1007/978-3-319-16465-6_10

agriculture. Water for irrigation was extensively developed and subsidized compared to other uses (Fietelson 2013). By the mid-1960s, Israel completed its National Water Carrier (NWC), conveying water from the Sea of Galilee (Lake Kinneret) in the wetter north to the drier central and southern regions. Two important features of a national water policy emerged: allocation of a considerable percentage of water for agricultural uses and a conscious plan to avoid use of marginal cost pricing to limit irrigation water demands (Menahem and Gilad 2013).

During the mid-1960s, the Jordan River supplied an average of 1.5 billion cubic meters per year from the Sea of Galilee to the Dead Sea. Since then, increased quantities of water have been appropriated by Israel, Jordan, and Syria from the Jordan River for irrigated agriculture and domestic use. The result was reduced flows of the river to a much lower average of about 10 million cubic meters per year. Since 2000, falling levels of the Dead Sea have created environmental problems, including sink holes. Various plans to address the declines of the flow in the Jordan River and the water level in the Dead Sea have been debated. Two of the more commonly discussed measures are an increased scale of seawater desalination and reduced quantities of water currently appropriated from the Jordan River (Becker et al. 2012).

By the mid-1990s, introduction of new demands stressed the existing water management system. These new uses included increased urban water demands, increased allocations of water delivered to neighboring countries brought on by negotiated peace agreements, and increased demands for in-stream flows to support key ecological assets. By 2001, urban and domestic use replaced irrigation as the largest users of freshwater in Israel. Israel's demand growth led to reduced flows of the country's coastal streams, wetlands, and other ecological assets like the Dead Sea.

Despite all efforts, water scarcity has plagued Israel since its establishment. Since then, a tenfold increase in population and considerable industrial growth and economic development have placed ongoing and increased demands on Israel's scarce water (Zeitoun 2011). Climatic fluctuations combined with recurrent drought continue to challenge the delivery of freshwater supplies.

While demand continued to rise throughout the 1990s and 2000s, supply shortages became more acute. Meanwhile, a series of studies concluded that in the face of climatic change, future rainfall supplying the country could fall by up to 30 % from historical patterns, with more frequent and prolonged droughts combined with increased evaporation (Lavee et al. 2011; Chenoweth et al. 2011; Sowers et al. 2012).

In the last decade, a series of actions both on the demand and especially on the supply side have created a more sustainable future in terms of water supply. However, little effort was devoted to understanding the importance of pricing in achieving the goals of the water sector in Israel. Water prices reflect laws and institutions. While Israel manages its water sector largely in a centralized way, others (e.g., the United States) approach it with a more decentralized pricing system. The main purpose of this chapter is to incorporate pricing within the history of Israel's water—a resource never intended to be run by economic rules, but one that finds it increasingly difficult to be diverted from the principles of the market system.

10.2 Setting the Framework: Supply and Consumption of Water

Freshwater: Israel has some particular characteristics that need to be revealed in order to understand its pricing mechanism and the way to analyze them. The most important aspect that does not exist in many other countries is that most of the country is connected through a single water system that supplies most of the users in urban centers and agriculture. This fact has a tremendous effect on pricing, because there may be many users in many regions but only one system that serves them all and needs to be efficiently operated.

In order to better understand this unique situation, it should be noted that Israel is a small and narrow country (Fig. 10.1); half of its area is a desert. Precipitation ranges from more than 700 mm per year in the north to less than 35 mm in the southern part of the country. Rainfall occurs only during the winter months. The core functions of the water sector have been to store water from winter for use in the summer and from rainy years to dry ones. In the 1960s, an additional task of the water system was to divert water from the north to the center and the south. Starting from the 1980s, two sources have been added to the country's water supply: When the population expanded and urbanization grew, treated and recycled sewage water was added to the supply, mostly for use in agriculture, but with smaller amounts allocated to aquatic habitats. More recently, desalinated seawater has become a significant source of water. Figure 10.2 presents the distribution of sources of freshwater.

Freshwater is stored in the Sea of Galilee in the north and in several groundwater reservoirs; the largest two are the mountain aquifer and the coastal aquifer. The mountain aquifer is located mostly under the West Bank. The coastal aquifer stretches along the Mediterranean from a site south of Haifa to Gaza. Fewer smaller aquifers add to the amount of freshwater reaching a total of 1,392 million cubic meters (MCM) (State of Israel 2012). Figure 10.1 also contains blue lines that are stretched in a north-south direction. They represent the national water carrier. It is a system of conduits running southwest from the Sea of Galilee, connecting most of the sources and users of water in the country in a single, closed system. The Mekorot Company, a government-owned utility company that supplies about two-thirds of the water in Israel, operates the national carrier. The other suppliers are private well owners, municipalities, and regional cooperatives.

Water Users: The main water users include agricultural, domestic, and industrial users. Agricultural water use has for a long time been using about 1,000 MCM. Domestic water consumption ranges around 100 MCM per capita, and industrial water use is about 100 MCM and has stayed roughly constant for many years. In addition, Israel took upon itself the responsibility of supplying water, according to treaties, to areas of the Palestinian Authority and the Kingdom of Jordan (about 50 MCM for each). Another sector that has been recognized as a legitimate user is nature. In an amendment to the water law, the water authority must take aquatic habitats into consideration and report the yearly allocation for this

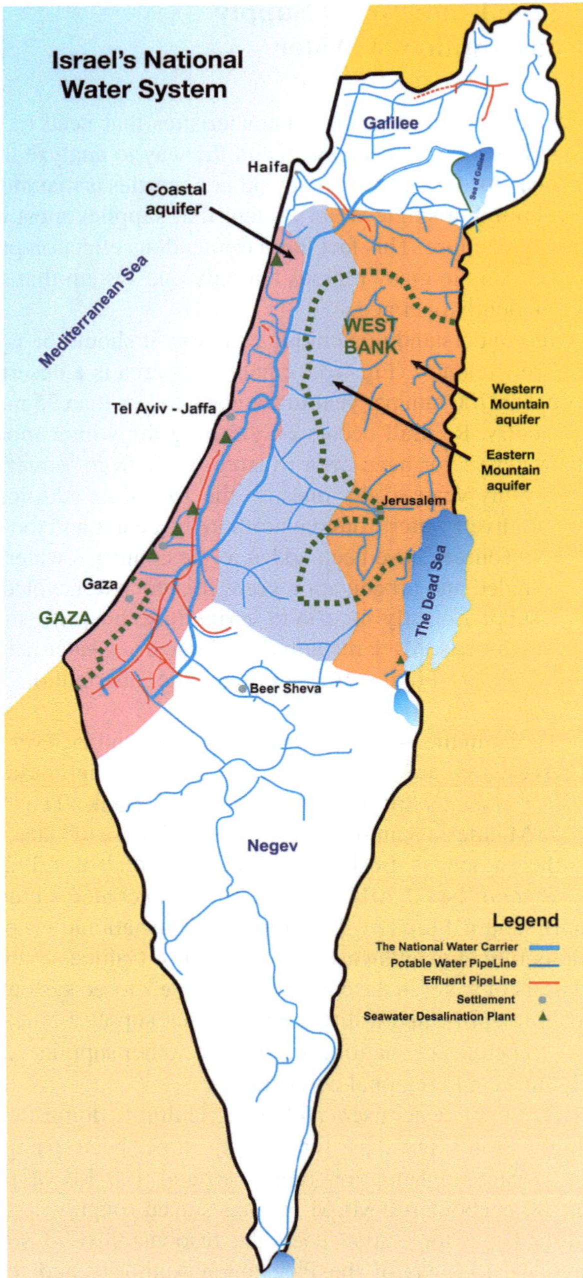

Fig. 10.1 Israel's national water system (Source: Mekorot and own editing)

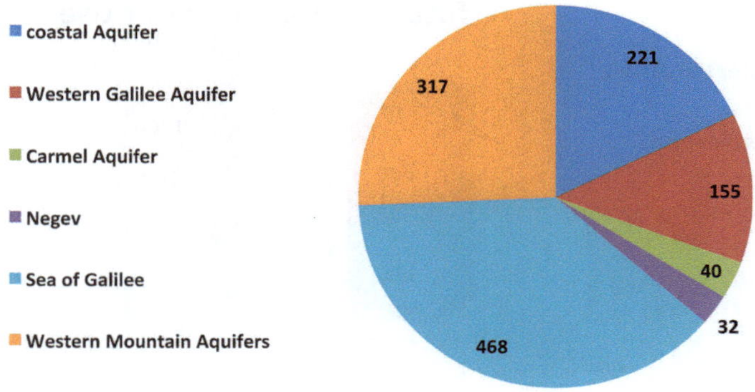

Fig. 10.2 Distribution of freshwater resources in Israel (Source: State of Israel Water Authority 2012)

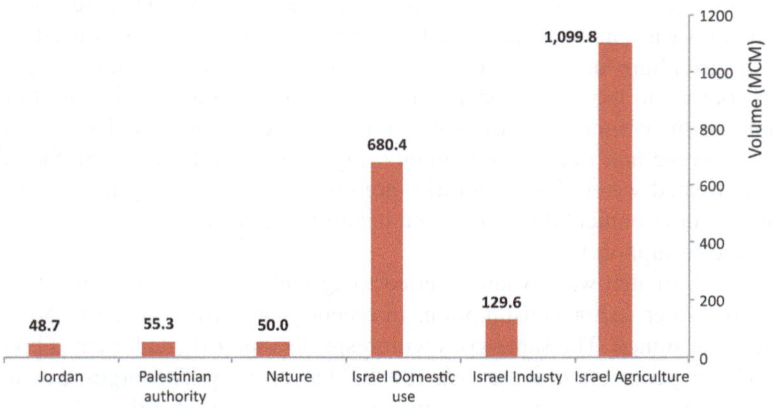

Fig. 10.3 Water main users (2012) (Source: State of Israel Water Authority 2012)

cause. The recommended amount was set to 50 MCM. A schematic representation of water consumption is shown in Fig. 10.3 for 2012.

Expanded Supply of Water, Wastewater and Desalination: Clearly, the fresh natural water supply cannot accommodate all water users, as Israel is characterized by a relatively high population growth (1.8 % annually). To help overcome this gap, marginal water (e.g., brackish and treated wastewater) and desalination plants have been playing a larger role during the last two decades, as can be seen in Fig. 10.4.

Treated Wastewater: In the past, either sewage leaked into the reservoirs or in places that had central sewage systems, the collected wastewater was sent to the sea or to nearby streams. Contamination of the beaches and two outbreaks of

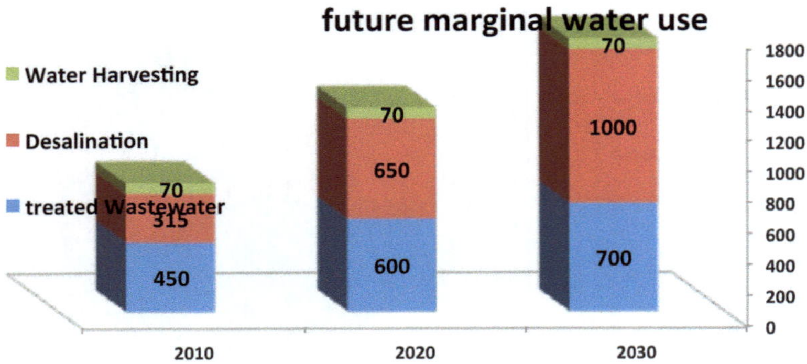

Fig. 10.4 Past, current, and future trends in marginal and desalinated water use (Source: State of Israel Water Authority 2012)

epidemics—cholera in 1970 and polio in 1988—helped raise awareness of the dangers of neglecting wastewater and the need to develop modern sewage systems and build wastewater treatment facilities. The treated wastewater is used for irrigation in agriculture and, in a few cases, was diverted to rivers (Kislev 2011).

In contrast to the water system built around the national carrier and linking the regions of the country, wastewater treatment resources are handled locally. Cities collect sewage and treat it, and neighboring farmers build recycling facilities and use the treated water. These facilities are costly, but since they help in mitigating important environmental damages and health hazards, significant public assistance is offered to support them.

In 1963, treated wastewater diverted to agriculture constituted only 4 % of the domestic water sector consumption. In recent years, it picked up to 55 % of the amount consumed. The same goes with respect to the irrigated area: It hiked from 1,500 ha 50 years ago to more than 100,000 ha currently. The largest treatment and recycling facility is Shafdan, collecting wastewater from 15 cities (Tel Aviv and its vicinity). The treated water, although not drinkable, was approved for unlimited irrigation. Currently, a more stringent standard is being adopted; however, its implementation makes wastewater treatment more costly.

Desalinated Water: The possibility of desalinating seawater was first mentioned as early as the 1960s, yet rejected due to technical and economic reasons. The first master plan that recommended seawater desalination on a large scale was submitted in 1997, but was rejected.

After the drought in 1998–1999, the government decided to begin practical preparations for seawater desalination. Today, there are five plants that produce about 550 MCM annually. In order to expedite the building of the plants, the government issued tenders, and four out of the five plants operate through private companies, which sell the water to Mekorot at an agreed price. The fifth plant is directly owned by Mekorot, after the company that won the tender went bankrupt.

10.3 Institutions and Reforms

It is essential to first understand the rationale for governmental intervention in the water sector. Three market failures characterize the water sector in Israel and probably many other countries as well: (1) Water is an open access resource, (2) treating sewage contains both positive and negative externalities, and (3) water infrastructure is characterized by increasing returns to scale. Added to these three economic reasons, there were also social reasons that formed the basis for the water law (Kislev 2011).

Accordingly, the water law (1959) states that all water sources in the country are publicly owned; there is no private ownership of water. A government agency is responsible for the utilization of the resource. The law requires measurement of all uses of water. Wells and pumps are monitored and consumers pay by the volume of water they use.

An important economic aspect of the law is that water for agricultural uses is set by quotas. On the contrary, water for other purposes, such as urban and industrial, is not restricted by quantity. The reason was that due to the heavy government subsidy in agriculture, this sector is able to bear consumption fluctuations.

Two far-reaching reforms took place in the water economy in the last 10 years. One was restructuring of the regulatory body, namely, the water authority, while the other was the removal of urban water provision from the control of municipalities. These are discussed below.

The Water Authority: The Israeli government is involved in the water sector in an active way, more than just setting the rules. There is thus a specific agency in charge of the sector. The original law gave the power to a water commissioner. Although this is a single person, he was assisted by the staff of the water commission, a government agency. The agency was responsible for the hydrological management, as well as planning and water allotments and permits (to the agricultural sector). The agency was also in charge of the development part of the sector (e.g., wastewater reuse and desalination plants).

The main problem was that many aspects/issues were left to the responsibility of other units (Kislev 2011). Those include quotas to farmers set by the ministry of agriculture and the ministry of the interior's control of the urban sector and that water prices were set by the ministry of treasury. The multiplicity of participants in decisions on water issues was seen as an obstacle to the efficient management of the sector. The government proposed to modify the water law and restructure the underlying institutional setting.

The new law (effective from 2007) abolished the position of the commissioner and established a governmental authority for water and sewage that was headed by a director.

The reformed law expanded significantly the area of responsibility of the new regulatory body. The water commissioner was responsible only for the regulation of the resources and their utilization. But his involvement in price setting was minimal. The water authority, on the contrary, is responsible for all aspects of the water sector. That includes setting the tariffs, as well as overseeing investments in Mekorot and the urban corporations.

Water Corporations: For years, the municipalities have been responsible for water and sewage services in their jurisdictions. They purchased water from Mekorot or pumped it from wells they owned and then supplied it to households. They were also responsible for sewage collection (which was transferred into treatment facilities). The residents paid for these services through a water charge.

The problem with that stemmed from the fact that water services were provided as part of the overall activity of the municipality; there was no separation between water-related revenues and other revenues. Hence, there was no full accounting of the water and sewage services. Political and other considerations made it easy for city mayors to postpone costly work needed for their water and sewage systems and instead diverted the accumulated funds to other more pressing and visible needs. The prevailing results were evident: water loss rates were high, and sewage was not collected and properly treated. This unsustainable situation caused a high water infrastructure depreciation rate without balancing it by financial reserves to support its maintenance and replacement. The estimated deficit was about $3 billion USD. In 1993, the government decided to transfer the administration of the urban water sector to independent companies and a committee. It was not until 2001 that the law was approved. By the original law, the municipalities were authorized to form corporations to supply water and sewage services, and they would be owned by the municipalities themselves. The water system's capital would be transferred to these corporations, which would then take over provision of services. To oversee them, the government created the water and sewage authority, which would be responsible for the quality of the services and the tariffs and to monitor the agreements between them and the municipalities as well as to approve their development plans. As of today, there are 55 such water corporations that serve almost six million residents.

The formation of the corporations also raises certain problems. The payments for water and sewage services helped the budgetary management. This source was now drying up. Further, the removal of the responsibility for water and sewage services from the local water and sewage utilities created a built-in tension between the local municipalities and the corporations with respect to tariffs.

10.4 Water Pricing: A Theoretical Framework

In order to understand the essence of the Israeli water policy, I will first start from the theoretical basis for water pricing and only then turn to their implementation in Israel.

Figure 10.5 is a schematic representation of demand and supply of the water sector in Israel. As can be seen from the figure, there are a stepwise supply function and three representative demand functions. Looking first at the supply curve, it can be seen as the marginal cost of producing water. That is, instead of asking, "How much water is to be produced at a given price?" we ask, "At a given amount of water produced, what is the marginal cost of producing an additional cubic meter (CM) of water?" The figure is a schematic representation, but it can be generalized to

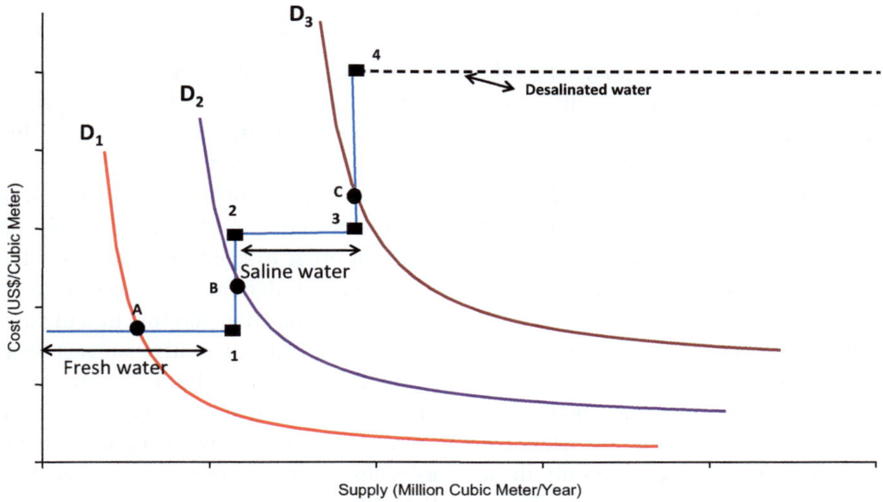

Fig. 10.5 A schematic analysis of demand and supply of water in Israel (Source: Own calculation based on the State of Israel Water Authority 2012)

represent a more detailed situation. As can be seen, the marginal cost of water production starts at the lower step, which is associated with withdrawing freshwater from the different reservoirs. However, this quantity is limited and if more water is to be produced, there is a need to treat different sources of brackish water. This is represented by the second step. This quantity of water is also bounded and if more water is to be produced, we finally reach desalinated water. In this case, the situation is different since that amount of water is not limited when the source of water is the Mediterranean Sea.[1]

In order to find the efficient price of water, we need to intersect the supply curve with the demand. We argue that the price at the intersection is efficient, since demand is also the marginal benefit of that extra unit of water. It is just at the intersection of the two curves that the marginal cost of the last unit of water is equal to the marginal benefit of that unit. Any other amount of water produced (and consumed) will cause marginal benefit to be either higher or lower than its marginal cost and therefore dictates inefficiency (Riegels et al. 2013).

The three demand curves[2] represent the dynamic idea of water pricing. It is dynamic, since water demand is changing rapidly and does not stay constant. It changes for different reasons. The two major reasons are associated with population growth, and the third reason is the emerging issues of water for peace and water for nature. Even without incorporating the impact of climate change (on the lower

[1] It may be bounded by other factors such as land, which is usually set on binding coastal areas.

[2] This is a total demand decomposed from agricultural, domestic, industrial, and nature and peace obligations.

stepwise curve), we see that the equilibrium price may dictate different policies and different pricing over the year.

The equilibrium point A, set by demand curve D_1 and the lower stepwise curve, dictates that there is no scarcity of water. In that case, water could be simply priced at the marginal cost equal to the height of the lower stepwise curve. However, this situation changes for the demand curve D_2. Actually, any point on the vertical line between the two switching points 1 and 2 creates scarcity of water. The reason is that if water continues to be charged at the previous cost, total demand would be larger than the available amount of freshwater. In order to keep the previous price effective, administrative quotas need to be set. However, that raises the issue of efficiency, because the central planner needs to know the exact demand function of each and every user to get it right. This is almost impossible. The alternative suggests that water price needs to rise until total demand will be equal to the total amount of freshwater. The difference between point B and switching point 1 is the scarcity rent. It is an additional cost users have to pay to allocate the water in an efficient way. It is efficient because only those users whose marginal benefit is higher than point B will purchase the water. These are the highest bidders. It is important to note that until demand reaches point 2, there is no need for a supply expansion. Any added amount of water would cost at the margin more than its marginal benefit.

The situation changes after point 2. Any intersection between demand and supply curves along the horizontal stepwise along the points 2 and 3 dictates supply expansion by starting to desalinate brackish water.[3] Even though the marginal cost of water is equal to a real marginal cost of producing water, still the scarcity rents exist for the freshwater resources, if one wants to hold the one-price criterion.

The same rationing can be implemented to the vertical line between the two switching points 3 and 4. Thus, the intersection point C raises the scarcity price above the cost of desalinated brackish water. When demand intersects point 4, the optimal time for desalination has arrived, and all users must pay that price. Scarcity rent should be set as the difference between any water source cost of production and the desalination cost.

10.5 The Evolution of Water Costs and Prices in Israel

Figure 10.6 depicts the major features of an estimation of what seems to be the situation in Israel, based on data gathered from the water authority and demand in which the only price-sensitive sector is agriculture (State of Israel Water Authority 2012; Lavee et al. 2011). I have used data for 2 years, namely, 2010 and 2030, to present the empirical point raised earlier.

As seen in Fig. 10.6, the cost of local production of water is \$0.31 per CM and reaches \$0.71 per CM for desalinated seawater. It is interesting to observe that even

[3] One could also think about the lower and middle stepwise curves as producing water from regional sources and the Sea of Galilee, respectively.

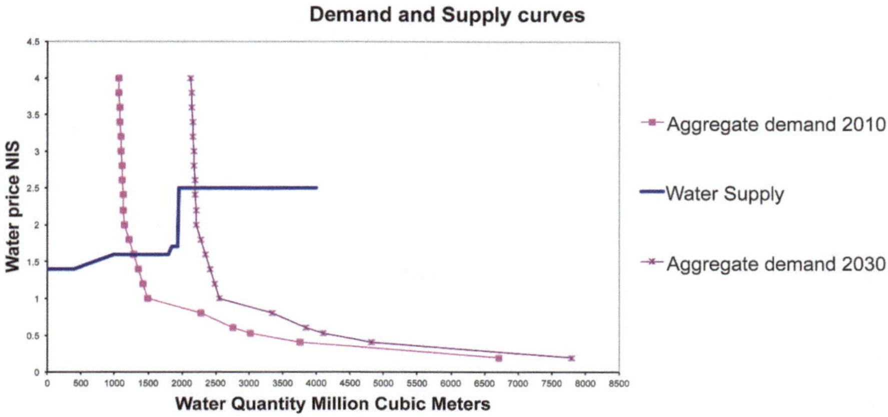

Fig. 10.6 Equilibrium points for 2010 and 2030 underestimated real conditions (Source: Own calculation based on the State of Israel Water Authority 2012)

in 2010, there was no need for desalination plants. Figure 10.6, however, ignores several points: First, climate change reduced the amount of available freshwater that is now assumed as sustainable yield. Second, water for peace and nature are not entered into the demand curve, and their insertion may create an incentive to start desalination earlier than what the figure suggests. On the other hand, wastewater treatment creates a new buffer that mitigates water shortage for farmers. All these new points are presented in Fig. 10.7 for the year 2010. There are four staircases in the supply function: treated wastewater, freshwater, saline water, and desalination. They are presented in an increasing trend to represent the increasing marginal cost of water production. The demand sectors are presented by boxes, and they sum up to an overall demand given by the red curve. It is worth noting that the lower staircase can be consumed only by agriculture and nature as is clearly shown in the figure.

As can be seen from the figure, there is an extra 230 MCM until desalination becomes economically efficient. One can look at it from another point of view that there is still place to give up more water for peace and nature at a lower price than what was thought before.

Another aspect of water pricing is the separation to its different structural components, fixed and variable cost. For domestic consumption, post-use costs are also added to sewage collection, treatment, and drainage where needed. The total cost is the sum of these components.

Covering the variable cost can be simple by charging the average cost per CM. The problem of fixed cost is a more complicated one. Since consumers do not necessarily cover total cost, fixed cost can be covered from two sources: The first is from the general budget of the water utility. The second goes back to water consumers and prices of water in a two-tiered method. This can be done either by an advance payment equal to the fixed cost component, a periodic fixed payment to cover the

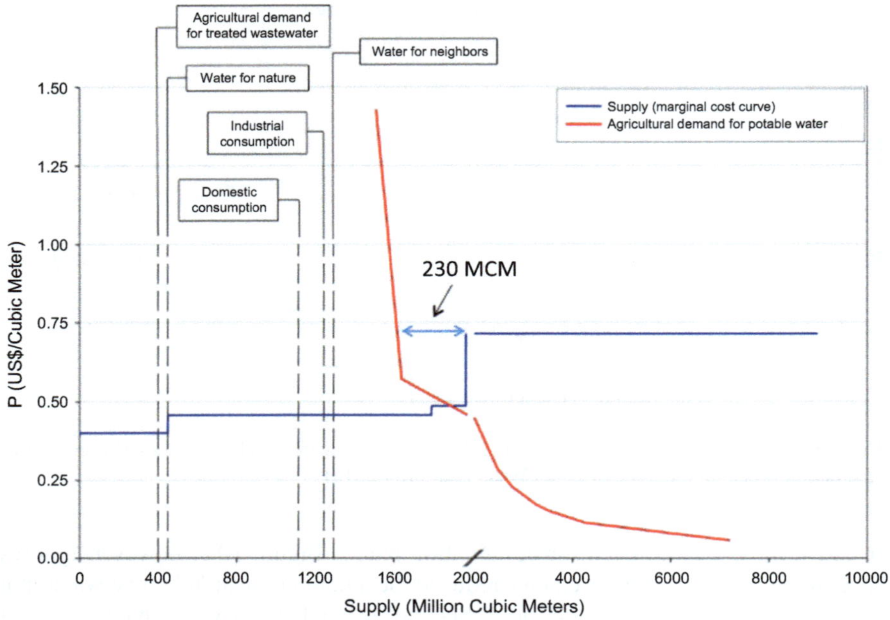

Fig. 10.7 Water supply and demand under full consideration of recognized sectors for 2010 (Source: Own calculation based on the State of Israel Water Authority 2012)

periodic fixed cost, or some integration of the fixed cost with the current volumetric charge for water. This last option is the one currently used in Israel. It is actually embodied in the water tariffs that, in turn, are based on forecasted consumption. They are thus subjected to trial and error and a balance of the net outcome over the years.

Based on Mekorot's financial report, as well as the water authority dataset, it is estimated that the average cost per CM from all sources and delivering to all destinations is \$0.52. The average cost per CM of desalinated seawater is \$0.6. Since desalinated water accounts for 50 % of the total water that Mekorot supplies, the average cost per CM without desalinated water is derived to be \$0.44. This estimate also accounts for treating brackish water; hence, the cost for CM of freshwater is about \$0.40.

10.6 Price Policy in Practice

Before I move to describe each sector and water use in depth, I present in Fig. 10.8 the price evolvement in the last 5 years (2009–2013). This figure describes price per CM for the agricultural, industrial, and domestic sectors. Because trends may shed

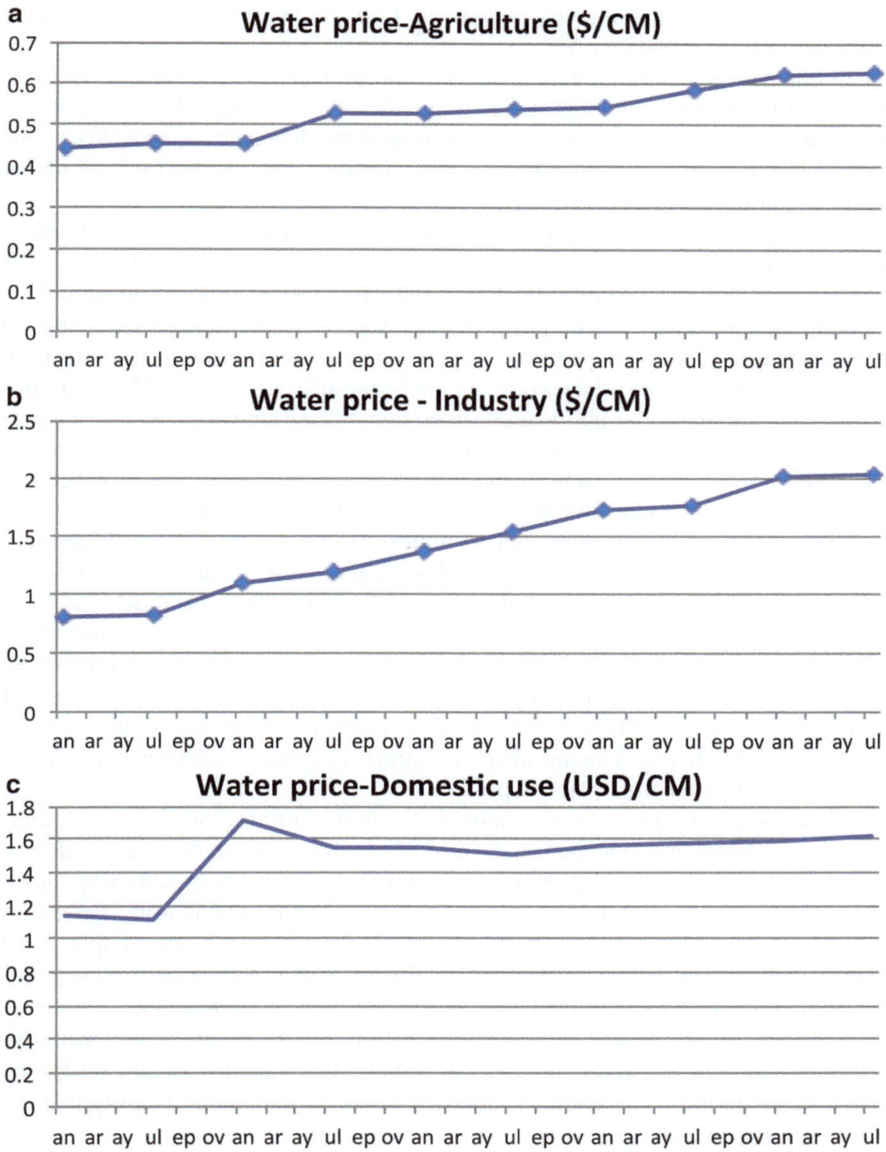

Fig. 10.8 Water prices for the three main sectors in Israel: 2009–2013 (Source: Private communication with the economics department at the water authority)

a light on evolution and not only magnitude, I use graphs here. Detailed prices for 2013 are given in Appendix A.

As can be seen from the figure, there is a fundamental difference among the three sectors. The water price for agriculture is significantly lower than the other two sectors. Another important aspect is the price jump in the domestic sector in 2010. This price jump is due to the change in water fee collection from the central government to the urban water corporations.

Prices in the Urban Sector: There are three major sets of water prices in Israel—prices for fresh and recycled water that is supplied by Mekorot, prices charged in the urban sector, and prices charged by private suppliers (mostly regional cooperatives). In addition, two other specific payments are paid for (1) sewage services and (2) scarcity rents on water pumped from the Sea of Galilee and the aquifers. In the past, the prices charged in the urban sector did not necessarily cover costs. Their determination was influenced by political considerations. Gaps between cost and revenue were covered for Mekorot by the state budget and in the urban sector from municipal sources. Since the creation of the water corporations, total cost of water and sewage services has been covered by the prices collected from users.

However, even though price support almost vanished, the sector still sees substantial government support. The state budget finances investment in recycling projects, in new urban corporations, and in sewage systems in poor localities.

Prices in Agriculture: About 60 % of water for agriculture is supplied by Mekorot. The water law distinguishes between the cost of water and water fees. *Cost* refers to the cost of extraction and supply. *Fees* are prices paid by the users of water, which the law allows setting based on various considerations, among them the users' ability to pay. The law also sets extraction levies to reflect water scarcity, and those may differ from place to place.

In the past, water prices were determined with the approval of Knesset committees with no explicit connection to the cost of provision. When the water authority was established, it was tasked with setting prices. In 2006, the government signed an agreement with farmers' representatives according to which water prices for agriculture would be set, based on the average Mekorot cost of water supply to the sector, including agriculture's share of desalinated water. As a result, water prices for agriculture have risen and will continue to rise in the coming years.

According to the water law, water for agriculture is supplied by administrative allotment, in quotas: Each consumer has a quota that was historically set by the planning authorities. The quota is supposed to be the maximum quantity that the consumer will receive.

Freshwater: The prices farmers pay to Mekorot for freshwater are of increasing block rate. Each agricultural consumer is allotted a quota, and, as of 2014, the prices are (calculated at the exchange rate of NIS 3.50 per $1.00) $0.43 per CM for the first block (50 % of the quota), $0.48 per CM for the second block (next 30 %), and $0.60 for the third block.

Since water allocation is also limited by quotas, in dry years, farmers may be limited to take only part of the quota, but the price structure does not change. In 2010, a contract was signed between the government and the farmers that dictates a gradual rise in water prices until they reach the average cost of supply, including the cost of purchased desalinated water. It is expected to be fully implemented by 2015. That is to reach about 0.7 USD/CM (depending on the true desalination cost and the exchange rate at that year).

Treated Wastewater: Treated urban sewage is mostly used in agriculture. Mekorot operates two large recycling plants, near Tel Aviv and near Haifa, and several smaller facilities. All others are owned and run by local operators, mostly regional agricultural cooperatives. Mekorot's price for recycled effluent is $ 0.24 per CM. The construction of private facilities is subsidized, aiming to set cost equal to Mekorot's charges so the individual farmers pay the same price, whether their treated wastewater are provided by Mekorot or by a regional cooperative.

Scarcity Rents: The prices farmers pay Mekorot are the same throughout the country. Consequently, the cost of water to well owners and those pumping directly from rivers or the Sea of Galilee may be significantly lower than those who receive water from the national system. In the past, the government operated an equalization fund: Low-cost water users paid into the fund and high-cost users were on the receiving end. Since Mekorot supplied water to remote areas, they were the prime beneficiaries of that fund. This policy was modified in 2000. Scarcity rents replaced payments to the equalization fund. In principle, these rents may change according to locality and source of water. In reality, the rents were not set as they should have been according to the theory presented in the previous section. They do differ geographically and by sector, however, as well as by quantity, season, and precipitation.

For example, farmers in the coastal area pumping coastal aquifer water pay block-rate rents: The first block, which consists of 25 % of the withdrawal license, will cost $0.02 per CM. The second block, for an additional 55 %, will cost $0.27 per CM, and the third block will cost an additional $0.41 per CM.

Farmers in other areas, particularly farmers who pump their water from rivers and the Sea of Galilee, pay lower rates. Some of these are higher in dry years and lower in rainy times.

The Policy Aspect: The gradual shift of freshwater from agriculture to other sectors is part of the growth process and the change in the structure of Israel's economy. The freshwater goes over to the urban sector and is replaced, though only partially, by marginal water. Since the management of water is in the hands of the state and the decisions of public agencies have been geared to reducing agricultural supply, the changes in water use have been perceived as coercive and arbitrary, generating sharp criticism on the parts of the farmers (Kislev 2011).

The gradual diminishment in the quantity of water supplied to agriculture was accompanied by another phenomenon, generating even harder criticism: that of repeated reductions in the water allocated to agriculture in dry periods. Agriculture has borne the burden of the crises in the water sector, and authorities sought to lay these burdens every time there was a shortage. This phenomenon stemmed from the structured price system in which agriculture was the marginal sector that had to be the buffer for hard times.

The Urban Sector: Since water corporations replaced municipalities in controlling water management for the urban sector, there are actually two independent water systems. The first is the national system containing the water sources, the national carrier, and the desalination plants. The second is the urban water sector containing the intra-urban water supply, sewage removal, and the treatment facilities. The local municipalities provide about 15 % of the urban water directly from reservoirs that are actually part of the nationwide economy. The other 85 % are purchased from Mekorot.

The two sectors deal with different problems and issues. The nationwide economy deals with questions of sustainable resource management and development of water supply, whereas the urban economy deals with water distribution and sewage collection and treatment.

Much of the work of the water authority in the last couple of years has been on tariffs for the urban water and wastewater corporations. Since by law these tariffs have to cover costs, the original intention of the authority was to set locality-specific prices to reflect local cost. However, these intentions encountered harsh criticism. Accepting the criticism, the new price schedules set for the urban corporations in 2010 are identical—the same prices for each and every corporation (Kislev 2011).

Price structure in the domestic sector is also characterized by block-rate tariffs. The first block is for a "basic quantity" of 2.5 CM per person, per month, and is priced at $1.62 per CM. The second block is for additional amounts and is priced at $3.18 per CM.

Figure 10.9 demonstrates the location of Israel in comparison with other Organisation for Economic Co-operation and Development (OECD) countries. These rates cover both water and sewage services, and they include an 18 % value-added tax. The estimate for Israel was calculated as a weighted average for the two price rates, according to consumption at each rate. As can be clearly seen, Israel's water charge for domestic use is below the OECD mean price (3.64 USD per CM).

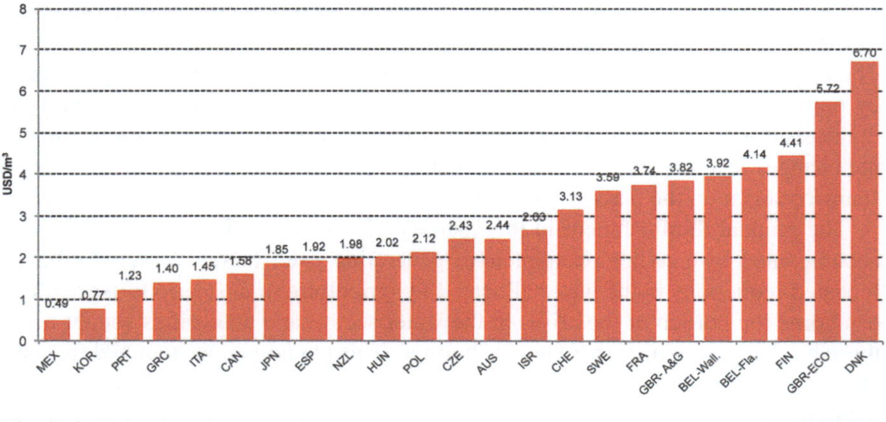

Fig. 10.9 Unit price of water and sanitation services to households (USD/CM) (Source: EU 2012)

To have the corporations cover just costs, not making profits or suffering losses, Mekorot prices of bulk quantities at the city gate are set differentially. These range from $0.74 to $2.01 per CM. Low-cost corporations pay Mekorot per unit of water more than do others.

Municipalities collect from homes and buildings one-time connection levies to cover investment in water and sewage infrastructure. The water authority council attempted to replace these levies with a capital component in the price per CM. The municipal water authority objected because they might lack funds for new developments. The inclusion of capital outlays in the price of water, if accepted, will add $0.21 per CM to the rates quoted above (Kislev 2011).

Fifteen percent of the water provided in the urban sector is drawn from wells owned by municipalities or by the water corporations. When deciding on new prices, the water authority council sets extraction levies for locally drawn water so that its cost is similar to the cost of water purchased by the corporations from Mekorot.

Cross-Subsidization: In the past, households paid for water charges by block rates as well. The help to big families came in the form of allowing them to charge more water in the lower block rate. Usually those families were poor as well. Hence, the smaller and richer families subsidized the big and poor families within each municipality. The potential problem with this kind of support is that some cities are characterized by a majority of the first kind of families (big and poor), while other municipalities are characterized by the latter kind of families. Given an equal cost structure of water supply, the nonpoor minority in poor municipalities have to pay more than similar families in nonpoor cities. Nonpoor families that live in wealthy municipalities may even be getting a subsidy. The new water corporations did not take that into account, and it was not the intention of the water authority to set municipal-specific prices for Mekorot water. This price structure could be avoided if tariffs were not set at block rates and only one price was charged. However, this simple solution is politically difficult to implement (Kislev 2011).

10.7 Summary and Conclusions

Water pricing is ambitious and must be dealt with great caution and care. It often attempts to reach too many goals, such as cost recovery, income redistribution, and efficiency to name a few (EU 2012). In Israel, water pricing reflects the changes that accompanied the state along its 66 years since its establishment in 1948.

In the first 30 years of its creation, the main goal of water pricing was to support social goals, such as population distribution in peripheral areas, and to keep agriculture as a main player. Water pricing was thus barely associated with efficiency and cost recovery criteria. This has changed gradually starting from the 1980s. The country became less socialist and more market oriented, and there was a significant decline in the role of the agricultural sector as one that needs special assistance relative to other needs. On the other hand, population increased at a relatively fast rate, which required diversion of water to cities. Awareness to climate change hiked after several severe drought years in the 1980s and 1990s. This situation is not unique to Israel (e.g., Fragoso and Marques 2013; Rivers 2013).

Pricing seemed to gain attraction, as it were, providing a crystal ball to the real cost and value of water in its different uses. This was true to farmers and to cities. For the latter case, the water budget was mixed with the general local government budget, and revenues from water use could not meet the costs of keeping the system in a sufficient reliability situation.

Currently, water prices reflect true cost much more than in the past, but pricing is also used to achieve goals of income redistribution as well as natural monopoly aspects and open access characteristics of the system. As a consequence, Israel has still some way to go toward full marginal cost pricing.

From the supply side, wastewater treatment and desalination costs decreased significantly, allowing the country to use about 50 % of its freshwater supply to be allocated to sources other than agriculture. Wastewater treatment also helped to restore some aquatic habitats, which became more and more important to the public as general environmental awareness has risen over time.

There is still a significant way to go toward treatment of water as an economic good that needs to be treated with caution and with the understanding of the true tradeoffs among different alternatives of managing this resource. However, the situation today is less severe, and this is an achievement for a semiarid country that was several times on the verge of water catastrophe. The problem lies with the fact that the country needed to undergo several crises to understand that its path is not sustainable.

The future holds many uncertain situations. What will be the final water settlement between Israel and its neighbors? What will be the future with respect to population growth? From the supply side, questions such as climate change and the future path of water production cost, especially marginal water, such as sea desalinated water, are important blocks in building a wise water policy. Pricing can play a vital role in accommodating each one of these issues.

Appendix A: Water Price Structure for 2013 (in USD per CM)

Agricultural water tariff*:
Lower – 0.55
Middle – 0.63
Higher – 0.78

Saline and Treated wastewater tariff:
For agricultural use – 0.27
For other uses (e.g., nature) – 0.23
Saline water – 0.35

Domestic use**:
Price paid to Mekorot-
Lower – 0.72
Higher – 1.98
Price paid by end users to the water corporations-
Lower – 1.60
Higher – 3.12

Industrial tarrif: 2.01

* Lower tariff is for the first 50 % of allotment. Middle is for the next 30 % and higher is for the rest.
** Lower tariff is for personal use up till 2.5 CM per capita.

References

Becker, N., Lavee, D., & Tavor, T. (2012). Desalinate or divert? Coastal non-market values as a decision tool for an integrated water management policy: The case of the Jordan River basin. *Ocean and Coastal Management, 64*, 27–36.

Chenoweth, J., et al. (2011). Impact of climate change on the water resources of the eastern Mediterranean and Middle East region: Modeled 21st century changes and implications. *Water Resources Research, 47*, W06506. doi:06510.01029/02010WR010269.

EU Water Initiative. (2012). *Pricing water resources to finance their sustainable management.* Finance Working Group (EUWI-FWG), GWP, Drottninggatan 33, SE-111 51 Stockholm, Sweden.

Fietelson, E. (2013). The four eras of Israeli water policy. Ch.2. In N. Becker (Ed.), *Water policy in Israel* (pp. 15–32). Dordrecht: Springer.

Fragoso, R., & Marques, C. (2013). *The economic impact of alternative water pricing policies in Alentejo region* (CEFAGE-UE Working Paper 2013/02).

Kislev, Y. (2011). *The water sector in Israel.* The Taub Center discussion policy report 2011.15. Jerusalem.

Lavee, D., Ritov, M., & Becker, N. (2011). Is desalination the most sustainable alternative for water-shortage mitigation in Israel? *International Journal of Sustainable Economy, 3*(4), 410–424.

Menahem, G., & Gilad, S. (2013). Israel's water policy 1980s–2000s: Advocacy coalitions, policy stalemate and policy change. In N. Becker (Ed.), *Water policy in Israel* (pp. 33–50). Dordrecht: Springer.

Riegels, N., Velazquez, M. P., Doulgeris, C., Sturm, V., Jensen, R., Møller, F., & Gottwein, P. B. (2013). Systems analysis approach to the design of efficient water pricing policies under the EU water framework directive. *Journal of Water Resources Planning and Management, 139*, 574–582.

Rivers, N. (2013). *Economic modeling of water pricing strategies* (GWF Discussion Paper 1335). Canberra: Global Water Forum.

Sowers, J., Vengosh, A., & Wrinthal, E. (2012). Climate change, water resources, and the politics of adaptation in the Middle East and North Africa. *Climate Change, 104*(3–4), 599–627.

State of Israel Water Authority. (2012). Master plan for the national water sector.

Zeitoun, M. (2011). The global web of national water security. *Global Policy, 2*(3), 286–296.

Chapter 11
Water Pricing in Italy: Beyond Full-Cost Recovery

Antonio Massarutto

Abstract This chapter provides an overview of the Italian water management system, which is segmented by sectors and characterized by a wide plurality of management systems, operators, and financing patterns. In the last 20 years, Italy has introduced far-reaching reforms of water management, which concerned in the first place urban water supply and sanitation. The most important aim was to create the basis for an autonomous and self-sufficient water industry, driving the sector out of the public budget. Financial equilibrium of water undertakings and access to market-based finance have dominated other possible aims of water pricing. Other sectors, and notably irrigation, continue to follow more traditional schemes. The chapter also discusses further reform opportunities in the search for using water prices as economic incentives for a more sustainable use of water resources.

Keywords Italy • Water abstraction charges • Price regulation • Increasing block tariff • Cost pass-through

11.1 Background: The Italian Water Management System and Pricing

11.1.1 Water Resources, Population, and Issues in Water Supply

Despite being located in the Mediterranean region, Italy is fairly rich in water resources on average. Regional variability is high: the northern region, located south of the Alpine chain, enjoys more regular and abundant flows that compensate for the relatively lower rainfall in summer. Central and southern Italy benefit from the presence of the Apennine mountain ridge; however, these provide far lower natural stocking opportunities (no glaciers, no big lakes, etc.). In southern regions and in the islands, the climate is more arid; nonetheless, the natural endowment is

A. Massarutto (✉)
DIES – University of Udine, Via Tomadini 30a, 33100 Udine, Italy
e-mail: antonio.massarutto@uniud.it

A. Dinar et al. (eds.), *Water Pricing Experiences and Innovations*,
Global Issues in Water Policy 9, DOI 10.1007/978-3-319-16465-6_11

normally sufficient on a regional scale, with the notable exception of Apulia, in the southeastern part of the peninsula.

On the other hand, the relatively good endowment of freshwater resources and low rainfall during the summer months have encouraged an intensive use of available resources. According to the recent Environmental Performance Review of Italy (OECD 2013), water stress in Italy derives more from the intensity of water use than from absolute scarcity of natural resources. National statistics are outdated and imprecise. The latest comprehensive survey dates back to 1999. Since then, more updated figures are available for public water supply. Statistics still report water abstractions and not net water consumption. It is misleading to compare uses in different sectors, given that in many cases (hydropower, cooling, and irrigation), they do not represent a real subtraction from the natural environment, because the resource returns to the environment immediately after its use. Other uses (e.g., urban) impact water supply on a qualitative rather than a quantitative way. With these caveats in mind, Table 11.1 may offer an overall picture of the structure of water uses.

Irrigation concerns 50 % of total withdrawals (two-thirds of which take place in the north). Urban uses spread out more uniformly and concern a further 20 %, and household uses are about 85 % of the total. More than half of industrial uses (20 % in total) and the totality of hydropower uses (11 %) are again located in the north.

With a stable population (nearly 60 million) and an overall diminishing per capita water demand (−9.2 % between 1999 and 2008 according to Istat 2012), no meaningful pressure seems to arise from the side of public water services. The main drivers of change are the completion of the infrastructural system, an improvement of service reliability (especially in the south, where many families still lament regular interruptions), renewal, and maintenance.

Industrial and agricultural uses, at the same time, exhibit an overall declining trend; however, the main cause of water stress in perspective lies in the concentration of water-demanding activities in areas with lower natural potential, whereas the demand (e.g., for irrigating the high-value-added crops that represent the backbone of the Italian agro-industry) becomes more rigid.

On the quality side, Italy is still far from reaching the target of "good ecological status" imposed by the EU directives (EEA 2012). The sewage collection and treatment system is still in the phase of catching up with the requirements imposed by the 91/271 directive, with respect to which Italy has a number of infringement procedures open (AEEGSI 2014).

We may expect that water quality—more than water quantity—issues will represent the main driver of water policies and water investments in the next years.

Groundwater depletion poses a further series of problems that are still poorly investigated (Civita et al. 2009). Evidence of a lowering water table is occurring in many areas of the country, with the related consequences (saline intrusion, land subsidence). It is still unclear, however, how different causes, either natural or human, are responsible for the phenomenon. Evidence of an increasing pressure of self-supplied irrigation suggests that this is a major explanatory factor, especially along the peninsula and in coastal areas. Diffused pollution also constitutes a challenge for groundwater quality.

Table 11.1 Breakdown of water abstractions by sector

	Urban		Irrigation			Industrial			Hydropower		Total	
	Total	%	Total	%	per ha	Total	%	per employed	Total	%	hm³/year	%
	hm³/year		hm³/year		m³/year	hm³/year		m³/year	hm³/year			
NW	2.251	6 %	8.193	20 %	11.223	3.520	9 %	1.486	1.863	5 %	15.827	39 %
NE	1.428	4 %	5.277	13 %	5.277	1.648	4 %	903	2.538	6 %	10.891	27 %
Center	1.539	4 %	970	2 %	4.409	1.482	4 %	1.176	72	0 %	4.063	10 %
South	1.746	4 %	3.506	9 %	7.627	879	2 %	798	36	0 %	6.167	15 %
Islands	876	2 %	2.191	5 %	7.547	457	1 %	1.130	–	0 %	3.524	9 %
Italy	7.840	19 %	20.136	50 %	7.458	7.986	20 %	1.147	4.509	11 %	40.471	100 %

Source: IRSA-CNR (1999)

Overall, the Italian water management system requires significant investments for completion, maintenance, and renewal. This urgency, combined with the difficult contingent situation of public accounts, forces the water sector toward the search of alternative financing models. The transition from public spending to market-based finance has been more remarkable in the case of residential water services and industry, while agriculture is still lagging behind.

11.2 Structure of the Water Management System and Financial Flows

The structure of the Italian water management system is rather complex. Many uses approach directly the natural resource through abstractions and self-operation of wastewater and drainage. Others, in turn, use a collective service, under separate arrangements and with dedicated institutions for each sector.

Although it is difficult to generalize, the following points summarize the main distinctive features in the Italian water sector:

Hydropower is in most cases independent, especially when large facilities with upstream storage and flow regulation are present. Power producers usually operate the whole hydropower production and delivery system, including dams, reservoirs, bypass channels, and all the concerned infrastructure works. Run-of-the-river plants are also usually independent. Sometimes, however, hydropower facilities are located along man-made artificial watercourses managed by third parties. This is, for example, the case of canals operated by reclamation boards (see below). Similarly, hydropower facilities may benefit from upstream water regulation (e.g., dams operated by third parties).

For *industrial uses*, self-supply is the general rule, especially when water is an important input in the production process (e.g., pulp, food, or textile industry). In a few cases, special-purpose industrial aqueducts are in place. Other industries generally rely on the main public water supply system. Approximately 15 % of water supplied by public aqueducts is destined to non-household uses.

Industrial sewerage is sometimes operated directly by individual companies, but more often it is managed by dedicated collective establishments, particularly when industrial discharge requires specific ad hoc treatment. These systems can later discharge into public sanitation systems or directly into watercourses, depending on the local situation and convenience.

Public water supply (PWS) systems reach nearly 100 % of residential population, the exceptions being small isolated rural premises and dwellings that traditionally rely on local individual or community systems. Sewage collection is converging toward the standards set by the EU Urban Wastewater Directive (UWWD), with still some failure especially concerning sewage treatment equipment. After the reform initiated by Law 36/1994, public water supply and sanitation belong to the so-called integrated water service (IWS), a responsibility of local authorities to be organized by inter-municipal entities, named ambito territorial ottimale (ATOs) (optimal management areas). Governance rules vary among regions. Originally, there were 91 ATOs, later reduced to 72, covering completely the national territory,[1] although some of them still exist only on paper and not all of them have completed all steps.

The largest share of *irrigation supply* derives from collective institutions (reclamation boards). Their creation dates back to the nineteenth century or earlier. These are private associations of landowners having a public status. Although regulated by the law (now devoted to regions), reclamation boards enjoy a substantial autonomy. Individual direct abstractions at the farm level increasingly integrate and often entirely replace collective irrigation, due to a more flexible, reliable, and timely water supply. Although estimates are rather imprecise, this management form concerns 10–20 % of irrigation water, but a far higher share in water-stressed districts, such as coastal areas or the southern part of the Po basin. Direct abstractions from groundwater concern in particular high-value-added cultures.

Reclamation boards also perform important tasks in the field of *land drainage* in rural areas and management of small watercourses. *Flood protection* and riverbed maintenance in all other cases are direct tasks of regions, which sometimes have created dedicated institutions, such as the AIPo (Interregional Agency for the Po River) in the Po basin.

Rainwater management is officially a task of municipalities. Since two-thirds of sewage collection networks are mixed (rainwater + wastewater), operation is very often delegated to IWS systems; in some regions, these are also allowed to recover the cost directly within the IWS bill. On top of this, reclamation boards may provide "bulk drainage" services, since their networks may receive the outflow of urban rainwater systems and/or of sewage treatment plants.

In sum, we can identify three different typologies of water service providers:

- Public water supply and sanitation (PWS)
- Reclamation boards
- Industrial water service operators

As for IWS operators, each AATO (autorità di ambito territoriale ottimale, that is the authority that rules the ATO) delegates operation of water services to professional companies, whose ownership can be either public or private. The law prescribed a single undertaking serving each ATO; however, it also allowed the possibility to have more than one operator without prejudice of efficiency and effectiveness. The exact number of operators is unknown, since many still operate on a provisional entrustment. The last report issued by the national regulator, AEEGSI, identifies 268 subjects operating in 61 ATOs, which represent only 55 % of the total population. These figures show that the process of concentration initiated by the reform 20 years ago is still far from the target.

As for reclamation boards, the last survey operated for ANBI (Associazione Nazionale Bonifiche e Irrigazioni) by Leone (2004) identifies 136 consortia, cover-

[1] The Autonomous Provinces of Trento and Bolzano, due to their special autonomy, have a different and specific organization.

ing a surface of 15 M ha. These boards associate 96 % of total irrigable land and 91 % of total irrigated land. More than 75 % of total irrigated land is located in the northern part of the country.

Many industrial water service operators have merged under the roof of IWS; some of them continue to operate independently. In most cases, these operators belong to mixed establishments in which industrial firms are involved with the participation of public bodies (local authorities, chambers of commerce), publicly owned financial intermediaries, and so on. In the lack of a systematic survey at the national or regional level, it is not possible to provide reliable figures about the number and the economic dimension of the sector.

Most undertakings, regardless of their sector, provide all phases of water management "from the source to the tap," eventually relying on neighboring operators for some services. In some cases, however, further institutions upstream manage water resources on a bulk basis. Again, we can identify different typologies.

In some cases, these are truly independent companies. Ownership may be public: this is the case of Romagna Acque, serving the coastal provinces of Emilia-Romagna, and Enas, managing reservoirs and bulk water transfers in Sardinia. Others are private concessionaires, particularly in the south (e.g., Sicilia Acque, Sorical, and Acquedotto Campano Occidentale).

A second category of bulk suppliers concerns entities created for the sake of administering upstream regulation and storage works and allocating available flows to entitled subjects. For example, all the big subalpine lakes are artificially regulated at their mouth, and consortia of entitled users manage the gauging works.

Finally, a few bulk water schemes operate in the agricultural sector and provide water to irrigation systems. Occasionally, they may also provide services to other water users, as in the case of CER (Canale Emiliano-Romagnolo), which provides complementary supplies to urban and touristic dwellings along the Adriatic coast of Emilia-Romagna.

The regulatory framework involves many government layers, whose interplay often lacks a precise allocation of tasks, causing overlap of competences and lack of jurisdiction (Oecd 2013). Water resource regulation is framed by the European Union and national legislation and implemented at the basin level through the "river district plan," elaborated by river basin authorities (RBA). This plan identifies the actions needed to guarantee the desired ecological quality targets. Following the plan, regions provide administrative tasks, such as water use licensing and pollution control.

Economic regulation of water services depends on the concerned sector. As for all public services ("services of general economic interest" in the EU jargon), their organization should follow general framework rules. National legislation has tried to introduce market-based orientation for IWS (such as compulsory competitive tendering), but this approach was finally rejected by a popular referendum in 2011. At present, competent authorities (in the field of water, these are normally local authorities) can choose among a range of solutions that include own enterprises ("in-house" delegation) and many types of public-private partnership, including full delegation.

Figure 11.1 illustrates the governance scheme that concerns IWS. Services are delegated to professional companies—either public, private, or mixed—based on a contract, which usually entails a concession scheme (i.e., operators are responsible for investments at own risk). However, contracts are not sovereign for any detail. In particular, price regulation and other aspects (such as definition of minimum standards) are ultimately the responsibility of an independent national authority (AEEGSI), which is also responsible for electricity and gas services.

For other segments of water use, regulatory responsibilities are usually devoted to regions. This applies, for example, to irrigation and drainage, since the framework governance of reclamation boards lies under regional jurisdiction.

The complex structure of the water management system outlined above reflects an analogously complex structure of financial flows. Figure 11.2 provides a simplified diagram of financial transactions between different levels.

Each final user sustains a cost that includes tariffs and charges paid to access water services and the costs sustained directly (e.g., for groundwater pumping). The positive difference between these costs and the value extracted from water (e.g., electricity or agricultural products sold to the market, direct utility obtained from final consumption) represents in economic terms a rent, namely, the extra price users would be willing to pay to continue using water.

Similarly, retail operators sustain some costs directly (labor, capital, goods, and services acquired on the market) and pay for water services they receive from bulk

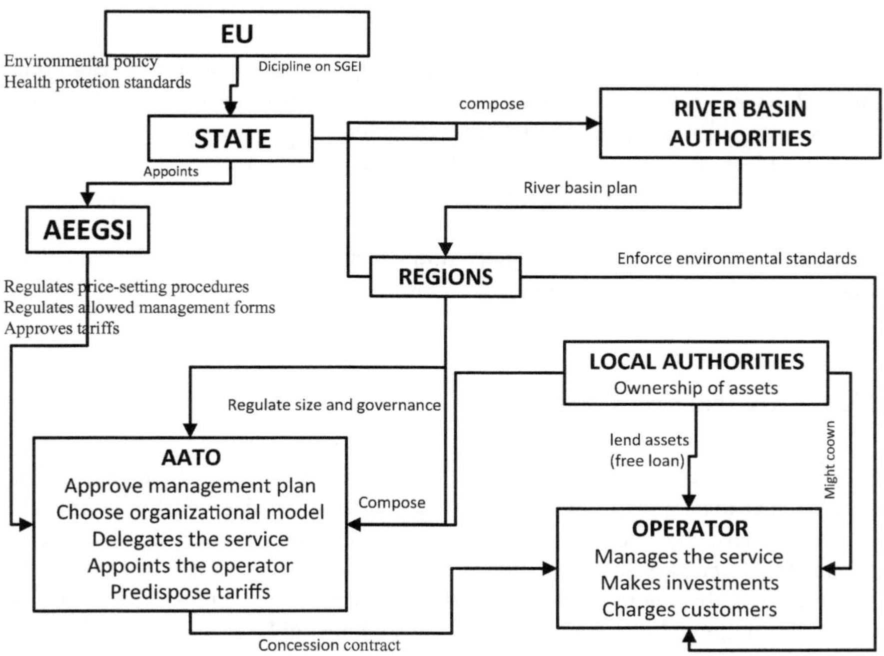

Fig. 11.1 Structure of the governance and regulatory system of the integrated water service (IWS) (Source: author elaboration)

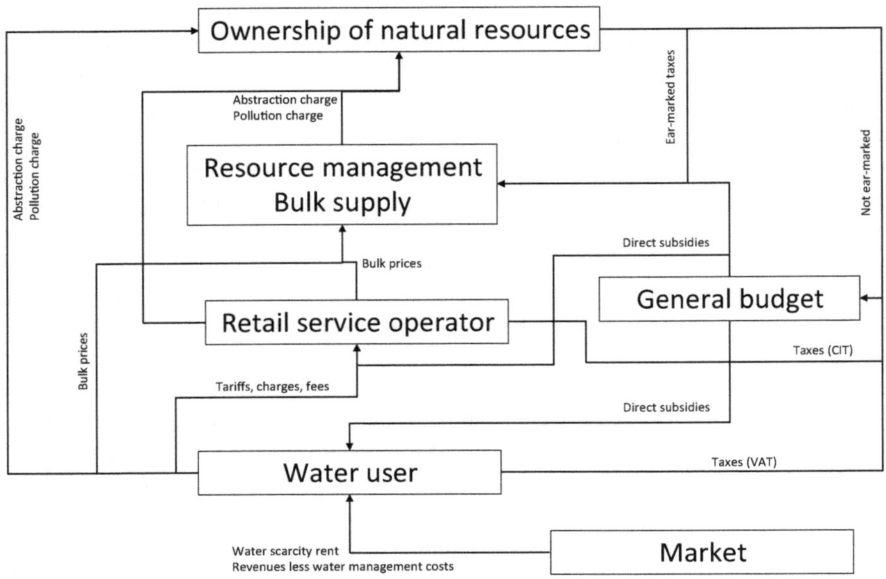

Fig. 11.2 The structure of financial flows characterizing the Italian water management system (Source: own elaboration)

suppliers or other retail operators. The same happens to bulk operators. Both retail and bulk operators have to recover their costs out of the revenues received from their clients, eventually complemented by state transfers.

The subject that extracts water from the natural environment (either the bulk supplier, the retail operator, or the final user directly) may be required to pay a charge to the resource owner (the state).

Finally, the state receives financial flows from taxation, but also finances the water sector through direct and indirect subsidies.

11.3 Abstraction Charges

All water abstractions—made either directly by users or by one of the intermediary subjects that manage bulk or retail systems—have to be licensed by the resource owner, namely, by the state. In the 1990s, resource ownership and related administrative powers have been devoted to regions.

Until 1994, this regime characterized surface waters only, while groundwater use was free and unregulated. Law 36/1994 extended the public domain to groundwater also. Therefore, at least nominally, groundwater abstractions have to follow the same licensing regime; however, the great number of individual abstractions—in the reach of tenths of thousands—makes the enforcement of this principle very difficult.

Abstraction licenses imply the payment of an abstraction charge. These are differentiated by sector. In the case of hydropower, charging principles reveal the clear intention of capturing at least a part of the economic rent. Rates are a function of nominal electricity generation capacity (a standard measure of the potential production), regardless of other characteristics of the site (quantity of water used, height of the dam, etc.). Hydropower producers have also to pay further fees to compensate local communities, which again depend on nominal capacity.

For all other uses, the reference unit of abstraction charges is the "module," corresponding in general to a volume of 100 l/s. Table 11.2 summarizes abstraction charges applied throughout the country—keeping in mind that each region can now set charge levels and application rules. Many regions have recently changed charging rules and rates or have announced the intention of doing so.

Translated in the correspondent amount per cubic meter, figures in Table 11.2 mean that the abstraction charge amounts to an overall negligible value, a fraction of a €c/m^3. On a national basis, our estimate of annual revenues provides a meaningful figure only for hydropower (in the range of 200–300 M€/year, which also includes local community compensations). Massarutto and Pontoni (2015) estimate that the share of the hydropower rent that accrues to regions and local communities lies in the range of 13–21 %.

Industrial charges generate another 40–50 M€. Revenues from other uses are negligible: both irrigation and public water supply generate less than 1 M€; still lower figures arise from other sectors of water use.

Overall, the abstraction charges do not represent at present neither a meaningful revenue source nor a serious incentive to water users. Proposals toward a comprehensive reform have arisen in many occasions, including recent reports of Oecd (2013). The consideration of environmental and resource costs of water use, which is foreseen by article 9 of the EU Water Framework Directive, but has never been implemented until now, offers a unique opportunity in this direction. The most promising option seems to be the adoption of a scheme that is similar to the French one, i.e., concerning a system of water taxes aimed at fueling the various public spending programs that concern water, e.g., in order to cofinance investments and avoid the need to rely entirely on market-based repayable finance.

In the IWS alone, for example, Massarutto (2015) calculates that a tax in the order of 0.10 €/m^3 could generate an annual cash flow of 600 M€, corresponding approximately to one-quarter of the annual investments actually planned. This tax could apply to abstractions from the natural resource and be passed-through only up to a standard level of allowed leakage, in order to provide an incentive to IWS operators. Moreover, its rate structure could take into account effluent quality and environmental costs of discharge, in order to penalize those with the lowest pollution abatement records.

Table 11.2 Abstraction charges in Italy (mod = 100 l/s)

						Irrigation		
	Hydropower	PWS	Hygienic	Fish farming	Industrial	No restitution	With restitution	Unmetered
	€/kW	€/mod	€/mod	€/mod	€/mod	€/mod	€/mod	€/ha
Average	13.4	1,664	2,824	264.2	9,646	35	8	0.33
Median	14.5	2,110	603		13,474	46	–	0.40
Max.	35.1	4,008	98,816		41,361	190	50	2.64

Source: Author elaboration on a direct inquiry

11.4 Experiences with Residential Water Pricing

11.4.1 Past Experiences with Residential Water Pricing

Prior to 1994, tariffs hardly allowed the recovery of operational costs (Malaman and Cima 1999). The public budget financed all investments. As a result, Italian residential water tariffs were extremely low and hardly noticeable in the family budget. On top of this, billing and revenue collection efficiency was often very poor, especially in the south, where most people actually did not pay anything and the very existence of updated records was questionable.

These circumstances reduced the practical importance of the pricing structure at the retail level. Metering was the rule—even if not everywhere. The pricing structure entailed a complex increasing block schedule, which is still in force today.

As for sanitation, sewage collection and treatment charges rely on a simple and uniform charge, proportional to the volume. The charge was set at the national level by budget law. At the launch of the reform, the sanitation charge was 170 and 500 ITL/m^3 (corresponding to 0.35 €/m^3 in total).

Law 36/94 introduced the principle of full-cost recovery for water and sanitation services supplied under the regime of the "integrated water service." As already mentioned, the IWS includes both water supply and sanitation for residential uses as well as business premises connected to the public service.

The reform left the tariff structure untouched; in turn, it aimed at an overall restructuration of the financial model: subsidies from the general budget should diminish drastically until disappearing, while the private capital market would sustain investments. In order to make this happen, tariffs would have to generate sufficient margins for repaying debt.

The state assumed the responsibility of designing a "normalized tariff method" (MTN) that each ATO would apply.

The MTN was adopted 2 years later (DM 1 August 1996). Its mechanism was rather simple. A first step concerned the calculation of *allowed costs*. These included both operational (Opex) and capital (Capex) expenditure in each ATO. The baseline Opex was set at the level reached by previous undertakings, possibly considering the effects of mergers and reorganization. Future dynamics depended on planned service extensions and efficiency improvements.

A price-cap mechanism required operational cost to diminish by at least 0.5 % per year, or more, according to the deviation from a standard cost obtained from an econometric formula. In case actual costs exceeded the benchmark for more than 20 %, higher price caps were imposed and a stricter authorization procedure was required. Some cost items were considered as exogenous and passed-through automatically (electricity, bulk water purchase, local taxes, and fees).

Municipalities were supposed to entrust existing facilities on a free-loan basis. Already existing financial obligations (e.g., pending loans) would be taken over by the operator (and passed on as operational costs) or reimbursed to municipalities. Concession fees were often introduced in exchange of the contribution of physical assets as IWS companies' equity and passed-through as well.

The regulatory asset base (RAB) included all investments made by the operator at the historical value, net of grants, and subsidies eventually received. This was intended on an ex ante basis, i.e., the basis was *planned investments*, saving an eventual negative ex post compensation in case actual investment was below schedule.

The RAB also included all assets originally owned by IWS companies (e.g., because they had been realized previously with their own funds or contributed as equity by parent municipalities). This provision created an uneven basis for departure: companies beginning their operation newly after the reform had in practice a zero asset base from the start, while already established ones could benefit from the cash flow generated by existing assets.

Allowable Capex included depreciation and remuneration of capital. Depreciation relied on accounting principles allowed by tax law. This provision made it possible—although not automatically—to adopt financial amortization schedules (i.e., with an economic life equal to the duration of the contract, regardless of the actual technical life of equipment). Capital remuneration consisted of a lump-sum pretax of 7 % applied to the RAB, net of depreciation; the MTN prescribed regular updates of this rate according to market conditions.

Once the allowed cost was defined, a tariff would be set in such a way to allow its full recovery, yet a smoothening mechanism would limit annual increases. In practice, the plan approved a "real average tariff" (TRM), on a €/m^3 basis. Operators would then translate it into the pricing structure, so total revenues would match.

The MTN required regular updates every 3 years (or earlier, in case that substantial unbalances arose). Reviews should assess the regular implementation of investment plans (and reduce tariffs accordingly in case of incomplete realization) and adjust allowed operational costs.

Seemingly, this scheme guarantees financial equilibrium. In practice, this did not necessarily happen. Substantial margins of discretional power were actually left to the implementation phase; regulators may force the interpretation of framework principles according to other priorities, and this was more likely to be the case when regulators are conditioned by political entities.

Massarutto and Ermano (2013) offer a thorough analysis of how discretional power of regulators could actually pledge tariff setting to other priorities.

First, the MTN allows adopting whatever depreciation schedule is in accordance with fiscal norms. In practice, this means that the chosen schedule may vary between the extreme of true economic life (which is actually very long for IWS assets) and the opposite extreme of financial amortization (i.e., following the duration of concession contracts).

In the former case, the actual impact on prices is low, but cash flows will be hardly high enough to cover financial expenses (given that bank loans will likely require shorter and tighter repayment schedules). Financial amortization, in turn, is more coherent with the requirement to repay loans (since the operator will be able to extinguish financial obligation before contract expiration); in turn, the impact on tariffs may be dramatic and difficult to sustain in political terms.

Just as an example, in Tuscany and Emilia-Romagna—two regions in which investment plans were actually more challenging and a coherent price regulation was adopted—prices reached rather soon the threshold of 2 €/m^3, more than double than prior to the reform.

Second, the allowed rate of return had been set provisionally at nominal 7 % pre-tax; the MTN prescribed regular updates following market conditions, but this never took place. In the first phase, it came out to be much lower than the market rate; after Italy joined the eurozone, however, the opposite became true. Investing in WSS assets would allow a "secure" return higher than borrowing rate. The boost of the global financial crisis in 2008 opened a new phase, with far higher rates than the allowed one.

Third, Opex revision soon proved a troublesome task. National benchmarking formulas were too imprecise and generic, unfit for the task of actually revealing efficiency levels. ATO plans relied on desktop calculations, often provided by external consultants using rules of thumb and replicating basic templates. Since the starting level was admittedly unreliable, the adoption of more realistic figures was postponed to the revision phase. Rather than a routine exercise, therefore, this became a crucial and delicate decision, for which the regulatory system was unprepared, in the lack of appropriate tools and procedures.

AATOs were in fact sovereign about the decision to review them (aligning to actual accounts) or force the implementation of initial calculations. In practice, the outcome could range from sticking to the initial calculation and refusing to review operational cost upwards to humbly adopting operators' accounts "out of pocket." Political opportunity, rather technical assessment, very often inspired the actual decision. The national supervisory committee, Conviri—a very weak office, understaffed, and lacking decisional autonomy—could only verify formal aspects, but not enter into the merits of figures.

Although price reviews were due every 3 years, many AATOs failed to do so because of the difficulty to reach the formal agreement of a large number of associated municipalities. As a result, reviews required a longer time, and substantial gaps between allowed and actual costs continued to occur.

According to the MTN, the outcome of the price regulation process was *a fixed average charge per cubic meter*. This corresponded to the total cost divided by the forecasted volumes. In case the latter were overestimated, a lower unit charge resulted, and this frequently was on purpose, in order to artificially soften the effective increase. Price reviews were supposed to provide an assessment of actual volumes: delaying the latter would thence allow postponing tariff increases.

Finally, the MTN was supposed to enter into operation once the reorganization of the IWS was complete. The prescribed time schedule of 12 months soon revealed to be unrealistic: it took in fact many years to see the first completed reorganizations, and more than one decade for the majority of ATOs. In order not to postpone cost recovery and to start at least with more urgent investments, an interim price regulation was approved in 1999. This was based on a simple price-cap rule to be defined each year. This incorporated an automatic incentive to efficiency complemented in some years by allowances aimed at sustaining urgent investment needs. Yet in most of the years, the national authority in charge of setting price limits decided to maintain a zero nominal price increase. This was technically argued through the definition of an X factor equal to the inflation rate, but it was clear enough that political willingness not to disturb electors with price increases was predominant.

11.4.2 From the MTN to the MTI

The MTN remained untouched until 2011, despite increasing evidence of its inadequacies. Paradoxically, its actual implementation managed to discontent everybody: low prices and unpredictable dynamics to sustain investments, but high enough to attract public concern and mobilize social protest. This culminated in the summoning of the 2011 referendum, triggered by alleged threats of privatization and successfully addressed abolishing the norm requiring IWS tariffs to include an "adequate remuneration of capital," which most voters understood as a guaranteed private profit over an essential service (Massarutto and Ermano 2013).

After the referendum, competences about water tariff regulation were transferred from the previous ministerial committee, Conviri, to an independent authority, already competent for electricity and gas supply (Autorità per l'energia elettrica, il gas e il Sistema idrico—AEEGSI).

Following a consultation phase, AEEGSI issued in December 2012 its regulatory norms valid for 2012–2013 (transitional period). The "transitional pricing method" (MTT) is in fact far less "transitional" than its label suggests, since it moves from the attempt to set up a general scheme able to fit most local situations, anticipating as much as possible the philosophy of the definitive rule.

The scheme is based (as for the MTN) on the identification of an allowed total revenue, corresponding to the sum of costs. The regulatory outcome, however, is expressed in terms of *allowed total revenue* and not on *unit charge*. Update of unit tariffs is automatic without a formal price review, therefore eradicating the bad habit of inflating planned volumes on purpose.

As far as operational cost is considered, they are divided into two categories (endogenous and exogenous), the former being considered as potentially influenced by operator's effort.

Endogenous costs are based on actual costs, as they appear in operators' 2011 accounts. This will represent the new starting basis. Afterwards, a systematic comparison between actual costs and benchmarking parameters will be adopted. Pending the calculation of new and more reliable standard costs, existing plans will serve as a reference.

For the first regulatory period, the AEEGSI gave up the attempt to use parametric formulas and decided to rely on previously forecasted costs as a basis of comparison. This allowed Opex to converge to the average between actual and forecasted cost. While this is reasonable, in the lack of more appropriate benchmarking instruments, it perpetuates the gaps eventually caused in the past by different attitudes of AATOs, since the planned cost has not necessarily been calculated in an appropriate way and gaps between actual and forecasted cost do not necessarily reflect operator's inefficiencies.

In the future, price reviews are expected to take place every 4 years. At the beginning of each regulatory period, actual operational cost of the previous period will be confronted with a benchmark, and efficiency-improving price caps will be introduced accordingly. A very detailed unbundled accounting system is being introduced, with the aim of allowing more effective comparisons between the costs of different service components and establishing more meaningful and reliable benchmarking formulas.

Exogenous costs (which include electricity, wholesale service local taxes, and contributions) are passed-through, yet a cap is placed on some items, based on national average market prices, such as, for example, in the case of electricity.

Overall, the MTT substantially maintains the previous approach about Opex, but provides a more predictable and automatic framework for regular updates, reducing the discretional power of AATOs.

The approach to capital cost regulation, instead, reversed completely the previous one (Table 11.3). The RAB is now based on *existing physical assets* calculated on an ex post basis, whatever their ownership and whatever the source of funding.

For this purpose, existing assets are stratified according to the year of realization, and values are systematically updated with inflation so as to correspond to their net reconstruction value. On the other hand, depreciation schedules are now calculated on the basis of true expected economic life. New investments enter the RAB with a 2-year time lag (i.e., an investment realized in year t will be considered in the regulatory cost starting from year $t+2$).

Therefore, depreciation costs are considered for all assets, including those that have not been financed by the operator. However, cash flows arising from public funds or from assets owned by municipalities will be set aside in a fund that can be used for new investments or social purposes (the so-called fund for new investments, FoNI).

The regulatory rate of return is based on a calculation that follows the capital asset pricing model (CAPM), namely, considering the risk-free rate plus a risk premium, which is calculated on the basis of market data. An extra bonus of 1 % is added, as a lump-sum compensation for the time lag of 2 years.

Referendum promoters, on the claim it would betray the results of the popular vote, challenged this provision, but, in April 2014, the court rejected their appeal.

In 2013, the AEEGSI issued the "definitive" price regulation package (MTI). It maintained most of the provisions of MTT, with some important differences. On the first place, the MTI recognizes the need to apply differentiated approaches to fit different situations. For this reason, it identifies four "regulatory menus," depending on two circumstances.

The first one concerns whether or not the already existing asset base is sufficient to sustain the planned volume of investments. In case it is not, the "ordinary" Capex is complemented by the possibility to generate accelerated cash flows through (1) the adoption of shorter depreciation schedules and (2) the provision of an "anticipation for new investments," which is set aside to the FoNI.

The second depends on whether the actual operational cost reflects or not the cost of the standard management system. In case it does not, AATOs can define a new level with substantial freedom. Additionally, the MTI introduces an automatic compensation scheme, in case actual revenues are different from allowed ones or in case of variation of exogenous costs.

Another important innovation consists in procedural aspects. While AATOs maintain (and even improve) their discretional powers, operators can bypass them

Table 11.3 Comparison of capital cost accounting criteria in MTN, MTT, and MTI

	Normalized tariff method (MTN)	Transitional tariff method (MTT)	Definitive tariff regulation (MTI)
Period	1996–2011	2012–2013	>2014
Operational costs (Opex)	Estimated through a desktop study	Based on 2011 accounts and Opex admitted by previous regulation	Opex = average between 2011 accounts and Opex admitted by previous regulation
	Revision admitted but not regulated in detail	Opex converge to 2011 accounts or Opex admitted by previous regulation depending on specific circumstances	Possibility to define a new OP in case of structural change in service
Efficiency gains	Price cap based on benchmarking formula	No incentives in the transition period	No incentives in the I period
		Announced for the future	Announced for II period
Pass-through costs	Electricity, bulk water, local charges, and taxes	As for MTN	As for MTN
			Electricity cost is passed-through, within the limit of average market price * actual consumption
Asset base	Assets already owned by operators at book value	Assets already owned at reconstruction cost	As for MTT
	New investment made by operator at historical cost, anticipated according to the contract (compensation ex post on a triennial basis)	New investment at reconstruction cost (only actually realized investments after a time lag of 2 years)	Additional provision (anticipation for new investments) foreseen in case RAB < than a certain fraction of investment needs
		Assets owned by municipalities at reconstruction cost (cash flow set aside to the FoNI)	
Grants received	Not included	Included (depreciation only), set aside to the FoNI	As for MTT
Depreciation	Any schedule admitted by tax legislation	True economic life	As for MTT
Financial amortization	Allowed	Not allowed	Allowed in case the RAB <50 % of investment needs
Rate of return	Lump-sum rate (7 %) on all investments sourced by the operator (on historical cost basis)	Market-based rate on all investments sourced by the operator (on revalued historical cost)	As for MTT
		Same rate applied to assets owned by municipalities (revenues set aside to the FoNI)	
		Further lump-sum (1 %) for new investments to compensate the time lag	

Source: Author elaboration

in case of inaction or disagreement and appeal autonomously to the national regulator. This provision will give AATOs the possibility to tailor solutions to the local situation, but at the same time to prevent the abuse of political discretion.

The MTI also introduces some norms with the aim of rationalizing the pricing structure. However, what it does in practice is to reduce the room for a discretional definition of parameters while confirming the increasing block structure. Fixed charges cannot imply free allowances and should not exceed 20 % of total revenues for each category. Subsidized rates can apply only to residential uses. The number of categories can diminish but not increase—with the clear attempt to promote a simplification of the whole structure.

11.4.3 Tariff Structure

The tariff structure for IWS remained substantially the same since it was first regulated in 1974.

The water supply charge includes a fixed charge, a subsidized block (for residential clients only), an average block (tariff base), and up to three upper blocks with an increasing unit charge. Dimension of blocks can vary, while different schedules apply to different use categories (e.g., domestic, second houses, commercial, industry, etc.). Essential water endowments and poor households are entitled to rebates and special subsidized charges. Public uses (e.g., fire protection, hospitals, street cleaning, public buildings) have dedicated (and subsidized) charges.

Although metering is the general norm, there are still cases of (individual and collective) unmetered customers, whose tariffs are calculated on a flat basis, possibly taking into account some indicator of water quantity, such as the diameter of the pipe.

It is difficult to provide a picture that summarizes the whole country, since these general rules apply in very different ways across the country. The number of different tariff schemes can be very high (up to 10–20 different types, according to the category of use). The size of blocks also varies significantly. Table 11.4 illustrates some basic figures, derived from a sample of ATOs.

Table 11.4 Structure of blocks and charges applied in 2013

	Subsidized		Base		I block		II block		III block
	€/m³	Up to (m³)	€/m³	Up to (m³)	€/m³	Up to (m³)	€/m³	Up to (m³)	€/m³
Average	0.44	88	0.81	166	1.33	233	2.00	339	2.81
Max.	1.12	131	1.49	274	2.89	390	4.08	520	5.15
Min.	0.00	20	0.24	48	0.41	96	0.84	144	1.14
n. ATO		41		40	40		33		15
Population		27.6		27.5	27.5		24.2		15.6

Source: Utilitatis (2014)

Charges for sewage collection and treatment follow a much simpler schedule, since they apply a uniform volumetric charge to all uses.

Recently, the AEEGSI has announced the intention of promoting a rationalization, by reducing the number of categories, more uniform block size, etc. The fixed charge should not exceed the 20 % of total revenues, while flat tariffs should progressively disappear.

Table 11.5 provides examples of tariff schemes applied to some categories of uses in four case studies that have been appositely analyzed for the present study.

11.4.4 Tariff Dynamics and Affordability

After the introduction of MTN in 1996, tariff dynamics have been rather impressive, moving from 0.97 (the average tariff in year zero) to 1.37 €/m^3 in 2010 for water supply and sanitation (Utilitatis 2014).

The growth of expenditure is much larger, since 0.97 €/m^3 already includes some of the increases introduced by interim tariff regulations during the transition phase. Actualized estimates of the aggregate industry annual revenues in the pre-reform era were 3.37 billion € (Malaman and Cima 1999); the same aggregate in 2010 came to 7.61 billion € (Utilitatis 2014). Hence, a first apparent outcome of the reform is that tariff revenues more than doubled, with a net increase of 4.14 billion €/year.

After the introduction of AEEGSI methods (MTT and MTI), the price increase trend continued. Since AEEGSI has not completed the procedure of approval of all tariff proposals submitted by AATOs, only partial results are available. Setting 2011 tariffs as the starting level ($t_{2011} = 1$), the average index grew to 1,024 in 2012 and 1,058 in 2013. The final figure will be probably higher, since the published data concern only ATOs with an annual increase below the threshold of 6.5 %, while those asking for higher increases are subject to a more detailed inquiry.

Future dynamics is expected to be rather impressive, as well. ATO plans foresaw an overall average tariff of 1.46 in equilibrium (after the full deployment of investment plans). Yet these were only the initial forecasts: after the first interim reviews, planned tariffs were revised and further increased in order to finance investments and guarantee balance-sheet equilibrium.

Although again no systematic data are available on a national basis, evidence from selected case studies shows that the MTI implies a much higher price increase—for the same planned investment—than the MTN, especially where the previous regulation had not opted for financial amortization (Massarutto 2015).

Table 11.6 illustrates the average expenditure of Italian households, calculated with the application of the pricing structure approved by each ATO. Two household typologies are considered, a single-person family (60 m^3/year) and a family of three persons (150 m^3/year). The national average is 99 and 242 €, respectively, with a minimum in the northwest and a maximum in the center. The variability is arguably

Table 11.5 Examples of tariff structures in some case studies in 2012

			Roma		Milano		Firenze		Sardinia	
			Tariff	Up to	Tariff	Up to	Tariff	Up to	Tariff	Up to
				m^3/person		m^3/apartm		m^3/person		m^3/person
Resident	Fixed charge	€/year	22.66		2.83–25.13		30.05		12–15	
	Subsidized block	€/m^3	0.17	92			0.37	60	0.36	70
	Baseline block	€/m^3	0.56	184			1.28	150	0.80	140
	First block	€/m^3	1.00	276	0.11	128	2.74	200	1.30	200
	Second block	€/m^3	2.03	368	0.23	274	4.08	∞	1.88	250
	Third block	€/m^3	3.98	∞	0.39	∞		∞	2.60	∞
	Sewage collection	€/m^3	0.16	∞	0.11	∞	0.46	∞	0.17	∞
	Sewage treatment	€/m^3	0.47	∞	0.28	∞	0.64	∞	0.41	∞
	Solidarity contribution	€/m^3	0.01							
Poor	Fixed charge	€/year			2.83–25.13					
	Subsidized block	€/m^3							0.17	70
	Baseline block	€/m^3			0.07	∞			0.37	140
	First block	€/m^3							1.30	200
	Second block	€/m^3							1.88	250
	Third block	€/m^3			0.11	∞			2.60	∞
	Sewage collection	€/m^3			0.11	∞			0.17	∞
	Sewage treatment	€/m^3			0.28	∞			0.41	∞

(continued)

Table 11.5 (continued)

			Roma		Milano		Firenze		Sardinia	
			Tariff	Up to	Tariff	Up to	Tariff	Up to	Tariff	Up to
				m³/person		m³/apartm		m³/person		m³/person
Nonresidents	Fixed charge	€/year	22.66						50	
	Baseline block	€/m³	0.56						0.80	140
	First block	€/m³	1.00						1.30	200
	Second block	€/m³	2.03						1.88	250
	Third block	€/m³	3.98						2.60	∞
	Sewage collection	€/m³	0.16						0.17	∞
	Sewage treatment	€/m³	0.47						0.41	∞
	Solidarity contr	€/m³	0.01							
Non-household	Fixed charge	€/year	4.7–41.96		2.83–25.13		30–360		50–200	
	Baseline block	€/m³	0.62		0.38	∞	1.28	∞	1.45	100
	First block	€/m³	1.00		–				1.88	200
	Second block	€/m³	2.03		–				2.60	∞
	Third block	€/m³	3.98		–					
	Sewage collection	€/m³	0.16	∞	0.11	∞	0.46		0.17	∞
	Sewage treatment	€/m³	0.47	∞	0.28	∞	0.64		0.41	∞
	Solidarity contr	€/m³	0.01							

Source: Direct survey

Table 11.6 Average expenditure for IWS (water and sanitation) in 2011

	% residential	Per capita consumption		Annual expenditure		Average expenditure	
		m³/year	l/day	€/year		€/m³	
				60 m³	150 m³	60 m³	150 m³
Northwest	72	75	205	85	208	1.41	1.39
Northeast	69	59	162	101	256	1.68	1.71
Center	71	57	156	111	262	1.84	1.75
South	79	49	134	103	247	1.71	1.65
Islands	77	52	142	103	253	1,72	1.68
Italy	74	60	164	99	242	1.65	1.61

Source: Our elaboration on Utilitatis (2014)

Table 11.7 Distribution of annual expenditure for a standard consumption of 150 m³/year

Annual expenditure (€/150 m³)		Number of systems	Population	
From	To		million	%
0	150	1	1.34	4.6 %
150	200	6	5.54	19.1 %
200	250	12	8.55	29.4 %
250	300	10	7.22	24.9 %
300	350	9	3.19	11,0 %
>350		6	3.19	11.0 %
		44	29.02	100.0 %

Source: Our elaboration on Utilitatis (2014)

Table 11.8 Indicators of affordability of water and sanitation services (IWS)

	% of IWS on average annual expenditure on total consumption	Incidence of IWS expenditure on the average income poverty line (%)
60 m³	0.47	1.39
150 m³	0.72	1.53

Source: Our elaboration on Utilitatis (2014)

influenced by technical features, such as the lower energy requirements in the north, thanks to gravity pumping, and the higher/lower density of customers along the network.

Table 11.7 provides further details by showing the distribution around the average. The 4.6 % of population (1 ATO) spends in the range between 0 and 150 €/year. The most numerous classes lie in the median range, with nearly 30 % of the population (12 ATOs) spending an average of 200–250 €/year. Further, 22 % of Italians (15 ATOs) spend no less than 300 €/year.

Table 11.8 finally illustrates the impact of water tariffs on families in terms of affordability. The first indicator (share of IWS expenditure on total family consumption) shows that IWS expenditure is still quite modest and far below the affordability thresholds that are commonly proposed in the international literature (3 % on average). In turn, the second indicator shows some disquieting information concerning the impact on the poor. Families whose income is equal to the poverty line spend on average 1.39–1.53 % of their income on IWS. This suggests the need to consider specific subsidies to poor families, which by no means should afford IWS alone, since the impact of price increases in other utilities (electricity, gas, transportation) is even more relevant (Miniaci et al. 2008).

11.5 Experiences with Other Sectors

11.5.1 Experiences with Irrigation Pricing

Reclamation boards (RBs), which supply collective irrigation water, are not precisely equal to "service providers." They are in fact private associations, ruled by boards that represent landowners. Charges paid by associations are more similar to "condominium fees" than to tariffs. On top of these, RBs may obtain further revenues from the market (e.g., from the sale of electricity produced by hydropower plants located along the distribution network).

Accounting rules generally follow cash flows rather than accrual criteria. Legislation obliges consortia to reach annually a balance between revenues and expenses, although public institutions may contribute grants and subsidies that are registered in the accounts. In the past, this allowed many RBs to elude cost recovery provisions, since public contributions constituted in practice systematic annual bailouts. Currently, budget equilibrium enforcement is stricter, especially for operational costs. Table 11.9 illustrates the result of an original study we conducted on a sample of 14 RBs, located in nine regions. Accounting data have been normalized and translated in a reclassified profit and loss account. A negative gross operational margin earnings before interest, tax, depreciation and amortization (EBITDA)

Table 11.9 Average normalized profit and loss accounts (operational cost = 100) for a sample of reclamation boards

	NW	NE	C	S + I	Italy
Fees paid by associates	91	93	101	56	87
Other revenues	19	13	10	13	13
Operational cost	100	100	100	100	100
EBITDA	9	6	11	–31	–0
Use of set-aside provisions	7	23	–	2	11
Depreciation and provisions	2	24	–	2	11
Net capital costs	1	0	0	3	1
EBIT	14	5	11	–34	–1

Source: Author elaboration on direct inquiry (*NW* Northwest, *NE* Northeast, *C* Center, *S* South, *I* Islands)

means that direct revenues (from associates and market activities) do not allow breakeven. This situation still occurs in the south and island regions, while in northern and central Italy, margins are positive, witnessing the capacity to self-finance at least a share of capital expenditure. User charges generally allow recovery of maintenance expenses, while public contributions fund new investments.

On the other hand, the construction of the Italian irrigation network took place along a period of many centuries; most of it is fully amortized now. New investments do not fund extensions of irrigated surfaces and, by no means, imply further abstractions; rather, they concern incremental improvements of water use efficiency (e.g., substitution of open-air canals with pumped pipelines, introduction of drip irrigation and sprinklers to replace submersion, maintenance of river corridors, greening of water infrastructure, and so on).

In fact, absolute water volumes actually used by agriculture are seemingly declining, in line with the overall reduction of agricultural activity. Estimates provided by past studies (IRSA-CNR 1999) considered theoretical requirements and licensed volumes rather than effective consumption and actual abstractions. Evidence from river district plans shows that consortia actually use only a fraction of licensed use rights.

Metering and volumetric charges are still exceptional in the north, while elsewhere they are more diffused than in the past, particularly where water resources are scarcer (e.g., in Emilia-Romagna and in the south). Elsewhere, associates pay a fee based on irrigated surface. However, this does not take into account water demand. Per ha fees can be differentiated according to cropping choices; guaranteed supplies and water on demand may imply extra charges. Although no systematic studies exist, evidence from case studies shows that the state of the art, although nonoptimal according to orthodox economic theory, is not completely unreasonable, given that significant investments would be required in order to adopt metering on a systematic basis, and these are not necessarily justified.

Unfortunately, there are no recent systematic surveys of water charges for irrigation. The latest published study dates back to the 1990s and is not representative of today's situation. Table 11.10 provides the result of an original study we have carried out using the database collected by INEA (the National Institute of Agricultural Economics). The database, still under construction, collects structural and economic information for each consortium. Although the survey is still incomplete, it is useful for a general overview. At present, it covers 92 consortia (out of 136) and an irrigated surface of 1.5 million ha (57 % of the total).

Where surface charges are applied, the average value is around 120–130 €/ha, with high fluctuations either among areas or within each area. Binomial charges typically entail a fixed charge (68 €/ha on average, again with significant fluctuations) and a variable charge, whose value is again quite variable. Only in the islands we have found values around 1.5 €/m^3, while elsewhere the typical charges are 0.2–0.3 €/m^3 or lower.

Massarutto (2003), for example, argues that most crops are actually not very responsive to marginal price, at the existing water price level, given the high-value-added crops. A case study in Friuli (northeast) shows that the frequency of drought

Table 11.10 Average, minimum, and maximum irrigation charges in 2012 – breakdown per macro-regions

	Northwest	Northeast	Center	South	Islands	Italy
Total surface associated to reclamation boards	949.410	3.805.119	2.362.702	3.916.712	1.148.181	12.182.124
Of which irrigated	58 %	15 %	6 %	5 %	5 %	13 %
Irrigation technology						
Submersion	80 %	40 %	17 %	14 %	12 %	48 %
Sprinklers	19 %	49 %	71 %	42 %	64 %	38 %
Drip	1 %	12 %	12 %	44 %	23 %	14 %
Water distribution technology						
Gravity	91 %	64 %	60 %	63 %	45 %	76 %
Pumped	9 %	36 %	40 %	37 %	55 %	24 %
Water use						
Average (m^3/ha)	8.226	4.078	3.765	4.823	5.555	4.931
Length of irrigation period (days/year)	141	164	188	208	196	180
Availability						
On demand	26 %	65 %	96 %	48 %	60 %	51 %
By turns	74 %	35 %	4 %	52 %	40 %	49 %

Charging method						
Surface	39 %	49 %	37 %	50 %	41 %	45 %
Volumetric (binomial)	39 %	49 %	37 %	50 %	41 %	45 %
Mixed	21 %	3 %	27 %	0 %	19 %	10 %
Charges per ha (surface only)						
Average	123	78	140	169	220	127
Min.	35	17	55	45	170	17
Max.	304	220	400	500	270	500
Charges per m^3 (binomial)						
Average fixed charge per ha	82	67	36	44	178	68
Average charge per m^3	0.12	0.24	0.14	0.20	1.57	0.31
Min. charge per m^3	0.00	0.02	0.01	0.01	1.56	0.00
Max. charge per m^3	0.24	0.86	0.22	0.40	1.57	1.57

Source: Our elaboration on INEA

events should be lower than one every 3–5 years to justify a systematic change of actual patterns of agricultural water use.

On the other hand, we must say that the use of economic instruments is still in its infancy. Many studies argue that incentive pricing for irrigation cannot automatically induce more sustainable patterns of use, while superior results could arise from a combined use of different economic instruments, such as water markets and insurance schemes (Mysiak et al. 2013; Cornish et al. 2004; Massarutto 2003).

In the Italian context, this is particularly true, especially if we consider that irrigation-driven water stress is not necessarily linked to high water consumption but rather to the intensive use of water in high-value crops in water-stressed subregions, as happens in the southern reach of the Po basin (Massarutto and de Carli 2009; Viaggi et al. 2010). Poor design and scant political acceptance hamper at present a more widespread use of economic instruments.

We can argue that agriculture—as for IWS—awaits a more widespread use of economic instruments, but it is more for the sake of increasing the level of self-financing than to provide incentive to a more efficient use of water for irrigation. On the other hand, the problem of unsustainable extractions and guarantee of environmental flows seem to require institutional instruments (stakeholders' cooperation) rather than exclusively using economic instruments (water pricing and markets). Nonetheless, economic instruments could have a further role to play in the design of compensation schemes that could alleviate the burden of measures aimed at improving sustainability and reallocating water endowments. Evidence shows that willingness to pay of farmers, especially in the high-value-added areas, is much higher than actual charges, whereas the capacity of the public budget to continue supporting investments is diminishing.

11.5.2 Experiences with Industrial Pricing

As discussed above, water services dedicated to industrial premises may be a part of the IWS or as separate activities. The latter case represents the least known part of the Italian water industry, with lack of systematic surveys. Evidence on a spot basis shows that these undertakings operate on a cost recovery base, even if they might have benefitted from some public funds in the past, especially at the time of the initial investment, through direct injection of subsidies, soft loans, etc.

Industrial premises connected to the IWS pose, in turn, a number of issues that have recently attracted the attention of the national regulator.

A first important issue concerns the case for cross-subsidies. This is generally not the case for water supply. We have already pointed out that industries for which water represents an input in the production process normally rely on self-supply from direct abstractions, for which they pay the abstraction charge, but do not receive a service. Industrial and commercial premises connected to the IWS are normally doing so for sanitary purposes. This justifies treating them as any other commercial premise. In turn, the national legislation explicitly foresees the possibility to introduce a cross-subsidy in favor of domestic uses and especially for low-income customers.

For industrial sewerage, the pricing structure is rather different from civil uses. According to Presidential Decree 24 May 1977, the formula for calculating industrial charges was a function of pollution potential (Eq. 11.1):

$$T_2 = F_2 + \left[f_2 + dv + K_2 \left(\frac{O_i}{O_f} d_b + \frac{S_i}{S_f} d_f \right) + da \right] V \tag{11.1}$$

with T_2 = tariff; F_2 = fixed charge; f_2 = unit cost of collection; dv = average cost of primary treatment; O_i, S_i = chemical oxygen demand (COD) and suspended solids (SS) of the concerned effluent; O_f, S_f = total COD and SS treated in the facility; and K_2, da = parameters capturing special features. Regions, which inherited regulatory functions, often introduced further parameters.

This scheme was supposed to apply to each treatment facility. This favored a wide differentiation of tariffs for the same effluents, even in the same territory. Figure 11.3 provides an example of the range of variability throughout the country. While a difference among sectors is normal, given the different polluting potential, differences within the same industry is entirely due to the variability of cost between different facilities.

While being originally inspired to the polluter pays principle, this formula has encountered criticism for many reasons. First, it does not take into account technological change that occurred since 1977, charging the same price regardless of the efforts aimed at reducing pollution (thence, contradicting the PPP). Second, charges are specific for each installation, with the result of generating rather different tariffs for similar effluents even in the same territory. Third, the structure does not include any fixed charge, resulting in an unfair pattern of cost allocation. Furthermore, the

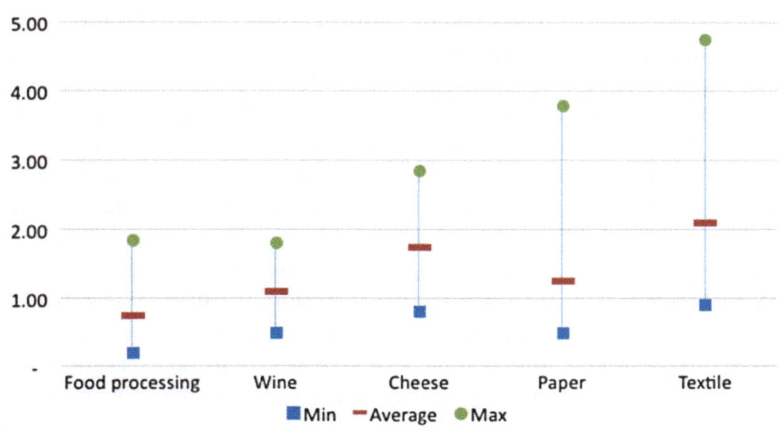

Fig. 11.3 Range of variation of industrial sewerage charges for selected industries in a sample of ATOs in 2010 (€/m^3) (Source: REF Ricerche 2014a)

same rate applies to collection and treatment, which is probably unfair (collection has the same cost, regardless of pollution).

The AEEGSI has proposed a uniform approach. The new tariff should apply the same rates within any ATO and will apply a uniform rate for collection and a specific one for treatment, considering pollution abatement costs in a more effective way.

11.6 Current Debates and Future Directions

The policy debate about water pricing in Italy has been dominated in the last 20 years by financial considerations, leaving the incentive effect of tariffs in the background.

Despite the efforts financial issues—namely, creating the base for a self-financing water industry, able to approach private capital markets—have not been definitively solved the long-run sustainability of the water industry, which faces an enormous investment effort, is still questionable, despite the undeniable progress allowed by the introduction of the MTI (Massarutto 2015).

With particular reference to IWS, the policy approach of the 1994 reform created self-standing undertakings, each able to sustain its own financial obligation, but with shortcomings implying industry concentration. This did take place, but too slowly and not necessarily in the right direction. The creation of ATOs aimed at unifying urban and rural areas under the same roof proved to be insufficient for allowing weak areas to sustain investments. The present debate considers alternative strategies, for example, the creation of mutual funds derived from earmarked water taxes.

Regarding the use of tariffs for providing incentives, the debate about reforming water pricing structures is still confined to academic audiences and, at best, informs the policy recommendations issued by multilateral institutions. Proposals have been made, for example, to introduce more explicit incentive schemes, such as lump-sum rebates on fixed charges in order to promote water saving or pollution abatement. Installation of household equipment has been found to be more sensitive to capital incentives than to marginal savings at variable cost (Conte et al. 2011).

Affordability and water poverty are not yet a real issue at present, since annual family expenditure is still rather low, compared with other EU countries, and one of the lowest in the OECD. However, projections of further increases show that this might not be true in the future, once all investment costs are transferred to consumers. Actual tariff arrangements foresee the possibility of adopting subsidized charges for the poor, but there will probably be a need for more effective action to address possible cases of water poverty. Financial schemes aimed at averaging out at least partially per capita costs among different territorial units may also become necessary (Massarutto 2015).

However, the regulatory priority seems to be that of influencing operators' choices and investment policies. For example, the AEEGSI has considered the possibility of rewarding priority investments and merit expenses with bonuses on investment remuneration (AEEGSI 2013).

Others have proposed to introduce penalties and rewards, based on the actual achievement of quality targets (REF Ricerche 2014b; Conte et al. 2011). An opportunity in this sense is offered by the consideration of environmental and resource costs. These could be charged to the operator, whereas their transfer in the water bill may be limited according to the policy objectives (e.g., transfer of only a given part of the abstraction charges, corresponding to the target level of leakage).

Another promising innovation concerns the use of water taxes, based on abstractions and/or pollution, either as an incentive to water users or as a complementary source of finance (Andersen et al. 2011; Oecd 2013).

References

AEEGSI. (2013). Fabbisogno di investimenti e individuazione degli strumenti di finanziamento per il raggiungimento degli obiettivi di qualità ambientale e della risorsa idrica, Primi orientamenti; Documento di consultazione 339/2013/R/IDR, www.autorita.energia.it

AEEGSI (Autorità per l'energia elettrica, il gas e il Sistema idrico). (2014). *Relazione annuale 2014*. www.autorita.energia.it

ANBI (Associazione Nazionale Bonifiche e Irrigazione). (2004). *Indagine sull'irrigazione nei comprensori di bonifica e irrigazione*, Roma.

Anderson, M. S., Speck, S., & Mautone, O. (2011, December 15). *Environmental fiscal reform: Illustrative potential in Italy*. Prepared for the conference: Environmentally related taxation and fiscal reform. Roma: Ministry of Economy and Finance. http://www.dt.tesoro.it/export/sites/sitodt/modules/documenti_it/eventi/eventi/EEA_Briefing_Note_for_ETR_Workshop_Rome_finaldraft.pdf

Civita, M. V., Massarutto, A., & Seminara, G. (2009). Groundwater in Italy: A review. *Atti dei Convegni Lincei, 248*, 13–74.

Conte, G., Bragalli, C., Branchini, S., de Carli, A., Lenzi, C., Masi, F., Massarutto, A., Pollastri, M., Principi, I., Bolognesi Andersen, M. S., Speck, S., & Mautone, O. (2011). Environmental fiscal reform. Illustrative potential in Italy. Innovative urban water management as a climate change adaptation strategy: Results from the implementation of the project Water Against Climate Change (WATACLIC). *Water 2012, 4*(4), 1025–1038. doi:10.3390/w4041025.

Cornish, G., Bosworth, B., Perry, C., & Burke, J. (2004). Water charging in irrigated agriculture. An analysis of the international experience. *FAO water reports* n. 28. ftp://ftp.fao.org/agl/aglw/docs/wr28e.pdf

EEA (European Environment Agency). (2012). *European waters: Current status and future challenges* (Technical report 9). http://www.eea.europa.eu/publications/european-waters-synthesis-2012

ISTAT. (2012). Giornata mondiale dell'acqua. Le statistiche dell'Istat, file:///C:/Users/Massarutto/Downloads/Le%20statistiche%20dell%E2%80%99Istat%20sull'acqua%20-%20 21-mar-2012 %20-%20Testo%20integrale.pdf.

Irsa-CNR. (1999). *Un futuro per l'acqua in Italia* (Quaderni di ricerca 109/1999). Roma: IRSA-CNR.

Malaman, R., & Cima, S. (1999). *L'economia del servizio idrico*. Milano: Franco Angeli.

Massarutto, A. (2003). Water pricing and irrigation water demand: Efficiency vs. sustainability. *European Environment, 13*(2003), 100–119.
Massarutto, A. (2015). *Devil and the details. Italian water pricing reform between technical rules and political will*. Paper prepared for the Jean Monnet module on water utility management, University of Pisa, 5 June.
Massarutto, A., & de Carli, A. (2009). I costi economici della siccità: il caso del Po. *Econ fonti di energia e dell'ambiente, 2*, 143–152.
Massarutto, A., & Ermano, P. (2013). Drowned in an inch of water. How poor regulation has weakened the Italian water reform. *Utilities Policy, 24*, 20–31.
Massarutto, A., & Pontoni, F. (2015). A parametric estimation of the hydropower rent in Italy. *Energy Policy, 78*, 31–40.
Miniaci, R., Scarpa, C., & Valbonesi, P. (2008). Distributional effects of price reforms in the Italian utility markets. *Fiscal Studies, 29*(1), 135–163.
Mysiak, J., Gòmez, C. M., Andersen, M. S., Vollaro, M., Defrance, P., Green, C., & Dworak, T. (Eds.). (2013). *Evaluating economic policy instruments for sustainable water management in Europe* (EPI-Water synthesis report). http://www.feem-project.net/epiwater/docs/epi-water_DL_5-1.pdf
OECD. (2013). *Environmental performance review of Italy*. Paris: OECD.
REF Ricerche. (2014a). *Tariffe per gli usi industriali: le questioni in attesa di soluzione* (Laboratorio Servizi Pubblici Locali, Quaderno n.18). www.refricerche.it
REF Ricerche. (2014b). *Incentivare la qualità del servizio idrico: premi, penalità, controlli* (Laboratorio Servizi Pubblici Locali, Quaderno n.17). www.refricerche.it
Utilitatis. (2014). *Blue Book. I dati sul servizio idrico in Italia*. Roma: Utilitatis. www.utilitatis.org
Viaggi, D., Raggi, M., Bartolini, F., & Gallerani, V. (2010). Designing contracts for irrigation water under asymmetric information: Are simple pricing mechanisms enough? *Agricultural Water Management, 97*(9), 1326–1332.

Chapter 12
Water Pricing in Mexico: Pricing Structures and Implications

Hilda Guerrero-Garcia-Rojas, Faustino Gómez-Sántiz, and J. Refugio Rodríguez-Velázquez

Abstract Mexican water price structure is set to reflect water availability and its economic value. Considering this framework, nine water availability zones have been established, in which the highest price is paid for zones with scarce amounts of water, and the lowest price is paid in zones with an abundance of water. Additionally, different tariffs have been established according to sectorial users, such as industry, households, and agriculture. This chapter develops a brief framework for water management in Mexico as a context for analyzing the pricing system of water actually used in Mexico. Also, the chapter briefly describes payment for environmental services—hydric (PES-H), as an instrument of environmental policy, because of its effects on pricing water from a forest conservation perspective. We conclude that although the water pricing system depends on water availability, the application of intra-regional tariffs for consumption (the largest water user being the agricultural sector) encourages irrational use due to subsidies applied to consumption.

Keywords Mexico • Payment for environmental services • Sustainable development • Water availability zones • Water management

12.1 Introduction

Water has influenced different aspects of Mexico's social and economic development. For the past 90 years, the increasing water use in cities, industries and agriculture has based its growth on expansion of the hydraulic infrastructure, as well as on the establishment of different policies to assure proper water management. The number of conflicts over water has already increased as a consequence of the continued population growth and urbanization. These conflicts occur between urban

H. Guerrero-Garcia-Rojas (✉) • F. Gómez-Sántiz
J.R. Rodríguez-Velázquez
Facultad de Economía, Universidad Michoacana de San Nicolás de Hidalgo (UMSNH),
Calle Juan de Villaseñor no. 100 Col. Nueva Valladolid, CP 58190
Morelia, Michoacán, Mexico
e-mail: hilda.guerrerogr@gmail.com

A. Dinar et al. (eds.), *Water Pricing Experiences and Innovations*,
Global Issues in Water Policy 9, DOI 10.1007/978-3-319-16465-6_12

and rural users, among neighboring cities and, more commonly, among neighboring states and regions.

The problems requiring a new approach to water management have been diverse; they include inefficient water use practices, deteriorating quality in water bodies, equity issues among those who have access to water services and those who have not, reduction in level of water services as consequences of inadequate maintenance, as well as a lack of organizational capacity for providing these services. In addition, water was, and still is, seriously underpriced carrying out inefficiencies in the resource allocation of its most beneficial use, as well as affecting the quantity and quality of water services expected by the population and their economic activities. In the early 1990s, the Mexican government introduced a number of structural reforms concerning the water sector and the management of the national water resources. Legal and institutional modification took place, and a series of strategies were implemented with a view of reverting negative trends. These transformations produced a significant impact on water consumers.

The purpose of this chapter is to develop a brief framework for water management in Mexico in order to analyze the water pricing system actually used, as well as the principal water uses. In the last section, we include the issue of payment for environmental services—hydric (PES-H) as an instrument for environmental policy, which has in the last 10 years allowed Mexico to develop negotiations between users of water resources, the state, and owners of the forest resources to preserve and/or improve water quality through the preservation of forests. Finally, we conclude that the system of water pricing depends on water availability and the application of intra-regional charges for its consumption, principally the agricultural sector, which reports the largest consumptive use of water, as well as encourages irrational use, due to subsidies it receives.

12.2 Water in Mexico

12.2.1 Previous Circumstances

Mexico has a long hydraulic tradition, dating back to pre-Hispanic times when water was not just for economic development but was associated with religious purposes. The hydraulic structure consisted of irrigation systems, aqueducts, chinampas (floating gardens of Xochimilco), and the hydraulic system of the Gran Tenochtitlan for both flood control and navigation (Guerrero 1995). The Tropic of Cancer cuts almost by half the country, giving it a specific climatic characteristic with arid climates in the north, warm-humid and subhumid in the south, and temperate or cold conditions in regions with greater elevation. Mexico has a vast diversity in its territory, making it a country with heterogeneous topographical characteristics, a wide variety of natural resources, and a wide range of climates. Two thirds of the territory is arid or semiarid, and the rest ranges from very humid to moderate.

The Mexican per capita water accessibility is around 4,263 m^3/hab/year (CONAGUA 2011, T2.2 pp. 21). According to the World Bank and United Nations, a per capita water availability lower than 1,000 m^3/year is a signal for severe water scarcity, while less of 2,000 m^3 means a significant water stress level, principally under years of low precipitation (CNA 2001). However, Mexico seems to be a country far from having water problems. And it would be true if the water availability were similar throughout the territory, but the climatic and topographic characteristics in Mexico are varied and have a significant influence on the economic activity, as less than a third of total runoff occurs within 75 % of the national territory, where most of the country's largest cities, industrial facilities, and irrigated lands are located. The estimated population of Mexico in 2014 is 119 million inhabitants (CONAPO 2014).

Guerrero (2002) highlighted how both population and economic activity are inversely related to water availability in Mexico, since 32 % of the runoff occurs where 77 % of the population resides, and 86 % of the GDP is generated (Fig. 12.1). These numbers have not changed in a significant way.

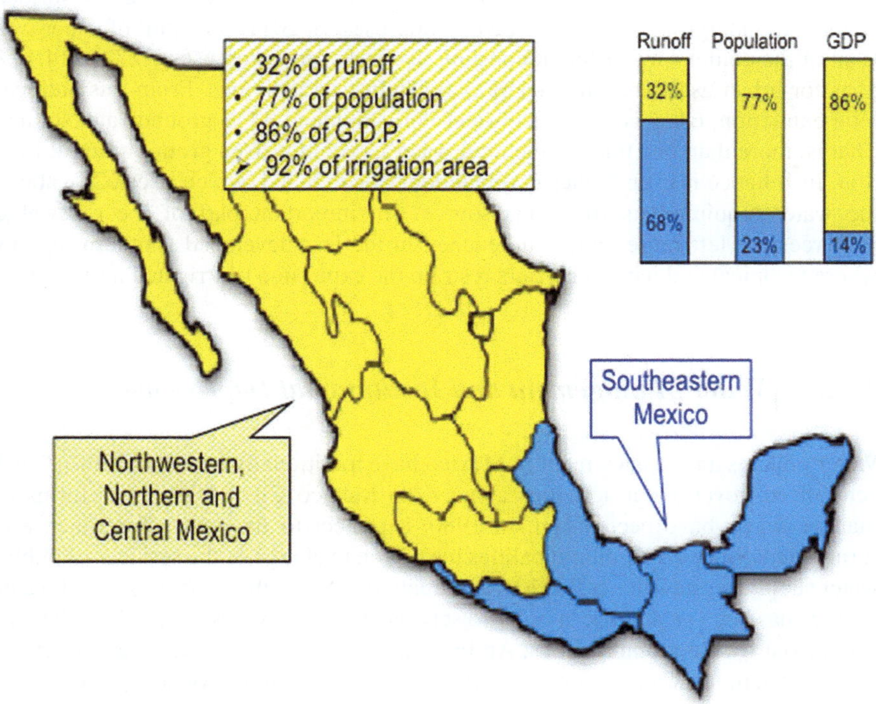

Fig. 12.1 Water availability and economic activity (Source: Guerrero 2002)

Table 12.1 Consumptive water uses for Mexico for 2011

	Surface water		Groundwater		Total	
Water use	km^3	%	km^3	%	km^3	%
Agricultural	40.9	81	20.9	69	61.8	76.7
Urban (domestic)	4.3	9	7.1	24	11.4	14.1
Industry (self-supplied)	1.6	3	1.7	6	3.3	4.1
Electric energy except hydroelectric	3.6	7	0.4	1	4.1	5.1
Total	50.4		30.1		80.6	

Source: CONAGUA (2011)

Conflicts for the use of water have increased causing important political and social effects. Most groundwater use takes place in the arid and semiarid areas of central, northwestern, and northern Mexico, where the pumping/recharge balances are negative with the consequent overexploitation of numerous aquifers. Therefore, groundwater has become fundamental for the Mexican economy and sustainable development. It represents the major (or even the single) source of water in the arid and semiarid regions of the country.

Table 12.1 summarizes water volume withdrawals for origin source, for surface water and underground, and for type of water uses. In 2011, 80.6 km^3 of water was used in the country for different consumptive uses. Irrigation uses 76.7 %, 14.1 % goes for urban uses, and only 4.1 % is for industrial activities. From this national total extraction, 63 % is of surface water derivation and 37 % groundwater source. That is, more than one third of the total water use comes from groundwater utilization. Its reliance is even higher in urban/domestic demand, which rises 62 % above the water requirements from this source. An important part of the renewable resources are left more or less untouched in the less-developed southern regions where technical and natural barriers restrain the expansion of irrigated agriculture.

12.2.2 Water Management and Institutional Improvements

Water policies and management in Mexico have traditionally been "top-down," and centralized government activities come from Mexico City.[1] Different reforms to manage water have been taken. In 1980, a project to fix quotas for water was formulated. Since 1983, municipalities have been in charge of the services of public water supply, wastewater collection, and treatment. Since then, the creation of water utilities has been promoted, in order to separate these activities from others that are carried out by the municipalities. An important breaking point was made in 1989, when the national water commission (Comisión Nacional del Agua—CONAGUA)

[1] The political administration in Mexico is organized, at the top by federal government, next the state governments (31), and last, the municipal government. The number of municipalities varies from one state to the other.

was created as an autonomous agency to become the sole federal authority dealing with water management. As early as 1990, necessary studies were carried out to design a new legal instrument. A national water law, authorized in December of 1992, promotes decentralization, stakeholder participation, more control of water withdrawals and wastewater discharges, efficient use of water, greater private sector participation, and establishment of economic instruments and fiscal policies related to the collection of water levies for both water use and water pollution control.

The agricultural sector deserves special consideration, since agriculture has been a traditional activity in Mexico. Lately, it has suffered a variety of essential changes, including the use of water. The agricultural sector administration has varied widely. The ejido, or communally farmed plot, emerged as a unique Mexican form of redistributing large landholdings. Confronted with a dysfunctional character of much of Mexican agriculture, the government in 1992 radically changed the ejido land tenure system, codifying some existing actions that were against the law but widely practiced and introducing new characteristics.

Currently, the concession of water rights in the agricultural sector may adopt one of next four forms: water rights settled through (1) concessions to single individuals for the use and exploitation of the water resources for farming purposes or to enterprises for the administration and operation of irrigation systems or the shared use and exploitation of common water sources for agricultural purposes; (2) ejidos and rural communities in coordination with legal dispositions derived from the new Agrarian Law; (3) irrigation units, as defined by the previous water law; and (4) public irrigation systems.

Since 1989, the federal government started the transference of operation and maintenance activities of irrigation districts to user associations. The water user associations, WUA,[2] are organizations whose main function is the operation, maintenance, and management of the irrigation infrastructure. The national water commission concedes volume water rights to the irrigation districts.

12.3 Water Economic Context

Water is just one of society's many scarce resources. In Mexico, this is not an exception. The water available in a basin is finite, and the consumers' demand for water for different uses, potable water, irrigation water, and water for industrial processes, principally increases. In addition, the water quality in Mexico has declined with increased water contamination—pollution is a real problem for surface water and groundwater sources. Many of the aquifers in Mexico are overexploited. The population has grown very fast. Taking all these elements together results in a serious water availability problem. The most commonly used instruments for improved

[2] They can be established as civil associations and granted certain fiscal privileges. The board of directors of these associations are selected by the assembly and composed of water users of the irrigation modules in the irrigation districts or units.

efficient water use are taxes, charges, subsidies, levies, and quotas. But the effectiveness of the results that any of them could generate is a function of the economic and political context under which they are applied.

The price of water has been utilized as an economic tool to enforce water users to become more efficient, since a commodity price is seen as a measure of its scarcity. Meaning, a price is supposed to be a sign of the right and of the whole social costs for supplying water, including resource depletion.

> In Mexico, water management rests on a delicate balance between government regulation and market mechanisms. Rather than going into a complicated scheme of calculating, if possible, opportunity costs or long-term marginal costs, Mexico's approach of pricing water has been pragmatic in nature, considering the political resistance associated with the introduction of any kind of new fiscal burden. That is why, the introduction of water levies in 1986 was motivated less for the purpose of assigning the 'right' price than for introducing the concept of water as an economic good with a specific value. (Guerrero and Thomas 2004, p. 1)

The basis for developing and consolidating Mexico's water financing system is established by a system of charges for water use and wastewater discharge. Besides providing an incentive to increase water use efficiency, which is already measurable, the collection of both kinds of charges—for water use and for water discharge—has resulted in the generation of financial resources to perform water programs and activities. Like in many other countries, Mexico's modern water management lies on a fragile balance between governmental regulation and market mechanisms. The country's legal and institutional reforms implemented in the early 1990s have the objective to reach this balance.

12.3.1 Water Pricing in Mexico

Water pricing in Mexico needs to be addressed from the way it was structured more than two decades ago. The Mexican water price structure is set to reflect water availability and the economic value of water. Water users have to pay abstraction charges, depending on their geographical situation, which is delineated according to relative water scarcity.

Water use charges and wastewater effluent charges are part of the Mexican legislation. In agreement with the legal framework, those who benefit from water use or those using the water courses to dispose of wastewater have to pay toward (a) the management and development of the resource and (b) the restoration and improvement of water quality, in proportion, respectively, to their water consumption or to the amount and characteristics of wastewater they discharge.

Taking into consideration that industrial water consumption comes, principally, from a self-supplied water, surface water, as well as underground sources, its exploitation is under a concession. That is, water right, or license granted by the national water commission, and the industry is under obligation to pay for a federal fiscal right for the use of water and wastewater discharge on national streams. These are one-time payments. In addition, self-supplied industrial water users have to pay quarterly abstraction charges per cubic meter, depending on their geographical location, which is determined according to relative water scarcity. For effluent emission discharge, industries also have to pay for contaminants as well as for the volumes discharged. Regarding the abstraction charges payment, there are some subsidies. Additionally, some municipalities are compensated for a given proportion of their water charges, resulting in an implicit subsidy scheme. The amount of all of these payments is set up in the federal law act (Ley Federal de Derechos en Materia de Agua) and is updated each year.

Since water tariffs and commercial efficiency of water utilities are very low, federal, state, and municipal governments provide financial support (subsidy). Actual fees are not sufficient to cover water utilities operation and maintenance costs. Tariffs would have to be increased at least by 100 % in order to promote self-financing water utilities. This situation is critical in rural areas where state and federal subsidies have to cover all the service costs.[3]

Regarding the agricultural sector, users of water for irrigation pay no abstraction charges. This policy has been the cause of intense discussions since irrigated agriculture accounts for most of the water abstraction (not considering hydropower) and water consumption. Pro and con argumentation goes beyond economic rationality; it has to do with social and political considerations. Nevertheless, government policies have been adopted to introduce efficiency in water use for irrigation through the irrigation management transfer program.[4] It was designed primarily to ensure that water user associations had adequate financial resources to be self-sufficient. This meant that the irrigation fees or water tariffs had to reach a level at which the cost of operation, administration, and maintenance (O&M) at the module (district) was covered. In addition, the water tariffs have to be sufficient to meet the module's share of the costs of operation, administration, and maintenance at the main canal and water source level as well.

In line with the policy of making irrigation districts more financially sustainable, it was recognized that users would have to pay the real O&M costs to their irrigation district. The general idea was to eliminate bureaucracy, reduce costs, and make those costs proportional to the benefits the farmers receive. Irrigation districts would, under this strategy, advance rapidly toward financial self-sufficiency.

Probably the most relevant issue in water pricing reform within the agricultural sector is associated with charging for abstraction and, to some extent, the establishment of pollution rates (nonpoint pollution rates). Under existing fiscal laws, no user is exempt from paying abstraction charges, including the agricultural water users. At present, the corresponding tariff for agricultural water users is set at zero. This is important, since no major legal modifications are required. The national congress sets water abstraction charges annually, and it is in this context that the issue will have to be discussed and resolved.

[3] To review the performance of water utilities in Mexico, see Guerrero (2008).

[4] Transferring the management of the irrigation districts to the users was foreseen as the proper strategy to create a different relationship between the government and water users.

12.3.2 Mexican Water Price Structure

Water has more value as it becomes less available, that is, as its scarcity rises. So, the water tariffs are determined as a function of the water availability. Before 1986, when the federal water law was modified substantially, the pricing system employed the same fixed price per cubic meter throughout the country. Since 1986, the water pricing system in Mexico has incorporated two kinds of tariffs: one is a fixed price per cubic meter of water used tariff, differing by water supply zone, which reflects the production and conveyance costs. The other type of tariff is an increasing block rate structure, reflecting the service zone, that is, the water tariffs are determined as a function of the water availability. Since 1986, up to now, the idea has been that the water price needs to be quantified by the magnitude of its four components: right, services, use, and preservation (Guerrero 1995), as is explained below.

- *Right*: represents the associated cost of disposable water regionally and locally.
- *Service*: represents the tariffs of the user connected to the water supply system. It addresses both irrigation water and potable water.
- *Use:* represents the charges to cover the management cost and operating and maintenance costs that are a function of the use level.
- *Preservation:* represents the wastewater treatment cost.

Prices were established taking into account the regional heterogeneity of the water availability. Up to 1996, four types of zones were defined from the hydrological point of view: Zone 1, where water is now scarce relative to demand; Zone 2, where supply and demand are in balance but only for the short term; Zone 3, where supply is enough to satisfy demand for the intermediate term; and Zone 4, where water is in abundance for the indefinite future. Pricing weights are assigned to each zone. Four principal water uses were also established: irrigation, hydroelectric generation, urban (potable), and industrial.[5] For each kind of user, a pricing weight is assigned. The industrial sector gets the highest weight; in second place is the water for urban use (potable); third is the water for irrigation; and the water for hydroelectric generation is assigned the lowest weight. The criterion for assigning these weights has not been made explicit, but it clearly includes considerations of return flow and ability to pay.

Table 12.2 displays the weights established in 1986 from the amendments in the federal water law for regional heterogeneity (up left), for water users (up right), and the way both are combined to build a sort of matrix that defines price of water in Mexico by users and water availability zones. The methods used to fix the prices appear to combine both the water availability weights and the sector weights, plus some political considerations. As noted above, there are four water supply zones. Thus, we can describe at least 16 different tariffs for the use of water. In fact, each

[5] In the first instance, this water price structure is not explicitly considering the environmental issue; certainly, there is not an evident tariff for environmental services, but, in fact, water price takes into consideration the care of the environment when regional heterogeneity (water availability zone) is part of the price.

Table 12.2 Water price structure in Mexico (1986)

Regional heterogeneity

Water availability		Weight
1	Scarce	1.00
2	Equilibrium	0.50
3	Enough	0.15
4	Abundance	0.05

Water users

Water use	Weight
Industry	1.000
Potable	0.800
Irrigation	0.013
Hydroelectric	0.001

Availability water zone	Weight	Efficiency use (weight) Industry 1.000	Potable 0.800	Irrigation 0.013	Hydroelect. 0.001
		- - - - ($/m^3)		- - - -	($/kwh)
1 Scarce	1.00	1.50	0.08	0.020	0.030
2 Equilibrium	0.50	0.75	0.04	0.010	0.015
3 Enough	0.15	0.50	0.02	0.005	0.008
4 Abundance	0.05	0.38	0.01	0.005	0.005

Source: Guerrero, H (2002)

municipality has the option of defining its own pricing steps beyond the first block. Regarding the kind of use, the industrial sector has the highest cost; in second place is the water for urban use (potable); in third is the water for irrigation; and the water for hydroelectricity is the cheapest.[6]

In the same way, using the four availability zones, the tariffs for wastewater discharge were established. Each group of four different tariffs is determined by the volume discharged and also by the sort and percentage of pollutants discharged (BOD, TSS, DO, and other pollutants). On the discharge side, we also have at least 20 different tariffs. Thus, the tariffs depend on both the regional hydrological characteristics for water use and the type of pollutants discharged.[7]

A specific zone will have different tariffs: one for water use and maybe more than one for discharges, which depends, principally, on the industrial activity that

[6] It is important to mention that there is not a specific equation that converts the regional heterogeneity and the water users' coefficients into efficiency use. In fact, there is not a rule. For water users, coefficients were established from a point of view of the estimated value that water produces into each economic sector. Regarding water availability zones, the procedures were similar, giving the highest coefficient where scarcity is the highest. It may look delicate, but the principle of this structure is to comply with economic performance of water users as well as the economic principle of scarcity. These coefficients have not changed over time, but tariffs indeed have had actualizations, principally in the form of inflation and taking into account social impacts.

[7] For more information regarding the determination of tariffs for pollutant discharges into water bodies, see CONAGUA (2013).

the zone has.[8] CPNH (1980) establishes that the principal objective to fix a price for water in Mexico is for one side, the efficiency for all the users and, for other, the equity in the cost that each one has to pay. Therefore, a domestic water user has to pay the minimum, given its low productivity, and the industrial user has to pay more, due to the aggregate value of the product generated from the water use and its productivity. But those prices should not have to be so small, since wasting water is feasible at a low water price.

The industrial water users have suffered the biggest impact, since many industries tap their water supply by themselves and CONAGUA applies the respective tariffs for the right to use water, according to the availability zone weights. As the industries tap their own water from groundwater that is a common source for other urban users, industries contribute to the overexploitation of the aquifers. In addition, industries also have to pay for the discharge of wastewater.

Regarding the other two main customers, urban water is being charged in a block structure with increasing prices that try to recover operating and maintenance (O&M) costs. Farmers pay the cost of O&M in terms of the irrigated area, based on the water consumption of the crop being raised. The irrigation districts have been transferred to the irrigators, and a recent change in the water law allows for the transfer of water rights. This mechanism attempts to generate an efficient use of the water in scarcity zones. The objective of the payment is to compensate the system cost, which includes both the investment in hydraulic structures (dams, canals, etc.), and the O&M costs. The amount of the fee depends on the regional hydrological characteristics. And it tries to reduce the subsidy on water, attempting to eliminate it altogether.

Since 1989, and due to CONAGUA's policy on implementing an efficient use of water that is becoming scarce, the water tariffs have risen substantially. Looking to induce a rational water use and efficient allocation, the water law has undergone transformations. This has produced a considerable fee collection increment. In 1992, the federal fee collection increased again because the discharges of wastewater on streams or sewerage were taxed strongly, where the policy "polluter pays" is applied. These fee augmentations have generated a more efficient water use, as well as a source of capital for the sector.

In 1997, the national water commission changed the number of availability zones, which were previously defined from a hydrological point of view, based on administrative concerns. The principal change is that zone 1—the scarcity one—was extended up to six zones, due to a number of practical difficulties handling water rights payment mechanisms, following problems like the application of subsidies, permissions, and exceptions for some industrial uses or municipalities. Nevertheless, zones 1 to zone 6 are still considered as scarcity zones. The other zones (7–9) retain the conditions as they were defined in 1986. Therefore, zone 7 is equivalent to zone 2 (equilibrium), zone 8 is equivalent to zone 3 (enough), and

[8] Also, there are prices for special cases and other uses as commercial and services, livestock, irrigation for sporting fields, and aquariums. But they are few and do not have incidence in the cost, because they comprise approximately 1 % of the water withdrawal.

Table 12.3 Water prices for 1993 and 2001 (Mexican $/m^3)

1993	Industry (M$/m^3)	Urban (M$/m^3)	Scarcity value
1 Scarce	1.30	0.0600	The highest
2 Equilibrium	0.90	0.028	
3 Enough	0.32	0.014	
4 Abundance	0.24	0.007	The cheapest
2001	Industry (M$/m^3)	Urban (M$/m^3)	
1 Scarce	11.4960	0.2277	
2	9.1966	✓	
3	7.6638	✓	
4	6.3228	✓	
5	4.9815	✓	
6	4.5022	✓	
7 Equilibrium	3.3890	0.1060	
8 Enough	1.2043	0.0530	
9 Abundance	0.9026	0.0264	

Source: Guerrero (2002). Prices are from "Ley Federal de Derechos en Materia de Agua" 1993 and 2001. M$/m^3=(Mexican pesos per cubic meter in current terms)

zone 9 is equivalent to zone 4 (abundance). Table 12.3 displays, using data from 1993 to 2001, the way the water structure changed from four to nine water availability zones.

Each year, CONAGUA updates water charges that each user should pay, as well as the catalog of the municipality's localization by water availability zone. In some circumstances, the municipality should change its water availability zone, for example, some of them could move from zone 9 to zone 8, or zone 7, and so on. These updates are published in the federal law act (Ley Federal de Derechos en Materia de Agua). Table 12.4 presents water prices for 2013.

The agricultural water user pays the same amount for water no matter where the water availability zone is located. When water is drawn from surface source or underground source, excepting from the sea, for agricultural use, it is paid the right for water use for each cubic meter only in the case which it exceeds the volume granted in concession to each irrigation district or per cubic meter that exceeds the volume allocated to the remaining agricultural users. Table 12.4 shows the amount paid by the agricultural user in 2013 for such a case of exceeding the concession.

The way that the Mexican water law has determined water prices per cubic meter as a function of the availability zone is very effective: the highest price for the scarcity zone and the cheapest for the abundant water zone. It is responding to the theoretical principle that a commodity price is seen as a measure of its scarcity. And also, there exists the difference between water users that respond to the knowledge that while water is an input for industrial processes, it also is a necessity and has a different value for industry, irrigation, and, consequently, for other uses.

Table 12.4 Water prices of first semester 2013 (Mexican $/m^3)

Water zone (quotas 2013)	Industry ($/m^3)	Urban ($/m^3)	Agricultural ($/m^3)
1 Scarce	20.5042	0.40620	0.14520
2	16.4028	✓	✓
3	13.6689	✓	✓
4	11.2770	✓	✓
5	8.8845	✓	✓
6	8.0297	✓	✓
7 Equilibrium	6.0437	0.18915	✓
8 Enough	2.1472	0.09446	✓
9 Abundance	1.6092	0.04702	✓

Source: CONAGUA (2013, pp. 14–17). Prices are in current terms. Banxico (2014)
Note: For reference 13.0088 Mexican $ = 1 USD in December 2013

The water price structure applied in Mexico is efficient in discouraging water wasters. But beyond that, the problem becomes knowing how the prices were determined and what they represent.

12.4 Environmental Considerations

The scheme of payment for environmental services—hydric (PSE-H), developed in Mexico, is an instrument that has not only allowed the establishment of a proportionate share of water rates but has also encouraged good practices in forest management that provide environmental services (mainly capture and carbon sequestration, and capture and infiltration of water). Therefore, its contribution to sustainable forest management is directly related to the climate change policies, in particular the REDD + (reducing emissions from deforestation and degradation) instrument, aimed at reducing carbon emissions.

Mexico has two antecedents of environmental policies that affect the pricing of water from forest conservation. The first is the creation of FIDECOAGUA, which was established in 2002 to preserve the forest and to establish the PES-H in the mountain area in Coatepec, Veracruz. Resources for the PES-H come from two sources: (1) the municipality of Coatepec, which allocates an annual amount to the trust under the contribution fund for municipal strengthening (FAFM), and (2) from citizen water users through a monthly fee included in the water bill of the city commission on drinking water and sanitation of Xalapa. So, domestic consumption pays 1 peso per serving, while merchants and industrialists pay 2 pesos per hydrant. During the period 2003–2009, it was possible to transfer more than $2,360,000 for the preservation of forests. From this experience (in collaboration with international agencies, such as FONAFIFO in Costa Rica and the World Bank), it was possible to establish (in 2003 as a national policy) a program for payment for environmental services, resources that most (about 70 %) are intended for the preservation of hydrological services (PES-H).

The second case is the establishment, in 2007, of the fund for payment for hydrological environmental services of the state of Mexico (FIPASAHEM) with a seed capital of 30 million pesos, approved by the state executive. In subsequent years, similar contributions from 30 million in each year (PROBOSQUE 2014) were collected. Also, by state decree, an additional fee for the water supply by way of payment of contributions improvement environmental services was created, through which water utilities of local councils deliver into the trust an equivalent of 3.5 % of the total water rates collected bimonthly (Decree 94 and 233 of the financial code of the state of Mexico). During 2007–2013, the collection of more than $467 million, of which over $324 million have been earmarked for the PES-H, was achieved.

In the last 5 years, the design process of the REDD+ program (ENAREDD + in Mexico, national strategy for reducing emissions from deforestation and degradation, which is being developed in pilot areas) provides for the integration of PES mechanisms in its operation structure, which would strengthen and promote, for example, local markets for hydrological services under an environmental governance perspective, as part of the measures for mitigation and adaptation to climate change.

In general, two instruments of the national policy of Mexico that relate to the management of forests and water are identified: (1) the payment for hydrological environmental services and (2) local payment mechanisms for environmental services through concurrent fund.[9] Both measures, particularly the latter, are emerging, at least for the case of Mexico, as economic instruments for watershed management and with an ecosystem vision, which seeks to revalue the water resources through construction of water prices, from institutional agreements between water users and owners of forest resources that provide water services.

12.4.1 *Payment for Environmental Services: Hydrological Service (PES-H)*

According to CONAFOR data (2012), potential areas for PES-H are estimated at 62.9 million hectares of forest, representing more than 45 % of the forest area in the country. These areas are categorized in three areas of differentiated payment ranging between $382 and $1,100 per hectare/year (Gómez 2012). Since the beginning of the program in 2003 until 2011, about 3.27 million hectares of forest land was incorporated or reincorporated for hydrological environmental services with a total amount allocated of around 5,411.10 million pesos (CONAFOR 2012).

[9] Both programs are supported by the national forestry commission (CONAFOR—Comisión Nacional Forestal), whose purpose is to "develop, support and promote productive preservation and restoration activities in forestry, and to participate in plans, programs and implementation of sustainable forestry development policies."

Table 12.5 States in the PES-H program (2007–2011)

State	Thousand hectares in PES-H (2007–2011)	State	Thousand hectares in PES-H (2007–2011)
Oaxaca	243.18	Veracruz	80.70
Chihuahua	158.35	Guerrero	80.61
Durango	149.32	San Luis Potosí	71.52
Quintana Roo	145.86	Estado de México	69.55
Chiapas	108.11	Michoacán	47.15
Yucatán	101.04	Tamaulipas	46.63
Jalisco	92.23	Nuevo León	42.10
Campeche	81.26	Otros estados	298.60
Total		**1,816.20**	

Source: CONAFOR (2011)

Areas that receive compensation for water environmental services are usually located in the upper basin. In many instances, residents are farmers and indigenous people, and those lands are "the richest in biodiversity (not more productive in the conventional sense) and the better preserved" (Hernández 2008). Inserted areas in the program include northern, central, and southern states (Table 12.5). This shows the high potential and interest in the preservation and/or improvement in quality and quantity of forest water resources. The funding mechanism of this program is through resources derived from charges for water use rights, according to the federal law act (article 223), as well as resources from the Mexican forestry fund.

12.4.2 Local Markets for PES-H

Under the matching fund scheme, CONAFOR has financed hydrological environmental service projects in areas where environmental service users and providers are clearly identified. Management for promoting this type of mechanism emerges with the intention of having the hydrological environmental consumers pay directly to land owners that provide the environmental services (Chagoya and Iglesias 2009).

By 2010, 30 agreements have been registered to boost these mechanisms, and 18 areas were valid. Each kind of financing was conducted under different institutional arrangements, according to the needs and capacities of those involved. There are schemes launched in Mexico with the participation of users such as Comision Federal de Electricidad (CFE), irrigation districts, and water utilities. As an example, the agreement among CFE, the state of Chiapas government, and CONAFOR (initiated in January 2012) with an established matching funds scheme in the amount of 16 million 500 thousand pesos for a 5-year period, where 2,500 ha of forest, distributed in 11 ejidos (social property), will be preserved.

Likewise, the participation of water resource users for agricultural irrigation has initiated PSA-H schemes, as in the case of "Fabricas de Agua Centro Sinaloa"

(FACES), whose main operational area is Tamazula River basin, where the preservation and restoration project of 2,807 ha in ejido Imala forestry areas that will diminish the siltation process of the dam and water gauge increase started. With support through the matching funds scheme, starting in 2009, FACES and CONAFOR committed themselves to provide equal shares of a 10 million pesos to be distributed over a 5-year period. The matching fund resource provided by FACES derives from a fee of 20 pesos per hectare for farmers through irrigation districts (Gómez 2012).

There are also experiences that reflect how institutions (e.g., basin commissions) influence the creation of this kind of local mechanism. For instance, in 2009, CONAFOR joined this effort by signing a 5-year agreement with the Comisión de Cuenca del Alto Nazas to conduct preservation works in an 8,622 ha area. CONAFOR is committed to a contribution of 10 million pesos, while the basin commission is committed to an equal amount for a total of 20 million pesos. As with FACES, resources provided by the basin commission come from the integration of funds from water users in Irrigation District 017 and direct contributions of well concession.

In addition, the Comisión de Agua Potable y Alcantarillado (potable water and sewerage commission) of Uruapan has joined this effort by signing an agreement with CONAFOR to improve and/or preserve forest areas in five indigenous communities of Michoacan. The total amount allocated is more than 10 million pesos for a 5-year period (Gomez and Guerrero 2013). Likewise, there are projects for designing these PES-H local schemes in urban areas, as in the case of the dynamics presented by the city of Morelia (capital of the state of Michoacan) and the El Calabozo microbasin (Guerrero et al. 2013).

Finally, it is important to emphasize that these local mechanisms are integrated from contributions of financial resources from CONAFOR and the stakeholders. Usually, CONAFOR provides 50 %, and its counterpart (companies, water utilities, NGOs, cities, etc.) provides the remaining half.

These schemes help preserve and appraise water resources as a basic input in the production process (for consumptive and nonconsumptive use) and, at the same time, encourage the participation of stakeholders needed for an integrated water management. The incentives presented by the schemes of PES-H promote forest preservation, as well as their capacity of retention, and capture of carbon.

12.5 Conclusion

If natural resource management cannot be measured in monetary terms, then it shall not be properly managed. Water management in Mexico is a resource whose distribution is not homogeneous across the territory and the reason for dispute in the way their allocation are disbursed to other sectors of the economy. Institutional development in water management in Mexico has allowed the conformation of agents, such as the watershed councils, water banks, and transfer of irrigation districts, which minimize social conflicts in the struggles for water use and encourage the participation of stakeholders in making decisions about integrated management of water resources.

With regard to the preservation of forests, the purpose of Mexico's national forestry commission is to develop, support, and promote productive preservation and restoration activities in forestry and to participate in plans, programs, and implementation of sustainable forestry development policies. Good practices in forest management provide environmental services (mainly capture and carbon sequestration, and capture and infiltration of water). Payment for environmental services is a management tool for forest and water, which encourages the participation of users and suppliers of water resources and, in the long run, can generate local markets from "PES-H" through cross cooperation between government agencies, NGOs, suppliers, and users.

Although the allocation of water pricing in Mexico is methodologically correct with regard to regional availability and the type of users, the subsidy application to users (agriculture) undermines the goal of the design of them, as well as the strategy of using the price of water as an economic tool for the efficient use of it. While the water policy of Mexico is strongly supported by its legislation, it should be adapted to the current context of the globalized Mexican economy. This is especially true in the case of agriculture in which water fees are highly subsidized. Compared to the other sectors, agriculture is the largest consumer of water, and there are agricultural users in areas of low water availability that report high economic value in production, mainly for export and which does not represent the national economy.

References

Banxico. (2014). Banco de México. *Tipos de cambio y resultados históricos de las subastas.* Tipo de cambio y promedio del periodo. http://www.banxico.org.mx, 8 July 2014.

Chagoya, Jorge Luis y Leonel Iglesias Gutiérrez (2009). "Esquema de pago por servicios ambientales de la Comisión Nacional Forestal, México". Capítulo 10. En Sepúlveda, Claudia y Ibrahim, Muhhamad (Eds.) Políticas y sistemas de incentivos para el fomento y adopción de buenas prácticas agrícolas como una medida de adaptación al cambio climático en América Central. Serie téc-nica, Informe técnico No. 377. Costa Rica: Centro Agronómico Tropical de Investigación y Enseñanza (CATIE), pp. 189–204.

CNA. (2001). *Programa Nacional Hidraulico 2001–2006. Resumen Ejecutivo* (2nd ed.). México: Comision Nacional del Agua.

CONAFOR. (2011). *Estadística Forestal Nacional.* Download 11 April 2012 in www.cnf.gob.mx:8080/snif/portal/economica/estadistica-forestalnacional

CONAFOR. (2012). Download 7 July 2012, in www.conafor.gob.mx/portal/.

CONAGUA. (2011). *Estadísticas del agua en México. Comisión Nacional del Agua.* www.conagua.gob.mx, edición 2011.

CONAGUA. (2013). *Ley Federal de Derechos en Materia de Agua.* Comision Nacional del Agua. www.conagua.gob.mx. edición 2013.

CONAPO. (2014). *Proyecciones de la Población de México, 2010–2050.* Mexico: Consejo Nacional de Población.

CPNH. (1980). *El Sistema del Precio del Agua.* Comision del Plan Nacional Hidraulico, SARH, Abril. Mexico.

Gómez, S., Faustino. (2012). *La Acción colectiva en dos comunidades y su incidencia en la gestión de los bosques en torno al esquema de pago por servicios ambientales hidrológicos.* El área de influencia de la cuenca del río Cupatitzio. El Colegio de Michoacán, A.C. thesis of Master in Human Geography.

Gómez, S., Faustino, & Guerrero-Garcia-Rojas, H. (2013). El esquema de pago por ser-vicios ambientales hidrológicos como instrumento para la gobernanza ambiental en la cuenca del río Cupatitzio: análisis y perspectivas. In H. Guerrero & R. López (Eds.), *Desarrollo Local y Sustentabilidad*. México: UMNSH y UNAM. ISBN 978-607-424-361-1.

Guerrero-Garcia-Rojas, H. (1995). Water pricing in Mexico in principles and in practice: The case of Mexico. MA. thesis, Department of Economics, University of Colorado, Boulder.

Guerrero-Garcia-Rojas, H. (2002). *Irrigation water pricing: The case of Mexico*. International conference irrigation water policies: Micro and macro considerations meeting. Agadir, 15–17 June (World Bank. org/agadirconference). Morocco.

Guerrero-Garcia-Rojas, H. (2008). Ch 4. El costo del suministro de agua potable en México: análisis y propuestas de política. In *El agua en México: Consecuencias de las políticas de intervención en el sector*. Coordinadores: Hilda R. Guerrero García Rojas (UMSNH), Antonio Yúnez-Naude (Colmex) y Josué Medellín-Azuara (University of California, Davis). Lecturas de El Trimestre Económico N° 100. Ed. Fondo de Cultura Económica ISBN: 9-789681-686130. Octubre.

Guerrero-Garcia-Rojas, H., & Thomas, A. (2004). Water pricing reforms in Mexico: The case of the manufacturing sector. In P. Koundouri (Ed.), *Econometrics informing natural resources management selected empirical analyses,* Cheltenham: Edward-Elgar Pub.

Guerrero-Garcia-Rojas, H., Gómez, S., Faustino & Rodríguez, J. Refugio. (2013). Gestión de bosques por pago de servicios ambientales en la Microcuenca El Calabozo: análisis del costo de oportunidad del uso de suelo agrícola. In H. Guerrero & R. López (Eds.), *Desarrollo Local y Sustentabilidad*. México: UMNSH y UNAM. ISBN 978-607-424-361-1.

Hernández, R. (2008). *El Programa de Pago de Servicios Ambientales y su aplicación en un núcleo agrario con manejo comunitario del bosque*. Universidad Nacional Autónoma de México, D.F. México. Tesis de doctorado.

PROBOSQUE Protectora del Bosque. (22 de agosto de 2014). Programa para el Pago por Servicios Ambientales. Obtenido de http://portal2.edomex.gob.mx/

Chapter 13
Water Pricing in the Netherlands

Marianne S. Schuerhoff and P. Hellegers

Abstract In the Netherlands, about 75 % of monitored groundwater extractions are used for the production of tap water. Water extraction for tap water production is the main consumer. The tap water rate of the various drinking water companies largely depends on the share of groundwater used, which requires lower treatment costs than surface water. In 2014, the tap water rate varied between 1.11 and 2.21 euro/m^3. There are various taxes in place on groundwater use, as well as on tap water use. Such taxes can aim to recover costs, trigger water-saving technologies, or reduce water demand for environmental purposes. In 1995, the national groundwater tax was implemented—a so-called "win–win, green" tax that aimed to reduce the income tax burden and to have an environmental impact in terms of reduced groundwater extraction. From 2012 onward, the Dutch government, however, revoked it, as it was fiscally inefficient and environmentally ineffective. It increased distortions by taxing only a narrow base and by interfering with groundwater management programs funded by an existing provincial groundwater fee. In 2014, the national tap water tax was increased. But given the fact that only 0.6 % of a household's budget on average is dedicated to tap water, it is not likely that it will substantially reduce water demand. This increase in the tap water tax contradicts the low rate of the value-added tax (VAT) on tap water.

Keywords The Netherlands • Income tax burden • Distortions • Provincial groundwater fee • Green tax

---

M.S. Schuerhoff (✉)
PBL, Netherlands Environmental Assessment Agency, Peterseliesteeg 2, 3512TL Utrecht, The Netherlands
e-mail: marianne.schuerhoff@gmail.com

P. Hellegers
Water Resources Management Group, P.O. Box 47, 6700 AA Wageningen, The Netherlands
e-mail: petra.hellegers@wur.nl

A. Dinar et al. (eds.), *Water Pricing Experiences and Innovations*,
Global Issues in Water Policy 9, DOI 10.1007/978-3-319-16465-6_13

13.1 Introduction

13.1.1 Water Resources, Population, and Issues in Water Supply

The Netherlands is a densely populated country with 16.8 million people living on 35,000 km^2, which is one of the highest population densities in Europe (Statistics Netherlands 2014a). In 2013, the number of households reached 7.6 million with 2.2 people per household (Statistics Netherlands 2014b) on average. Nearly half of the country lies below sea level, with many of the main urban hubs at risk of floods. The country is famous for its dikes, dams, and reclaiming of land. In terms of drinking water supply, the country is practically 100 % connected to drinking water facilities (WHO 2010). The quality of supplied water is high (Accenture 2010).

Both surface and groundwater are sources for tap water in the Netherlands. Surface and groundwater are also self-extracted by household and industries. Table 13.1 shows water use for different water users in the Netherlands. The majority of the tap water is used by households; groundwater is mainly extracted by drinking water companies, agriculture, and several food industries. Surface water is also extracted for tap water, but mainly used for cooling purposes by energy companies and the chemical industry (Statistics Netherlands 2014c).

Several taxes and fees are levied on the extraction of groundwater, delivery of tap water, and discharge wastewater. Surface water extraction is not taxed probably because there is no scarcity of surface water in the Netherlands and its extraction is not directly damaging. Most surface water that is extracted is used for cooling and is returned after use. Groundwater is the preferred alternative to tap water for companies that need high-quality water. The taxes can serve various purposes. It can trigger adoption of water-saving technologies, recover costs, or affect behavior of

Table 13.1 Water use in 2011, million m^3

Use 2011[a]	Tap water[b]	Raw groundwater	Raw surface water
Households	783		
Farming, forestry, and fishery	43	89	31
Mining	2	0	1
Manufacturing	143	141	3,391[c]
Energy companies	12	5	10,928[c]
Drinking water companies	0	756	473
Others	97	1	476
Totals	1,080	992	15,300

Source: Statistics Netherlands (2014c)

[a]Preliminary figures

[b]Drinking water companies extract raw surface and groundwater—adding all three categories of water presented here would result in double counting

[c]Mostly water used for cooling purposes and as process water

Table 13.2 The various taxes and fees on groundwater and tap water in the Netherlands

Kind of tax/fee	Since	Size (€/m³)	Purpose	Aims to give an incentive for
Provincial groundwater fee	1986	0.014[a]	To finance research and to manage groundwater	Reduce diffuse groundwater extraction by drinking water companies and industry
National groundwater tax	1995–2011	0.196[b]	Revenue, environment	Change behavior of drinking water companies, industry
National tap water tax	2000	0.33[c]	Revenue, environment	End users
VAT on tap water		6 % of the rate	Revenue	–

[a]Average in 2010
[b]Rate 2011
[c]Rate in 2014

consumers (i.e., reduce water demand for environmental purposes). In the Netherlands, there are various taxes on groundwater extraction and tap water delivery; see Table 13.2 for an overview of the kind, purpose, and aim of the various taxes. The delivery of tap water has been taxed on a national level since 2000, and this tax is levied to generate revenue for the government and to encourage economic use of water. The tap water tax is a uniform rate per m^3 for the whole country and is only charged over the first 300 m^3 of delivered tap water. The extraction of groundwater is charged on a provincial level and was also taxed on a national level from 1995 to 2011. The national tax was installed to generate revenue for the government and, simultaneously, to provide an incentive to reduce groundwater extraction. The national groundwater tax was charged to extractors of fresh groundwater, with several exemptions: smaller capacity pumps, ice ranks, fire pumps, etc. While it was still in place, two rates were applied to the whole country: one for normal extraction and a reduced rate when water was infiltrated afterward. This tax was revoked, as it was fiscally inefficient and environmentally ineffective. It increased distortions by taxing only a narrow base and by interfering with groundwater management programs funded by an existing provincial groundwater fee (Schuerhoff et al. 2013). The rates for this provincial groundwater fee differ per province, and while the revenue of the provincial tax is used for provincial groundwater protection, the revenue of the national tax was dedicated to the general budget.

Lower-level governments, such as provinces and municipalities, also have the right to levy water-related charges, all of which are earmarked and should only recover costs. Municipalities charge a sewage charge to households and other users that discharge wastewater, based on "pollution units" for discharging of wastewater. Water boards charge a fee for the treatment of discharged water and for water quantity (i.e., water level) management.

In the Netherlands, drinking water companies produce and distribute tap water. In 2012, the drinking water companies served over 7,700,000 administrative connections, of which 96 % are metered (Vewin 2012). Water prices of tap water consist of a fixed rate per connection and a variable rate per delivered cubic meter of water. The rate further depends on the size of the connection and the amount of delivered water. The cost of tap water is largely determined by the mixture of sources used. Because groundwater is of relatively high quality, treatment costs are lower and drinking water companies using mainly groundwater have on average lower cost per connection (Accenture 2013). The system of determining tap water prices has not changed much over the past years, but the number and level of the various rates have changed. Customers also pay VAT over the rates, including taxes charged by the drinking water companies. For delivered tap water, the rate is reduced to 6 %, compared to the normal rate of 21 %.

As of 1997, every 3 years, a benchmark is commissioned by the drinking water company association (Vewin 2012). This benchmark is used to learn from best practices, to reduce costs and prices, and is obligatory since 2010 (Ministry of Housing, Spatial Planning and the Environment 2009). The majority of drinking water companies are public limited companies and municipalities, and provinces are shareholders (Vewin 2012). In 2012, the water sector had a revenue of €1.5 billion and profits added up to €135 million.

13.2 Past Water Pricing Experiences

13.2.1 Past Experiences with Irrigation Water Pricing

Agriculture uses both tap water and self-extracts of surface and groundwater. Water is mainly used for irrigation and to water cattle. As agriculture is mainly rainfed, water extraction by agriculture especially peaks in dry summers on the sandy soils in the southern and eastern parts of the country (Statistics Netherlands 2013). Farmers face the same tap water price as all other users (see section on experiences with industrial water pricing). Farmers also face the tap water tax, and in case they self-extract groundwater, they need to pay the provincial groundwater fee and the national groundwater tax.

Water scarcity issues in the Netherlands were a reason to introduce the provincial groundwater fee in 1986, which aimed to reduce diffuse groundwater extraction. Provinces charge a fee on groundwater extraction since then. Some provinces introduced fee-free thresholds, ranging from 10,000 to 100,000 m^3 (IDWG 2007), that effectively exempt all smaller farmers. The revenue is used to finance research and to manage groundwater. The revenue is also used to compensate farmers facing land-use constraints due to groundwater protection. The rates of the provincial groundwater fee varied between 0.85 and 2.75 eurocents per m^3 in 2010. The rate was 1.39 eurocents on average, which is less than one-tenth of the rate of the national groundwater tax. This tax was levied on groundwater extraction from 1995 to 2011. At the start, small-capacity pumps were exempt, which caused small-scale users like farmers to install multiple small-capacity pumps (Iwaco 1997). Groundwater extracted to water cattle was exempted from groundwater tax, and near the ending

Table 13.3 National taxes on water (€/m^3)

	Groundwater tax	Tap water tax
1995	0.154	–
2000	0.16	0.129
2005	0.181	0.146
2010	0.1951	0.157

of the tax, groundwater extracted for irrigation was exempted as well. As of 2012, the groundwater tax has been revoked. Table 13.3 provides rates for a few years for both the national groundwater tax and the tap water tax. The rate of the groundwater tax gradually increased over time until it was revoked in 2012. The tap water tax was introduced in 2000 and also gradually increased.

In periods of extreme droughts, irrigation is prohibited, which is a far more effective instrument than water pricing. This will not always affect farmers' behavior as some crops have high revenues from well-timed and secure irrigation, or users are unable to act in a rational manner.

13.2.2 Past Experiences with Residential Water Pricing

Residential use is around 70 % of tap water in the Netherlands (Graveland and Baas 2013). Since the 1990s, the total tap water use has gone down (Fig. 13.1), despite population growth, mainly due to water-saving equipment. Household use remained rather constant, while the reduction of tap water use is mainly due to business (Statistics Netherlands 2013).

The tap water rate of the various drinking water companies depends on the share of groundwater used, which requires lower treatment costs than surface water (Table 13.3). Drinking water companies using surface waters avoid the groundwater tax and provincial groundwater fee, but they have higher treatment costs. Groundwater-dependent drinking water companies pass taxes through to customers, but they pay less for treatment. By dropping the groundwater tax, groundwater becomes an even cheaper source than surface water. The cost per m^3 includes the provincial groundwater fee (and previously the groundwater tax), but excludes the tap water tax and VAT of 6 %. The rate charged to customers is therefore higher than the cost presented in Table 13.4. The abolition of the groundwater tax is reflected in the rate; in 2012, rates decreased, both per connection and per m^3 (Accenture 2013).

Average water rates decreased over the last years (Accenture 2013). The companies charge a flat rate for delivery under 150 or 300 m^3, which results in a flat rate at household level, as a family consumes less than 100 m^3 on average. Table 13.5 shows the fixed and variable parts of the rate charged by households for three cities. These three cities are served by three different types of water companies: Waternet

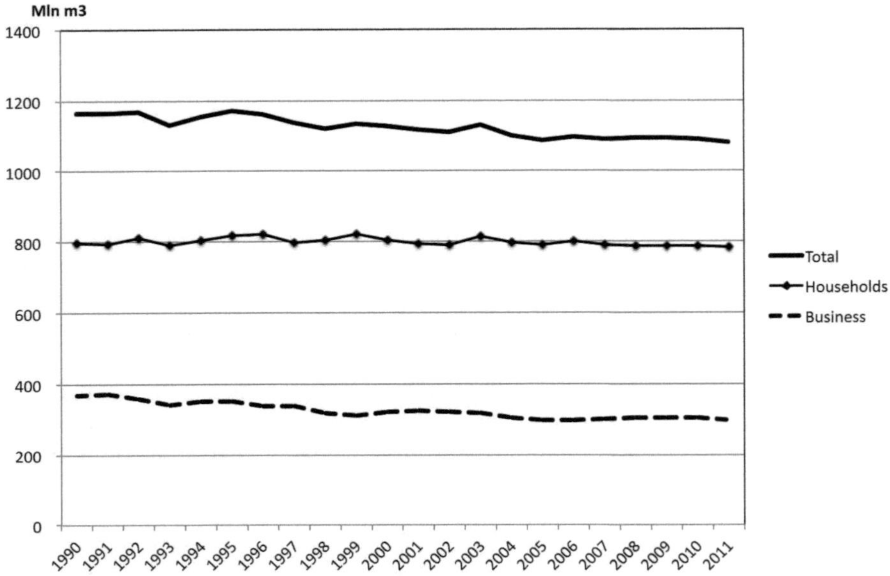

Fig. 13.1 Total tap water use by households and businesses

is a surface water company in the western part of the country, Vitens is a groundwater company serving most of the middle and northern part of the country, and WML is a mixed company in the south (Accenture 2013). Surface water companies generally have higher fixed rates, because treatment costs are higher.

The tap water tax is only levied on the first 300 m^3 of delivered water per connection. This means that most households pay the tax over all their water consumption (see Table 13.3 for rates). Drinking water companies function as intermediaries and collect the tax from customers. The tap water tax is reported separately on the water bill to provide insight to consumers. Customers also pay 6 % of VAT. Table 13.6 shows the components of the final customer prices for three cities for a household with average water consumption.

13.2.3 Past Experiences with Industrial Water Pricing

Tap water is delivered to all industries. The food and beverages industry and the (basic) metal industry are among the main users. Raw surface water is mainly extracted by energy companies for cooling purposes. This is, however, nonconsumptive use. Raw groundwater is mainly extracted by the food and beverage

Table 13.4 Percentage of water sources from groundwater, and cost of water produced, and its various cost components by drinking water company in 2012

Names of drinking water companies	Groundwater (%)	Cost (€/m³)	Taxes (€/m³)[a]	Operation expenditure (€/m³)	Capital and deprecation (€/m³)
Brabant Water	96	0.97	0.02	0.60	0.35
WMD	100	1.02	0.01	0.75	0.27
WBGR	88	1.05	0.02	0.73	0.30
Vitens	97	1.11	0.02	0.57	0.52
Evides	7	1.27	0.03	0.60	0.64
WML	69	1.45	0.01	0.74	0.70
Waternet	0	1.52	0.01	0.90	0.62
Oasen	9	1.58	0.05	1.10	0.43
PWN	5	1.69	0.00	1.01	0.68
Dunea	0	1.76	0.13	1.00	0.63
Average	56	1.27	0.03	0.71	0.53

Source: Accenture (2013, figure 39 and 79)

[a]Including provincial groundwater fee, fees for laying pipes over municipal land, excluding tap water tax and VAT

Table 13.5 Rates for metered households, fixed and variable rate, and total price charged by three drinking water companies in 2011

City and water company	Type of company	Fixed rate (€/year)	Variable rate (€/m³)	Total price[a] (€/m³)
Amsterdam (Waternet)	Surface water	42.15	1.26	1.69
Arnhem (Vitens)	Groundwater	25.00	1.055	1.31
Maastricht (WML)	Mixed	67.20	1.0184	1.71

Vewin (2011)

[a]Total price per m³ (including the fixed and variable rate) for an average households with a water consumption of 97.3 m³ a year

Table 13.6 Calculations for average household prices in three cities in 2011

City and water company	Type of company	Total price for average use[a] (€/household)	Total tap water tax[b] (€)	VAT of 6 % (€)	Final customer price (€)
Amsterdam (Waternet)	Surface water	164.37	15.37	10.79	190.60
Arnhem (Vitens)	Groundwater	127.46	15.37	8.57	151.41
Maastricht (WML)	Mixed	166.38	15.37	10.91	192.66

[a]Based on the total rate charged by the drinking water company per m³ (shown in the last column of Table 13.5) and an average use of 97.3 m³

[b]Based on the tap water tax of 0.158 €/m³ and an average use of 97.3 m³

Table 13.7 Average rate per m^3 per type of connection in 2009

Type of user	Delivery m^3	Connection size m^3	Average rate € per m^3
Small	1,500	3	1.09
Average	10,000	5	1.01
Large	25,000	10	1.04

Source: Accenture (2010, figure 32, 2013, figure 37)

companies, metal, manufacturer of chemicals, and manufacturer of rubber and plastic products. Up until 2011, the industry had to pay both the national groundwater tax (Table 13.3) and the provincial groundwater fee on self-extracted groundwater. The prices charged to industrial users depend on the capacity of connection and the amount of water delivered. Table 13.7 shows average rates charged to different types of users.

Figure 13.1 shows a relatively fast decrease in tap water use since 1990, notwithstanding population and per capita economic growth. For bulk users with high-capacity connections, reduced rates are available. For businesses, the average rates showed an increase in 2000 (compared to 1997) when the tap water tax was introduced (see Table 13.3 for rates of tap water tax).

13.3 Present Water Pricing Practices

The rates vary per drinking water company and are largely dependent on the mixture of sources used. This mixture, in return, depends on the region, as the use of surface or groundwater is largely determined by local circumstances. Furthermore, operations and maintenance costs determine the fixed rate. The present year differs from previous years for two reasons: First, the groundwater tax was revoked in 2011, and the rate of the tap water tax was doubled in 2014. This means customers served by groundwater companies saw their prices decline, while all users saw their prices rise because of the doubled tap water tax rate a few years later. The regional spread in rates charged by drinking water companies increased from 2011 to 2012. While the initial plans in 2011 were to drop the tap water tax, along with the groundwater tax, this tax still exists to date. As of July 2014, the rate of the tap water tax even increased from 0.17 €/m^3 in 2013 to 0.33 €/m^3 in 2014, both to raise more revenue to close a governmental budget gap and to reduce water use. This increase is supposed to yield an extra 205 million € each year (Ministry of Finance 2013). Initially, it was proposed to start charging for tap water above 300 m^3 as well, but in June 2014, this proposed change was canceled (Ministry of Finance 2014). This proposed change could have had the following effects: First, large-scale users would have an incentive to reduce tap water use. This could lead to absolute water saving with lower need for raw water extraction, including groundwater, or, on the contrary, may also cause users to switch to self-extracting groundwater, which is even cheaper since the groundwater tax was dropped (Vewin 2014b). Second, the difference in rates between household and industrial users would have been reduced with this change. The Ministry of Finance consulted different parties affected by the change and decided not to release the tax-free threshold. Releasing the threshold would have negative environmental effects, increase administrative complexity, discourage innovation, and have random effects for specific companies only (Ministry of Finance 2014). Now only the rate is increased, which affects households most. Given the fact that only on average 0.6 % of a household's budget is dedicated to tap water, and the increase in the tap water tax will only cost 10–20 € annually per household, it is not likely that it will substantially reduce water demand.

13.4 Present Experiences with Irrigation Water Pricing

Depending on the size of a connection, tap water is charged according to general business rates. Besides tap water, agriculture uses surface and groundwater. Extraction of surface water is free of charge. A fee is levied for groundwater extraction, which differs per province. No special rates for agriculture exist, but in most provinces, groundwater fees have fee-free thresholds or exemptions that effectively exempt farmers. Table 13.8 provides rates and exemptions for three provinces. The revenue of the provincial groundwater fee is earmarked and used to manage and protect groundwater levels.

Drinking water companies do not have specific rates for agricultural use, but the information in the section on industrial water also applies to agriculture.

13.5 Present Experiences with Residential Water Pricing

Rates vary per drinking water company, and the companies serve only a specific region. This indicates that residents pay the rate of the region they live in. For households, drinking water companies charge a fixed price per connection, varying from 35 €/year to 117 €/year (in 2014). The variable part of the rate[1] per m^3 delivered varies from 0.46 €/m^3 to 1.24 €/m^3. Table 13.9 shows an example for current (2014) prices for three different cities.

Table 13.8 Provincial groundwater fee rates and exemption threshold for three provinces

City (province)	€/m^3	Exemption threshold (m^3)
Amsterdam (Noord-Holland)	0.0085	12,000
Arnhem (Gelderland)	0.013	100,000
Maastricht (Limburg)	0.015	10,000

Source: Overheid.nl (2014)

[1] Including provincial groundwater fee and excluding tap water tax and VAT.

Table 13.9 Rates for households (metered), fixed and variable parts, and total rates charged by three drinking water companies in 2014

City and water company	Type of company	Fixed rate (€/year)	Variable rate (€/m^3)	Total rate[a] (€/m^3)
Amsterdam (Waternet)	Surface water	42.15	1.24	1.68
Arnhem (Vitens)	Groundwater	40.00	0.73	1.15
Maastricht (WML)	Mixed	86.64	0.7655	1.67

Vewin (2014a)
[a]Total price per m^3 for an average family with a water consumption of 95.5 m^3 a year

Table 13.10 Calculations for average household prices in three cities in 2011

City and water company	Type of company	Price for average use[a] (€)	Tap water tax (0.33 €/m^3)	VAT of 6 % (€)	Final customer price (€)
Amsterdam (Waternet)	Surface water	160.44	31.52	11.52	203.47
Arnhem (Vitens)	Groundwater	109.83	31.52	8.48	149.82
Maastricht (WML)	Mixed	159.49	31.52	11.46	202.46

[a]Based on price per m^3 in Table 13.9 and an average use of 95.5 m^3

On top of the price that the drinking water companies charge, the customers pay tap water tax and VAT. Table 13.10 shows calculations for an average family using 95.5 m^3 water in a year.

Fixed rates vary per region, depending on the sources used and the municipal distribution refund levies[2] (Vewin 2014a). Average water usage is 45 m^3 per person per year in 2013. For an average family with a water use of 96.4 m^3 per year, the price per m^3 ranges from 1.09 to 2.09 € in 2013. When water delivery is not metered, companies set either fixed rates per household or estimate the water use based on household size etc. (Vewin 2013). In 2014, the average prices (total of fixed and variable) per m^3 range from 1.11 to 2.21 € (Vewin 2014a). Tap water tax and VAT are levied on top of the rates set by the drinking water companies.

13.6 Present Experiences with Industrial Water Pricing

Businesses can request higher-capacity pumps and, again, rates vary per company and region. In general, it holds that the higher the capacity of the connection, the higher the fixed rate (see Table 13.11). Per cubic meter prices, however, decline with the size of the connection capacity.

[2]Levy paid by drinking water companies for having pipes in public land, the rate may differ per municipality. Not all municipalities levy distribution refunds. And municipalities that levy distribution refunds cause a greater spread at the top. These distribution refunds are charged to the residents of the municipalities in question and lead to local increases in the annual drinking water invoices of up to € 40 (Accenture 2013).

Table 13.11 Average rate per m³ per type of connection in 2012

Type of user	Delivery m³	Connection size m³	Average rate € per m³
Small	1,500	3	0.88
Average	10,000	5	0.82
Large	25,000	10	0.81

Source: Accenture (2013, figure 37)

Table 13.12 Example of fixed rates for business users by Waternet 2014

Capacity	€/year
Qn 1,5 m³/h	€ 42,15
Qn 2,5 m³/h	€ 42,15
Qn 3,5 m³/h	€ 49,00
Qn 6 m³/h	€ 59,00
Qn 10 m³/h	€ 75,00
Qn 15 m³/h	€ 118,00
Qn 20 m³/h	€ 135,00
Qn 30 m³/h	€ 150,00
Qn 40 m³/h	€ 164,00
Qn 50 m³/h	€ 195,00
Qn 60 m³/h	€ 213,00
Qn 100 m³/h	€ 276,00
Qn 150 m³/h	€ 341,00
Qn 250 m³/h	€ 470,00
Qn 400 m³/h	€ 634,00
Qn 600 m³/h	€ 793,00

Vewin (2014a, 36)

The drinking water companies determine the fixed and variable rates in different ways. To provide an example, Table 13.12 shows the different fixed rates, depending on the capacity for a water company in the western part of the Netherlands, which is classified as a surface water company. This company has a single variable rate of 1.26 €/m³ (excluding tap water tax and VAT). Most drinking water companies charge increasing block rates for the fixed rates, but the steps of the blocks and the rates may differ per company. Most drinking water companies charge decreasing block rates for the variable rates, one for small-scale users (up to 150 m³) and one for users that use more than 150 m³.

The variation in rates across regions increased in 2012, compared to previous years, due to the ending of the groundwater tax. These average rates exclude the tap water tax and VAT.

13.7 Current Debates and Future Directions

The quality of both groundwater and surface water is of concern, due to various kinds of pressures. The current (plan to) search for shale gas might be a risk for groundwater quality (Warner et al. 2013). The RIVM (2013) also worries that, in the future, the quality of groundwater might be under the norm due to pressures from agriculture, sewerages, industries, and buildup of soil pollution. Surface water is further threatened by the discharge of medicines through human excreta, insecticides, etc. Pathogens in water are becoming a serious problem (Vermeulen and Hofstra 2014; Hofstra et al. 2013). Also the recycling of phosphorus from wastewater is getting more and more attention (Senguptaa and Panditb 2011). As a result of seawater rise, the importance of in situ groundwater to avoid saltwater intrusion is increasing. Besides pressures on water quality, there are also various pressures on water quantity, mainly from climate change and economic development.

13.8 New Approaches for Water Pricing Under Consideration

It was recently proposed to simplify the VAT system and levy only one rate of 15 %. For tap water, this would mean an increase of 9 % in VAT; however, this plan has not yet been executed (CPB 2014).

13.9 Conclusion

An interesting aspect of the Dutch drinking water sector is the degree of transparency. Since 1997, a benchmark is executed every 3 years which gives insights in the water pricing mechanisms. One of the issues that stand out from this benchmark is that households and industries are bound to the rates of the drinking water company in their region. These rates can vary significantly based on the mixture of available raw water sources per drinking water company.

Taxes and fees can, however, also serve other purposes, such as revenue generation and cost recovery, and promote adoption of water-saving technologies. More advanced technologies, such as water-saving toilets and showers, reduce water use per person over time. Whether the adoption of such modern water-saving technologies has been triggered by the taxes is, however, hard to confirm. Given the fact that only on average 0.6 % of a household's budget is dedicated to tap water, this is not very likely.

The government sends mixed messages, charging the lower VAT rate on tap water and simultaneously taxing water to generate revenue and influencing behavior. While local fees on groundwater have been stable for nearly 30 years, plans for national taxes have, however, been subject to change in recent years. The Dutch government revoked the national groundwater tax after 16 years for being fiscally inefficient and environmentally ineffective on 31 December 2011. The rate of the tap water tax has recently increased, but a proposed drop of the tax-free threshold was canceled after consultation with stakeholders. Households will continue to pay the highest rate, while business is spared.

Acknowledgment The authors would like to acknowledge the useful comments received by Cor Graveland.

References

Accenture. (2010). *Reflections on performance 2009, benchmarking in the Dutch drinking water industry, Vereniging van waterbedrijven in Nederland*. Almelo: Association of Dutch Water Companies.

Accenture. (2013). *Reflections on performance 2012, benchmarking in the Dutch drinking water industry, Vereniging van waterbedrijven in Nederland*. Almelo: Association of Dutch Water Companies.

CPB. (2014). *Bouwstenen voor een moderne BTW* (CPB Policy Brief 2014/02). The Hague. http://www.cpb.nl/publicatie/bouwstenen-voor-een-moderne-btw. Last accessed 21 Apr 2014.

Graveland & Baas (2013). *Watergebruik steeds efficiënter*. Den Haag http://www.cbs.nl/nl-NL/menu/themas/natuur-milieu/publicaties/artikelen/archief/2013/2013-3768-wm.htm

Hofstra, N., Bouwman, A. F., Beusen, A. H., & Medema, G. J. (2013). Exploring global cryptosporidium emissions to surface water. *Science of the Total Environment, 442*(1), 10–19. ISSN 0048-9697.

IDWG. (2007). *Interdepartementale Werkgroep Grondwaterbelasting, Grondwaterbelasting nader bezien; Aspecten verbonden aan overheveling van de grondwaterbelasting naar een provinciale bestemmingsheffing en aan afschaffing van de grondwaterbelasting*. The Hague: Interdepartementale Werkgroep Grondwaterbelasting.

Iwaco. (1997). *Deelrapportage Milieu-effecten Grondwater Belasting: Evaluatie wet Belastingen op Milieugrondslag* (Report 1062060.002). Rotterdam: Ministry of Finance.

Ministry of Finance. (2013). *Herfstakkoord 2013/Belastingplan 2014*. The Hague: Ministry of Finance.

Ministry of Finance. (2014). *Belasting op Leidingwater, 4 juni 2014, IZV/2014/0089M*. The Hague. http://www.rijksoverheid.nl/documenten-en-publicaties/kamerstukken/2014/06/04/belasting-op-leidingwater.html. Last accessed 28 June 2014.

Ministry of Housing, Spatial Planning and the Environment. (2009). *Drinkwaterwet*, 18 juli 2009, BWBR0026338, Ministry of Housing, Spatial Planning and the Environment. The Hague.

Overheid.nl. (2014). *Overheidsinformatie, Regelingen van gemeenten, provincies en waterschappen*. http://zoekdienst.overheid.nl/CVDR/CVDRResultaat.aspx?searchtype=Advanced&term=grondwaterheffing&datumgeldendop=17-08-2014&provincie=&sortby=relevance%2fdescending. Last accessed on 17 Aug 2014.

RIVM. (2013). *Zorg over kwaliteit van bronnen voor drinkwater*. National Institute for Human Health and the Environment. http://rivm.nl/Documenten_en_publicaties/Algemeen_Actueel/Nieuwsberichten/2013/Zorg_over_kwaliteit_van_bronnen_voor_drinkwater

Schuerhoff, Weikard, & Zetland. (2013). The life and death of the Dutch groundwater tax. *Water Policy, 15*, 1064–1077.

Senguptaa, S., & Panditb, A. (2011). Selective removal of phosphorus from wastewater combined with its recovery as a solid-phase fertilizer. *Water Research, 45*(11), 3318–3330. doi:10.1016/j.watres.2011.03.044.

Statistics Netherlands. (2013). *Environmental accounts of the Netherlands 2012*. The Hague: Statistics Netherlands.

Statistics Netherlands. (2014a). *Bevolking, Bevolking, Huishoudens en bevolkingsontwikkeling*. http://statline.cbs.nl/StatWeb/publication/?DM=SLNL&PA=37556&D1=0&D2=l&VW=T. Last accessed 22 June 2014.

Statistics Netherlands. (2014b). *Maatwerktabellen watergebruik/wateronttrekking door bedrijven en huishoudens 1976–2011*. http://www.cbs.nl/nl-NL/menu/themas/natuur-milieu/publicaties/milieurekeningen/cijfers/archief/2014/2014-watergebruik-bedrijven-huishoudens-t-mw-1976-2011.htm. Last accessed 1 June 2014.

Statistics Netherlands. (2014c). *Environmental accounts, water use and abstraction*. http://statline.cbs.nl/StatWeb/publication/?DM=SLEN&PA=81398ENG&D1=0-1,4&D2=0,3,5-6,39,41&D3=l&LA=EN&VW=T. Last accessed 22 June 2014.

Vermeulen, L. C., & Hofstra, N. (2014). Influence of climate variables on the concentration of Escherichia coli in the Rhine, Meuse, and Drentse Aa during 1985–2010. *Regional Environmental Change, 14*(1), 307–319. ISSN 1436–3798.

Vewin. (2011). *Tarieveoverzicht drinkingwater per 1 januari 2011, Verening van waterbedrijven in Nederland*. Rijkswijk: Association of Dutch Water Companies.

Vewin. (2012). *Drinkwaterstatistieken, Vereniging van waterbedrijven in Nederland*. Rijswijk: Association of Dutch Water Companies.

Vewin. (2013). *Tarievenoverzicht 2013, Vereniging van waterbedrijven in Nederland*. Rijswijk: Association of Dutch Water Companies.

Vewin. (2014a). *Tarievenoverzicht 2014, Vereniging van waterbedrijven in Nederland*. Rijswijk: Association of Dutch Water Companies.

Vewin. (2014b). *Belasting op leidingwater, impact groter dan voorzien*. http://www.vewin.nl/nieuws/Pages/Waterspiegel_Update_impact_BoL_groter_dan_voorzien.aspx. Last accessed 10 Mar 2014.

Warner, N. R., Christie, C. A., Jackson, R. B., & Vengosh, A. (2013). Impacts of shale gas wastewater disposal on water quality in western Pennsylvania. *Environmental Science and Technology, 47*, 11849–11857. doi:10.1021/es402165b.

WHO. (2010). *Water, sanitation and hygiene: Exposure data by country*. http://apps.who.int/gho/data/node.main.606. Last accessed 1 June 2014.

Chapter 14
New Zealand Water Pricing

Bryan Jenkins

Abstract Methods of charging, water use, and cost comparisons were made for municipal, irrigation, and hydropower generation uses of water. For municipal use, city size and water metering influenced per capita use, with larger cities and metered use being associated with lower per capita use. Drinking water quality (for smaller councils), demand management (for growing cities), and long-term asset management are the developing issues for municipal water supply. For irrigation, the cost of entitlements related to the age of the scheme (older schemes with capital paid off had lower costs), recent capital investment, and operating costs. Investment in irrigation schemes was being undertaken to improve reliability of supply (through storage) and water use efficiency (through conversion of flood to spray irrigation and replacing open distribution channels with pipes). Water used for hydropower generation was driven by electricity markets. Water values were imputed with rivers with multiple hydro stations, capturing more of the head in the river system having higher values.

Keywords Water cost comparison • Water demand management • Methods of water charging • Water use comparison • Water metering

14.1 Introduction

14.1.1 Water Resources in New Zealand

While New Zealand is endowed with an abundance of water, increasing demand, particularly for irrigated agriculture, is putting pressure on available supplies using current means of abstraction and application, especially in the dry east coast of the country. New Zealand also derives a significant proportion of its energy from hydroelectric power generation. This contributes a significant proportion to the country's

B. Jenkins (✉)
Waterways Centre for Freshwater Management, University of Canterbury and Lincoln University, Private Bag 4800, Christchurch 8140, New Zealand
e-mail: bryan.jenkins@canterbury.ac.nz

A. Dinar et al. (eds.), *Water Pricing Experiences and Innovations*,
Global Issues in Water Policy 9, DOI 10.1007/978-3-319-16465-6_14

energy supply, although the proportion is declining because of little new development.

New Zealand receives about 611,000 million m^3 of rainfall per year: 2.3 m average over its area of 268,000 km^2 or 137,000 m^3 per person for its 4.47 million population (Statistics New Zealand 2011, 2013). There are marked regional variations with the west coast with high rainfall (up to 10 m/year) and the east coast with low rainfall (around 0.6 m/year).

Use for hydroelectricity generation is very high with an allocation of 160,000 million m^3/year (about 36,000 m^3/year per person). With an agricultural export economy, irrigation is the dominant consumptive use and is also high with 5,800 million m^3/year allocated to irrigated pasture, crops, and horticulture (about 1,300 m^3/year per person). Irrigation use is highly seasonal with summer allocation of 348 million m^3 per week. Allocation to municipal use (i.e., residential, commercial, and industrial) is 1,800 million m^3/year (about 410 m^3/year per person). Actual consumptive use (excluding hydro) is estimated to be 51 % of allocated use (Aqualinc 2010). With an estimated annual consumptive use in 2007 of 3,925 million m^3, the Organisation for Economic Co-operation and Development (OECD) places New Zealand second among OECD countries in terms of per capita consumptive use at 940 m^3 per person (OECD 2010). However, allocated water for consumptive use comprises less than 5 % of its renewable freshwater resource (Ministry for the Environment 2014). In contrast to its high per capita use figure, at 1.2 %, New Zealand was the third lowest OECD country in terms of water withdrawal as a percentage of gross annual availability (OECD 2009).

14.1.2 Historical Background

New Zealand has a highly devolved system of water and water infrastructure management. There was a major reform of natural resource management in the late 1980s. The number of local and regional government units was reduced from 625 to 94. Municipal water supply had been the responsibility of city and district councils, and this remained after the reforms, albeit with larger organizations and greater areas of supply. The provision of water infrastructure for hydropower generation and much of the irrigation had been the responsibility of central government through the Ministry of Works and Development, which was abolished in 1988. The change was associated with new legislation, the Resource Management Act, with a shift in philosophy for government's role to be one of regulatory activities rather than planning activities. While there are environmental and agricultural policy agencies, there is now no natural resource or water resource agency in the central government. Regional councils with boundaries based on catchments play a major role in water resource regulation.

From 1912 to 1987, the Ministry of Works and Development had the responsibility for the design, construction, and operation of government-owned irrigation schemes, as well as the responsibility for recommending annual water charges to the minister for approval. Farmers frequently stated that the charges were set too high for farms to remain viable, while officials believed the charges to be too low to recover capital costs associated with schemes (Farley 1994).

Of the economic reforms beginning in 1984, the primary focus was on the agricultural sector, and virtually all agricultural subsidies were removed. Between 1988 and 1990, 49 government-owned irrigation schemes in New Zealand were sold to private irrigators. Very few of the schemes yielded a high sale price for the government; many sold for $1 or less. Nearly $60 million of capital investment by the government was unrecovered (Farley 1994).

Also, as part of the reforms, the New Zealand Electricity Department was corporatized as the Electricity Corporation of New Zealand in 1987. Until then, New Zealand had a centrally run system of providers of generation, transmission, distribution, and retailing. There were further reforms, including in 1996 the establishment of a wholesale spot electricity market and generation assets split into several state-owned enterprises (Electricity Authority 2011). Subsequently, all the electricity generation state-owned enterprises were fully or partially privatized.

14.1.3 Recent Review of National Infrastructure

In a recent strategic review of national infrastructure (transport, telecommunications, energy, water, and social infrastructure), each sector was analyzed against guiding principles as "effective/could be further developed/ineffective" (New Zealand Government 2011). The water infrastructure sector ranked poorly. In terms of the guiding principles, *investment analysis* (i.e., investment is well analyzed and takes sufficient account of potential changes in demand), *funding mechanisms* (i.e., maintaining a consistent and long-term commitment to infrastructure and utilizing a broad range of funding tools), and *regulation* (i.e., regulation enables investment in infrastructure that is consistent with other principles and reduces lead times and uncertainty), the water infrastructure sector was ranked as "ineffective". The water infrastructure sector was considered that it "could be further developed" in relation to *resilience* (i.e., national infrastructure networks are able to deal with significant disruption and changing circumstances), in relation to *accountability and performance* (i.e., it is clear who is making decisions and on what basis and what outcomes were being sought), and in relation to *coordination* (i.e., infrastructure decisions are well coordinated across different providers and are sufficiently integrated with decisions about land use).

Strategic opportunities were seen for (a) better demand management practices and consistent performance criteria for water infrastructure, (b) the promotion of partnerships and activities within the sector, and (c) ensuring that water management assets contribute to improved social, economic, environmental, and cultural well-being of communities (New Zealand Government 2011).

14.1.4 Current Water Pricing Arrangements

A large number of municipal councils, irrigation companies, and hydrogenerators were approached to provide data and insights into their approaches to water supply for this chapter. Effective responses were received from 13 councils and 4 irrigation companies. Commercial constraints limited the responses from irrigation companies. Furthermore, for hydrogenerators with the emphasis on electricity spot prices, a different approach to volumetric valuations of water was needed.

This means more data can be presented on municipal supply in terms of per capita use and cost of supply. There is also more information on common challenges, including the need to upgrade infrastructure to meet relatively new drinking water standards for water quality, managing demand to reduce consumption, and improving information available for forecasting, planning, and asset management.

Comparative analysis of types of irrigation schemes shows marked variations in water pricing, depending on age and level of recent investment in scheme infrastructure. As water availability is becoming scarce, investments are occurring in storage for improved reliability and piping and irrigation technology to improve water use efficiency. Limited water trading occurs, and the range of traded prices is presented.

For hydroelectricity, where a spot market determines electricity prices at half-hourly intervals, short-term water pricing is problematic. However, the market structure is based on the premise that average prices over time will provide for the incentive for investment in new installed capacity when it is warranted economically. This enables a calculation of the long-term average value of water for a hydropower generation scheme.

Environmental flow requirements in rivers and lake level requirements for ecological purposes are set as legal constraints through regional plans and consenting rules in relation to the use of water under the Resource Management Act. There are restrictions on takes from a river. These include the specification of "minimum flows," the flow at which abstraction must cease, and "allocation limits," limits on the total volume that can be taken. Regional councils measure river flows and monitor compliance with restrictions. For example, the Waimakariri irrigation scheme is a run-of-river scheme, but can only take its full allocation of 10.5 m^3/s when the flow in the Waimakariri River is greater than 63 m^3/s at the gauge at the Old Highway Bridge. If the river flow drops to 41 m^3/s, then only a stockwater provision can be accessed. Thus, environmental services are protected by regulation rather than pricing.

14.2 Municipal Water Pricing

14.2.1 Methods of Charging

Of the 13 councils reviewed, the majority sought cost recovery for water supply infrastructure as a targeted rate (i.e., a charge) per residential property (seven councils) or as a targeted rate on property value (two councils), whereas commercial

Fig. 14.1 Locations of councils used in survey

properties were usually metered and charged on a volumetric basis. Four councils had universal metering for their urban areas. The locations of the councils surveyed are shown in Fig. 14.1.

14.2.2 Management of Infrastructure

In most cases, the water supply infrastructure was managed by a department within the council. With the creation of the Auckland "super" city by amalgamating seven city and district councils and the regional council in 2010, a separate council-controlled organization (Watercare Services Limited) was created to manage all water infrastructure (water supply, wastewater, and stormwater). For Greater Wellington, bulk water supply is provided by the regional council to four city councils. Water distribution for two of the councils (Wellington and Hutt City) is undertaken by a council-controlled organization (Capacity Infrastructure Services), and the other councils (Upper Hutt and Porirua) are in the process of becoming shareholders in Capacity.

14.2.3 Water Use Data

Data on water usage for 13 councils, ranging in size from Auckland at 1.4 million to Kaikoura at 4,000 people, was obtained. Table 14.1 sets out the estimated population served and the water production[1] for each council. Where data on water consumption were available, these are also included. Such data were available from councils with universal metering and from those where there had been recent surveys of distribution leakage and other unaccounted-for water use (e.g., fire-fighting use). The overall council use per capita is also shown: this includes residential and nonresidential use for the council area.

Figure 14.2 plots the results for per capita production in liters per person per day (l/p/d) against the population served (on a log scale). The graph shows the reduction in per capita usage with city size. Auckland has the lowest per capita production at 274 l/p/day, and Kaikoura has the highest at 685 l/p/day. Rural residential use is expected to be higher than for urban settings.

In Fig. 14.2, councils with metering for residential cost recovery are shown with squares, and those with property charges are shown with circles. It is noteworthy that metered supplies have lower per capita use for cities of comparable size. Tauranga, at 296 l/p/day, has lower per capita production, compared to other cities of around 100,000 population (338–356 l/p/day). Nelson (450 l/p/day) has lower per capita production than other urban areas of 30,000–50,000 population (495–619 l/p/day). Tasman (531 l/p/day) has comparable per capita production to Gisborne (495 l/p/day) but lower per capita consumption (296 compared to 431 l/p/day).

14.2.4 Water Cost Data

Table 14.2 sets out the cost of production for water (in \$NZ[2]/m^3) for the 13 councils surveyed. The costs vary from \$0.47/m^3 for Christchurch to \$1.91/m^3 for Tasman. Where metered charges for residential and commercial use are in place, these are

[1] Water production is the annual amount of water supplied into the water distribution system from treatment plants, bore pumping, or supplied from another council.

[2] Exchange rate \$NZ = 0.84 \$US in August 2014.

Table 14.1 Available data on annual water production and per capita use for 13 councils (2012/2013)

Council	Estimated population served	Water production $m^3/year \times 10^3$	Water consumption $m^3/year \times 10^3$	Per capita production l/p/day	Per capita consumption l/p/day
Auckland[a]	1,375,893	137,000	119,000	273	237
Christchurch[b]	360,411	47,000	–	357	–
Wellington[b]	199,280	26,601	–	366	–
Dunedin[c]	119,500	15,119	12,095	347	277
Tauranga[a]	117,000	12,654	10,931	296	256
Hutt City[c]	103,000	12,700	–	338	–
Palmerston North[c]	75,000	9,800	–	360	–
Nelson[a]	45,000	7,400	5,000	450	304
Timaru (urban)[c]	32,000	7,794	–	667	–
Gisborne[c]	30,600	5,532	4,810	495	431
Tasman[a]	27,777	5,383	3,000	531	296
Ashburton[c]	24,000	5,425	–	619	–
Kaikoura[c]	4,000	1,000	–	685	–

Main form of residential charging
[a]Metered volume
[b]Capital value rate
[c]Property charge

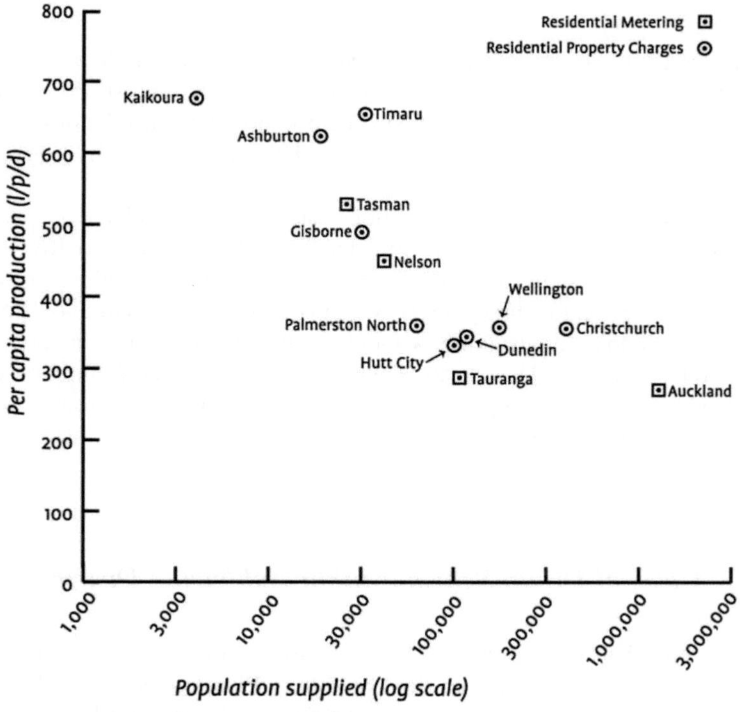

Fig. 14.2 Per capita production of water compared to size of population supplied

Table 14.2 Cost of water production and metering charges for council supplies

Council	Cost of water production ($/m^3)	Residential metering charge ($/m^3)	Commercial metering charge ($/m^3)
Auckland	1.09	1.343	1.343
Christchurch	0.47		0.62
Wellington	1.42		1.797
Dunedin	1.72		1.33
Tauranga	1.37	1.73	1.73
Hutt City	1.10		1.69[a]
Palmerston North	0.70		0.966
Nelson	1.47	1.914	1.914[b]
Timaru urban/rural	0.52		0.57
Gisborne	0.91		1.04
Tasman	1.91	1.87	1.87
Ashburton	0.84		0.72
Kaikoura	0.95		1.00[c]

Notes:
[a]$1.69/m^3 up to 100,000 m^3; $1.21 >100,000 m^3
[b]$1.914/m^3 for 0–10,000 m^3/year; $1.493/m^3 for 10,000–100,000 m^3/year; $1.179/m^3 for >100,000 m^3/year
[c]For use >365 m^3/year

indicated for the main component (i.e., administrative and other charges associated with metering are excluded). Figure 14.3 shows the cost data for the councils in ascending order of population supplied. There is no apparent relationship to city size. Rather, differences are more likely to be explained by hydrogeological setting.

Christchurch, with the lowest production cost, has the most advantageous setting for urban water supply. It receives snowmelt from the Waimakariri River, which supplies an aquifer system that acts as a natural sand and gravel filter. The aquifer system flows underneath Christchurch, which is on flat terrain. Groundwater bores bring water suitable for drinking without treatment under artesian pressure to Christchurch's doorstep. In contrast, Dunedin ($1.72/m^3) and Wellington ($1.42/m^3), two of the higher cost suppliers, receive most of their supply from distant surface water storages. There is also variable terrain over which water has to be distributed.

14.2.5 Costs of Urban Schemes in the Ashburton District

There are ten urban water supply schemes in the Ashburton District (population 24,000) in a wealthy rural area on the Canterbury Plains centered on the town of Ashburton. The district's accounting system allows for the identification of costs of the different community water supplies. Table 14.3 shows the number of

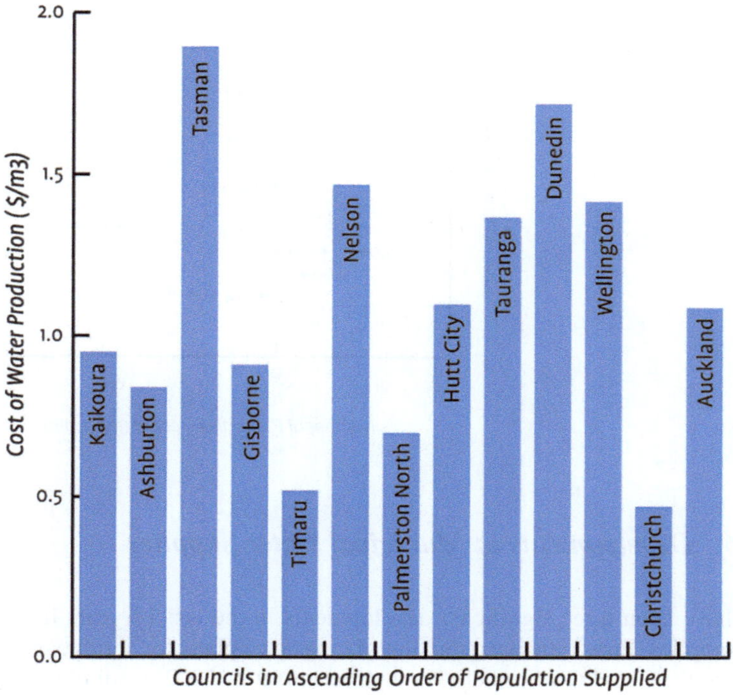

Fig. 14.3 Water production costs for councils in ascending order

Table 14.3 Unit costs of urban schemes in the Ashburton district

Scheme	No of properties connected to scheme	Unit cost ($/m^3)
Ashburton	7,693	0.85
Methven	938	1.55
Rakaia	525	0.81
Hinds	129	0.73
Mt Somers	101	2.07
Lake Hood	86	3.03
Chertsey	81	0.98
Fairton	73	1.25
Mayfield	60	1.25
Hakatere	59	3.83

properties connected to the schemes for the ten urban areas and the unit cost for each scheme.

Figure 14.4 plots the cost of water supply (in $/m^3) against the number of properties connected to urban water supplies. While there is some variation, it shows economies of scale, with costs varying from $3.90/m^3 for the smallest community Hakatere (59 properties) to $0.85/m^3 for the largest community Ashburton (7,693 properties).

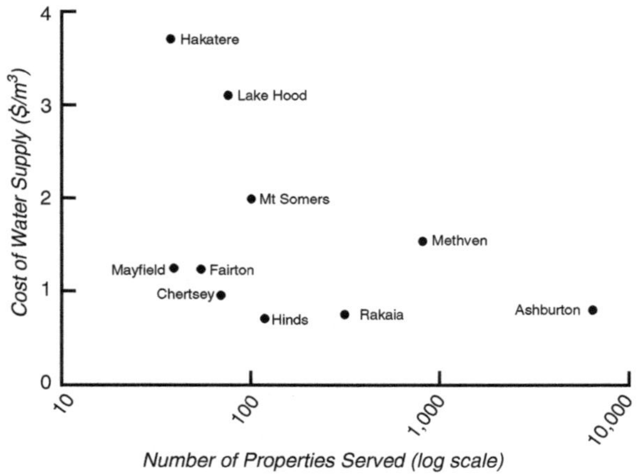

Fig. 14.4 Cost of water supply for urban areas in Ashburton District

14.2.6 Developments in Municipal Water Supplies

There have been three significant developments in the last 12 years in municipal water supplies. First is the introduction of long-term plans for significant activities as part of the Local Government Amendments Act of 2002, leading to water supply activity management plans being prepared by all councils. Second is the introduction of the Health (Drinking Water) Amendment Act 2007, requiring the preparation of public health risk management plans (PHRMPs)[3] that identify threats to drinking water quality and how they should be managed. Third, there is a growing interest in water demand management as issues associated with water availability in New Zealand have intensified. These issues were highlighted in a recent review by the Office of the Auditor General, which noted: "Common challenges included the need to upgrade infrastructure to meet the drinking water standards for water quality, managing demand to reduce consumption, and improving the information available for forecasting and asset management" (Office of the Auditor General 2010).

14.2.7 Drinking Water Quality Management

New Zealand has relatively high rates of largely preventable enteric and gastrointestinal disease. For example, the campylobacteriosis rate is twice that of England and three times that of Australia and Canada, which is partly attributable to contamination of drinking water (Ministry for the Environment 2007). While the risk

[3] Retitled "Water safety plans" in legislative changes in December 2013.

management approach considers both source protection and water treatment, many of the risks of contamination come from runoff from high rainfall events in unprotected catchments or from leaching of contaminants from land use intensification over unconfined groundwater. Two examples below highlight the costs associated with these issues.

14.2.8 Hurunui District Council

The Hurunui District is one example of a community that has relatively poor water drinking water quality. There are 13 water schemes run by the council that extract water from 23 different water intakes. Eight of the supplied communities are on permanent "boil water notices"[4] (Hurunui District Council 2011b). Out of the 23 water intakes, six are deep bores (more than 70 m deep) that are less at risk of contamination, while the other 17 are shallow supplies with a higher risk of contamination (Hurunui District Council 2011b). This rural community has had multiple temporary boil water notices for water supplies in the area, due to the detection of *Escherichia coli* (*E. coli*) in the shallower bores (Hurunui District Council 2014).

The Hurunui District Council is producing PHRMPs for its drinking water sources as part of its work program for meeting its water supply responsibilities. The work to be completed to improve the water supply has been estimated at $14 million for capital works, plus $484,000 per annum for operational costs. Legislation requires the Hurunui District Council to achieve compliance within 3 years of the PHRMP becoming operative. However, since the Hurunui is a small, rural council funding of the upgrades is challenging, and, with the agreement of the Ministry for Health, the compliance period has been extended to 10 years. Hurunui has planned the installation of nine new mixed oxidant plants that will improve the quality of the drinking water supply so that boiling notices can be removed from water schemes (Hurunui District Council 2012).

The Cheviot water scheme in the Hurunui District is one water supply that has been on a permanent boil water notice since 2007 (Hurunui District Council 2011a). Although the PHRMP has been approved, for this scheme to be in compliance with the New Zealand drinking water standards, there will need to be significant changes in the system. These changes include either treating the whole water supply or finding a completely new water source. Treating the water in Cheviot would cost $1.1 M (capital costs) plus an additional $175,375 (operating costs) and would cause a 31 % increase in rates. This equates to an additional $389 to the property rates for each dwelling connected to the network.

[4] Boil water notices are a formal requirement of the supply authority to notify residents of the need to boil drinking water owing to the ongoing risk of microbial contamination.

14.2.9 Timaru District Council: Downlands Water Scheme

Through the PHRMP process, the Timaru District Council has identified areas in its water schemes that need additional attention and upgrading. The majority of the land in the supply catchments is used for farming. One example is the Downlands water scheme, which supplies part of three district councils: Timaru, Waimate, and Mackenzie. The water within the Downlands water scheme requires improvement to prevent illness in humans and stock. Currently, it is treated by chlorine at three treatment plants and five reservoirs, but upgrading of the system is needed to meet New Zealand drinking water standards. At present, risks to the water scheme include contamination of the source (the Waitohi River), the presence of *E. coli* and protozoa after water treatment, high turbidity after treatment, and taste and odor complaints due to the need to dose with high levels of chlorine (Downlands PRHMP). The risks could occur due to multiple causes, including contamination due to agricultural activities, growths of cyanobacteria, poor raw water quality or leakage, and backflow.

Improvement to the Downlands water supply will require upgrading existing infrastructure and possibly additional water storage. These improvements as well as others discussed in the Downland's PHRMP will most likely increase rates paid by consumers. According to Timaru District Council forecasts, the annual water charges for domestic users would increase from current levels of $392.00 (2012–2013) per property to $510.00 from 2015 to 2022 (Timaru District Council 2014).

14.2.10 Water Demand Management

Three examples are provided in relation to water demand management. The first relates to Tauranga, a city of about 120,000 people, 200 km southeast of Auckland, which introduced water metering as a demand reduction measure and to defer expenditure on new water supply infrastructure. The second is provided by Greater Wellington (population nearly 400,000), where water demand is declining through leakage reduction, water-saving devices, and behavioral change, but without metering. The third comes from Auckland City, with 1.4M people—40 % of New Zealand's population and increasing at about 1.1 % per annum. Universal metering has been in place since the early 1990s, so more sophisticated demand reduction techniques are now needed, as access to further water supplies is becoming increasingly constrained.

14.2.11 Tauranga City Council Water Metering

In 2001, Tauranga implemented universal water metering, including residential water supply. This approach was taken to reduce water demand to delay the need for a new water supply scheme. Tauranga has a population of 120,000 people, which

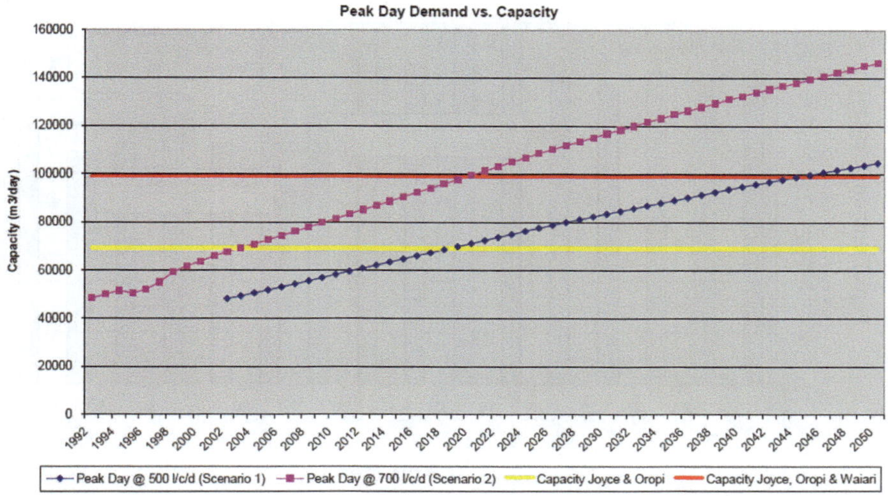

Fig. 14.5 Demand/capacity projections for Tauranga water supply (Sternberg and Bahr 2011)

has doubled in the last 20 years and is projected to double again by 2050. Demand projections in the mid-1990s predicted that the capacity of the existing plants would be able to cope with a peak demand of up to 700 l/p/day until approximately 2004–2005. Building a new water scheme (Waiari) was estimated to have a capital cost of \$75M (Sternberg and Bahrs 2011).

The implementation of universal water metering resulted in a reduction in peak demand of approximately 30 %, with average demand reducing by about 25 %. This enabled the proposed Waiari water scheme to be delayed by at least 10 years. Figure 14.5 shows the existing system capacity of nearly 70,000 m^3/day (with the Joyce and Oropi schemes), which is capable of being increased to nearly 100,000 m^3/day (with the addition of Waiari scheme). With a peak day demand of 700 l/p/day, existing capacity is exceeded in 2004, and even the expanded capacity is exceeded in 2020. However, with peak demand reduced to 500 l/p/day, existing capacity can meet demand until 2018 and expanded capacity to 2045 (Sternberg and Bahrs 2011).

Prior to the introduction of metering, water restrictions were required during the summer peak demand even after a plant upgrade to increase peak capacity (the Joyce Road plant upgrade). Since the introduction of metering, there has been no requirement for restrictions, despite a population increase of about 27 % during this period (Fig. 14.6).

As water demand reduced, there was a corresponding reduction in wastewater volumes generated. This decreased the volume that needs to be conveyed and treated, resulting in operational savings as well as deferral of expenditure for increased capacity.

Based on a financial analysis over 30 years of a "with meters" and "without meters" comparison, there were cost savings of \$141M (net present value of \$83M at a 5 % discount rate). This includes the installed value of meters at \$9.9M for

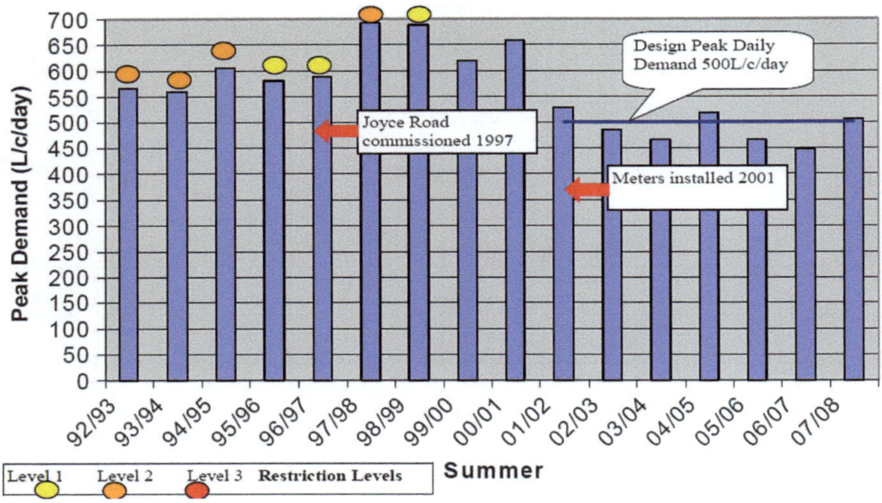

Fig. 14.6 Change in peak demand and restrictions since metering in Tauranga (Sternberg and Bahr 2011)

39,000 meters, i.e., an installed cost of about $250 per meter and an annual cost of $19 per year, including meter replacement every 15 years (Sternberg and Bahrs 2011).

14.2.12 Greater Wellington Water Usage

Greater Wellington supplies bulk water to a population of nearly 400,000, which is growing at around 1 % per annum. Greater Wellington has seen a reduction in per capita consumption over the last 7 years, which has offset the increased demand from population growth. Figure 14.7 shows a 20 % decline in annual gross water consumption from 425 l/p/day in 2006 to 340 l/p/day in 2013. While the population supplied has increased from 377,000 to 396,000 (a 5 % increase), the average daily supply has declined from 158,000 to 136,000 m^3/day (a 14 % decrease) (Greater Wellington Regional Council 2013).

This has occurred in the absence of metering of residential water supply. Indicators point to behavioral change and lower rates of leakage from distribution pipes as the main reasons for declining per capita usage.[5] There are trends toward higher-density housing and more efficient water-using appliances, toilets, showers, and taps.

[5] Between 2011 and 2012 and 2012 and 2013, there have been reductions in residential, commercial, and unaccounted-for water. There has been an education program for water conservation and efficiency (Wellington City Council 2011) and leak reduction measures of zone metering, pressure management, and leak detection (Capacity Infrastructure Services 2013).

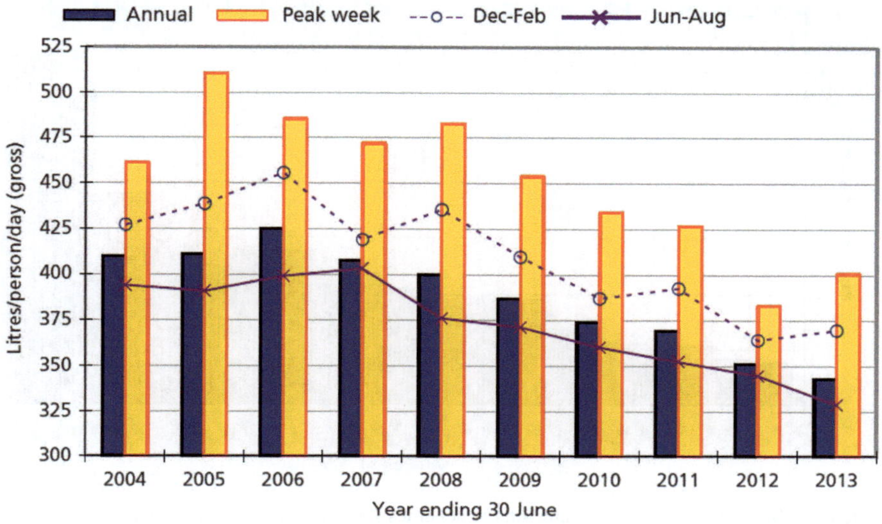

Fig. 14.7 Water consumption in Greater Wellington (Greater Wellington Regional Council 2013)

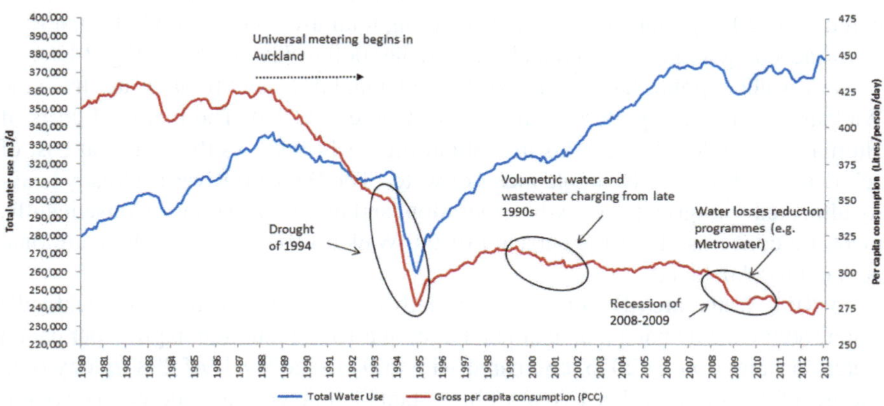

Fig. 14.8 Total and per capita water use for Auckland 1980-2013 (Watercare 2013)

14.2.13 Auckland Demand Management

Auckland implemented universal metering in 1990–1992, which saw a reduction in gross per capita consumption from 425 to 350 l/p/day (18 % reduction) before the drought of 1994 lowered the consumption to 275 l/p/day. This had rebounded to 320 l/p/day by 1999. However, ongoing water demand reduction programs have lowered per capita consumption to 275 l/p/day—the lowest for an urban area in New Zealand (Fig. 14.8) (Watercare 2013).

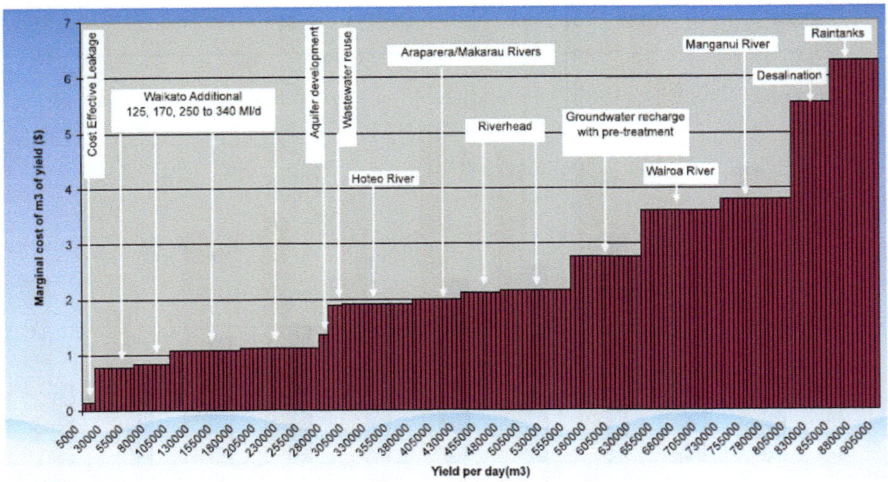

Fig. 14.9 Marginal cost of new water supply schemes for Auckland (Corneby 2010)

However, with Auckland's population growth, total water use for the city is still increasing. By 2051, Auckland's population is expected to increase by 57 %, compared to its 2011 population. The pre-metering total use volume of 335,000 m^3/day was reached again in 2003 and had reached 380,000 m^3/day in 2013 (Fig. 14.8).

New water supplies have been investigated with taking additional water from the Waikato River as the preferred alternative (Corneby 2010). The marginal costs of alternatives are double the Waikato option. Figure 14.9 shows the marginal cost of achieving each successive cubic meter of water. The Waikato River is already heavily allocated for agriculture, hydrogeneration, and municipal supply. Consequently, there are concerns about the capacity of the Waikato River to meet the additional demand for Auckland.

In 2011, Watercare (the council-controlled organization that supplies Auckland's water) adopted a demand management target of a 15 % reduction in gross per capita water consumption by 2025, compared with 2004 usage levels of 298 l/p/day (i.e., a target of 253 l/p/day). The two most important factors were considered to be population growth and peak demand. The Demand Management Plan is based on the *Guide to Demand Management* developed by the Water Services Association of Australia (Institute for Sustainable Futures 2008).

In its Demand Management Plan, Watercare has targeted leakage reduction (6,000 m^3/day), residential education (4,000 m^3/day), water efficiency in new buildings (3,000 m^3/day), and improved efficiency in nonresidential high water users, such as food processing (2,000 m^3/day), and is investigating source substitution, such as gray water and rainwater tanks, as well as further wastewater volumetric charging to achieve its water reduction target (Watercare 2013).

14.3 Irrigation Schemes

14.3.1 Methods of Cost Recovery and Management of Infrastructure

Only a limited number of responses were received from irrigation companies. The private schemes that provided information were all shareholder-owned cooperatives. One response was from a district council scheme.

Share ownership in a private irrigation company entitled farmers to a water allocation for an irrigable area. The share price covered the capital costs, and an annual charge per share covered the operational costs with an allowance for a small surplus. The district council scheme provided the option of either a lump-sum entry payment to cover capital costs or an annual land value rate against the irrigable area of the property. Operational costs (including depreciation and overheads) are recovered by a metered charge.

14.3.2 Cost of Water Entitlements

The nature of irrigation schemes is that irrigation is a supplement to rainfall, so the quantity used is highly variable depending on the combination of crop requirements and rainfall. Irrigators who are shareholders in a scheme purchase an entitlement to access water. Table 14.4 sets out the arrangements between shares and entitlements for four different schemes.

Converting the entitlements to weekly equivalents provides a basis for comparing water use requirements. It is interesting to note the higher allocation (37 mm/

Table 14.4 Scheme entitlements

Scheme	Irrigated area	Share requirement	Water entitlement	Weekly equivalent
Old border dyke scheme	2,300 ha	30 shares per allocation, 10 shares per ha	1,200 mm over 227 day irrigation season	37 mm/week
Recent run-of-river scheme	18,000 ha	0.075 l/s per share, typically 7 shares per ha	0.525 l/s/ha	31.8 mm/week
Recent scheme with storage	16,000 ha	Infrastructure share plus annual water charges	0.41336 l/s/ha with 5,625 m^3/ha annual cap	25 mm/week
Horticulture scheme with pressurized supply	4,500 ha	Up-front payment (or capitalized charge) plus annual charge	1.8 mm max per day	12.6 mm/week

Source: Survey responses from irrigation companies

Table 14.5 Cost of irrigation entitlements

Scheme	Share price ($/ha)	Annualized share cost ($/ha/year)	Operating cost	Entitlement	Capital cost contribution ($/m³)	Operating cost contribution ($/m³)	Total cost ($/m³)
Old border dyke scheme	2,500[a]	406.90[d]	$376,194 for 2,300 ha	1,200 mm for season	0.0339	0.0136	0.0475
Recent run-of-river scheme	7,350[b]	1,196[d]	$15.55/share $108.85/ha	31.8 mm/week	0.111	0.0101	0.121
Recent scheme with storage	5,250[b]	854[d]	$0.048/m³	5,625m³/ha annual max	0.152	0.048	0.200
Horticulture scheme with pressurized supply	4,850[c]	324.75[e]	$0.2553/m[c]	1.8 mm/day	0.0661	0.2553	0.3214

Notes:
[a]Price for new quota created by improved efficiency
[b]Current market price for shares expressed as $/ha
[c]Council entry price to the scheme
[d]Annualized share cost at 10 % pa over 10 years
[e]Targeted rate set by council

week) needed for border dyke (flood) irrigation, compared to a more recent run-of-river scheme, which is predominantly spray irrigation (31.8 mm/week). A lower entitlement (25 mm/week) is made in which storage provides a higher reliability of supply. Horticulture requires less than pasture irrigation with only 12.6 mm/week needed. The cost of these entitlements consists of a capital and an operating component as set out in Table 14.5.

The low capital cost of the old border dyke scheme reflects not only the greater time for capital repayment but also the effective subsidy of the capital write-off by the government when the scheme was transferred to farmer ownership in the late 1980s. The higher capital component of the recent scheme with storage reflects the high capital up-front investment in storage. The high operational cost of the pressurized supply reflects the pumping cost involved in maintaining the irrigation system under pressure.

Actual use averaged over the whole scheme is much lower than the entitlements for all schemes. Over the last 5 years, irrigation demand for the storage scheme varied between 31.5 and 51.7 Mm^3 (i.e., 5.8 and 9.5 mm/week, compared to the weekly entitlement of 25 mm). Over the last 4 years, water use for the horticultural scheme varied between 1.99 and 2.91 million m^3 (i.e., 1.13 and 1.66 mm/week, compared to the weekly entitlement of 12.6 mm/week).

14.3.3 Changes in Irrigation

The rapid expansion of dairying into dryland farming areas has led to an unprecedented demand for water for irrigation. Dairying is now New Zealand's main export earner. The OECD examined trends in irrigated areas in member countries for ten-year periods between 1990–1992 and 2001–2003 (OECD 2008). In aggregate OECD area irrigated rose by 8 % over those 10 years. New Zealand exceeded all other OECD countries with a 90 % increase in irrigated area in that time. It continues to grow at 6 % pa with 70 % of the growth in the Canterbury region (Aqualinc 2010).

This has led to a heightened interest in increased storage, increased efficiency of water use, and improved reliability of supply for consistent dairy pasture growth (Jenkins 2013). An off-river storage at Arundel is nearing completion to supply 14,000 ha of new irrigation at a cost of \$82M. Consent approval has been received for a tributary storage on the Waitohi River, with diversions from the Hurunui River, with a possible storage capacity of 210 million m^3 to irrigate 60,000 ha. Another consent approval for varying lake levels in a hydro storage at Lake Coleridge will make more water available for the Central Plains Irrigation Scheme with a stage 1 capacity of 20,000 ha proposed to be irrigated.

In relation to irrigation efficiency, there continues to be a marked shift from border dyke (flood) irrigation to spray. For example, the Ashburton-Lyndhurst scheme, originally designed for border dyke irrigation, now has 67 % spray irrigation with a current conversion rate of 7 % a year. In mid Canterbury, there have been projects to upgrade the original open channel network to reduce conveyance losses. The Ashburton-Lyndhurst scheme has completed the first stage of a piped delivery system and is proceeding with a second stage. The initial scheme (at a cost of \$8M) replaced 31 km of open channel with pipe servicing 3,500 ha of irrigated land and enabling a further 550 ha to be irrigated with improved efficiency. A second stage (estimated to cost \$95M) involves more than 200 km of pipe to supply the remaining 21,000 ha of the scheme with the ability to supply a further 4,000 ha. This would also restore 100 ha of land currently in channels to productive farmland. With the use of a pressurized pipe system, there is reduction in energy requirements for pumping irrigation water. Similar "pipe replacement of open-channel" projects is in progress for the Valetta scheme (13,000 ha) and the Mayfield-Hinds scheme (32,000 ha) (Jenkins 2013).

There have also been storages at a smaller scale to improve reliability of supply. There have been many on-farm storages, e.g., on a 779 ha dairy farm milking 1,600 cows, a 2 ha storage pond capable of holding 40,000 m^3 of water has been constructed as insurance against weather and water restrictions. This provides enough water to irrigate pasture with a 585 m center pivot for 10 days. Irrigation schemes are also putting in storage to offset run-of-river restrictions. Mayfield-Hinds Irrigation is constructing a 6.1 Mm^3 capacity pond at Carew to offset a 20 % river restriction for 21 days. Waimakariri Irrigation Limited is seeking approval for an 8.2 Mm^3 storage at Wrights Road. This will hold enough water for 9 days of full irrigation to 18,000 ha of farmland. Water will be stored when river flows are high and irrigation demand is low and used when abstraction is on restriction at times of low river flow. The additional storage would have made the scheme fully reliable for 27 of the past 42 years. Without storage, the scheme would have been fully reliable for 1 year in 42 years. In the dry conditions of last summer (2012–13), an estimated $30M of production was lost because of restrictions to irrigation supply (Jenkins 2013). These infrastructure investments will significantly increase the annual charges to irrigators, reflecting the increased value being placed on reliable water supply. Mayfield-Hinds estimates an increase in its annual charge from $53/ha to $343/ha and Waimakariri Irrigation from $107/ha to $217/ha (Irrigation New Zealand 2012).

An emerging issue for irrigation development is the need to recognize and make provision for environmental flows. Reviews of river flow regimes indicate the need to maintain and enhance instream flows to protect instream environmental services, thereby constraining or reducing water availability for out-of-stream uses. A second issue is water quality impairment from land use intensification, which is constraining further irrigation expansion due to nitrate concentrations and bacterial contamination affecting drinking water quality, as well as nutrients (nitrates and phosphates) and sediments adversely affecting aquatic ecosystems.

14.3.4 Water Trading

The Resource Management Act allows transfer of water permits. Transfers are governed by the same consenting rules as new applications for water in relation to approval requirements and assessment of effects. Thus, while water was available to be allocated, few transfers occurred. Now that allocations are constrained in Canterbury and Otago, there is greater interest in transfers of allocations. A company, HydroTrader, has been established to facilitate the buying and selling of leasing of water permits.

A summary of the prices of water trading facilitated by HydroTrader is set out in Table 14.6. This shows a maximum of $1.59/$m^3$, a minimum of $0.25/$m^3$, and an average of $0.88/$m^3$. This is from 24 comparable trades between May 2008 and September 2013 for permits greater than 100,000 m^3/year or 1,000 m^3/day (HydroTrader 2013). This represents a low level of trading activity.

Table 14.6 Summary of prices of water trading (HydroTrader 2013)

Price range	Price ($/m³)	Timing
Top price	1.59	June 2010
Average price	0.88	May 2008–Sept 2013
Bottom price	0.25	April 2011

14.4 Hydroelectricity

New Zealand is a major user of hydropower, with about 54 % of its electricity generated as hydroelectricity (Electricity Authority 2011). New Zealand has a competitive market for electricity generation, regulation of the natural monopolies of transmission and distribution, and a competitive market for retail. Generators make offers to supply electricity at 59 grid injection points, while retailers and major users buy electricity at 226 grid exit points. Auctions are every half hour and set prices at each node, with differences determined by the combined effects of losses and network congestion (Alvey et al. 1998). But, at the simplest level, the cheapest generation offers are accepted, and the highest-priced offer accepted sets the spot price for that half hour period—the market clearing price. Spot prices can vary significantly from period to period (Electricity Authority 2011).

Imputing a value for a cubic meter of water on the basis of electricity spot prices is problematic. Furthermore, in a hydro-dominated market, with limited storage, electricity prices are very sensitive to wet and dry years (Read 2009). For example, 2012 was a record low year for catchment runoff to hydro storages: the average wholesale price received for South Island generation production by the largest hydro producer was \$98.8/MWh (Meridian Energy 2012). By comparison, 2010 was an average year in terms of catchment runoff: the average wholesale price was \$48.3/MWh (Meridian Energy 2012). Tipping and Read illustrate the wide range of spot prices produced by the market and use a hybrid top-down/bottom-up model to impute a marginal value curve for water, as a function of relative storage level, on the basis of those spot prices (Tipping and Read 2010).

That curve reflects the fact that the opportunity cost of water in storage is highly variable. If a storage is full and further runoff is coming into the storage, then the opportunity cost of water is zero as further water will spill. However, if a storage is low, then the opportunity cost will be high. The situation is even more complex when there is a series of storages on the one river, since the potential energy value of stored water decreases as it flows downstream (Read 2009). But while the dynamic internal valuation guides short-run reservoir management, it is not really comparable with the average economic valuations discussed above, except in very long-run average terms.

The market design for NZ electricity only provides for a market in energy, i.e., there is not a separate market or provision for installed capacity as occurs in some market designs in other jurisdictions. The market design is based on the premise that average prices over time will provide the incentive for investment in new installed capacity when it is warranted economically (Read 2009). Thus, to provide an indication of the current average value of water for hydro generation, it is possible to consider a value based on the average generation multiplied by the average price received from generation divided by the average flow.

Table 14.7 sets out the typical generation from some of New Zealand's hydro schemes (in GWh per year) and the average flow of the river at the site of the final hydro station in the river sequence (in m^3/s). Based on the average wholesale gen-

Table 14.7 Water values for hydro schemes

Hydro scheme	Typical power generation (GWh/year)	Average flow downstream of last hydro scheme (m^3/s)	Average wholesale price for 2010 ($/MWh)	Water value ($/$m^3$)
Waitaki (8 stations)	7,665	385	56.5	0.036
Waikato (9 stations)	3,935	240	62.1	0.032
Clutha (2 stations)	3,700	490	56.1	0.013
Opuha (1 station)	25	8.28	59.5	0.0057

eration price for the relevant grid injection point for the year 2010 (Electricity Market Information 2014), which is a mid-range year for generation prices, a calculation of the value of water per cubic meter ($/$m^3$) for the hydro schemes has been made.

The value difference reflects the difference in the head in each river captured by hydro generation—potential energy is based on head and flow. With multiple power stations on the Waitaki and Waikato, higher water values ($0.036/$m^3$ and $0.032/$m^3$, respectively) are achieved compared to the Clutha ($0.013/$m^3$) and the much smaller hydro station on the Opuha Dam ($0.0057/$m^3$), which is a joint hydro and irrigation storage.

With the increasing demand for water for irrigation and municipal uses, and the increased attention being given to environmental flows, there is now consideration being given to the use of hydro allocations for other purposes. One example is whether it is cost effective to divert water from the upper Waitaki to provide irrigation supply in South Canterbury (URS 2014). The study compared the transfer costs, in the range of $20,000–$30,000 per hectare with replacement energy generation costs for the hydropower generation lost, with a long-run marginal cost of $122.4/MWh. The economic analysis showed that none of the transfer concepts produced a positive economic benefit nor were they affordable for any land use likely to utilize the water (URS 2014).

A second example evaluated the potential impact on the electricity sector from increasing minimum flow requirements on eight hydro schemes (Comet 2013). Increasing minimum flows can reduce the flows diverted to hydropower generation or reduce the flexibility to store water at low-value generation times for use at high-value times. The economic cost of increased minimum flows was based on replacing lost hydropower generation with non-hydropower generation—assumed to be at a cost of $85/MWh, which was estimated to be the approximate long-run marginal cost of new baseload generation. The effect of setting minimum flows at 80 % of the natural minimum flow had cost impacts between 4 % and 64 % of the scheme's total value (Comet 2013).

14.5 Concluding Comments and Future Pricing Directions

Despite its relative abundance in New Zealand, water availability is reaching sustainability limits of its current methods of abstraction and use.

Municipal water supply is managed by city and district councils. Water pricing plays a limited role in municipal water supply with many councils still using rates on property for cost recovery. However, councils that have universal metering and charge according to usage have lower per capita water use, compared to those of similar size using property rates. Drinking water quality (particularly for smaller councils), water demand management (particularly for growing cities), and long-term asset management are the developing issues for municipal water supply.

Irrigation companies are mainly shareholder cooperatives, after the government sold its schemes in the late 1980s. Share price covers annual costs, and an annual charge covers operational costs. For irrigation, the issues of water availability are increasing in significance, leading to interest in storage, increased efficiency of water use, and improved reliability of supply. Water trading has only a minor role. Also of importance are the maintenance of instream flows in rivers and water quality impacts of land use intensification.

Hydroelectric production is a major user of water. Competitive markets have been established for electricity generation and retail. Prices for water can only be imputed indirectly.

As noted above, for municipal supplies, the current concerns are with meeting drinking water standards and meeting future population demand. In relation to water pricing, the water industry association, Water NZ, is piloting a benchmarking process for financial and nonfinancial indicators for municipal water supplies. It is reinforcing the concept of universal metering to provide a direct financial incentive for users to conserve water (Water New Zealand 2013).

Driven largely by the growth in the dairying industry, there is an increased demand for irrigation water whose availability is reaching or has reached sustainability limits. There is now increased competition between alternative uses and for environmental services. In addition, water quality impairment from increased land use intensification is becoming a constraint. Allocation of nutrient capacity is becoming contentious. A market in nitrogen discharge allowances has been recently created for Lake Taupo.

Based on advice from the Land and Water Forum,[6] the central government has introduced a national policy statement on freshwater management (New Zealand Government 2011, 2014). This is focused on directing regional councils to setting limits for water quality for water bodies and for water quantity to protect environmental flows. The Land and Water Forum also recommended collaborative approaches for the development of any land and water strategy similar to the suc-

[6] The forum was a multi-stakeholder group (58 participating organizations) of water interests established with the support of the government "in the belief that the stakeholders needed to engage directly with each other if we were to find a way forward" (Land and Water Forum 2010).

cessful Canterbury Water Management Strategy (Land and Water Forum 2012). In relation to improving water allocation from the current first-come, first-served administrative system, it recommended the government "consider three options for efficiently allocating water after instream limits have been set: continuing existing consents but using consent expiry as an opportunity to make changes to conditions; using a different administrative system, based on efficiency criteria and community considerations; payment, including through tendering, auction or regular re-tendering of permits" (Land and Water Forum 2010).

The recent partial sales of state-owned electricity generators have raised the issue of who "owns" water under the Treaty of Waitangi. Under the current legal paradigm, water is considered to be a public good that no one owns.

While projections of climate change are being generated for New Zealand, it is only recently being considered in policy terms. The general pattern for the west coast is for higher temperatures, fewer frosts, and more rain and, therefore, more favorable agricultural conditions. However, the east coast is projected to have increased potential evaporation deficit, leading to greater demand for irrigation, lower winter rainfall, and, hence, less groundwater recharge. While the headwaters of the alpine rivers are projected to receive more rain and less snow, thereby shifting river peaks from late spring/early summer to winter making them less reliable for run-of-river irrigation in summer.

Acknowledgments This chapter would not have been possible without the help from many contributors of information. The initial work and surveys were greatly assisted by Sarah Hayman, who worked on data collection as part of a university summer scholarship. Contributions through survey responses and follow-up reviews came from Tim Joyce, Sarah Stewart, Jeff Cuthbertson, David Burn, Russell Holden, Chris Stanley, Liberty Gunto, Piotr Swierczynski, Nick Urlich, Gerry Essenburg, Phil Ruffell, Marcus Koll, Rob Green, Peter Bahrs, Ramari Young, Shane Adams, Catherine Thompson, Andrew Kerr, David Payne, Andrew Curtis, Paul Reese, David Hendrikz, Ian McIndoe, Brian Smith, Chris Ellington, Brent Walton, Tony McCormick, and Gary Kelliher. Guidance from Grant Read on hydroelectricity and reviews of the draft text by Ken Taylor and Martin Ward were also most helpful.

References

Alvey, T., Goodwin, D., Xingwang, M., Streiffert, D., & Sun, D. (1998). A security constrained bid-clearing system for the New Zealand wholesale electricity market. *IEEE Transactions on Power Systems, 13*, 340–346.

Aqualinc. (2010). *Update of water allocation data and estimate of actual water use of consented takes 2009–2010*. Christchurch: Ministry for the Environment.

Capacity Infrastructure Services. (2013). *Annual report 2012–2013*. Wellington: Capacity Infrastructure Services.

Comet. (2013). *Evaluation of potential electricity sector outcomes from revised minimum flow regimes on selected rivers*. Wellington: Ministry for the Environment.

Corneby, D. (2010, February 5). *The Auckland region's water supply: Past, present and future*. Paper presented at the New Zealand Society for Sustainability Engineering and Science. Auckland.

Electricity Authority. (2011). *Electricity in New Zealand.* Wellington: Electricity Authority.
Electricity Market Information. (2014). Prices $/MWh. http://www.emi.ea.govt.nz
Farley, P. J. (1994). *Privatisation of irrigation schemes in New Zealand* (Short Report Series on Locally Managed Irrigation). Columbo: International Irrigation Management Institute.
Greater Wellington Regional Council. (2013). *Water supply annual report for the year ended 30 June 2013.* Wellington: Greater Wellington Regional Council.
Hurunui District Council. (2011a). Cheviot Ward Committee Report.
Hurunui District Council. (2011b). Hurunui Waiau Zone Committee. http://files.ecan.govt.nz/public/cw-hw-zip-reports/HDC%202011%20Drinking%20Water%20Report%20D%20Edge.pdf
Hurunui District Council. (2012). Hurunui Community Long Term Plan 2012–2022. Amberley: Hurunui District Council.
Hurunui District Council. (2014). *Water and sewerage services.* http://www.hurunui.govt.nz/services/water-and-sewerage-services
HydroTrader. (2013). *Trade history.* http://www.hydrotrader.co.nz
Institute for Sustainable Futures. (2008). *Guide to demand management.* Sydney: Water Services Association of Australia.
Irrigation New Zealand. (2012). *Survey of costs of surface water irrigation schemes in Canterbury.* Christchurch: Irrigation New Zealand.
Jenkins, B. R. (2013, August 28–30). *Progress of the Canterbury water management strategy and some emerging issues.* Paper presented at the New Zealand Agricultural and Resource Economics Society conference Lincoln University.
Land and Water Forum. (2010). Report of the Land and Water Forum: A fresh start to fresh water.
Land and Water Forum. (2012). Second report of the Land and Water Forum: Setting limits for water quality and quantity, and freshwater policy- and plan-making through collaboration.
Meridian Energy. (2012). *Meridian energy annual report 2012.* Wellington: Meridian Energy.
Ministry for the Environment. (2007). *Proposed national environmental standard for sources of human drinking water: Resource Management Act section 32 analysis of the costs and benefits.* Wellington: Ministry for the Environment.
Ministry for the Environment. (2014). *Freshwater demand (allocation).* http://www.mfe.govt.nz/environmental-reporting/fresh-water/freshwater-demand-indicator/index.html. Accessed 3 June 2014.
New Zealand Government. (2011). National Infrastructure Plan 2011. Wellington: New Zealand Government.
New Zealand Government. (2014). *National policy statement for freshwater management 2014.* Wellington: New Zealand Government.
OECD. (2008). *Environmental performance of agriculture in OECD countries since 1990.* Berlin Centre: OECD.
OECD. (2009). OECD in Figures 2009.
OECD. (2010). Water consumption. In: *OECD Factbook 2010: Economic, Environmental and Social Statistics.* OECD Publishing. doi:http://dx.org/10.1787/factbook-2010-61-en
Office of the Auditor General. (2010). *Local authorities: Planning to meet the forecast demand for drinking water.* Wellington: Office of the Auditor-General.
Read, E. G. (2009). *Economic behaviour in a hydro-dominated electricity market.* Christchurch: EGR Consulting.
Statistics New Zealand. (2011). *Water physical stock account: 1995–2010.* Wellington: Statistics New Zealand.
Statistics New Zealand. (2013). *National population estimates: At 30 June 2013.* Wellington: Statistics New Zealand.
Sternberg, J. & Bahrs, P. (2011). Water metering—The Tauranga journey. Tauranga: Tauranga City Council.
Timaru District Council. (2014). Timaru District rates resolution 2013/2014. Timaru: Timaru District Council.

Tipping, J. & Read, E. G. (2010). Hybrid bottom-up/top-down modelling of prices in hydro-dominated power markets. In: Rebennback, S., Pardolos, P. M., Pereira, V. F., Iliadis, N. A. (Eds.), *Handbook on power systems optimisation* (Vol. II, pp. 213–238). Berlin: Springer.
URS. (2014). *Tekapo transfer review 2014 with water quality and cultural aspects*. Christchurch: Environment Canterbury.
Water New Zealand. (2013). *2011/2012 National performance review of water utilities*. Wellington: Water New Zealand.
Watercare. (2013). *Auckland regional water demand management plan 2013–2016*. Auckland: Watercare Services Limited.
Wellington City Council. (2011). *Water conservation and efficiency plan*. Wellington: Wellington City Council.

Chapter 15
Water Pricing: The Case of South Africa

Barbara Schreiner

Abstract South Africa is a water-scarce country with a high level of income inequality, based largely on race. The issue of water pricing for water services and raw water has been shaped over the years to try to address both of these issues and to ensure a revenue stream that, with the parliamentary appropriation, is sufficient to fund the management and infrastructure-related costs of providing water and protecting water resources. This chapter deals with the key aspects of water pricing in South Africa for irrigation, municipal and industrial use, and power generation. It outlines the legal framework for water pricing and how this has been interpreted since the current legislation was promulgated in the late 1990s. It also outlines some of the key debates currently being addressed, such as how to deal with irrigation subsidies, how to address issues of equity, the possible adoption of a national charge for water, and how best to structure infrastructure-related charges.

Keywords South Africa • Operation and maintenance costs • Subsidies • Financial viability • Bulk water supply

15.1 Introduction and Background

Charging for water use was introduced in South Africa in different areas and circumstances across the country, usually at the local level. Gradually, over time, a more coherent approach to pricing of water was introduced at the national level for raw water. The pricing of water supplied by municipalities, however, has always been the individual purview of local government.

There has, thus, been a continual evolution of pricing of water across the value chain, until the most recent introduction, which is looking at the application of a charge for discharging wastewater into a water resource.

B. Schreiner (✉)
Pegasys, 1209 Frances Baard St, Hatfield, Pretoria 0083, South Africa
e-mail: barbara@pegasysinstitute.org

A. Dinar et al. (eds.), *Water Pricing Experiences and Innovations*,
Global Issues in Water Policy 9, DOI 10.1007/978-3-319-16465-6_15

15.1.1 Water Resources, Population, and Issues in Water Supply

South Africa is a middle-income country, with a population of around 50 million. The country has a huge wealth disparity, with a Gini coefficient among the highest in the world—0.63 in 2009 according to the World Bank.[1] There is also a huge disparity in access to water, due to the discriminatory policies of apartheid, which saw most of the nation's water being concentrated in the hands of the white minority.

While around 95 % of the population has access to improved water sources for domestic use, there are significant challenges with this provision, which include poor operation and maintenance, high levels of water wastage through leaks, low levels of payment for services in many areas, and lack of technical capacity to manage the water services.

On the macroscale, South Africa is a water-scarce country, and most of the country's basins are closed or approaching closure. Per capita water availability, despite significant infrastructure development, falls in the water stress level, at less than 1,100 m^3 per capita, per annum. Around 60 % of the country falls into shared river basins.

The country experiences high levels of interannual climate variability, with recurrent floods and droughts. Water is also unevenly distributed across the country, with around 65 % of the country receiving less than 500 mm of rain per annum. Rainfall decreases as one moves west across the country. The western and northwestern part of the country is semiarid, with very low rainfall (less than 200 mm per annum) and high levels of evapotranspiration (Fig. 15.1) (DWA 1986).

Although groundwater in South Africa is most common in hard rock aquifers that limit the quantity that can be easily abstracted for use, it is an important source of water for outlying communities. Where groundwater is found in dolomitic and sand aquifers, large volumes of water are abstracted, mainly for irrigation (DWA 1986; DWAF 2004).

The quality of the surface water is generally good in natural conditions, except for some areas where the water is salty under low-flow conditions due to the local geology. However, the country has high levels of pollution from industry, mining, and poor or failing sanitation systems.

15.1.2 Past Experiences with Irrigation Water Pricing

In the 1970, *Report of the Commission of Enquiry into Water Matters* (Government of South Africa 1970), it was recognized that the price charged for water on irrigation schemes was insufficient to cover annual expenditures on the schemes. The report also recognized that the cost of water formed such a low percentage of the farmers'

[1] http://data.worldbank.org/indicator/SI.POV.GINI accessed 17/05/2014.

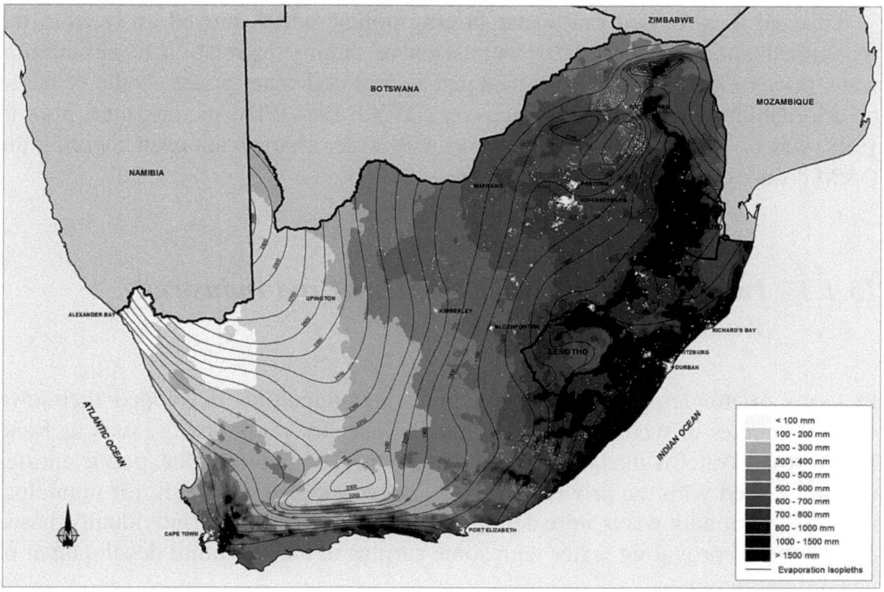

Fig. 15.1 Average annual rainfall across South Africa (Source: DWA 1986)

total costs that it had little impact on irrigation practices. It further stated that "it is generally accepted that water rates should be increased to cover at least the annual operating and maintenance (O&M) costs of schemes. It would not be unreasonable to fix water rates at such a level that on the average about 3 % of the farmers' gross returns had to be paid in water taxes. This would probably leave a small surplus to pay part of the interest on the capital" (Government of South Africa 1970).

Despite this, the political power of the agricultural sector prevented the implementation of the recommendations of the 1970 commission. By the time the new pricing regime was implemented (after the promulgation of the National Water Act in 1998) (RSA 1998), most irrigators were still paying less than the annual O&M costs of schemes.

There were also subsidies available for irrigators, with up to 30 % of new off-farm infrastructure costs being subsidized by the Department of Water. The Department of Agriculture was responsible for subsidies for on-farm infrastructure.

In 1980, a committee of inquiry into the price policy with regard to the determination of water tariffs (DWAF 1984) found that over a 6-year period, the tariffs for irrigation had risen by 120 %, but were still too low to cover the full operation and maintenance costs of government water schemes. It also found that small farms would not be able to absorb further water price increases without impacts on financial viability and that a different tariff policy for the agricultural sector in relation to the industrial sector was appropriate, because of the significant differences between the impact of water charges on their financial viability.

This led to a revised raw water pricing policy, which moved away from the recommendations of the 1970 commission of inquiry in relation to agricultural water prices. Government determined that agricultural water prices should be based on affordability, rather than cost recovery (DWAF 1984). The picture, then, prior to 1998, was of a highly subsidized sector, with water charges not even covering the O&M costs of the schemes.

15.1.3 Past Experiences with Municipal and Industrial Water Pricing

In terms of municipal water supply, many municipalities developed their own water resources or were provided with bulk water by water boards, such as Rand Water. Rand Water is the largest of 13 water boards in South Africa, public entities that are tasked with the provision of bulk water supplies primarily for municipal use. Tariffs for bulk water were determined by the water boards individually, based on the cost of providing water with some surplus to enable future development of new infrastructure.

The commission of inquiry (Government of South Africa 1970) found that water was not a significant cost to industry. In the manufacturing industry, the cost of water was calculated to be 0.17 % of the cost of materials used. However, some subsidies were provided to municipalities that were paying particularly high costs for water, and the recommendation was made by the commission to subsidize sewage treatment with local authorities, making the water available for reuse (Vawda et al. 2011).

In the 1980 review of the pricing policy, it was recommended that for municipal and industrial water use, full cost recovery should apply, based on a marginal cost approach and including the capital cost (DWAF 1984).

15.2 Present Water Pricing Practices

Water pricing is currently governed by the White Paper on a National Water Policy for South Africa (RSA 1997a), and the National Water Act (NWA) (Act 36 of 1998) for the setting of raw water charges, and the Strategic Framework for Water Services (RSA 2003), and the Water Services Act (WSA) (Act 108 of 1997) for charges for potable water. The sources of funding from the fiscus, and the different charges across the value chain, and the purpose they serve, are set out in Fig. 15.2.

The NWA provides the basis for a pricing strategy for raw water, which was first promulgated in 1999 and revised in 2007. A further review was initiated in 2012, but had not been completed by mid-2014.

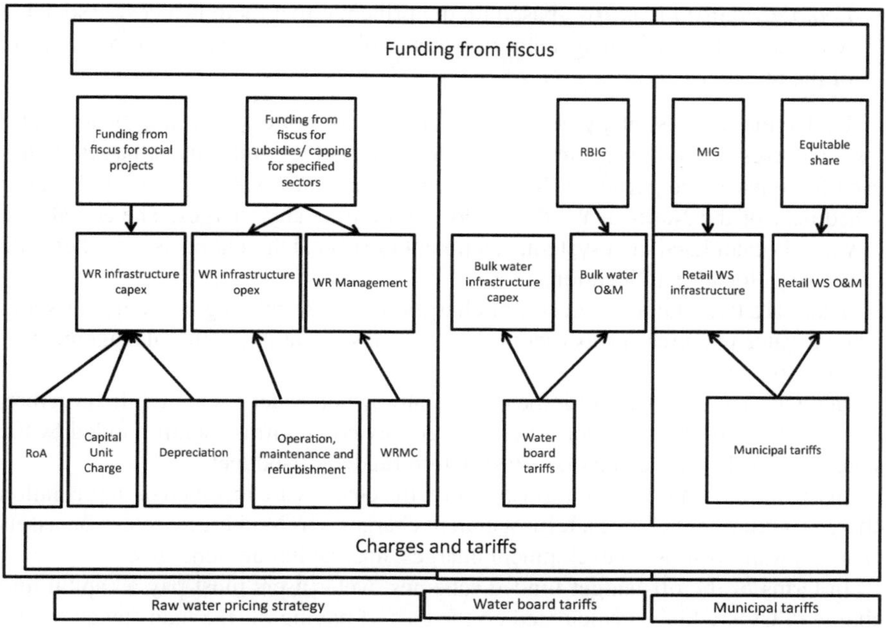

Fig. 15.2 Funding sources and charges across the value chain

The provision of water supply and sanitation is constitutionally a local government function, and municipal tariffs are governed by the WSA and determined by each municipality in accordance with regulations promulgated under section 10 of the WSA.

15.2.1 Raw Water Pricing Strategy

The raw water pricing strategy is governed by four key principles:

- *Social Equity*—to be achieved through financial assistance to provide affordable water to those who were excluded from the mainstream economy under apartheid
- *Financial Sustainability*—ensuring that sufficient funds are generated either from water use charges or from the fiscus to cover the costs of development, operation, maintenance and refurbishment of water resource infrastructure, and effective water resource management
- *Economic Efficiency*—setting the price of water to reflect its scarcity value in order to improve the economic efficiency in the use of this scarce resource

- *Ecological Sustainability*—based on the principle that there is a cost associated with the ecological management of water, which should be paid for by all users of the resource

Under the 1999 strategy, and the 2007 revised strategy, all significant water resource uses, including commercial afforestation,[2] face water use charges. Water used for subsistence purposes (food gardening and stock watering) that falls under schedule 1 of the National Water Act does not attract any charges. The actual cost of water is calculated on a system, catchment or sub-catchment basis, and therefore varies from location to location.

There are two major categories of charges: (1) those relating to the use of state-owned water resource infrastructure and (2) those relating to the management of water resources.

In relation to the charges for the use of state-owned water resource infrastructure, there are two different categories: (1) charges levied on infrastructure funded by the state and (2) charges levied on infrastructure funded off-budget.

In relation to state-funded infrastructure, the charges are made up of depreciation charges, betterment charges, refurbishment charges, O&M charges, return on assets charges, water resource development charges, and capital unit charges.

In terms of the off-budget funded schemes, water users must pay a capital unit charge (CUC), which is calculated on the cost of paying off the loans taken out by the state to build the infrastructure. This approach is only used for that portion of any new scheme that is intended to serve commercial interests or domestic users who can afford to pay this charge. If there are social users of the water from the scheme (e.g., poor communities that require water for subsistence or household purposes), this capital portion of the scheme is paid from the fiscus, and the CUC is not charged to these users. When the water is provided to a municipality, the overall financial viability of the municipality is considered in relation to whether the capital should be funded by the state, from the markets, or through a mixture of both.

15.2.2 Charges for State-Funded Infrastructure

The charges relating to state-funded infrastructure are described in Fig. 15.3. Operation and maintenance charges include direct costs of administration and operation and maintenance and indirect costs that are not specific to one scheme, but contribute to the overall management of infrastructure, such as regional and head office overheads. These costs are recovered on a scheme or system basis, either on actual cost or forecast cost.

Charges also include a return on assets charge intended to provide the state with some return on the value of the infrastructure. In 1999, this figure was set at 4 % of

---

[2] Commercial afforestation is deemed to be a “streamflow reduction activity” under the law, significantly reducing the amount of runoff in a catchment, and therefore draws certain water use charges.

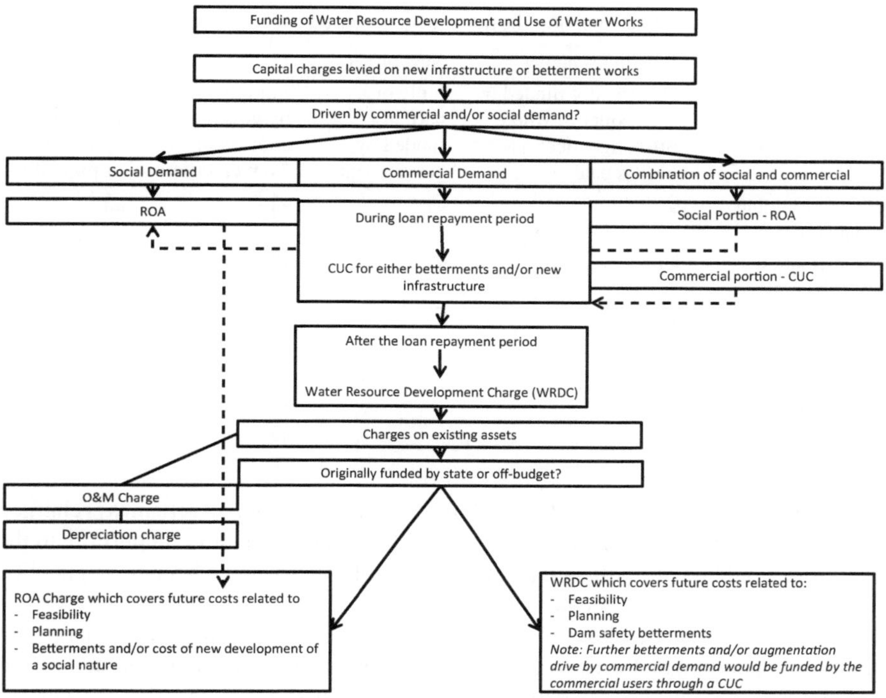

Fig. 15.3 Components of infrastructure-related charges under current pricing strategy

the depreciated replacement cost of existing infrastructure, or the completion cost of new infrastructure, and remained at that level in the 2007 revision of the strategy. This was considered a fair rate of return on capital employed by the government to finance the development of water infrastructure. This component of the charge is set on a scheme-related basis and is only applied to those sectors with increasing demands. These include local government, industry, mining, and energy, but exclude agriculture. The intention of this charge was to support the development of new infrastructure to be funded by the state.

A charge for depreciation and refurbishment is also levied and calculated on a straight line basis over the useful life of the asset as the annual depreciable portion of the replacement value of the assets with revaluation of the assets to be carried out every 10 years. The depreciation charge is intended to pay for the refurbishment of existing assets. This charge is only applicable on off-budget funded schemes, once the loans have been paid off. If refurbishment is required prior to this, a "refurbishment charge will be arranged by agreement between the parties" (DWA 2007). This has not yet been necessary.

A betterment charge on commercial schemes funded off-budget may be levied in consultation with end users, based on either actual cost recovery or taking into account the need to smooth charges over time. Betterment of social schemes is funded through the return on assets (RoAs) charge.

Table 15.1 Application of elements of the raw water charges on under different funding conditions

Charges to be levied	Existing schemes	New projects		
	Historically funded by Exchequer or where off-budget debt has been repaid	Fully or partially funded by government	Initially funded by government and recouped from end users	Off-budget funding applied fully or partially
Operations and maintenance	☑	☑	☑	☑
Depreciation/ refurbishment	☑	☑	☑	☑
Return on assets	☑	☑		
Water resource development	☑			
Betterment			☑	☑
Capital unit charge			☑	☑

Additionally, a water resource development fee for off-budget schemes is charged once the loan on the scheme has been repaid. This charge is similar in purpose to the RoA and would be promulgated by the minister of water when it became applicable. It has not yet become applicable on any scheme.

Table 15.1 indicates how the different elements of the raw water charge are applied, depending on the source of financing used to develop the infrastructure and the status of payment of the capital. Thus, operation and maintenance charges are levied on all schemes, while other elements of the charges are applied specifically under different funding conditions.

The water resource management charge (WRMC) is calculated on the basis of the costs associated with managing a geographically defined area called a water management area (WMA). A WMA is based on catchment boundaries and can be either a sub-catchment or a collection of smaller catchments. In 2014, nine water management areas were declared, covering the whole of South Africa. These charges cover the cost of inter alia, planning, and implementation of catchment management strategies; monitoring and assessment of water resource availability and quality; flood and drought management; management of raw water allocations; evaluation and processing of water use authorization and registration applications; and water resource protection and pollution control. These functions are currently performed by the Department of Water and Sanitation, except in two WMAs where some functions have been delegated to the two existing catchment management agencies (CMAs). In due course, nine CMAs will be in place, one in each WMA, and will take over these functions from the national department.

The WRMC is levied on registered use and was capped in 1999 at a maximum of ZAR 0,02/m^3, increasing with inflation to ensure that costs were contained. Unfortunately, the original calculation of the WRMC was too low, and this constraint in increasing the charge has resulted in constant underfunding of water resource management activities.

The 2007 pricing strategy introduced the concept of a waste discharge charge system, based on the polluter-pays principle, but did not set out any specific mechanism for calculation of the charge. Subsequent work has been done on how to calculate this charge, but it has not yet been implemented.

The strategy also creates the possibility for an economic charge to be applied on a scheme or system basis and to be determined either administratively or via market-based mechanisms. However, the strategy also recognizes that the economic charge would not be applied prior to a process of compulsory licensing. Since this process has not yet been completed anywhere in the country, the economic charge has not been applied. The specific application of these charges to the various sectors is dealt with under the sections below.

15.3 Calculation of Charges

The calculation of the various charges is set out in the sections below.

15.3.1 Water Resource Management Charges

The department, or CMAs where they have been established, develops an annual budget for the costs of activities to be performed in each WMA, based on a schedule of functions listed in the pricing strategy. This includes functions related to the management of abstraction of water and functions related to the discharge of wastewater. These costs are apportioned differently to abstractors or dischargers. Some costs are allocated to both, since they are necessary to manage both functions. The budgeted costs are divided by the registered water use in the WMA to derive a cost per m^3 to be paid by all registered water users. Due to the issue of affordability, a cap has been placed on the charges to be paid by the agricultural and afforestation sectors.

The WRMC on water transferred into a WMA through an interbasin transfer must be transferred to the donor catchment. Where poor water quality from an upstream WMA to a downstream WMA results in additional costs to the downstream WMA in managing the poor water quality, this additional cost should be paid for through increased WRM charges levied on waste dischargers in the upstream WMA paid across to the downstream WMA. In practice, however, this has not yet happened, largely because there are only two CMAs in place, and apart from these two WMAs, the WRMC is paid to the department that apportions the money as required to make up for any shortfall through their parliamentary appropriation.

15.3.2 Determination of Annual Sectoral Use Volumes for Pricing Purposes

For charging purposes, the registered water use of each user and each sector is used. This is the volume of water that users register with the department and is not a directly measured volume. For agriculture, the volume of water is calculated using the SAPWAT program to convert hectares under irrigation to an annual volume. SAPWAT, now in its third version, is a program specifically designed for calculating the irrigation water requirements of crops, farms, and drainage or administrative regions for planning purposes in South Africa.

For afforestation, modified tables based on work done by the Water Research Commission are used to determine the average annual use of water by plantations. The total volume of registered water use in each WMA is compared with the total yield of resources in the WMA at 98 % assurance of supply. The allocable volume must exclude water set aside to meet the environmental flow requirements (the reserve) for international obligations or for transfer to another WMA.

In some case, registered water use exceeds the allocable yield. In this case, the total volume of registered water use is used to determine the charges. Where the registered water use is less than the allocable amount, the volume of allocable water is used to determine the volumetric charges, and any shortfall in income is subsidized through the fiscus.

The volumetric use for each sector reflects the assurance of supply for that sector as follows:

- Irrigation sector, 91 % (100 %@70 % + 70 %@30 % of the time)
- Domestic, industrial, and mining; 97 % (100 %@70 % + 90 %@30 % of the time)
- Strategic industrial sector, 100 % (no water restrictions)

Percentages may be applied to determine the price differential on the CUC based on the assurance of supply. The assurance of supply is applied as follows. If, for example, a scheme has 100 million m^3 of available water per annum and if 30 % is allocated to domestic and industry (30 million m^3) and the balance of 70 % is allocated to agriculture, then the long-term average use of allocations will be calculated as follows:

- Domestic and industry, 30 million $m^3 \times 0.97 = 29.1$ million m^3.
- Irrigation, 70 million $m^3 \times 0.91 = 63.7$ million m^3.
- Total, 92.8 million m^3.
- Domestic and industry allocation of cost will be 29.1/92.8 = 31.36 %.
- Irrigation allocation of cost will be 63.7/92.8 = 68.64 %.
- Total cost allocation, 100 %.

Under the current example, domestic and industry will pay a premium of 1.36 % as a result of a greater assurance of supply, while irrigation will receive a discount of 1.36 %, as a result of a smaller assurance of supply (DWAF 2007).

15.4 Capping on Charges

15.4.1 Water Resource Management Charge

Due to the issue of affordability, caps were placed on the water resource management costs to the afforestation and irrigation sectors. WRM charge for afforestation is capped at R10 per hectare, plus producer price index (PPI) rate (%) at April of each year, with 2002–2003 financial year as the base year. Resource-poor foresters do not pay this charge.

WRMC to the irrigation sector is capped at 1.5 cent per m^3, plus the PPI rate (%) at April of each year, with 2007–2008 as base year. The WRM charges for resource-poor farmers and resource-poor forest growers are phased in over a period of 5 years, from the first registration of their water use to enable them to build up sufficient capital to be able to pay the charges.

15.4.2 Infrastructure Charges

On the infrastructure-related charges, the following limitations are imposed in the pricing strategy:

Domestic/industrial/mining/energy sector: Annual increases for existing state-funded schemes are limited to 10 %+ PPI (rate taken in April), until full cost recovery is reached.

Agricultural sector:

Commercial farmers: Full operation and maintenance costs will be recovered annually, with an annual increase limited to 50 %. Depreciation charges for existing schemes will be capped at 1.5 cents per m^3, plus PPI (rate), with 2007–2008 as base year, with annual increase limited to 20 % of the previous year's charge. Full financial cost recovery (including ROA) will be charged on new schemes.

Resource-poor farmers: Operation and maintenance charges will be phased in over 5 years from the date of registration of the relevant water use. Depreciation charges will be waived for 5 years from the date of registration of water use, and then charges will be capped at 1.5 cent per m^3, plus PPI (rate) with 2007–2008 as base year, with annual increases limited to 20 % of the previous year's charge. The capital cost of new infrastructure will be subsidized by the state. Further, waiving of charges can also be considered for a limited time period, if requested by other relevant departments in order to support governmental initiatives, such as land and agricultural reform.

15.5 Summary of Present Experiences with Raw Water Pricing

15.5.1 Irrigation Water Pricing

Despite the intention of the 1970 commission of inquiry to move toward greater cost recovery on irrigation water charges, the cost of water for agricultural purposes remains heavily subsidized, with charges only covering a portion of the O&M charges, with capped increases allowed per annum. The return on assets charge is not charged to irrigation users, and the water resource management charge is also capped and does not cover full costs of water resource management.

A singular challenge that has been faced in relation to the irrigation charges is the low level of cost recovery. There are several reasons behind this. The first is that some water user associations (WUAs) have expressed a concern that they are not being provided with the services for which they are paying. As a result, some WUAs have decided to withhold their payments from the department and to hold them until such time that evidence of the services being provided has been given. This results in something of a vicious cycle, since the department does not have the funds to provide the services that they should be providing, and the WUAs are not prepared to hand over the funds until the services are provided. It is, at this point, unclear exactly how much money has been withheld in this way, but it appears to amount to millions of dollars.

The second reason lies in the weakness of the registration system for water users run by the Department of Water and Sanitation and weaknesses in the billing system. All water users should be registered on the Water Authorisation Registration and Management System (WARMS) of the Department of Water and Sanitation. However, a failure to align the work of the deeds office in registering changes in land ownership with the registration process of the department means that change of land ownership has not necessarily been registered with the Department of Water and Sanitation. As a result, the registration of water users is outdated, and some bills have been sent to previous rather than current landowners.

The third reason lies in the weakness of the billing system run by the department and the failure to identify and follow up on nonpayment. Considerable work has been done over the recent years to turn this around, and overall recovery of charges for irrigation water and other water uses has improved but is still well below what it should be.

The failure to charge the full O&M costs to irrigators, combined with poor cost recovery, has meant that maintenance of state-owned, irrigation-related infrastructure (dams, canals, etc.) has been undermined, and there is now a significant backlog in maintenance that needs to be funded and implemented.

15.6 Present Experiences with Afforestation Water Pricing

South Africa is one of the few countries, if not the only, that consider commercial afforestation to be a water user and require afforestation companies/individuals to apply for water use licenses and to pay water use charges. They do not generally pay the infrastructure charges, except for one case in which the license for afforestation was issued on the condition that they contribute to the cost of the new infrastructure that was going to be required to offset the water use by the afforestation.

Due to an agreement made during the drafting of the first pricing strategy, timber growers have capped water resource management charges and do not pay the water quality management portion of the WRMC, since they argued that they do not impact on water quality. This is, indeed, a moot point, as timber growing in South

African conditions can, if not extremely well managed, increase siltation of water courses and does reduce streamflow, thus reducing the dilution factor of water pollution in the catchment. This is, therefore, an aspect of the WRMC that should be revisited.

15.7 Present Experiences with Municipal Water Pricing

There are two types of water charges that municipalities face—the charges described above, for raw water supplied from state-owned infrastructure, and charges for bulk potable water from water boards. Not all municipalities are supplied with water either by the national department or by a water board. A number of municipalities have their own surface or groundwater infrastructure for providing water. In this case they only pay the WRM charges to the national department.

Water board charges are calculated on the basis of the cost of the water provided to the municipality. Water boards are, as state-owned entities, not allowed to make a profit, but can make a surplus in order to provide for future infrastructure development, refurbishment, or betterment. The water board tariffs therefore take into account capital costs, O&M costs, and costs of future infrastructure development.

The most significant challenge in terms of providing water to municipalities is the significant backlog in payments from municipalities to both the department and water boards. This issue is dealt with in Sect. 15.9.

15.8 Present Pricing Experiences of Environmental Services

The only references to the pricing of environmental services in the current pricing strategy relate to the water for ecological purposes, which are specifically excluded from water charges, and the cost of controlling invasive alien plants (IAPs) that have an impact on water availability.

In relation to the latter, the full cost of the control of certain IAPs may be charged to water users in a particular area, but only in consultation with the affected water users and only where the control of IAPs is the most cost-effective method of making more water available for use or increasing the reliability of supply.

15.9 Present Experiences with Water Services Tariffs

It is worth noting that the provision of water supply and sanitation is constitutionally a local government function in South Africa, and, as such, municipal tariffs are determined by the municipalities themselves. Although there are now intentions to establish a national economic regulator that will regulate, inter alia, municipal tariffs, to date, the only requirement for municipalities is that they comply with the regulations promulgated under section 10 of the Water Services Act.

In 1994, when South Africa achieved democracy, nearly 40 % of the population lacked access to a basic supply of safe drinking water, and more than 50 % did not have access to adequate sanitation (RSA 1994). The access to water and sanitation was worst in the rural areas and the ex-homeland areas (Eales 2011).

In 1994, the White Paper on Water Supply and Sanitation (RSA 1994) was published, with a particular focus on ensuring effective water supply and sanitation services to the poor and the historically marginalized in the country. A basic water supply of 25 l per person per day within 200 m of the household was determined as the minimum that should be provided. The white paper argued that water services should be paid for by everyone except poor communities that were unable to afford basic services, in which case the state would subsidize the construction costs of the basic minimum services but not the operating, maintenance, or replacement costs. A social tariff that covered only the operating expenses would be charged for communal water sources, while higher levels of service would attract tariffs covering the full cost of supply (Eales 2011).

The white paper also proposed a rising block tariff with a minimum of three blocks—the first is a lifeline or social tariff, the second is a normal tariff, and the third is a marginal cost tariff for high levels of consumption. This was then incorporated into the Water Services Act of 1997.

In 2000, the principle of free basic water was introduced in South Africa, in order to give effect to the constitutionally guaranteed right of access to sufficient water, on the argument that the ability to pay for water could not be allowed to prevent poor South Africans from accessing this right. This meant that the first block, of 6 kl per household per month, should be provided free of charge.

Interpretation of this principle has, however, varied from municipality to municipality, with some providing a free basic water allowance of 6 m^3 per household, per month to all households, while others only provided the 6 m^3 per month to indigent households registered on an indigent register. In addition, all municipalities were required under the regulations promulgated under section 10 of the Water Services Act to introduce a stepped tariff with a minimum of three steps. In some municipalities, the first step after the free basic water was extremely high, while in other municipalities, it was relatively low.

The section 10 regulations require that "A water services institution must, when determining its revenue requirements on which tariffs for water services are based, take into account at least the need to: (a) recover the cost of water purchases; (b) recover overhead, operational, and maintenance costs; (c) recover the cost of capital not financed through any grant, subsidy, or donation; (d) provide for the replacement, refurbishment, and extension of water services works; and (e) ensure that all households have access to basic water supply and basic sanitation" (RSA 1997b).

15.9.1 State Contribution to the Costs of Municipal Water Services Provision

In order to overcome the legacy of apartheid, and to ensure the provision of water services to poor black communities in particular, the government has introduced a range of funding support mechanisms for water supply and sanitation, as follows:

- *The Equitable Share* (*ES*): The constitution requires that local government must receive an equitable share of national revenue and a portion of this financing water targeted for the operation and maintenance of water supply and sanitation.
- *The Municipal Infrastructure Grant* (*MIG*): The MIG is funded from the national revenue and is intended to support the capital costs of the provision of basic services, including water supply and sanitation, to the poor. The MIG was introduced in 2004 and was based on a formula of how many people in a municipality still lacked basic services.
- *The Regional Bulk Infrastructure Grant* (*RBIG*): The RBIG was introduced in order to fund the capital costs of regional bulk infrastructure, i.e., infrastructure that served more than one municipality and/or other users.

A significant challenge in the setting of municipal water services tariffs is that few municipalities have accurate information on the status of their infrastructure and the actual costs of providing water. It is, therefore, hard to know whether municipalities are undercharging or overcharging users. In addition, the revenue from water services is not ring fenced, despite a legal requirement for this to happen, with the result that revenue from water services goes into the general municipal budget and is not necessarily spent on water services, with preventative maintenance being shortchanged as a result.

In addition, in many municipalities, the recovery of charges is poor, contributing to the underfunding of water services functions and the deterioration of the infrastructure.

A further challenge is that while the national treasury calculates ES on the number of indigent households in a municipality, based on census data, municipalities do not necessarily have the same measures of indigency. The cutoff level for being defined as an indigent household in Cape Town, for example, is three times higher than that set by National Treasury (Eales 2011). This results in a gap between the funds provided through the ES and the requirements of the municipality. In addition, actually identifying the indigent households on the ground in order to provide free basic water to them is difficult, relying on municipal indigency registers, with significant inclusion/exclusion errors.

15.10 Current Debates and Future Directions

15.10.1 Addressing Subsidies, Water Quality, Infrastructure Funding, and Equity Issues

A revision of the pricing strategy for raw water was instituted in 2012 and had not reached conclusion by June 2014. However, this section encapsulates some of the critical issues driving the revision of the strategy.

Key challenges were identified that drove the need for a review of the pricing strategy (other than the legal requirement that it should be reviewed every 5 years and which has not yet been met). The current pricing strategy provides a blanket subsidy to agricultural water users, regardless of the nature of the crop or the financial viability of the farming activity. As a result of these blanket subsidies, the Department of Water and Sanitation is making subsidy decisions that ought to be made by the Department of Agriculture, Forestry and Fisheries, in consultation with the Department of Economic Development.

The current pricing strategy does not provide sufficient protection for the poor against rising water prices, resulting from new infrastructure development. It does not provide a robust enough method of generating revenue for the development of infrastructure intended for social or economic stimulus purposes in areas where the user base will only be able to afford the charges after the infrastructure has been developed. The RoA is determined as a percentage of asset value, with little justification as to why it is set at 4 % as opposed to any other value.

The current pricing strategy does not make it possible for department of Water and Sanitation to set charges that reflect the full cost of delivering water, resulting in insufficient revenue for water resource management and sustainable infrastructure asset management. In addition, the waste discharge system has been further refined and needs more details incorporation into the pricing strategy.

The revision of the pricing strategy has been led by the Department of Water and Sanitation, with a stakeholder committee that was used during the process to advise on and test various models and approaches. This committee included representatives of the national treasury, water user bodies, WUAs, CMAs, water boards, the South African Association of Local Government, and other key government departments, such as the Department of Agriculture, Forestry and Fisheries. The key issues under consideration in this process are discussed below.

15.11 New Approaches for Raw Water Pricing Under Consideration

15.11.1 National vs. Hybrid Model

One of the major concerns in the pricing of raw water in South Africa is how to ensure that black South Africans, who were largely excluded from the benefits of state-funded infrastructure during the apartheid era, do not pay more for water than their white counterparts as a result of higher infrastructure costs for newer infrastructure. This is particularly true in the rural areas and the ex-homeland areas. One option that has been tabled for addressing this is the development of a set of national sectoral charges, for raw water.

The rationale for such an approach is that everyone within one sector would be paying the same for water, regardless of when the infrastructure was built or where. This would result in those with relatively cheap water from old infrastructure that was built in prime dam sites cross-subsidizing water users with more recent, more expensive infrastructure.

The challenge is that this approach removes any correlation between the cost of water in a particular area and water use, with the potential for resulting in suboptimal water use in water-short areas or areas in which infrastructure is particularly expensive due to the terrain or distances that it must be conveyed.

15.11.2 Introduction of Targeted Subsidies

A second issue under consideration is how to ensure that blanket subsidies, such as that currently provided through the caps on charges to the irrigation sector, are transformed into targeted subsidies, aimed at achieving specific ends in relation to economic development, job creation, and transformation. The intention is that such subsidies should be provided by the Department of Agriculture, Forestry and Fisheries, rather than the Department of Water and Sanitation. This requires the development of a policy and a process to enable this to be done, which is not yet in place.

Currently, the irrigation sector receives a subsidy of almost US$30 million per annum. This is excluding the implicit subsidy derived from the irrigation sector not being charged the RoA. The objective of such a large subsidy has not been made clear, and as a result, its level of success or failure in achieving its objective cannot be measured (Table 15.2).

15.11.3 Ecological Infrastructure

The third issue under consideration is how to deal with the funding of the protection and restoration of ecological infrastructure. In many catchments, the affordable engineering solutions to water management challenges have been largely exploited, and different approaches are required. Investing in the rehabilitation and maintenance of ecosystems can be a cost-effective addition to traditional infrastructure options in relation to water availability and water quality. It has therefore been proposed that in the revised strategy, "infrastructure" should be redefined to include natural infrastructure refurbishment and "betterment" when this is the cost-effective option.

Table 15.2 Current irrigation subsidy through caps in 2012–2013

	RoA (US$ million)	Depreciation (US$ million)	O&M (US$ million)	Total (US$ million)
Full cost	±118.3	±20.5	±42.9	±181.2
Revenue (capped)	±0	±9.8	±24.8	±34.6
Revenue loss due to capping	±118.3	±10.7	±18.1	±147.1

Source: DWAF

The proposal, therefore, is that elements of the costs of rehabilitating and maintaining natural infrastructure, in order to secure the water-related ecosystem services provided by the infrastructure, may be charged to water users in the catchment. This refers to the need to protect and restore aquatic ecosystems and habitat that produce and deliver water-related services that are of value to society, such as water quality enhancement, flood attenuation, reduction of sedimentation of dams, aquifer recharge, and streamflow regulation.

This might include (1) rehabilitation and maintenance of wetlands, riparian zones and watersheds through erosion control, sediment stabilization, re-saturation of drained areas, and re-vegetation; (2) fire management to prevent over-frequent or high-intensity fires that might result in high soil loss and damage to soil structure, with accelerated runoff, erosion, and reduced infiltration; and (3) the initial clearing and ongoing control of invasive alien plants that have a significant impact on water quantity and quality.

15.11.4 Future Infrastructure Build Charge

The RoA charge is intended to contribute toward funding of future social infrastructure development and the betterment of existing infrastructure. Section 56(2) (b) (v) of the NWA makes provision for the use of an RoA charge "for funding water resource development." However, the calculation of the RoA was based on 4 % of asset value, rather than on actual costs of the development and betterment of waterworks.

Under consideration, therefore, is the removal of the RoA charge and the introduction, instead, of a future infrastructure build charge (FIBC). While the purpose of the two charges is essentially the same, it is the calculation of the charge that is completely different. The FIBC would fund the activities listed under section 56(2) (b)(i, ii, and iii) of the National Water Act: the costs of investigation, planning, design, construction, and prefinancing of new infrastructure and the betterment of already existing infrastructure.

The FIBC would only be used to fund social and economic development stimulus infrastructure, which includes schemes in which there is a supply to municipal users that is associated with basic water requirements, whether this is the entire scheme in a rural area or a portion of a municipal supply system, and infrastructure that will provide for future economic water use for which there are currently no users or for which the existing users cannot afford the water supply (such as historically disadvantaged individuals (HDI) farmers), but where the water supply is necessary to provide for future economic development.

The proposal is that the FIBC would be calculated on the basis of a 10-year infrastructure plan, with the first 5 years being used to calculate the annual funding requirement. This will be divided by the water use volumes of all the included categories of water use to get a rate per m^3. The FIBC would be levied on all water use, other than irrigation and hydropower.

Any water use for municipalities, and all registered water use by nonnatural persons and other enterprises, will have to pay the FIBC, excluding hydropower. In line with the decision to keep irrigation charges capped, the FIBC will also not be charged to the irrigation sector.

15.11.5 Three-Year Charge Setting

Currently, water use charges are determined on an annual basis. Under consideration is the determination of multiyear charges, set for a period of 3 years at a time. The proposal for introducing this system is that, for the first 3 years, the charges will be reviewed annually on a rolling 3-year basis to ensure that the mechanisms and tools work effectively. Thereafter, i.e., in year 4 after the implementation of this approach, the charges would be set for 3 years, every 3 years.

15.12 The Proposed Waste Discharge Charge System

This section describes the proposed waste discharge system as incorporated into the draft pricing strategy revision by the Department of Water Affairs in 2013.

Section 56 (5) of the National Water Act (NWA) enables the minister to establish a system for charging waste discharges in terms of the pricing strategy, in order to promote the sustainable development and efficient use of water resources and the internalization of environmental costs by waste dischargers, create financial incentives for waste dischargers to reduce waste and use water resources in a more optimal manner, and recover costs associated with mitigating the water quality impacts of waste discharge.

The intention is that the WDCS will be implemented as one element of an integrated approach in a catchment or sub-catchment as part of a water resource management process that includes regulatory, economic, and other instruments, particularly where the water quality impact derives from the cumulative impacts from a number of dischargers, and the dischargers authorized water users under Section 21 of the NWA.

In order to keep the implementation relatively simple initially, the WDCS will only apply to surface water and may include, but not be restricted to:

- *Nutrients*: phosphate, nitrate, and ammonium
- *Salinity*: Total dissolved solids, electrical conductivity, chloride, sodium, and sulfate
- *Heavy metals*: arsenic, cadmium, chromium, copper, mercury, lead, nickel, and zinc
- *Organic material*: Chemical oxygen demand

The appropriate water quality variables will be determined, based on the water quality issues in a specific catchment.

The WDCS is proposed to consist of two distinct water use charges, either or both of which may be applied in a specific catchment: (1) a waste mitigation charge, intended to cover the costs of administratively implemented measures for the mitigation of waste discharge-related impacts, and (2) a waste discharge levy that will act as a disincentive or deterrent to the discharge of waste to water resources.

The proposed waste mitigation charge is expanded below. The waste discharge levy charge would have to be promulgated through a parliamentary money bill tabled by the minister of finance.

15.12.1 Principles for the Waste Mitigation Charge

The waste mitigation charge is intended for situations where mitigation measures provide a more economically efficient approach and the achievement of water quality objectives in a catchment than waste discharge reduction at source.

It is proposed that the following principles will apply to the waste mitigation charge: (1) it will be based on load discharge to avoid dischargers diluting effluent to reduce costs; (2) load, not concentration, will determine the charge; (3) only registered waste discharge-related water use under Sections 21 (e), (f), (g), and (h) of the NWA will be liable for waste mitigation charges, and the state will bear the costs associated with pollution loads that do not derive from registered water users; and (4) the load or concentration of pollutants in water abstracted by or supplied to the discharger may be deducted from the waste discharge charge.

The mitigation charge may be used in order to cover the costs of (1) developing and operating regional mitigation schemes, initiatives, or projects that will reduce pollution loads in the water resource; (2) reduced system yield resulting from the need to operate water systems to reduce the impact of water quality problems (i.e., using water for dilution purposes); (3) developing and operating treatment works to meet the requirements of downstream users; and (4) reducing waste load from a specific source, including regional schemes that collect and treat waste from a number of sources.

15.12.2 Calculating the Mitigation Charge Rate

The waste mitigation charge rate is proposed to be calculated as:

- Total annual mitigation cost ÷ total annual waste discharge load, where
- Total annual cost = annual operational cost (operations and maintenance) + any amortized capital cost over the design life of the measure (capital, interest, and depreciation).

The total discharge waste load in the catchment will be based on an assessment that distinguishes the contribution from point sources, and nonpoint sources, and that excludes background loads.

15.13 Economic Regulation Across the Value Chain

The Department of Water and Sanitation has established an economic regulator to regulate charges, tariffs, and related services standards across the value chain—from raw water to water services and discharge of water back to the resource. The proposed role of the economic regulator has been defined as:

> setting the rules to control, monitor, enforce and/or change tariffs/charges, tariff/charge determination structures and service standards for the water sector whilst recognizing and supporting government policy and broader social, environmental and economic imperatives.

The charges and tariffs that the economic regulator will regulate are outlined in Fig. 15.4.

The regulator will have to regulate a large number of bodies that are involved in the setting of charges and tariffs for water, including:

- The Department of Water and Sanitation Water Trading Entity (DWS WTE)
- Water boards (WBs)
- Catchment management agencies (CMAs)
- Trans Caledon Tunnel Authority (TCTA)
- Water services authorities (WSAs)
- Water user associations that are managing state-owned infrastructure (WUAs/local WMI)
- Water Research Commission (WRC)
- Private sector companies acting as water services providers (WSPs)
- Municipal entities acting as water services providers
- Water services intermediaries
- Water provided through international agreements/entities (e.g., KOBWA)

It is recognized by the department that the full suite of regulatory functions is large, and covers a large number of organizations, and that an incremental approach will have to be taken in building not only the capacity of the department to regulate

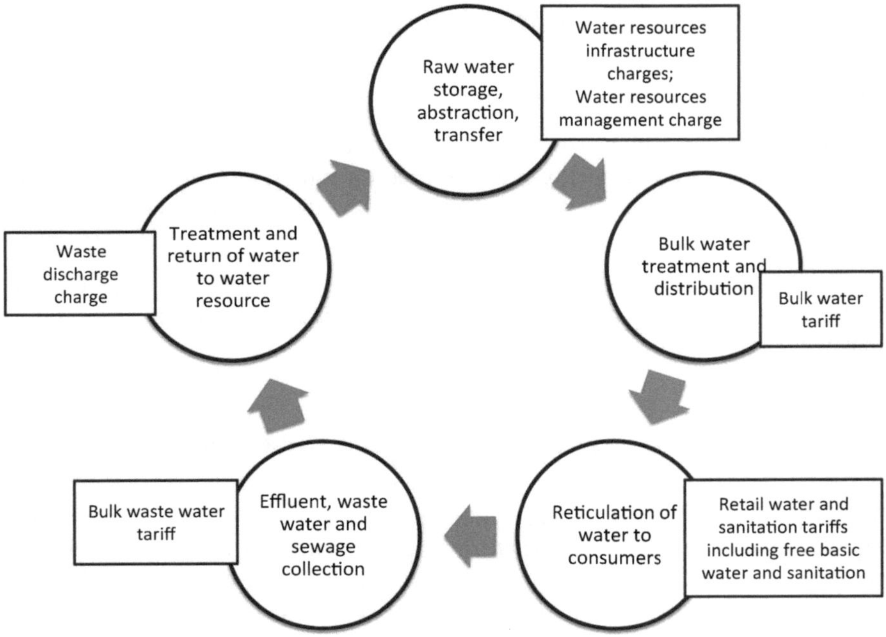

Fig. 15.4 Charges and tariffs to be regulated by the economic regulator

these entities, but, equally, the capacity of all of the entities to be effectively regulated through being able to provide reliable information on assets, services, and costs, which many, particularly at the local government level, are not currently in a position to do.

15.14 Conclusion

South Africa has a strong and well-considered pricing framework in place for both water resources and water services charges, with further work underway to improve the framework. The principles underpinning the pricing framework are largely sound and include a balance between the need to pay for the costs of services provided and the need to ensure that the cost of water does not interfere with the rights of the poor to water.

The major challenge in the South African context has been in the implementation, with several major factors impacting effective implementation, including poor billing systems, poor asset management systems resulting in an inability to calculate true maintenance and replacement/refurbishment costs, and, in some cases, lack of political will to implement water charges or to raise charges to appropriate levels. The result is that, despite the effectiveness of the water pricing system across the value chain from water resources to water services, there is an underfunding and under recovering in the system that, in many areas, is resulting in lack of maintenance of infrastructure and poor operation of services and the slow degradation of infrastructure and the services provided.

This challenge has, however, been recognized by the government, and considerable effort is being put into improving billing and cost recovery systems in order to support better provision of sustainable services over time.

References

DWA. (1986). *Management of the water resources of the Republic of South Africa*. Pretoria: Department of Water Affairs.

DWA. (2007). *Raw water pricing strategy*. Pretoria: Department of Water Affairs.

DWAF. (1984). *Report of the Committee of Enquiry into the price policy with regard to the determination of water tariffs*. Pretoria: Department of Water Affairs and Forestry.

DWAF. (2004). *National water resource strategy*. Pretoria: Department of Water Affairs and Forestry.

Eales, K. (2011). Water services in South Africa 1994–2009. In B. Schreiner & R. Hassan (Eds.) *Transforming water management in South Africa: Designing and implementing a new policy framework* (Series: Global issues in water policy, Vol. 2). Dordrecht/New York: Springer.

Government of South Africa. (1970). *Report of the commission of enquiry into water matters*. South Africa: Government of South Africa.

RSA. (1994). *White paper on water supply and sanitation*. South Africa: Government of South Africa.

RSA. (1997a). *White paper on a national water policy for South Africa*. South Africa: Government of South Africa.

RSA. (1997b). Water Services Act 108 of 1997. South Africa: Government of South Africa.

RSA. (1998). National Water Act 36 of 1998. South Africa: Government of South Africa.

RSA. (2003). *Strategic framework for water services*. South Africa: Government of South Africa.

Vawda et al. (2011). Pricing of water for cost recovery, economic efficiency and social equity. Vawda, M, King, N, Muller, M. In B. Schreiner & R. Hassan (Eds.), *Transforming water management in South Africa: Designing and implementing a new policy framework* (Series: Global Issues in Water Policy, Vol. 2). Dordrecht/New York: Springer.

Chapter 16
Water Pricing in Spain: Following the Footsteps of Somber Climate Change Projections

Javier Calatrava, Marian García-Valiñas, Alberto Garrido, and Francisco González-Gómez

Abstract As many other countries, Spain has to cope with, and be prepared to address, major water challenges: climate change, growing demand, and water pollution. Climate change projections indicate significant reductions of runoff and water recharge and more unstable climate regimes. Improving water allocation has become an urgent need. Water demand management is now one of the most relevant issues in the Spanish water policy agenda. The chapter discusses the controversial topic of water pricing, focusing on Spain. The Water Framework Directive (WFD) foresees that, in order to ensure an efficient and sustainable management of water resources, prices should be fixed according to the principle of cost recovery. But our analysis of all policy-relevant drivers and likely scenarios suggests that reforming water-pricing policies is likely to face numerous obstacles and to raise strong opposition from most water users. And yet, pricing policies in Spain are already innovative and fully implemented for all sectors. So the way to reform is already paved, and we expect that more progress will be made in next WFD planning period (2015–2020).

J. Calatrava (✉)
Escuela Técnica Superior de Ingeniería Agronómica, Universidad Politécnica de Cartagena, Paseo Alfonso XIII, 48, 30203 Cartagena, Spain
e-mail: j.calatrava@upct.es

M. García-Valiñas
Facultad de Economía y Empresa, Departamento de Economía, Universidad de Oviedo, Avda. del Cristo s/n, 33007 Oviedo, Spain
e-mail: mariangv@uniovi.es

A. Garrido
ETS Ingenieros Agrónomos, Avenida Complutense 3, 28040 Madrid, Spain
e-mail: alberto.garrido@upm.es

F. González-Gómez
Departamento de Economía Aplicada, Facultad de Ciencias Políticas y Sociología, Universidad de Granada, C/ Rector López Argüeta s/n, 18071 Granada, Spain
e-mail: fcojose@ugr.es

A. Dinar et al. (eds.), *Water Pricing Experiences and Innovations*,
Global Issues in Water Policy 9, DOI 10.1007/978-3-319-16465-6_16

Keywords Spain • Binomial water tariff • Climate change • Desalinization • Effluent control levy • Equity

16.1 Introduction

The general framework for water pricing in Spain was established in the 1985 Water Law (Garrido and Calatrava 2009). Since then, only partial amendments have been implemented, the most important of which resulted from the transposition of the EU Water Framework Directive (EC 2000) into Spanish national legislation.

When Spain became a member of the European Community in 1986, its lower per capita income relative to all member states entitled it to funds from the European Union that have been used primarily to construct infrastructures. Many of these funds were invested in water-related projects, including dams, canals, wastewater treatment plants, irrigation modernization projects, and water supply systems. These projects were heavily subsidized. The beneficiaries of these projects were thus serviced at a lower cost than if they would have financed them entirely. So one driver of water pricing in Spain in the past two decades has been the subsidization of water projects. Because of the expansion of the EU to Eastern Europe, Spain's relative position in per capita terms has improved, reducing the amount of European Union's Structural and Investments Funds that Spanish regions are eligible for.[1] As a result, most water users, together with the Spanish administrations, must bear all the costs of new water infrastructures.

The end of the era of massive funding for regional projects from the EU has coincided with the economic crisis, which severely hit the Spanish economy from 2008 onward, and with the first planning period of the Water Framework Directive, which establishes the obligation that 6-year basin plans should be approved in 2009. Spain, together with all other member states, embarked on a thorough assessment of, first, the ecological status of all water bodies, and second, of all water uses, including economic costs, pricing schemes and cost-recovery rates. This evaluation made clear that Spain requires a major water reform with significant changes in all fronts: planning criterions, projects' financing schemes, allocation mechanisms, serious environmental restoration projects, participatory processes, and water-pricing reforms.

Basin water plans were submitted to the European Commission after years of delay (some have not yet been finished and approved in June 2014, 5 years after the WFD deadline of December 2009). A reflection on its contents and approaches

[1] European Union's Structural and Investment funds finance actions targeted at economic development. There are five funds, namely, the Cohesion Fund, the European Regional Development Fund, the European Social Fund, the European Agricultural Fund for Rural Development, and the European Maritime and Fisheries Fund. Support from these funds depends on the level of economic development. However, only the less-developed regions (one in the case of Spain) can receive support from the first one.

revealed that they are too ambitious and complex and that they require large budgets, even after being cut by half with respect to the first drafts (De Stefano and Garrido 2013). There is a marked mismatch between the country's water policy goals (that include social and environmental objectives) and the inadequacy of pricing and financial schemes to achieve these goals. But pricing schemes would not be significantly changed in the absence of a water law reform, and the Spanish central and regional administrations could not, by any means, bear the costs of all the programs, actions, and projects defined in the basin plans.

Two other factors complete the scenario. First, the completion of the modernization reform of irrigation districts involved in most cases the replacement of open-air canals and conveyance systems with pressurized networks with tubes and valves (Lopez-Gunn et al. 2012). As a result of the reform of the electricity sector, energy prices for irrigators increased by 30–70 % (Hardy et al. 2012). Water costs for 80 % of the irrigators (surface and groundwater users) in Spain have increased by at least 100 %, not in the water rate itself but in the energy component. So "cheap water" before 2006 is now far from cheap but not because of a pricing reform.

Lastly, the need to improve, upgrade, and repair both urban wastewater treatment plants and water supply systems has already been identified as a top priority. Required investments have been estimated at €19 billion[2] (Aldaya and Llamas 2012). This could only be financed either by issuing public bonds (local, regional, and national), an option that is severely limited by the European Commission's overseeing of Spanish public finances, or by the private sector, taking on franchises or some other form of private contractual arrangement.

16.2 Key Issues in Water Supply in Spain: An Overview

Spanish population has increased notably during the present century, from 40.5 million in 2,000 to 46.7 million in 2013,[3] which created additional pressures on domestic water supply systems. However, current projections of population growth predict a population decrease of up to 2.6 million during the next decade. Even if such trend would be reversed if the Spanish economic situation improves in the coming years, population growth will be a major challenge in terms of treating wastewater, rather than in terms of satisfying tap water needs.

Probably, the major threat for water supply systems in the future arises from climate change. According to the CEDEX report (CEDEX 2011), by 2040 mean annual temperature in Spain could increase between +1.4 and +1.9 °C, and annual precipitation could decrease between 5 and 6 % (depending on the GHG emission scenario considered). In the case of the Canary Islands and the southern basins, precipitation will decrease between 7 and 14 % between 2010 and 2040, while the

[2] The 2013 yearly average Euro/US$ exchange rate was 0.783 Euros per US dollar, i.e., 1.277 US dollars per Euro, according to data from the US Internal Revenue Service (http://www.irs.gov).

[3] Data from the Spanish National Statistics Institute (http://www.ine.es).

eastern and northern basins are not expected to experience large changes in rainfall patterns (Garrido et al. 2013).

In terms of water availability, the CEDEX report predicts greater impacts in the southern half of the country. Southern basins (Guadalquivir, Mediterranean Andalusian basins, Guadiana, and Segura) could suffer reductions of up to 13 % of surface water runoff and 15 % of aquifers' recharge, while the eastern Mediterranean basins (Ebro, Catalonia Inland basins, and Jucar) would experience reductions in water availability of less than 10 %. Northern basins will suffer similar reductions, but their abundant resource availability will likely mitigate the impact (Garrido et al. 2013). However, some detailed studies, such as the one by Quiroga et al. (2011) in the Ebro basin, predict even larger reductions in runoff.

Regarding water quality, the implementation process of the WFD has forced Spanish authorities to collect a vast amount of data to characterize the ecological and chemical status of water bodies. Willaarts et al. (2014) have reviewed all related assessment reports provided by Spanish basin authorities. Despite significant flaws that have been detected in such reports (e.g., lack of data about relevant indicators and about 43 % of surface water bodies), Willaarts et al. (2014) conclude that almost 50 % of all surface waters in Spain are in poor ecological status, being the southern Atlantic basins in the bottom list in terms of percentage of water bodies in poor status.

16.3 Past Experiences with Water Pricing

16.3.1 General Features of Water Pricing in Spain

There is a wide variety of water-pricing systems in Spain, with sectorial, regional, and even local differences. However, there is a general pricing model that stems from the application of the 1985 Spanish Water Act and that remained mostly untouched after the 2001 Spanish Water Act. This model applies only to interregional hydrological basins that are managed by the basin agencies that depend on the Spanish national government. The fees and tariff systems are different for intraregional basins managed by regional governments. It also applies only to surface water, as most groundwater developments are legally privately administered.

Water pricing in Spain is based on four fees and tariffs that are paid by water users to river basin authorities (RBA) depending on the water services that they receive (Fig. 16.1): First, users of the public hydraulic domain are charged a levy to protect and improve the domain's conditions. It is charged on the occupation or use of land belonging to the public hydraulic domain, riverbeds, and river flows but not on water consumption. Second, urban and industrial users pay an "effluent control levy" (*Canon de Vertido*). A basic or reference value for this levy on point source pollution is set annually (0.01683 €/m^3 for urban sewage and 0.04207 €/m^3 for industrial wastewaters in 2014), which is multiplied by a coefficient, ranging between 1 and 2.5, depending on the contamination level of the discharged effluents.

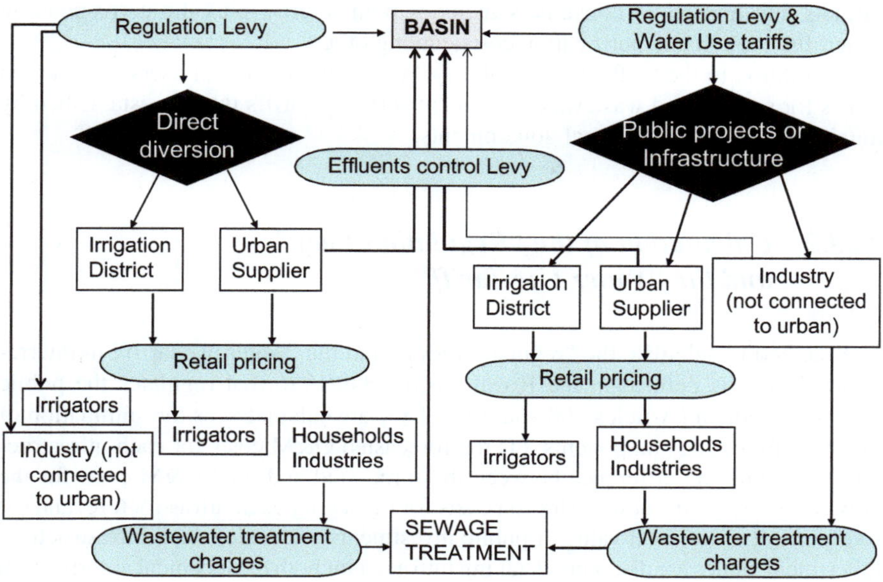

Fig. 16.1 The general model of water pricing in Spain (Source: Garrido and Calatrava 2009)

Third, users of surface resources pay a "regulation levy" (*Canon de Regulación*) to compensate the basin authority for the costs associated with building, operating, and maintaining public water regulation infrastructures, especially dams. Last, the "water use tariff" (*Tarifa de Uso del Agua*) aims to pay investment, operation, and maintenance costs of specific infrastructures, such as large canals, water transfers, etc., that are not regulation works. Only the users of such infrastructures pay it.

The 1999 Spanish Water Act introduced a multiplier factor, ranging from 0.5 to 2.0, to be applied to levies and tariffs charged to irrigators when their level of water consumption was below or above a reference consumption level. However, there is no documentation of this factor ever being applied.

Most water users benefit from both public regulation works and specific infrastructures and, thus, are charged both the "regulation levy" and the "water use tariff." Those users who abstract water directly from the surface water bodies must pay the "regulation levy," as they do not use any specific infrastructure but benefit from the general water regulation of the basin.

Groundwater users are not usually obliged to pay any levy or tariff, as they do not use public infrastructures to divert the resources they use, and it is assumed that they do not benefit from public water regulation infrastructures. Although users must have concession rights for groundwater resources, most of them are still under private ownership. They are responsible for the cost of drilling their wells and the O&M costs.

In most cases, irrigation districts and urban suppliers pay the "regulation levy" and the "water use tariff" to the basin agencies. The final retail water price paid by farmers and household consumers includes both figures, plus the corresponding water distribution and purification costs, among others.

In addition to the "effluent control levy," urban and industrial users are charged tariffs for sewage and wastewater treatment services, tariffs that are established by municipalities and/or regional governments.

16.3.2 Calculation of the "Regulation Levy" and the "Water Use Tariff"

The method to calculate the "regulation levy" and the "water use tariff" in interregional basins is defined in the Royal Decree 849/1986 that regulates the public hydraulic domain (Articles 300 and 307). They are calculated as the summation of the three following components: (1) the forecasted O&M costs for each infrastructure, including the difference between the forecasted and final O&M costs for the previous year; (2) the administration costs for managing each infrastructure; and (3) four percent of the total value of public investments to develop each infrastructure. Investment values are discounted, taking into account both the technical amortization of the infrastructure and inflation rates.

The only difference in their calculation is that the amortization period for infrastructure projects considered for the calculation of the investment cost component (3) is 50 years, starting from the year after the infrastructure became operational, in the case of the "regulation levy," and 25 years in the case of the "water use tariff." The valuation of existing infrastructures is done using the "historical value" criteria. A detailed explanation of the formulae used can be consulted in Calatrava and Garrido (2010).

Several authors have criticized the system used to calculate fees and tariffs for both the regulation levy and the water use tariff because, in most cases, it results in cost-recovery rates below 100 % (Pérez and Barreiro 2007; Bielsa et al. 2009). A number of studies (Berbel 2005; MMA 2005; Groot and Sánchez Chóliz 2006; Bielsa et al. 2009) have assessed the implications of this system under different assumptions and for different areas and basins. The general conclusion is that the computed capital costs are lower than those obtained using alternative standard accounting systems, including the one that existed before 1986. A detailed review of these studies is presented in Calatrava and Garrido (2010).

In sum, users pay all O&M and administration costs of wholesale water services (diversion, regulation, and transportation of surface resources) and a share of its capital costs. However, in general, wholesale water services only represent a small share of total costs of water services, whereas the main cost component corresponds to the purification and/or distribution phases.

In the calculation of the "regulation levy" and the "water use tariff," costs of wholesale water services are allocated among the different water users using specific infrastructures. Articles 301 and 308 of Royal Decree 849/1986 establish the system used for sharing the calculated annuities for both the regulation levy and water use tariff among the different water users (Calatrava and Garrido 2010). This system is based on a set of "equivalence coefficients" that are linked to the users' presumed paying capacity, which is based on the benefit that each user obtains from water use.

The system of "equivalence coefficients," which is summarized for all basins in MMA (2007), differs between basins and even within basins. It is based on a "stakeholder's agreement" at the basin level and takes into account the relevance of the different private and public users in each basin. For instance, it considers that urban uses have priority over other uses, and thus also a high supply reliability. As a result, average profit per cubic meter for domestic uses ranges between three to five times the values considered for irrigation (Table 16.1).

Users include not only agricultural, domestic, industrial, and energy generation but also the state as a beneficiary of public services, such as flood control and environmental services (pollution control, environmental river flows, etc.). For example, the share of total water regulation and transportation costs that are attributed to flood control services ranges between 0 % for the Duero basin and 50 % in the Segura basin, with a national average of 15 % (2007b). As a general rule, this share increases as we move from north to south and from the Atlantic toward the Mediterranean.

16.3.3 Exceptions and Regional Pricing Systems

There are some exceptions to the general model described above. First, the main water-pricing scheme described above is applicable only to those basins that are the responsibility of the national government. The fees and tariffs systems are different

Table 16.1 Equivalence coefficients for the main water uses in interregional basins

Basin	Irrigation	Domestic	Industrial	Nonconsumptive	Hydropower
Duero	1 l/s	5.41 l/s	5.41 l/s	0.1 l/s	0.1 l/s
Ebro	2 m^3	10 m^3	10 m^3	1 m^3	4 m^3
Júcar	1 m^3	2.5–4 m^3	2.5–4 m^3	1 m^3	0.96 kWh
Guadiana	1 m^3	1–5 m^3	1–3 m^3	0.6 m^3	0.6 m^3
Guadalquivir	0.25–3 m^3	0.75–5 m^3	0.75–5 m^3	0.3 m^3	0.96 kWh
Norte	2 m^3	10 m^3	10 m^3	1 m^3	3.6 kWh
Segura	1 ha	3 ha	3 ha	–	9.600 kw
Tajo	1 m^3	3 m^3	3 m^3	0.2 m^3	15 % of the price of kWh

Source: MMA (2007). These coefficients are used to share the costs of water supply among water users. Water costs are shared based on water consumption but corrected using these equivalence coefficients. For example, in the Duero basin, domestic and industrial users in the Duero basin are charged per m^3 5.4 times what irrigation is charged and 54 times what nonconsumptive users and hydropower generation are charged for intraregional basins managed by regional governments. As regional governments develop their own environmental policies, some of them have specific water charges. Catalonia, for instance, charges all final consumers since 2000 a water levy (*canon del'aigua*), irrespective of the type of agency servicing end users or if it is a public-private institution (ACA 2008). Rates vary across sectors, and farmers are exempt from the levy. Other regions, such as Galicia, have enacted similar levies. In the internal basins of the Basque Country, several private and public consortiums of municipalities are responsible for water resources management in different geographical areas, which results in a wide variety of water-pricing schemes (Gobierno Vasco 2005).

In the case of the Andalusian Mediterranean basins, the regional government of Andalusia maintains the general system of fees as tariffs but have created two additional levies by the 9/2010 Andalusian Water Law: the "improvement levy" (*Canon de Mejora*) that charges urban water use in order to finance wastewater treatment facilities and the "general services levy" (*Canon de Servicios Generales*), which is intended to recover the regional government's administration costs of water conservation services. This latter levy modifies the "regulation levy" and the "water use tariff" by not including their second component (administration costs for managing each infrastructure) that is now recovered through this new levy.

Another interesting case is the Canary Islands, where the absence of surface flows and a traditional water culture that appreciates the importance of water have given rise to numerous institutional arrangements. The private sector has been the major actor in these institutions, though more recently a number of governmental agencies have begun to participate actively in various water services (Aguilera Klink 2002). The 1990 Canary Islands' Water Act (Ley 12/1990 de Aguas de Canarias) considers three main fees and levies that charge the use of the public domain, the discharge of effluents, and the use of water from public infrastructures.

Other exceptions are groundwater users and historical users, which in general are not charged with any water levy. Groundwater users only pay for their extraction and distribution costs. However, in Catalonia, nonagricultural groundwater users are charged the above regional levy that is also charged to surface water users. Historical users are users that can provide evidence of having used water before major modern infrastructures were built in their area. Based on a favorable legal claim, they are currently exempted from the regulation levy (Calatrava and Garrido 2010). However, it remains to be seen whether the full application of the WFD results in the need to charge resource and environmental costs to these users.

Lastly, another relevant exemption is the Tajo-Segura Transfer (TST), a large canal connecting central and southeastern Spain, whose financing system is based on a specific law (Act 52/1980). Its tariffs are among the most expensive currently paid in Spain for bulk water, leaving desalinized water aside. Its relevance arises not from its size (2 % of water users in Spain), but from the relevance of the revenues collected by the TST managing authority, which are about a third of all funds collected from water services in Spain. More detail about its financing system is given in Calatrava and Garrido (2010).

16.3.4 Irrigation

The cost of water provision borne by farmers depends on the type and origin of the water resources used. Simply speaking, four main types of charges exist: (1) the regulation levy; (2) the water use tariff; (3) the "derrama" tariff, charged by the water users association (WUA) for water distribution costs met by the irrigation district (ID) or WUA which farmers belong to, a tariff that can include the WUA costs for pumping groundwater from the WUA's communal wells; and (4) costs farmers pay for abstracting water directly from rivers or aquifers, including maintenance of pumping equipment, energy, and labor costs.

Again, on a simplified manner, several situations can exist. First, farmers using surface water from specific infrastructures pay the regulation levy and the water use tariff to the river basin authority via the irrigation district's administration and an additional "derrama" tariff to cover the costs of the irrigation district itself (point 3 in the above paragraph) (Fig. 16.2). Second, in case the ID abstracts water directly and uses public regulation infrastructures (i.e., the ID does not uses specific infrastructures), farmers pay only the regulation levy and the "derrama" tariff to cover the district's own pumping, transport, and application costs.

Third, farmers not belonging to a WUA and abstracting water directly from a surface body of water pay the regulation levy directly to the RBAs and bear the abstraction costs themselves. Fourth, farmers using groundwater only pay their own abstraction and distribution costs. Last, there are historical users that are presently exempted from tariffs and levies.

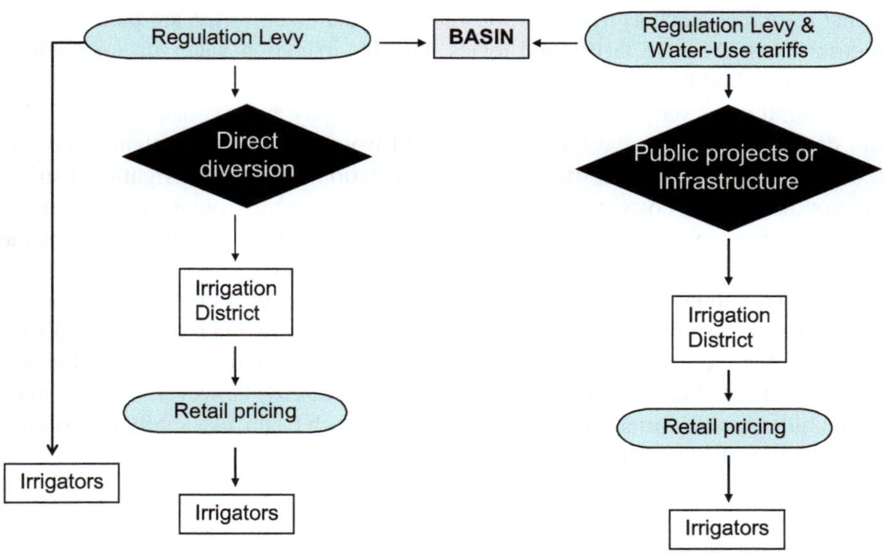

Fig. 16.2 Tariff structures for agricultural surface water users (Source: Own elaboration from Garrido and Calatrava 2009)

Table 16.2 Farmers' payments for irrigation water services in interregional Spanish basins (2001–2002)

Basin	Surface water resources			Groundwater		Surface and groundwater	
	Per ha						
	Distribution (paid to WUA)	WUA and basin tariff	WUA and basin tariff per m³ (*)	Cost per ha (**)	Cost per m³	per ha (***)	per m³ (****)
Duero	20	46	0.012	500	0.095	231	0.044
Ebro	49	120	0.030	829	0.150	113	0.020
Tajo	36	67	0.020	541	0.100	199	0.038
Júcar	81	16	0.020	383	0.074	283	0.055
Guadiana	19	102	0.025	232	0.048	188	0.039
Guadalquivir	101	70	0.035	744	0.150	400	0.081
Segura	34	151	0.038	789	0.163	464	0.096
Total	50	56	0.021	500	0.090	264	0.051

Source: Adapted from MMA (2007); all figures expressed in euros. (*) is the volumetric equivalent of the per-hectare surface water tariffs. (**) is the per-hectare equivalent of the volumetric cost of groundwater; (***) is the per-hectare conversion of the total costs of surface and groundwater; (****) is the per cubic meter conversion of the total costs of surface and groundwater resources

Table 16.2 summarizes the average farmer's payment for water services in some Spanish basins (MMA 2007). The regulation levy and the water use tariff are predominantly paid to the basin authorities on a per-hectare basis for irrigation uses and on a volumetric basis for urban and other uses. Groundwater and nonconventional resources, such as treated wastewater or desalinized water, are paid on volumetric terms. Average tariffs and levies paid for irrigation water in areas where water is supplied by river basin authorities are equivalent to 0.021 €/m³, except for agricultural users serviced by the TST project who pay approximately 0.09–012 €/m³, depending on the year,[4] and those served from desalinization plants who pay 0.40–0.45 €/m³ (all these tariffs include transportation costs to the irrigation districts but not distribution costs within the district). In areas using groundwater resources, recipients pay an average volumetric price of 0.04–0.16 €/m³, depending on the basin, which is based on extraction and other O&M costs.

Table 16.2 shows that average surface water tariffs are greater in the most water-stressed basins, such as Guadiana, Guadalquivir, or Segura. On the contrary, differences in average groundwater costs among basins are relatively smaller than for surface water and are not directly related to water scarcity. However, these average tariffs hide notable variations among areas within each basin, especially for groundwater and for the less-endowed basins.

[4] Current tariffs for raw water from the Tajo-Segura Transfer are 0.098445 €/m³ for agricultural users and 0.115768 €/m³ for urban users.

For example, in the southeastern Segura basin, which is one of the most water-scarce areas in the country, water costs for irrigation raw water are on the range 0.03–0.34 €/m^3, depending on the source of water, with an average of 0.11 €/m^3 (CHS 2007). The vast majority of farmers and irrigation districts in this basin pay volumetric tariffs and rely on several water sources with different prices to meet their water allotments. According to CHS (2007), average costs for the basin's own surface water is 0.038 €/m^3, whereas a 0.09–0.12 €/m^3 tariff is paid for water from the Tajo-Segura Transfer, 0.03 €/m^3 for treated sewage water, and 0.34 €/m^3 for desalinized water. The cost of groundwater resources ranges from 0.102 to 0.33 €/m^3, with an average of 0.115 €/m^3. These prices for raw water must be increased for the distribution costs (WUA's costs or derrama), which are on the 0.06 to 0.08 €/m^3 range (own estimate based on data from several WUAs), and for the transportation and distribution costs in the case of desalinization.

In the Valencia Region (Júcar basin), where groundwater is the main source of supply, the unit cost of water for farmers ranges from 0.04 to 0.22 €/m^3 for surface and groundwater, respectively, the average price for all water consumption being 0.11 €/m^3 (García 2002; García et al. 2004).

If we look at irrigation districts' prices, most Spanish water users associations have opted for one of the following water-pricing schemes: a fixed per-hectare tariff, a volumetric tariff, or a binomial tariff. Fixed per-hectare tariffs are calculated as the total costs attributable to farmers, divided by total irrigated area. It is the most common option in traditional districts (those built before 1950) served from surface resources, while volumetric tariffs are more frequent in districts served by groundwater or incurring significant energy costs. The third pricing system, binomial tariffs, combines a volumetric rate to cover variable costs (water and energy) and a fixed per-hectare rate for investment and management costs. According to MAPA (2001), fixed rates are applied across 82 % of the national irrigated acreage, whereas volumetric rates are applied in 13 % and binomial tariffs in 5 % of the national irrigated acreage. There is not more recent data at the national level about the penetration of volumetric and binomial tariffs, but a majority of Spanish irrigation districts have increased their tariffs because of the increasing energy prices rather than changing their tariffs structures. Quotas, rather than prices, remain the main allocation system.

16.3.5 Urban Users: Households and Industries

Spanish urban water prices are among the lowest in the European Union (OECD 2010, 2013). The Spanish National Statistics Institute calculates an index that shows the average revenue per cubic meter from urban water services.[5] Figure 16.3 shows

[5]The index is calculated dividing the total revenues from water services by the distributed water volume.

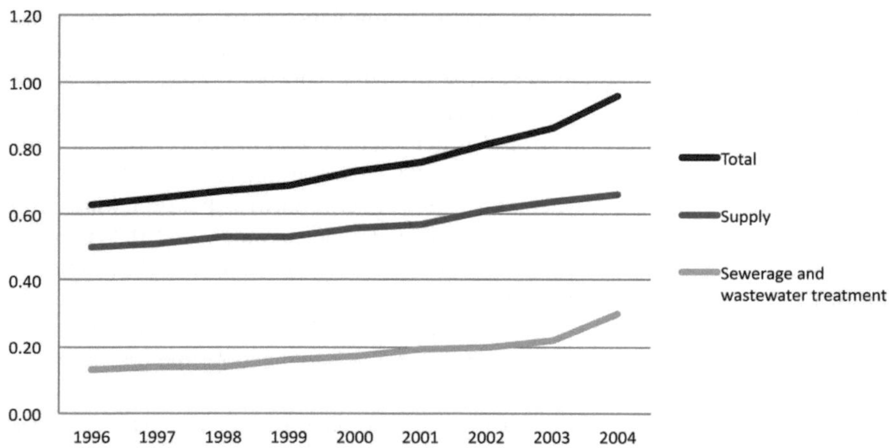

Fig. 16.3 Average revenue from urban water services in Spain: 1996–2004 (in €/m3)

the evolution of that index during the period 1996–2004. It also shows the evolution of the index disaggregated by service. In 8 years, the average revenue has increased by nearly 50 %. The increasing trend is more stressed for sewerage and wastewater treatment services. In fact, a significant raise is registered during the period 1996–2005, when those tariffs increased by nearly 170 %. This is explained by the extension of those services to additional areas. Although supply service is broadly extended to the population, sewerage and wastewater treatment services are lacking. Adapting the requirements of the European Water Framework 91/271/CEE, the Spanish National Plan of Sanitation and Sewerage 1995–2005 established to broaden those services to all the municipalities bigger than 2,000 inhabitants. In fact, in 1995 only 41 % of the population had access to sewerage services. At the end of the plan in 2005, the percentage was about 76 %. Then, new infrastructures were needed, so municipalities and autonomous communities started to charge sewerage prices looking for financial support in order to achieve the objective of costs' recovery.

Among the regions, Canary and Balearic Islands, Murcia, and Cataluña have the highest index levels, always above the national average. Most of them are regions characterized by scarcity problems, and they have applied desalinization technologies to obtain additional water to satisfy several users' demands. Something similar has happened in the Valencia Region, where suppliers have started to use similar technologies, so prices have increased during the last years. As it is expected, some regions located on the north of Spain registered the lowest prices. However, it is surprising to observe that Andalucía, which is a region frequently affected by drought episodes, has set prices below the national average during the whole period. Additionally, regional prices do not have a very high dispersion around the average. All those features are shown in Table 16.3.

Table 16.3 Average revenue from urban water services in Spain by regions: 1996–2004 (in current €/m³)

	1996	1997	1998	1999	2000	2001	2002	2003	2004
National average	0.63	0.65	0.67	0.69	0.73	0.76	0.81	0.86	0.95
Andalusia	0.53	0.55	0.57	0.58	0.59	0.64	0.69	0.79	0.94
Aragon	0.44	0.46	0.51	0.55	0.59	0.59	0.62	0.66	0.82
Asturias	0.36	0.41	0.42	0.45	0.51	0.55	0.59	0.65	0.65
Balearic Islands	1.12	1.16	1.16	1.24	1.32	1.45	1.48	1.42	1.31
Canary Islands	1.51	1.52	1.52	1.55	1.58	1.66	1.67	1.68	1.64
Cantabria	0.41	0.41	0.44	0.46	0.53	0.52	0.55	0.60	0.69
Castilla and Leon	0.41	0.41	0.44	0.42	0.42	0.45	0.49	0.53	0.61
Castilla – La Mancha	0.35	0.38	0.39	0.35	0.44	0.48	0.52	0.57	0.63
Catalonia	0.76	0.80	0.86	0.9	0.94	0.91	0.98	1.04	0.92
Valencia	0.62	0.60	0.62	0.62	0.66	0.72	0.78	0.83	1.07
Extremadura	0.44	0.49	0.49	0.60	0.72	0.74	0.76	0.73	0.72
Galicia	0.41	0.41	0.48	0.50	0.54	0.60	0.61	0.62	0.78
Madrid	0.64	0.65	0.66	0.68	0.69	0.76	0.81	0.86	1.00
Murcia	0.94	0.95	0.99	0.99	1.12	1.02	1.08	1.08	1.41
Navarra	–	–	–	0.45	0.60	0.59	0.63	0.73	1.11
Bask Country	0.98	1.02	1.04	1.06	1.12	1.09	1.14	1.15	0.83
La Rioja	–	–	–	0.30	0.41	0.42	0.44	0.54	0.96

Source: Authors elaboration from http://www.ine.es

Table 16.4 Average water price, 2002 (in current €/m³)

	Domestic	Industrial
Supply	0.660	0.870
Sewerage and wastewater treatment	0.350	0.430
Total	1.000	1.300

Source: AEAS (2003)

Tables 16.4 and 16.5 present some disaggregated statistics linked to different kind of users, elaborated by AEAS.[6] Table 16.4 refers to the level of prices in 2002. In Table 16.5, some general features are shown. Those statistics have been provided by AEAS, a private Spanish organization that conducts biannual surveys in a group of municipalities.

Table 16.5 shows some general features of water tariffs in Spain for different users. At the beginning of the century, the most usual structure was composed of a

[6] AEAS is a nonprofit professional association that integrates large groups of public and private operators of the Spanish water supply service. For residential users, the formula to calculate the average price is the following: (0.15*P7)+(0.75*P15)+(0.10*P25), where P7, P15 and P25 are water bills corresponding to 7, 15, and 25 m³ per month, respectively. In the case of industrial users, the average is not weighted, using the following formula: (P10+P150+P1,500)/3. In a similar way, P10, P150, and P1,500 are water bills corresponding to 10, 150, and 1,500 m³ per month, respectively.

Table 16.5 Urban water tariffs structure by users, 2002 (% population)

	Residential			Industrial		
	Supply	Sewerage	Wastewater treatment	Supply	Sewerage	Wastewater treatment
Fixed charge + increasing blocks	92	71	63	67	67	3
Fixed charge + constant price	3	18	27	24	23	90
Free allowance	5	4	2	9	3	0
Constant price	0	7	3	0	7	2
Flat fee	0	0	4	0	0	6

Source: AEAS (2003)

fixed charge and a variable charge, both for residential and industrial users. The variable charge is usually based on an increasing block structure. The number of blocks varies, depending on the municipality or/and region, but the most frequent number is three. Increasing block tariffs is the most common way to charge for consumption. They are adequate from an environmental perspective, helping to solve scarcity problems. They are highly recommended by international institutions, such as the World Bank or the OECD, as tools to help achieve efficiency objectives. However, increasing block tariffs results in serious equity effects for households in both developed and developing countries (Whittington 1992; Barberán and Arbués 2009). Then, it would be adequate to design per capita water tariffs, but usually their application requires high-quality information.

Additionally, it is important to mention that there is a percentage of the population that is charged using free allowances or, even worse, a flat fee. That percentage is even higher in the case of industrial users' supply services. A free allowance is a minimum water consumption that is charged at zero marginal prices. Although it is a low percentage, some international organizations have strongly advised against that kind of structure, because they can lead to water resources overuse. Moreover, free allowances lead to significant efficiency losses. Users fail to reveal their preferences, since they do not face a marginal incentive to conserve water and usually consume more than they need (Castro et al. 2002). Additionally, it is possible to observe a strong heterogeneity in the size of free allowances. This fact suggests that there is no uniform equity criterion to set that minimum amount of water. Actually, there is a clear trend to reject these practices, because they are not environmentally efficient (they generate overconsumption) and show lower levels of equity than expected (OECD 2003).

In general, the complexity of tariffs decreases when it comes to sewerage and wastewater treatment. In the case of supply service, a fixed charge is set in all the municipalities in the AEAS sample. However, not all the municipalities are charged a fixed component for those services. At the same time, it is more usual to find a constant price instead of increasing blocks.

In order to ensure affordability and equity, some tariff-related policies are applied.[7] Thus, several discounts are applied, especially at the household level. Among those discounts that have a clear redistributive aim, we find those applied to retirees and/or people over 65, those with large families, and those households whose income is below some preestablished level. Sometimes, the previous adjustments are combined with efficiency issues, rewarding low consumption levels (García-Valiñas et al. 2010). However, using prices to achieve redistribution goals does not result in cost-recovery goals. It is highly likely that social water tariffs do not allow recuperating total service costs (Bös 1985). Actually, at the beginning of this century, urban water prices in Spain did not comply with the European Water Framework Directive 2000/60/CE (González-Gómez et al. 2012; European Environmental Agency 2013). Additionally, when "social" criteria are not well defined, usually people in the medium-income class receive the higher benefits (Estache et al. 2001, 2002).

16.4 Present Water Pricing Practices

16.4.1 Irrigation

The charges paid by farmers in a selection of irrigation districts and RBAs are reported in Table 16.6. It presents a range of values of the water use tariff and the regulation levy for seven Spanish basins and shows some examples of prices paid by farmers.

Irrigation water pricing has barely changed since the 1985 Water Law was enacted. Wholesale water pricing is still guided by the Royal Decree 849/1986, and minor changes to the general pricing scheme made in the 1999 Water Act have not been applied in practice. A similar setting is found when looking at retail prices paid by farmers. Water prices have generally increased because of the rising energy costs, especially in the most modern districts. But, beyond that, most tariffs structures remain untouched.

[7] In this respect, two different types of policies can be used: income-support policies and tariff-related policies. Income-support policies focus on the income side when attempting to solve the consumers' affordability problem. In this group, it is posible to consider direct income aid or water service vouchers from the public sector, water utilities or other private or charitable sources, payment aids in the form of easier payment plans, special loan facilities, and arrears forgiveness. On the other hand, tariff-related initiatives consist of changing water charges (level and structure) in order to reduce the size of the typical water bill faced by low-income users. We include, among other measures, subsidizing utility prices, designing tariff structures ("social tariffs") to get cross subsidization, or capping metered tariffs for low-income users (OECD 2003).

Table 16.6 Water tariffs for a selection of Spanish irrigation districts and RBAs

Area/region	Source of water	Price	Comments	Source
RBAs regulation levies and water use tariffs				
Tajo basin	Surface	Regulation levy: 5,22–130,36 €/ha	Range of 2013 wholesale prices at basin level	www.chtajo.es
		Water use tariff: 33,17–300,11 €/ha		
		Summation of both: 5,22–309,72 €/ha		
Guadalete-Barbate basin (Andalusia)	Surface	Regulation levy: 21,69–136,49 €/ha	Range of 2011 wholesale prices	http://www.juntadeandalucia.es/medioambiente
		Water use tariff: 19,47–369,7 €/ha		
		Summation of both: 19,47–369,7€/ha		
Duero basin	Surface	Regulation levy: 15,99–73,06 €/ha	Range of 2013 wholesale prices at basin level	www.chduero.es
		Water use tariff: 14,9–245,97 €/ha		
		Summation of both: 14,9–261,46 €/ha		
Guadiana basin	Surface	Western basin:	Range of 2014 wholesale prices at basin level	www.chguadiana.es
		Regulation levy: 10,82–48,39 €/ha; Water use tariff: 0–295,14 €/ha		
		Eastern basin:		
		Regulation levy: 8,98–31,59 €/ha; water use tariff: 0–295,14 €/ha		
Mediterranean Andalusian basins (Almería province)	Surface	Regulation levy (0,0262–0,0547 €/m^3); water use tariff (0,0136–0,1523 €/m^3); RL + WUT: 0,0683–0,1786 €/m^3	Range of 2011 wholesale prices	http://www.juntadeandalucia.es/medioambiente
Mediterranean Andalusian basins (rest)	Surface/groundwater	Regulation levy (33,78–270,24 €/ha); water use tariff (29,99–147,55 €/ha)	Range of 2011 wholesale prices	http://www.juntadeandalucia.es/medioambiente

Final prices paid by farmers				
Segura basin	Various sources	0.03 €/m³ (surface); 0.09–0.12 €/m³ (Tajo-Segura interbasin transfer); 0.03 €/m³ (treated sewage water); 0.38–0.45 €/m³ (desalinization, incl. transport); 0.10–0.33 €/m³ (groundwater)	Wholesale prices at basin level	CHS (2007) and own data
Mazarrón (Segura basin)	Desalinized groundwater	0.6 € /m³	Pumping cost for private farmers	Puerto et al. (2013)
Several irrigation districts (Segura basin)	Price of pool of water sources (surface, groundwater, desalinized, recycled)	0.19 ± 0.04 €/m³ (Campo de Cartagena)	Volumetric prices paid by farmers in the 2002–2011 period	Soto (2013)
		0.22 ± 0.04 €/m³ (Lorca)		
		0.16 ± 0.03 €/m³ (TST Cieza)		
		0.21 ± 0.02 €/m³ (Pantano de la Cierva)		
		0.16 ± 0.04 €/m³ (Miraflores District)		
Castilla La Mancha Region (Júcar basin)	Groundwater	0.15 € m³	Pumping cost for private farmers	Domínguez et al. (2012)
Riegos del Alto Aragón (Ebro basin)	Surface	Average districts' charges: 0.013 €/m³ + 62.50 €/ha (sprinkler irrigated districts); 0.005€/m³ + 46.50 €/ha (surface irrigated districts)	Public irrigation development formed by 53 districts, some of them modernized	Lecina et al. (2010)
Bembézar MD district (Guadalquivir basin)	Surface	293 €/ha flat rate (equivalent to 0.04 € /m³)	Public recently modernized district	Carrillo Cobo et al. (2014)
Palencia province (Duero basin)	Surface	Regulation levy: 24 €/ha;	Several public irrigated districts	Gallego-Ayala et al. (2011)
		Water use tariff: 10–30 €/ha		
		Districts' charges: 20–30 €/ha		
		Total fees paid: 54–84 €/ha (equivalent to 0.015 €/m³)		
Western and Eastern La Mancha aquifers (Guadiana basin)	Groundwater	0.08–0.11 €/m3 (Western La Mancha aquifer)	Pumping costs form water farmers at current water table levels	Esteban and Albiac (2011)
		0.06–0.11 €/m3 (Eastern La Mancha Aquifer);		

Source: Own elaboration

One factor behind the predominance of flat rates across the country is the fact that a majority of surface resources are charged by basin authorities on a per-hectare basis. Public development initiatives during the twentieth century are behind an important share of Spanish irrigation districts. Served from public surface water infrastructures, they have always used flat-rate pricing. In fact, in the most water-scarce basins (Segura, Almanzora), where retail volumetric pricing predominates, tariffs and levies are also paid volumetrically to the basin's authority. As long as per-hectare wholesale pricing persists, districts will have fewer incentives to change their retail pricing schemes.

As commented, volumetric or binomial tariff structures are found in districts where groundwater is a relevant source of supply or with high levels of energy consumption. We could expect that the vast modernization process through which many Spanish irrigated areas has passed would promote a shift from per-hectare to volumetric or binomial structures also in districts served with surface water. However, no data is available regarding the impact of modernization of irrigation schemes on the districts' tariffs. In general, the WUA charges (derramas) have increased to pay for the associated investments, and there are cases of districts going through serious financial trouble to pay back their modernization loans, but to our knowledge no study has analyzed whether modernized WUAs have changed from flat to volumetric water rates.

A major shift in water pricing at districts' level happened in many areas of the Segura and Júcar basins in southeastern Spain during the process of creation of formal water users' associations after the 1985 Water Act. Many traditional associations using surface and/or groundwater resources changed from auction-based allocation mechanisms to volumetric tariffs and water quotas. This reflects a move toward more formal associations with greater infrastructures and technical and administrative staff that requires more stable revenues. However, despite the large investments in distribution infrastructures that have been made, there are many districts in which no fixed rate component has been set to finance such investments, which are fully recovered through the volumetric tariff. The traditional view that water users, rather than community members, should pay for all water costs, a notion that is embedded in the auction system, still prevails. Attempts by irrigation boards to introduce binomial tariffs have been frequently voted against by the WUA members. This is reinforced by the fact that new water supply sources in these areas are paid volumetrically.

Another shift in agricultural water pricing on the Mediterranean Coast could come from the increased availability of desalinized resources. Despite their high prices and the resulting increase in water prices, less-endowed WUAs are resorting to their use to complement their unreliable pool of water sources and to improve the quality of degraded groundwater. The Spanish government, owner of most of the desalinization plants, is proposing new supply contracts that guarantee a fixed volume of water to the districts, but these must pay a share of the prearranged water price for any unused volumes. It is yet to be seen whether this type of arrangement would result in changes of the districts' tariffs structure.

16.4.2 Urban Users: Households and Industries

Present water-pricing practices in Spain keep some of the elements mentioned in the previous section, with slight changes. Tables 16.7, 16.8, and 16.9 show the main figures about price level and structure in 2012. Price levels are presented in Table 16.7. Important increases have been registered for both kinds of users. Following the past experiences trend, industrial water prices are higher than those at the residential level, showing the presence of cross-subsidization strategies.

Despite those price increases, cost-recovery aims have not been fully achieved. The level of compliance with this aim is heterogeneous and depends on the region. Looking at the figures set in different river basins' hydrological planning from 2009 to 2015, the percentages of urban services cost recovery oscillates between 39 and 93 %.[8] However, those estimations should be interpreted with caution. The lack of homogeneous methodology to estimate environmental or resource costs could explain the registered differences. This is probably the most difficult issue to be incorporated into a full cost-recovery analysis (Ministerio de Medio Ambiente 2007; European Environmental Agency 2013).

In Table 16.8, we observe strong similarities with those set in 2002. Increasing blocks and a fixed charge is the preferred structure used for both residential and industrial users. Industrial structures seem to be simpler. Free allowances use has slightly decreased.

However, discounts in water tariffs have been strongly generalized. According to AEAS (2012), almost 70 % of Spanish municipalities set discounts in residential water tariffs. Those discounts are applied to a significant percentage of the Spanish population, as far as the probability of finding them increases with municipality size (higher probability in bigger cities). Around 50 % of them consider some adjustments, depending on the family size. They have become quite popular during the last few years. Table 16.9 shows different models for designing these kinds of discounts, applied in Barcelona and Granada in 2014.

In the previous table, the first model of family size adjustment is focused on solving the equity problems linked to increasing block tariffs. Thus, blocks are adjusted in order to adapt prices to the number of members in the house. That is a simple adjustment, in the sense that it does not take into account the presence of economics of scale in water

Table 16.7 Average water price, 2012 (in current €/m³)

	Residential	Industrial
Supply	0.924	1.232
Sewerage and wastewater treatment	0.672	0.837
Total	1.596	2.070

Source: AEAS (2013)

[8] Those figures have been taken from different river basin websites: www.chcantabrico.es (Cantábrico); www.chduero.es (Duero); www.chebro.es (Ebro); www.chguadalquivir.es (Guadalquivir); www.chguadiana.es (Guadiana); www.chtajo.es (Tajo).

Table 16.8 Urban water tariffs structure by users, 2010 (% population)

	Residential			Industrial		
	Supply	Sewerage	Wastewater treatment	Supply	Sewerage	Wastewater treatment
Fixed charge + increasing blocks	90	76	79	68	77	68
Fixed charge + constant price	2	8	8	15	7	14
Free allowance	4	5	3	4	2	3
Constant price	3	9	8	13	12	13
Flat fee	1	2	2	2	2	2

Source: AEAS (2012)

Table 16.9 Residential water supply tariffs: family size adjustments

Barcelona		Granada	
Block	Euros/m^3	Block	Euros/m^3
0–6 m^3	0.6188	0–2 m^3	0.4053
7–9 m^3	1.2376	3–10 m^3	0.6763
10–15 m^3	1.8564	11–18 m^3	1.3996
16–18 m^3	2.4752	>18 m^3	1.9171
>18 m^3	3.0940		
Eligible households: families with 4 or more members		Eligible households: families with 3 or more children	
Discount → blocks will be extended as follows:		Discount: 50 % in the variable charge corresponding to 10 m^3/month (two first blocks)	
First block: 2 m^3/month per additional person			
Second block: 3 m^3/month per additional person			
Third block: 5 m^3/month per additional person			
Fourth block: 6 m^3/month per additional person			

Source: Authors own elaboration

consumption (Whittington 1992; Arbués and Barberán 2012). However, it is a good method in order to avoid higher average prices that large families bear. The second model is oriented toward both equity and efficiency, limiting the discount to low consumption levels. However, it does not adjust the tariff according to the family size. Definitively, current urban water-pricing practices in Spain could be improved with more ambitious aims in terms of efficiency, equity, cost recovery, and environment.

Regarding industrial users, water tariff prices are more uniform, and penalties are set lower for water overuse. This feature makes more difficult to comply with environmental aims. In some cases, we observe some extensions of the block size based on the industry size. Table 16.10 shows the industrial water tariff set in 2013 in Madrid,[9] an area with high levels of economic activity. An increasing block tariff is set, but the block size is adjusted, depending on meter size. A similar scheme is applied to commercial users.

[9] This tariff is applied in the Madrid region, in those municipalities supplied by the Canal de Isabel II water company.

Table 16.10 Industrial water supply tariffs in Madrid: industry size adjustments

Block size (m^3/bimonth)										
	Meter size (mm.)									
Block	≤15	20	25	30	40	50	65	80	100	>100
1	<90	<150	<200	<350	<400	<550	<800	<800	<900	<900
2	90–180	150–300	200–400	350–700	400–800	550–1,100	800–1,600	800–1,600	900–1,800	900–1,800
3	>180	>300	>400	>700	>800	>1,100	>1,600	>1,600	>1,800	>1,800

Prices (euros/m^3)		
Block	Price	
	Summer	Rest of the year
1	0.407	0.407
2	0.687	0.550
3	1.460	0.973

Source: Author own elaboration

The positive effects of a peak-load pricing on the protection of water resources is neutralized by the extension on the block size, depending on meter size (which is directly linked to the industry size and the potential water consumption). Additionally, the progressivity of the increasing block scheme is lower than that of residential users, especially in setting the price for the third block.[10] This particular case study and other similar pricing schemes at the industrial level show that industrial water tariffs in Spain are not adequate at achieving the sustainability of water resources.

16.4.3 Environmental Aspects

The ecological status of Spanish water bodies shows mixed results. Based on information from the recent river basin management plans,[11] it seems that at the national level, only 42 % of surface water bodies are in good status. In general, problems related to quantity and quality for groundwater bodies are concentrated in the most arid regions in Spain (south and southeast of Spain).

Only regional governments charge levies for environmental purposes, in addition to the one charged by the basin agencies (effluent control levy, see Sect. 16.3). In principle, domestic water charges would suffice to cover wastewater treatment costs, so that urban spills are returned to the water bodies at the required standards. However, as Aldaya and Llamas (2012) indicate, many wastewater treatment plants do not operate correctly, and most that do only perform secondary treatment (not tertiary). The cost of revamping all poorly working treatment plants and upgrading those with secondary treatment has been evaluated to be in the range € 19 billion. A rough figure of this amounts to 4 € per m^3, which when annualized in 10 years would be about 0.45–0.5 € m^3, a surcharge that would represent an average increase of 20–40 % of the urban tariff. While this is not an extremely large increase, it is still politically sensitive.

On the other hand, nonpoint pollution is not addressed using charges or levies. Agricultural contamination is combated using cross compliance (see footnote 13) and zoning and controls in the nitrogen balance.

16.5 Current Debates and Future Directions

16.5.1 New Approaches for Water Pricing Under Consideration

As we stated at the beginning of this chapter, climate change projections have already been integrated in the water planning documents. Urban supply systems and even farmers associations are aware of the somber outlook for water resources and

[10] For the supply service to residential users, the price of the third block is 1.979 and 1.319 for the summer and the rest of the year, respectively.

[11] Or drafts, when final plans are not yet available.

availability. And yet, very few voices have emphasized the potential of pricing instruments to cope with the climate change challenge, other than reinforcing the financing mechanisms of infrastructures.

The problem of unsustainable groundwater uses will not likely be addressed with pricing mechanisms, although voluntary exchanges have been used in the past (Rey et al. 2014). The best solutions involve strong cooperation among all parties, robust online information systems, and a leading role of head organizations (generally users associations).

There are two avenues for change in irrigation water pricing. One, based on trading options, would in theory facilitate water scarcity signaling and internalization. However, as Rey et al. (2014) describe, trading has been limited and mainly concentrated in specific regions and across the Tagus-Segura interbasin aqueduct. Trading only occurs during the initial stages of a drought cycle and hardly ever in normal hydrological years. But clearly, market mechanisms have served to make transparent the market value of water at least for irrigation.

Another avenue of reform would require passing a new water law. Although current fees and tariffs do not cover all of the capital costs or environmental and resource costs, they are set on strict accordance with Spanish legislation. However, as of June 2014, it is very unlikely that the government, which is now beyond the equator of its term in office (2012–2016), will have time to make a proposal to be discussed and approved.

With water markets as limited instruments and in the absence of a thorough law reform, the most plausible scenario is that irrigation water pricing will not change in levels, approaches, and implementation in the near future.

In the case of urban and industrial uses, an alternative way to promote equity consists of offering users a "menu" of structures. Thus, some customers could choose, depending on their economic level (Estache et al. 2001). These "optional" tariffs lead to users' self-classification. That procedure makes it possible to obtain relevant information when it is scarce. In water supply, it is possible to find alternatives to choose between a flat/variable tariff and usual/low consumption pricing (OCDE 2003). In the same way, Barberán et al. (2006) proposed to offer two kinds of tariff structures, in order to correct the overcharging of large families. In a context of increasing block tariffs, they suggest the possibility of offering two alternative tariffs: the first one would consider total household water consumption, and the second one would consider the per capita water household consumption. Large-size families could be allowed to choose between them.

But the most important challenge in the near future is to meet earnestly the cost-recovery objective. Especially when it comes to estimating environmental and resource costs, important deficiencies have emerged. We have mentioned the investment needs that are required to modernize and upgrade the wastewater infrastructures. Increasing the component of the tariffs for sewage and wastewater treatment would be the closest to an internalization of the environmental costs. As to the resource costs, market signals have already been provided in the regions where exchanges have taken place (the areas suffering more water stress) and where a resource component is clearly more needed.

16.5.2 Interaction with Other Policy Instruments

The literature on water demand management deals extensively with market-based policies aimed at moderating consumption, in particular water pricing (Arbués et al. 2003; Worthington and Hoffman 2008). However, non-pricing policies have also been considered as complementary tools to control water consumption (Ferrara 2007; Worthington and Hoffman 2008). The installation of water-efficient devices has been shown to potentially be an effective way of reducing water consumption. Additionally, it has been shown that the level of educational attainment increased pro-environmental behaviors in the water field (OECD 2011).

In this respect, both regional and local governments in Spain have carried out different initiatives, such as subsidizing the adoption of low-water and energy-consuming technologies[12] and promoting educational campaigns oriented to modify water use habits. Such interventions can also achieve significant water savings. In several regions, new buildings or renovated buildings must be equipped with individual meters for each apartment/dwelling, low-pressure showers and faucets, and dual-flow toilets.

However, the literature has analyzed the issue of a possible rebound effect, in which water use increases after the installation of water-efficient equipment (Campbell et al. 2004). That is, after installing water-saving technologies, users adapt their water use practices and behaviors in such a way that the overall effect is an increase in water use. Thus, it is necessary to control this rebound effect when it comes to designing subsidization policies (European Commission 2012). A combination of technological improvement and pricing could be a useful instrument in order to reach this goal.

In the case of irrigation, the interaction with other policies is a missed opportunity. As we stated at the beginning of this chapter, the new common agricultural policy will provide Spanish agriculture and rural areas funds totaling €38 billion for a 7-year period starting in 2015. About 70 % of these will be given to growers in the form of direct payments, which have two parts. The first part, denominated, basic payment comprises about 70 % of the total direct payment of which eligible farmers are entitled. Eligibility is based on cross compliance[13] with a number of environmental and best agricultural practices. The remaining 30 %, denominated "greening component," is

[12] The *Plan Renove* is a subsidy program launched in Spain as part of the 2005–2007 Energy Saving and Efficiency Action Plan and was followed by a second wave of subsidies in 2008–2012. Its primary purpose was to provide financial incentives to households to replace some electrical appliances (fridges, freezers, washing machines, dishwashers, etc.) by others with a class A or A+ or A++ label. The subsidy was aimed at compensating for the price differential between the conventional appliance and the efficient one. The level of the subsidy, which was determined by each autonomous region (or autonomous community, AC), varied from 85 to 125 euros, depending on the appliances.

[13] Cross compliance: A mechanism that ties direct payments to farmers and a number of rural development payments to compliance with a series of legislative acts relating to the environment, food safety, animal and plant health, and animal welfare and to maintaining agricultural land in good agricultural and environmental condition (GAEC). Cross-compliance rules relate to 18 statutory management requirements and 15 GAEC standards. Noncompliance with these standards and requirements can lead to a reduction in CAP payments to the farmer.

optional and requires that farmers (a) rotate three crops, (b) leave a section (buffer) not cultivated of 8 % of the farm for biodiversity and wildlife conservation, and (c) conserve and tend pastures. Despite initial proposals for the basic payment or for the greening component, eligible irrigated farms will be required to fulfill the WFD requirements, including paying charges at or close to cost-recovery levels. An audit from the European Court of Auditors reveals the partial and limited success in integrating the objectives of EU water policy into the new CAP, due to a mismatch between the ambition of the policy objectives and the instruments used to effect change (Court of Auditors 2014). This audit highlights weaknesses in the two instruments currently used by the commission to integrate water concerns into the CAP (cross compliance and rural development) and pointed out delays and weaknesses in the implementation of the WFD.

16.6 Conclusions

Spain needs a full revamping and resetting of its water-pricing policies and schemes. Presently, the country is far from achieving full cost recovery and environmental objectives set at the WFD, although it has a solid base and an established culture for paying for water. An ambitious effort will require significant political willingness to make serious reforms, which would require also a water law reform.

For a start, the irrigation sector now faces the increasing cost of energy and electricity, which leaves less room for charging higher water rates. Inefficient pricing structures, such as those based on irrigated surface should be fully replaced by volumetric tariffs, following up the significant effort made to modernize about 30 % of the irrigated area.

The urban water supply and wastewater treatment system requires significant funds to be upgraded, improved, and repaired. Residential, industrial, and commercial rates should be increased, but Spain does not have a common regulatory framework. In fact, water companies and franchises have requested that there should be a unique regulator for the water sector in Spain. Looking for improving equity issues at the residential level, it recommends that regulators generalize some per capita-based formulas. Additionally, environmental efficiency should emerge as a key issue in this sector. Thus, users living in areas with higher levels of water stress should bear higher progressivity levels. Scarcity should be considered in the design of water prices, and markets should be improved and better regulated.

References

ACA. (2008). *Análisis económico del uso del agua* (Documento de presiones e impactos, y análisis del riesgo de incumplimiento de los objetivos de la DMA. Agència Catalana de l'Aigua). Madrid: Ministerio de Medio Ambiente.

AEAS. (2003). *Encuesta de Tarifas 2002*. Madrid: Asociación Española de Abastecimientos de Agua y Saneamiento.

AEAS. (2012). *Suministro de agua potable y saneamiento en España 2010. XII Encuesta nacional*. Madrid: Asociación Española de Abastecimientos de Agua y Saneamiento.
AEAS. (2013). *Tarifas 2012. Precio de los servicios de abastecimiento y saneamiento en España*. Madrid: Asociación Española de Abastecimientos de Agua y Saneamiento.
Aguilera Klink, F. (2002). *Los mercados de agua en Tenerife*. Bilbao: Bakeaz.
Aldaya, M. M., & Llamas, M. R. (Eds.). (2012). *El agua en España: bases para un pacto de futuro*. Madrid: Fundación Botín.
Arbués, F., & Barberán, R. (2012). Tariffs for urban water services in Spain: Household size and equity. *International Journal of Water Resources Development, 28*(1), 123–140.
Arbués, F., García-Valiñas, M. A., & Martínez-Espiñeira, R. (2003). Estimation of residential water demand: A state of the art review. *Journal of Socio-Economics, 32*(1), 81–102.
Barberán, R., & Arbués, F. (2009). Equity in domestic water rates design. *Water Resources Management, 23*(10), 2101–2118.
Barberán, R., Arbués, F., & Domínguez, F. (2006). *Consumo y Gravamen del Agua para Usos Residenciales en la Ciudad de Zaragoza. Evaluación y Propuesta de Reforma*. Zaragoza: Ayuntamiento de Zaragoza, Servicio de Cultura.
Berbel, J. (2005). *Análisis económico del agua en la Directiva Marco. Su aplicación a la Cuenca del Guadalquivir*. Conferencia ISR, Córdoba, Spain, 28/abril/2005. Available at: http://www.isrcer.org/jornadas.asp
Bielsa, J., Cazcarro, J., Groot, E., & Sánchez Chóliz, J. (2009). El coste financiero de la DMA: Tarifas sobre el uso del agua en agricultura. In J. A. Gómez-Limón, J. Calatrava, A. Garrido, F. J. Sáez, & A. Xabadia (Eds.), *La economía del agua de riego en España*. Almería: Fundación Cajamar.
Bös, D. (1985). Public sector pricing. In A. J. Auerbach & M. Feldstein (Eds.), *Handbook of public economics* (Vol. I, pp. 129–211). North Holland: Elsevier Science.
Calatrava, J. & Garrido, A. (2010). *Measuring irrigation subsidies in Spain: An application of the GSI method for quantifying subsidies*. Geneva: International Institute for Sustainable Development (IISD). Available at http://www.iisd.org
Carrillo Cobo, M. T., Camacho Poyato, E., Montesinos, P., & Rodríguez Díaz, J. A. (2014). New model for sustainable management of pressurized irrigation networks. Application to Bembézar MD irrigation district (Spain). *Science of the Total Environment, 473–474*, 1–8.
Castro, F., Da-Rocha, J. M., & Delicado, P. (2002). Desperately seeking θ's: Estimating the distribution of consumers under increasing block rates. *Journal of Regulatory Economics, 22*(1), 29–58.
CEDEX. (2011). *Evaluación del impacto del cambio climático en los recursos hídricos en régimen natural*. Madrid: CEDEX.
CHS. (2007). *Estudio General sobre la Demarcación Hidrográfica del Segura* (versión del 2 de julio de 2007). Confederación Hidrográfica del Segura, Murcia, Spain.
De Stefano, L. & Garrido, A. (2013). *Síntesis y conclusiones del 12 Seminario Nacional del Observatorio del Agua de la Fundación Botín: Reflexiones sobre el primer ciclo de planificación hidrológica de la Directiva Marco del Agua en España*. http://www.fundacionbotin.org/file/49645/
Domínguez, A., Martínez, R. S., de Juana, J. A., Martínez-Romero, A., & Tarjuelo, J. M. (2012). Simulation of maize crop behavior under deficit irrigation using MOPECO model in a semi-arid environment. *Agricultural Water Management, 107*(2012), 42–53.
EC. (2000). Directive 2000/60/EC (2000) of the European Parliament and of the Council of 23 October 2000. "Establishing a framework for community action in the field of water policy". *Official Journal of European Communities, L 327*, 1–72.
Estache, A., Gomez-Lobo, A., & Leipziger, D. (2001). Utilities privatization and the poor: Lessons and evidence from Latin America. *World Development, 7*, 1179–1198.
Estache, A., Foster, V., & Wodon, Q. (2002). *Accounting for poverty in infrastructure reform. Learning from Latin America's experience, studies in development series*. Washington, DC: World Bank Institute.

Esteban, E., & Albiac, J. (2011). Groundwater and ecosystems damages: Questioning the gisser-sánchez effect. *Ecological Economics, 70*, 2062–2069.

European Commission. (2012). *A blueprint to safeguard Europe's water resources. Communication from the Commission to the European Parliament, the Council, the European Economic and Social Committee and the Committee of the Regions.*COM (2012) 673 final. Brussels: European Comission.

European Court of Auditors. (2014). *Special Report Integration of EU water policy objectives with the CAP: a partial success*. LUXEMBOURG. http://www.eca.europa.eu/Lists/ECADocuments/SR14_04/SR14_4_EN.pdf. Accessed 15 May 2014.

European Environmental Agency. (2013). *Assessment of cost recovery through water pricing*. European Environmental Agency. Technical Report N°16/2013, Luxembourg: European Environmental Agency.

Ferrara, I. (2007). Residential water use: A literature review. In *Household behaviour and environment: Reviewing the evidence*. Paris: Organization for Economic Cooperation and Development (OECD).

Gallego-Ayala, J. J., Gomez-Limon, A., & Arriaza, M. (2011). Irrigation water pricing instruments: A sustainability assessment. *Spanish Journal of Agricultural Research, 9*(4), 981–999.

García, M. (2002). *Análisis de la influencia de los costes en el consumo de agua en la agricultura valenciana. Caracterización de las entidades asociativas para riego*. Doctoral Dissertation. Departamento de Economía y Ciencias Sociales. Universidad Politécnica de Valencia, Valencia.

García, M., Carles, J., & Sanchís, C. (2004, November 22–26). *Características institucionales y territoriales y su influencia en el costo del agua como input en la agricultura*. VII Congreso Nacional de Medio Ambiente. Madrid (Spain).

Garcia-Valiñas, M. A., Martínez-Espiñeira, R., & González-Gómez, F. J. (2010). Affordability of residential water tariffs: Alternative measurement and explanatory factors in southern Spain. *Journal of Environmental Management, 91*(12), 2696–2706.

Garrido, A., & Calatrava, J. (2009). Trends in water pricing and markets. In A. Garrido & M. R. Llamas (Eds.), *Water policy in Spain* (pp. 131–144). Leiden: Taylor & Francis.

Garrido, A., & Calatrava, J. (2010). *Agricultural water pricing: EU and Mexico. Consultant report for the OECD* (Document COM/TAD/CA/ENV/EPOC/RD(2008)49). Paris: OECD.

Garrido, A., Willaarts, B., López-Gunn, E., & Rey, D. (2013). Considerations on climate variability and change in Spain. In L. de Stefano & M. R. Llamas (Eds.), *Water, agriculture and the environment in Spain: Can we square the cire?* (pp. 191–202). Leiden: CRC Press/Balkema, Taylor and Francis Group.

Gobierno, V. (2005). *Recuperación de costes de los servicios del agua en la Comunidad Autónoma del País Vasco*. Vitoria: Departamento de Ordenación del Territorio y Medio Ambiente del Gobierno Vasco.

González-Gómez, F., García-Rubio, M. A., & Guardiola, J. (2012). Urban water service policy and management in Spain: Pending issues. *International Journal of Water Resources Development, 28*(1), 89–105.

Groot, E., & Sánchez Chóliz, J. (2006). *Valoración financiera de los costes asociados al uso del agua de riego: El caso del sistema de Riegos del Alto Aragón (RAA)*. Huesca: Comunidad General de Riegos del Alto Aragón.

Hardy, L., Garrido, A., & Juana, L. (2012). Evaluation of Spain's water-energy nexus. *International Journal of Water Resources Development, 28*(1), 151–170.

Lecina, S., Isidoro, D., Playán, E., & Aragüés, R. (2010). Irrigation modernization and water conservation in Spain: The case of Riegos del Alto Aragón. *Agricultural Water Management, 97*, 1663–1675.

López-Gunn, E., Mayor, B., & Dumont, A. (2012). Implications of the modernization of irrigation systems. In L. De Stefano & M. R. Llamas (Eds.), *Water, agriculture and the environment in Spain: Can we square the circle?* Leiden: CRC Press/Balkema.

MAPA. (2001). *Plan Nacional de Regadíos. Horizonte 2008*. Madrid: Ministerio de Agricultura, Pesca y Alimentación.

MMA. (2005). *Memoria Económica del Proyecto de Reforma del Real Decreto del Reglamento del Dominio Público Hidráulico*. Madrid: Ministerio de Medio Ambiente.

MMA. (2007). *Precios y costes de los servicios de agua en España. Informe integrado de recuperación de costes de los servicios de agua en España*. Madrid: Ministerio de Medio Ambiente.

OECD. (2003). *Social issues in the provision and pricing of water services*. Paris: OECD.

OECD. (2010). *Pricing water resources and water sanitation services*. Paris: OECD.

OECD. (2013). *Environment at a glance 2013, OECD indicators*. Paris: OECD.

Pérez, L., & Barreiro, J. (2007). Una nota sobre la recuperación de costes de los servicios del agua en la cuenca del Gállego. *Economía Agraria y Recursos Naturales, 7*(13), 49–56.

Puerto, P., Domingo, R., Torres, R., Pérez-Pastor, A., & García-Riquelme, M. (2013). Remote management of deficit irrigation in almond trees based on maximum daily trunk shrinkage. Water relations and yield. *Agricultural Water Management, 126*, 33–45.

Quiroga, S., Garrote, L., Iglesias, A., Fernández-Haddad, Z., Schlickenrieder, J., de Lama, C., & Sánchez-Arcilla, A. (2011). The economic value of drought information for water management under climate change: A case study in the Ebro basin. *Natural Hazards and Earth System Sciences, 11*, 643–657.

Rey, D., Garrido, A., & Calatrava, J. (2014). Water markets in Spain: Meeting twenty-first century challenges with twentieth century regulations. In K. W. Easter & Q. Huang (Eds.), *Water markets for the 21st century: What have we learned?* (pp. 127–147). Dordrecht: Springer.

Soto, M. (2013). *Estudio de la evolución de los regadíos mediante técnicas de benchmarking. Aplicación en la Región de Murcia*. Doctoral Dissertation. Universidad Politécnica de Cartagena, Spain.

Whittington, D. (1992). Possible adverse effects of increasing block water tariffs in developing countries? *Economic Development and Cultural Change, 41*(1), 75–87.

Willaarts, B. A., Ballesteros, M., & Hernández-Mora, N. (2014). Ten years of the water framework directive in Spain: An overview of the ecological and chemical status of surface water bodies. In P. Martinez-Santos, M. M. Aldaya, & M. R. Llamas (Eds.), *Integrated water resources management in the 21st century: Revisiting the paradigm* (pp. 99–120). Leiden: CRC Press/Balkema.

Worthington, A. C., & Hoffman, M. (2008). A state of the art review of residential water demand modelling. *Journal of Economic Surveys, 22*(5), 842–871.

Part II
Innovations in Water Pricing

Chapter 17
Introducing New Mechanisms into Water Pricing Reforms in China

Dajun Shen, Xudong Yu, and Jian Shi

Abstract This chapter analyzes the water pricing structure, reform process, and case studies in China and presents a overall picture of pricing water resources and its services during the past 60 years, particularly after 1980. China now implements a comprehensive water pricing framework and develops it step by step. The water resources fee was introduced in the 1980s, and the wastewater treatment and collection fee was developed in the late 1990s. By the 2000s, a comprehensive system was developed. Two case studies, involving Beijing and Shanxi Province, are discussed, which demonstrate increasing tariff standards in both regions. In the future, China will continue struggling with its water sector's increasing tariff levels in order to meet its multi-objective water pricing.

Keywords China • Water resources fee • Wastewater collection and treatment tariff • Comprehensive water-pricing system • Water shortage

17.1 Introduction

Water plays a critical role in social and economic development in China. More than 2,200 years ago, it was recognized that "water is with benefits and harms," by Maqian Si, the author of *The Records of the Grand Historian*. In 2011, the State Document No. 1 defined water as "the source of life, the element for production and the basis for ecosystem" (State Council 2011: 1).

The critical importance of water arises from the water resources problems in China. China is 1 of 13 countries in the world that is lacking in water resources. Although total water resources in China is up to 2.8 trillion m^3—the sixth in the world in absolute terms—water resources per capita in China is only 2,200 m^3, equivalent to a quarter of the world level. Moreover, the uneven distribution of water resources in time and space, and the lack of reasonable water management led to

D. Shen (✉) • X. Yu • J. Shi
School of Environment and Natural Resources, Renmin University of China, Zhongguancun Dajie No. 59, Beijing 100872, China
e-mail: dajunshen@ruc.edu.cn

A. Dinar et al. (eds.), *Water Pricing Experiences and Innovations*, Global Issues in Water Policy 9, DOI 10.1007/978-3-319-16465-6_17

many serious water problems in China. From 50 to 80 % of China's annual precipitation occurs over a period of 4 months during flood season. The area south from the Yangtze accounts for 80.4 % of the water, but for only 53.6 % of the population, 35.2 % of the arable land, and 55.5 % of gross domestic product (GDP). The annual water shortage in China is more than 50 billion m^3, two-thirds of 600 cities are lacking an adequate water supply. At the same time, pollution and degradation of the freshwater ecosystem is severe. In 2012, according to the Surface Water Environmental National Standard (GB3838-2002), water quality of about 36 % of the river length was assessed as worse than Grade IV and could not be used as drinking-water sources, spawning, fishing, or recreational swimming. The water quality compliance rate in water function zones, an indicator reflecting waters meeting use requirements, was only 46 % (Ministry of Water Resources 2013).

In order to tackle such water issues, China has adopted several water resource management instruments, including command and control, incentives, and self-regulation. In recent decades, with the development of a market-based economy, water pricing is the key instrument for promoting water resources reallocation and conservation. The decision on key issues to fully deepen reform in 2013 required promoting water-pricing reform and establishing competitive pricing regulation (Central Committee of Chinese Communist Party 2013). Therefore, the chapter will analyze the water-pricing structure, reform process, and case studies in China in order to present an overall picture of pricing water in terms of framework and practices.

17.2 Water Pricing Structures in China

Five types of fees and charges are included in water-pricing framework in China, which are related to resources, services, and environmental issues: (1) water resources fees, (2) water supply tariffs from hydraulic engineering, (3) urban water supply tariffs, (4) wastewater collection and treatment tariffs, and (5) pollutant discharge fees.

The water resources fee belongs to the resources charge. The fee is collected according to the water resources payment for use system, which was defined in the 2002 water law. At present, water resources fees are collected for industrial, domestic, and hydropower uses with an abstraction permit and based on the actual water use volume. Agriculture is exempted from fees. This discriminatory fee standard is designed for different purposes and water sources, and normally industrial water users are charged more than domestic users; groundwater is charged more than surface water (State Council 2006).

The hydraulic engineering water supply tariff is a service charge for water supplied from hydraulic engineering projects to users, such as cultivated lands, water supply companies, etc. The 2003 Management Methods for Water Supply Tariffs from Hydraulic Engineering, established by the Ministry of Water Resources (MWR) and the National Development and Reform Commission (NDRC), states that the tariff shall be formulated according to the principles of cost-recovery, reasonable profit, higher price for better quality, and fair-affordability and that water should be regulated according to changes in costs, expenditures, and water supply and demand. The tariff is grouped into agricultural water supply and non-agricultural water supply. The agricultural water supply could be formulated to cover costs and expenditures, but without profits and taxes. The non-agricultural water supply tariff is designed to cover costs and expenditures, as well as taxation and profits. The profit is based on the net water supply asset at a rate of 2–3 % more than the long-term commercial bank loan rate (MWR, NDRC 2003).

The urban water supply tariff, or tap water tariff, is a service fee that is charged for supply services by the urban water supply company. According to the 1994 Urban Water Supply Regulation (State Council 2004), the urban tariff is formulated based on the principles of cost-recovery and low profit for domestic supply and rational charge for production and commerce supply. The 1998 Notice on Improving Urban Water Supply Tariff Management, by National Development and Planning Commission (NDPC) and Ministry of Construction, grouped urban water supply tariffs into three groups: household, non-household, and special water uses. The tariff consists of water supply cost, expenditures, taxes, and profits.

The wastewater collection and treatment tariff is a service fee that is charged by the wastewater treatment company for collection and treatment services. The 2013 Urban Drainage and Wastewater Treatment Regulation of the State Council states that the wastewater tariff shall not be lower than the normal operation cost of wastewater treatment facilities (State Council 2013).

The pollutant discharge fee is the environmental charge. The 2003 Regulation on Pollutant Discharge Fee Collection and Management states that the polluter who discharges pollutants directly into the environment shall pay a discharge fee (The Polluter Pays Principle). This fee is calculated based on the concentration and volume of the key pollutants, with a fixed unit fee system applied all over the country. The concentration and volume is set based on the observations reported by the polluter and the calibration of the environmental protection agency (State Council 2003). The 2013 Urban Drainage and Wastewater Treatment Regulation states that those who discharge wastewater into urban wastewater treatment facilities and pay for the wastewater treatment tariff need not pay the pollutant discharge fee.

17.3 Water Pricing Reform Process After 1949

The history of charges for irrigation services in China is quite long. If a water project is operated sustainably, a rational water pricing system must be developed. The famous Dujiangyan Project, built in the second century BC, charged water for 75 Kg rice each ha (Dept. of Water Project Management, MWR 1991). Water pricing in China has faced a long reform process. The reform evolved with the water resources development and issues, such as water pollution and ecosystem degradation, as well as the understanding of water resources management. During this process, China has passed several stages, from "no charge" to "lower charge," introducing "resources charge" and "wastewater treatment charge" gradually. Even after the development of the comprehensive pricing structure and framework, China is still facing critical problems on how to charge for water in an equitable, sustainable, and effective manner.

17.4 No Charge Era (1949–1965)

After 1949, water resources development focused on constructing projects, rather than improving management. For a long time, all of the water supply costs and expenditures were fully covered by government subsidies. Therefore, water was supplied free of charge from hydraulic engineering between 1949 and 1965.

During this period, many water projects had been developed. The operational costs and maintenance expenditures were covered by governmental resources. But in some regions, very few tariffs were collected, or replaced by collecting some grains, or by labor inputs by the farmers for maintenance. Generally, water was supplied free. Only after 1964, when the Ministry of Water and Hydropower held the first water project management meeting, the methods to collect and manage water tariffs was developed. Free water supply had ended.

17.5 Lower Service Charge from 1965

In 1965, the Water Tariff Collection, Use and Management Methods for Reservoir Project was issued. The methods defined that the management, maintenance, and renewal of infrastructures and facilities of water projects with benefits should be funded by water tariff collection from beneficiaries; the water tariff should be formulated according to the principles of "self-financing" and "with reasonable accumulation (of profit)," while considering the benefits of beneficiaries and economic circumstance (ability to pay) at the same time. But due to the poor ability of the users to pay, the tariff was very difficult to collect. In most regions, water tariff revenues were too small to cover the management, operation, and maintenance costs because of low tariff payment standards.

In 1979, after implementation of the reform policy in China, the meeting on reservoir fishing and comprehensive operation by the Ministry of Water and Hydropower discussed how to strengthen operation and management to increase economic benefits of water projects, as well as water tariff collection and management. The 1982 State No. 1 document required the ministry to "re-assess water pricing for urban and rural industrial, and agricultural water supply" (State Council 1982). In 1985, the State Council issued the Water Tariff Assessment, Collection and Management for Water Project. The methods revealed the need to assess water tariff standards, based on water supply costs, which included operation and management costs and maintenance, with depreciation and other expenditures accounted for. The tariffs were formulated according to water supply types: the agricultural water supply tariff for grain production would be calculated according to water supply costs, and that for cash crops could be more than the supply costs for grains. The agricultural water supply costs excluded the depreciation from fixed assets formed by farmer labor input, while the industrial water tariff would be formulated according to full water supply costs and 4–6 % investment profit rate.

But due to the fact that water tariffs are impacted by many factors, the collection standard has never met the requirements of these regulations and methods in China. In 2003, the Management Methods for Water Supply Tariffs from Hydraulic Engineering was issued to replace the existing management methods.

17.6 Introducing Resources Charge in the 1980s

At the end of the 1970s and beginning of 1980s, the governments promoted a plan to save water in order to deal with the emerging water shortage problems in cities in northern China and coastal regions. Among three water supply systems – urban tap water system, water supply systems from water projects, and self-supplying systems – the self-supplying systems for larger and middle industrial and mining enterprises had a large water supply, which was 50 % more than tap water systems at that time, but was not included in the management plan and, thus, resulted in waste of water. Therefore, in order to strengthen these self-supplying sources, water administrative departments started to collect water resources fees in some provinces in North China.

The 1988 Water Law, the first water law in China, stated that the unit in the urban area directly withdrawing groundwater should be charged water resource fees; other units abstracting water directly from aquifers, rivers, or lakes should levy water resource fees that are decided upon by provincial governments. The 1997 Water Sector Industrial Policy stipulated that the state charges water resources fees for the direct abstraction from aquifers, rivers, or lakes. The 2002 Water Law reiterated the collection of water resources fees.

So the water resources fee is gradually being expanded from urban groundwater to all water abstractions, from some provinces to all provinces. In 2009, with Tibet starting to collect that fee, all provinces are now collecting water resources fees. In 2011, 13.6 billion RMB (the exchange rate between US$ and RMB is about 6.11 in 2014) fees were collected nationally (MWR 2011). The total amount of fees collected are increasing fast, due to the continuing regulation and adjustment of the fee standard.

17.7 Charging Wastewater Collection and Treatment Fees Since the Late 1990s

In the 1990s, the fast-growing economic development brought increased wastewater discharge into water bodies and caused serious water pollution. The pollution was exacerbated due to lagging urban wastewater collection and treatment infrastructure investments. The 1996 Water Pollution Control Law promoted the building of urban centralized wastewater treatment facilities and charged wastewater collection and treatment fees in order to sustain the normal operation of the facilities. The 1998 Urban Water Supply Pricing Management Method indicated that the standard of wastewater treatment fees shall be formulated according to operation, maintenance, and construction costs of urban wastewater collection networks and plants (National Planning Commission, Ministry of Construction 1998).

In a few cities, wastewater treatment fees were collected, together with urban water supply tariffs, but the fee standard was low, which could not compensate the operation and maintenance costs of the facilities. In order to strengthen fee collection and promote wastewater treatment, in 1999 NDPC decided to combine collection of wastewater treatment fees with water supply tariffs. The fees should cover the operation and maintenance costs of the collection network and treatment facilities to allow given profits. The charges could be regulated step by step, but should meet the operation and maintenance requirements in heavily polluted regions in 1999 (National Development and Planning Commission (NDPC), Ministry of Construction, State Environmental Protection Bureau 1999).

The 1999 Notice was not implemented fully, although the collection of fees was gradually promoted. In 2000, NDPC still required to charge for wastewater treatment fees for the cities not collecting them and to regulate the standard to cover cost for the cities that did collect (NDPC 2000). In 2009, the government further focused on alleviating the low collection standard of wastewater treatment fees in order to promote wastewater treatment sector development (NDRC, Ministry of Housing and Urban and Rural Development (MOHURD) 2009).

The 2013 Urban Drainage and Wastewater Treatment Regulation stipulated that the collection of standard wastewater treatment fees should not be lower than the normal operation costs of the urban wastewater treatment facility. If the collected revenue could not cover the normal operation costs, the local government should subsidize the difference. Therefore, the government finally recognized the difficulty in cost-recovery in the wastewater tariff and provided the option to subsidize the difference.

17.8 Implementing Comprehensive Water Pricing System After the 2000s

In fact, water pricing is a very complicated issue, not only decided upon by the water sector itself, but it is also affected by many factors external to the sector. Therefore, China tried to develop a water pricing framework incorporating these external factors. During this process, a water pricing system in China was gradually developed, and a comprehensive system is inevitable.

In 2000, NDPC issued the "Guidance for Reforming Water Pricing and Promoting Water Saving" in order to develop water pricing mechanisms and institutions that meet socialist-market economy and improve water tariffs from hydraulic engineering projects and urban supply (NDRC 2000). Thereafter, the comprehensive water pricing system started to develop.

The guidance realized that there were key problems in water pricing: (1) the mechanism was not reasonable. Water was not regarded as an economic good. The water supply tariff from hydraulic engineering was lower than water supply cost, resulting in operation and maintenance problems. The supply and collection network and wastewater treatment costs and expenditures were not covered. Water tariffs are not regulated in time according to changes in water supply and demand and costs. (2) The management institution was not reasonable. The end-use irrigation canal systems were poorly managed by village collectives, the tariff collection was not formal and based on a metering system. The urban supply company was not efficient and high cost. (3) The water tariff was low. All components of the water tariff was lower than costs. (4) The end-use water tariff was not in good order. In one respect, it was difficult to collect the agricultural water tariff. In another respect, the additional costs and charges in the water tariff increased the farmers' burden.

Therefore, comprehensive reforms were required to: (1) Increase the hydraulic engineering water supply tariff to a reasonable level by clarifying costs and expenditures of water projects based on its functions, such as flood control, water supply, hydropower, fishing, tourism, etc. The costs and expenditures with public interest could be covered by governmental resources and with non-public interest should be covered by tariffs. (2) Reform the urban water supply management institution and regulate supply and treatment tariffs. The water supply and treatment enterprises should be self-financing. The reform on network and plant separation could be introduced in middle and large cities. The wastewater treatment fee should be collected with supply tariffs and gradually regulated to a rational level, according to cover reasonable costs with some profits. (3) Reform rural water supply systems, improving canal systems by more investments to reduce leakage, improve metering facility to implement volumetric charge, and strengthen rural water supply tariff management by reducing the additional charges. (4) Develop a water pricing system to promote water savings by increasing water resource fees, implementing quota-exceeding, increasing block tariff structure and capacity, and a two-part volume tariff (NDRC 2000).

After the guidance, China continued to improve the comprehensive water pricing system. The 2009 Notice on Improving Related Issues of Urban Water Supply Tariff Management required an improvement in water pricing regulation procedures, such as cost auditing and public hearing; improved metering by introducing the block tariff for household use and quota-exceeding, and increasing the block tariff for non-household use; simplified the water tariff for household use, non-household use, and special uses; and fully consider the ability-to-pay of low-income families (NDRC 2009).

In terms of agricultural water tariffs, the 2011 "Decision on Strengthening Water Resources Development and Reform" promoted the agricultural water tariff comprehensive reform, according to the principles of promoting water saving, reducing famer's expenditure on water, and guaranteeing better operation of irrigation and drainage infrastructures. The operation and management cost for agricultural irrigation and drainage infrastructure could be subsidized by governmental resources. The probe of subsidized water tariffs within quota, the defined volume for a crop per ha, and increasing water tariffs for quantities exceeding the quota were encouraged (State Council 2011).

In terms of household water use, in 2013 NDRC and MOHURD requested to hasten the development of the block tariff structure and to develop the tariff structure before the end of 2015 for all cities. The formulation of block volume should clarify the basic and nonbasic requirements (NDRC and MOHURD 2013). Now, many cities have developed the block tariff structure, such as Beijing and Shanghai (in 2014).

17.9 Water Pricing Experiences in Various Locations in China: *Beijing*

Beijing is the capital of China. Nearly 21.15 million people were living in Beijing at the end of 2013, with an urbanization rate of more than 80 %. Beijing faces serious water shortage. The annual average precipitation is 585 mm, and the whole water resources is only 3.95 billion m^3. But the water usage in 2012 was 3.59 billion m^3, although less than those in the previous years (Beijing Water Affair Bureau 2012). Beijing has built 19 water treatment plants with a supply capacity of 3.4 million m^3 daily and 8 wastewater treatment plants with capacity of 2.6 million m^3 daily.[1]

Beijing regulates water pricing often by changing user groups, reforming tariff structures, and, most important, increasing tariff standards: charging water resources fees for self-supplying wells in the 1980s and extending it to all abstraction in 2002 (He and Ren 2004) and introducing wastewater treatment in 1997 (Beijing Pricing Bureau 1997). In 2014, the increased block tariff was introduced in the urban household water supply. Normally, urban water pricing in China is grouped into household and non-household water tariffs.

17.9.1 Household Water Tariff

The household water tariff in Beijing was adjusted several times in the past 20 years, in terms of tariff components (from urban tap water tariffs to water resource fees and wastewater treatment fees) and structure (from single charge to block tariff). The household water tariff increased from 0.12 RMB/m^3 to 0.30 RMB/m^3 in 1991 (exchange rate with US$ was about 3.70), to 0.50 RMB/m^3 in 1996, to 0.70 RMB/m^3 (including 0.1 RMB/m^3 wastewater treatment fee) in 1997, to 1.00 RMB/m^3

[1] Introduction of Beijing water affairs. http://www.bjwater.gov.cn/pub/bjwater/zfgk/znjj/swzs/201009/t20100908_49674.html

(including 0.1 RMB/m^3 wastewater treatment fee) in 1998, to 1.30 RMB/m^3 (including 0.3 RMB/m^3 wastewater treatment fee) in 1999, to 1.60 RMB/m^3 (including 0.4 RMB/m^3 wastewater treatment fee) in 2000, to 2.00 RMB/m^3 (including 0.3 RMB/m^3 water resources fee and 0.5 RMB/m^3 wastewater treatment fee) in 2002, to 2.30 RMB/m^3 (including 0.6 RMB/m^3 water resources fee and 0.3 RMB/m^3 wastewater treatment fee) in 2003,[2] to 3.70 RMB/m^3 (including 1.10 RMB/m^3 water resources fee and 0.90 RMB/m^3 wastewater treatment fee) in 2004, to 4.00 RMB/m^3 (including 1.10 RMB/m^3 water resources fee and 0.90 RMB/m^3 wastewater treatment fee) in 2008, and to 4.30 RMB/m^3 in 2009 (including 1.26 RMB/m^3 water resources fee and 1.04 RMB/m^3 wastewater treatment fee). In 2014, the block tariff structure was implemented in May, with the first block tariff of 5.00 RMB/m^3 (including 1.57 RMB/m^3 water resources fee and 1.36 RMB/m^3 wastewater treatment fee) (Table 17.1, Fig. 17.1).

Table 17.1 Beijing household water tariff (Unit: RMB/m^3)

Water source	Blocks	Annual household water use (m^3)	Total tariff	Components		
				Tap water fee	Water resource fee	Wastewater treatment fee
Tap water	1	0–180	5.00	2.07	1.57	1.36
	2	181–260	7.00	4.07		
	3	260	9.00	6.07		
Self-supply well	1	0–180	5.00	1.03	2.61	1.36
	2	181–260	7.00	3.03		
	3	>260	9.00	5.03		

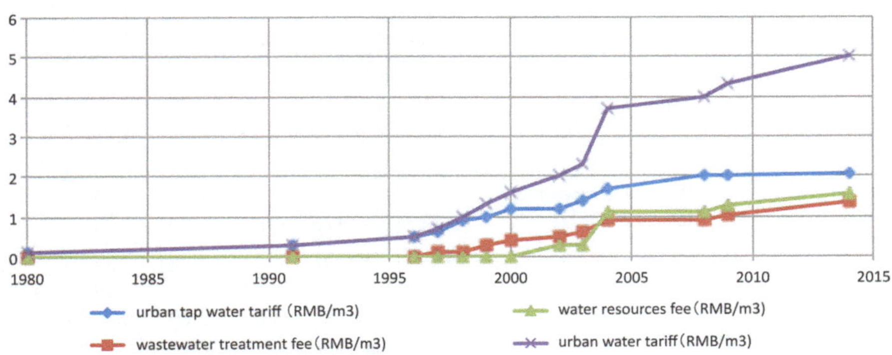

Fig. 17.1 Urban water tariff in Beijing

[2] http://news.xinhuanet.com/zhengfu/2004-06/03/content_1506586.htm

So, in Beijing, the household water tariff increased very quickly in the past 20 years (almost 40 times), compared to the beginning of 1990s. The charging structure became complicated, and there is a trend from service charge to resources charge and to wastewater treatment charge.

17.9.2 Non-household Water Tariff

The non-household water tariff charges for the water supply of public agencies (such as schools, hospitals, government agencies, etc.), industry and commence, hotels and catering, and special use, including car washing and bathing in China. The grouping had been changed. For example, public agencies, industries and commerce, hotels, and catering had been added as individual groups. But for the convenience of administration, it was merged in a 2014 water tariff regulation in Beijing.

The same as household water pricing, industrial water prices increase very quickly in Beijing. The non-residential water tariff is 7.15 RMB/m^3 in 2014, and the special water use is higher, at 160 RMB/m^3 (exchange rate with US$ was about 6.11) (Table 17.2). Additionally, the quota-exceeding increasing block tariff is applied according to the 2012 Beijing Water Saving Methods (Beijing Municipal Government 2012).

Table 17.2 Beijing non-household water tariffs (Unit: RBM/m^3)

Type	Tariff	User group	2004	2009	2014
Non-residential	Water resources fee	Agency	1.10	1.32	1.63
		Industry and commence	1.10	1.32	
		Hotel, catering	1.10		
	Tap water tariff	Agency	2.50	2.50	3.52
		Industry and commence	2.91	2.91	
		Hotel, catering	2.91		
	Wastewater treatment fee	Agency	1.50	1.68	2
		Industry and commence	1.50	1.68	
		Hotel, catering	1.50		
	Sum	Agency	5.40	5.80	7.15
		Industry and commence	5.60	6.21	
		Hotel, catering	6.10		
Special use	Water resources fee	Car washing and bathing			153
	Tap water tariff	Car washing and bathing			4
	Wastewater treatment fee	Car washing and bathing			3
	Sum	Car washing and bathing			160

17.9.3 Agricultural Water Tariff

According to 2007 Interim Measures for Agricultural Water Resources Fee Management in Beijing, water resources fees will be charged for usage exceeding the quota, the grain crops are charged 0.08 RBM/m^3, and other crops are charged 0.16 RBM/m^3. But, in fact, it is not implemented.

At present, there are four types of water charging in Beijing agricultural water use, including volumetric charging, such as a flat fee of 0.04 RBM/m^3 in Xinhe Irrigation District in Tongzhou District; area charging, such as 300–450 RBM/ha annually in East Xiaying village in Tongzhou district; electricity consumption charge in well; and no charge (Gu et al. 2008).

17.10 Water Pricing Experiences in Various Locations in China: Shanxi Province

The Shanxi Province is located at the western part of North China. The province is short on water resources, with only 351 m^3 per capita, one-sixth of the national average. But the province is plentiful in coal and is the national energy base.

17.10.1 Water Resources Fee

The province is one of the first among provinces to collect water resources fees. At the beginning of the 1980s, the province started to collect fees and regulate its use often in the following years in order to improve water safety and protect groundwater resources.

The latest regulation of the fee was in 2008 (Shanxi Provincial Pricing Bureau 2008) (Table 17.3). In terms of standard, the province has the highest level in the country, even compared to the provinces in North China. In terms of collection rates, the province is behind, because household water use is exempted.

17.10.2 Water Supply Tariff from Hydraulic Engineering Projects

In order to reduce farmers' burden on water use, and encourage the use of surface water from Yellow River, in 2009 the Energy Price and Water Price Compensation Management Methods for Large and Medium Pumping Stations implemented the compensation energy price of 0.06 RMB/Kwh and the irrigation tariff of no more than 0.25 RMB/m^3 for surface water irrigation pumping stations without consideration

Table 17.3 Water resources fees in Shanxi Province

Water source	Abstraction type	Purpose	Fee standard (RMB/m^3)									
			Non-over-exploitation zone					Over-exploitation zone				
			In quota	Exceeding quota				In quota	Exceeding quota			
				Less than 20 %	Between 20 % and 40 %	Between 40 % and 60 %	Over 60 %		Less than 20 %	Between 20 % and 40 %	Between 40 % and 60 %	Over 60 %
Groundwater	Self-supply	Special use	10.00	20.00	30.00	40.00	50.00	15.00	30.00	45.00	60.00	75.00
		General use	2.00	4.00	6.00	8.00	10.00	3.00	6.00	9.00	12.00	15.00
	Urban public supply and supply from hydraulic engineering	Special use	4.00	8.00	12.00	16.00	20.00	6.00	12.00	18.00	24.00	30.00
		Industry, commence, service	1.00	2.00	3.00	4.00	5.00	1.50	3.00	4.50	6.00	7.50
		Public agency	0.50	1.00	1.50	2.00	2.50	0.75	1.50	2.25	3.00	3.75
	Mining drainage	Mining (according to drainage volume)	1.20									
Surface water	Self-supply		1.00	2.00	3.00	4.00	5.00	–				
	Urban public supply and supply from hydraulic engineering	Special use	2.00	4.00	6.00	8.00	10.00	–				
		Industry, commence, service	0.50	1.00	1.50	2.00	2.50	–				
		Public agency	0.25	0.50	0.75	1.00	1.25	–				
	Thermal power generation (tubular cooling) (RMB/kw·h)		0.002									
	Hydropower generation (RMB/kw·h)		0.005									

for pumping waterhead (Office of Shanxi Provincial Government 2009a). In Sept 2009, the Notice to Clarify Energy Price and Water Price for Irrigation Pumping Stations implemented the requirement that a tariff of no more than 0.30 RMB/m^3, after adding end-use irrigation management costs, can be charged to water end users.

In terms of other water supply tariffs from hydraulic engineering projects, since 1996 tariffs have been increasing. At the same time, two-part tariffs and seasonal pricing are being tested. At present, the tariff for surface water supply from hydraulic engineering projects to agriculture is about 0.17 RMB/m^3 and that to industry is between 1.20 and 2.40 RMB/m^3.

17.10.3 Urban Water Supply Tariff

At present, the urban water supply tariffs of 11 cities in the province are grouped into household, industry, public agencies, services, and special use. The household tariff is between 2.30 and 3.40 RMB/m^3, and the block tariff structure has been introduced (Table 17.4) (Taiyuan Pricing Bureau 2008). The tariff for public agencies is between 2.50 and 5.65 RMB/m^3. The tariff for industry is between 2.90 and 5.65 RMB/m^3; for service, the tariff is between 3.90 and 5.65 RMB/m^3. The tariff for special use is between 13.00 and 49.00 RMB/m^3.

17.10.4 Wastewater Treatment Fee

As early as in 1999, "the Notice of Fastening Urban Wastewater Treatment Infrastructure and Collecting Wastewater Treatment Fee" required services to collect treatment fees of 0.20–0.80 RMB/m^3 to cover the O&M costs for networks and plants (Shanxi Provincial Government 1999). In 2009, the provincial government issued "the Notice of Fastening Urban Wastewater Treatment Infrastructure and Guarantee the Normal Operation" requesting all counties to collect before the end of 2009 and raise the standard to no less than 0.80 RMB/m^3 (Office of Shanxi Provincial Government 2009b).

By the end of 2011, the collection standard of wastewater treatment fees in the province was between 0.50 and 0.80 RMB/m^3, which did not reach the requirement of the policy. The lower standard resulted in most counties not accomplishing cost-recovery for wastewater collection and treatment.

Table 17.4 The urban water tariff in Taiyuan City, Shanxi Province

Structure	Block	Monthly volume per household (m^3)	Tariff (RMB/m^3)	Note
The block tariff	1	0–9	2.30	The user with "one household, 1 meter outside of apartment"
	2	9–13.5	4.60	
	3	>13.5	6.90	
The single tariff			2.40	Not with above conditions

17.11 Outlook

Water pricing reform is a long process, impacted by many factors. Although China has developed a general framework, it is just beginning to develop a rational water pricing formulation mechanism. In the future, the following issues will be the key driving forces to promote pricing reform.

17.11.1 Struggling for the Position of Water Services in Society and Economy

This is the out-of-sector factor impacting water pricing but is a dominant one. In the past 30 years, the position of the water sector in society has not yet been clearly defined. More seriously, it changes from one side to another side, from market mechanism to public interest, or vice versa. Reflecting on water supply, the issue is how to define water supply and wastewater treatment. Is it a public good, an economic good, or a mixed good (and how mixed)? The lesson experienced in the past 30 years resulting from this confusing position is that China has never realized a cost-recovery for water services, which is the fundamental principle for an economic good.

The confusion comes from social and economic aspects. Because of the critical water shortage, China wants to use pricing and realize cost-recovery as a key instrument to promote water saving and improve water resources allocation in a more efficient and optimal way. But, because of the importance of water resources to improve livelihood and secure grain production, China would like to subsidize agricultural production and provide lower tariffs to water supply services for domestic use. And the difficulty is that the policy has fluctuated in the past 30 years.

Until now, China has not yet clearly defined the position water services has in its society and economy, and the struggle for position of the water sector will continue and will be affected by social and economic reform. Only after is it clarified will China have a clear pathway to promote water pricing reform.

17.11.2 Increasing Charges for Environmental Protection

Due to the severe pollution in China, more investment will be required to control its effects. In terms of water pollution control, collecting tariffs from wastewater production is a right and reasonable source. So, in the next decade, China will continue to regulate environment control charges. The wastewater fee will be increased from a level of covering most costs, to covering all costs and expenditures, to result in profits. Only after China has successfully controlled its water pollution will increasing tariffs and charges for environmental protection be stopped.

17.11.3 Valuing Water Resources Shortages by Water Resources Fee

Since China is short on natural resources for its development, the country's water resources is and will continue to be a strategic interest. Therefore, to use water resources fees as a method to address the shortage is a rational mechanism for promoting water resources allocation. At present, water resources fees have increased to a relatively higher level. But China is shifting its governmental revenue from products to resources. The water resources fees, particularly for groundwater and water-short regions, will be the focus of water pricing regulation.

17.11.4 Increasing Tariff Standard

Whatever the water pricing component, change will affect the tariff standard. At present, given the lower tariff standards for water services, increasing shortage of water resources, and severe water pollution, the water tariff standard will continue increasing.

17.11.5 Improving the System to Meet Multi-objectives

Water pricing is a complicated issue, related to social, economic, and environmental circumstances out of the water sector, as well as water resources development, use, and protection in the sector. Among these conflicted targets, a comprehensive pricing mechanism is required to meet multi-objectives. So the investigation into improving the water pricing system in China will continue in the year to come.

Acknowledgment The research was supported by the Social Science Foundation of China (Project No. 11AZD007) and the Major Science and Technology Program for Water Pollution Control and Treatment (Project No. 2013ZX07603-003-005).

References

Beijing Municipal Government. (2012). *Beijing water saving methods.*

Beijing Pricing Bureau. (1997). *The notice of regulating water pricing and wastewater treatment fee.*

Beijing Water Affair Bureau. (2012). *Beijing water resources bulletin.*

Central Committee of Chinese Communist Party. (2013). *The decision on key issues to fully deepen reform.*

Department of Water Project Management, MWR. (1991). *Theory and calculation methods for water supply pricing of water project*. Beijing: China Water and Power Press.

Gu, Y., Zhang, F., Hao, Z., & Wu, W. (2008). The survey and analysis of agricultural water use tariff in Beijing. *Beijing Water Affairs, 2*, 26–28.

He, H., & Ren, J. (2004). Suggestions on dynamic collection on water resources fee in Beijing. *China Water Resources, 15*, 18–19.

Ministry of Water Resources. (2011). *Water resources management annual report of China 2011.*

Ministry of Water Resources. (2013). *Water resources bulletin of China 2012.*

Ministry of Water Resources, National Development and Planning Commission. (2003). *The management methods for water supply tariffs from hydraulic engineering.*

National Development and Planning Commission. (2000). *The guidance on reform water pricing to promote water saving.*

National Development and Planning Commission, Ministry of Construction, State Environmental Protection Bureau. (1999). *The notice to strengthen wastewater treatment fee collection, development the better operation mechanism for urban wastewater discharge and centralized treatment.*

National Development and Reform Commission, Ministry of Housing, Urban and Rural Development. (2009). *The notice on improving related issues of urban water supply tariff management.*

National Development and Reform Commission, Ministry of Housing, Urban and Rural Development. (2013). *The guidance on fastening to develop, improving urban household block tariff system.*

National Development and Reforming Commission. (2009). *The notice on improving related issues of urban water supply tariff management.*

National Planning Commission, Ministry of Construction. (1998). *Urban water supply pricing management method.*

Office of Shanxi Provincial Government. (2009a). *Energy price and water price compensation management methods for large and medium pumping stations.*

Office of Shanxi Provincial Government. (2009b). *The notice of fastening urban wastewater treatment infrastructure and guarantee the normal operation.*

Shanxi Provincial Government. (1999). *The notice of fastening urban wastewater treatment infrastructure and collecting wastewater treatment fee.*

Shanxi Provincial Pricing Bureau. (2008). *The notice of regulating water resources fee in order to promote water-saving.*

State Council. (1982). *The profile of national rural works conference.*

State Council. (2003). *Regulation on pollutant discharge fee collection and management.*

State Council. (2004). *Urban water supply regulation.*

State Council. (2006). *Regulation of management of water abstraction permit and water resources fee collection.*

State Council. (2011). *The decision on fastening water resources reform and development.*

State Council. (2013). *Urban drainage and wastewater treatment regulation.*

Taiyuan City Pricing Bureau. (2008). *The notice to regulate tap tariff and wastewater treatment fee of Taiyuan City.*

Chapter 18
How to Integrate Social Objectives into Water Pricing

Bernard Barraqué and Marielle Montginoul

Abstract The social dimension should be addressed in the sustainability of water services provision, but it is less well studied than the economic and environmental ones. The debate between pros and cons of water privatization led the Organisation for Economic Co-operation and Development (OECD) to publish a seminal paper on social issues in water pricing, back in 2003. Relying on this document and other literature review, we successively present various solutions to support "water-poor" people in the payment of their charges: reducing bills for targeted populations (rebates, increasing blocks), supporting the income of targeted populations, reducing bills for all customers, and reintroducing taxation as a source of income. A general outcome is that social tariff design entails administrative costs that may offset the benefits it is supposed to generate. Lastly, we advocate the development of new software to assess the redistributive effects of ongoing tariffs, and tariff changes between categories of residents and with the water utilities' capacity to invest.

Keywords Social tariffs • Water and sanitation • Tariffs taxes transfers • Macro-affordability • Micro-affordability

18.1 Introduction

While at the end of the twentieth century the issue of the "water-poor" people[1] and the right to water was considered "solved" in developed countries and only an ongoing problem in developing ones, a sort of backlash occurred after the debate roared between supporters and opponents to "water as an economic good," privatization, and full cost pricing; starting with England and Wales and, soon afterwards, in

[1] Water-poor here refers not to countries or regions but to people who experience difficulties in paying their water charges.

B. Barraqué (✉)
Agroparistech-Engref, 19 Avenue du Maine, F75015 Paris, France
e-mail: bernard.barraque@agroparistech.fr

M. Montginoul
UMR G-Eau – Irstea, 361 rue JF Breton BP5095, F34196 Montpellier cedex 5, France
e-mail: marielle.montginoul@irstea.fr

A. Dinar et al. (eds.), *Water Pricing Experiences and Innovations*,
Global Issues in Water Policy 9, DOI 10.1007/978-3-319-16465-6_18

France, the question of water charges/bills affordability was raised. However, this area remains the least studied among the issues of sustainable water management. Only in 2002, in the aftermath of full privatization of water services in England, Fitch and Price started to calculate how many people as a result would pay more than 3 % of their revenues to access water. This chapter aims at presenting the various solutions adopted in developed and emerging countries to support cheaper access to potable water and sanitation for low-income families or targeted social groups; it deals only with residential water charges. The review of previous publications on this subject shows that it is a much more complex issue than what water services managers and local politicians have thought.

18.2 The Emerging Social Issue in Water Supply and Sanitation (WSS) Services

In the second half of the twentieth century, an increasing number of cities in the developed world opted for volumetric pricing of potable water, considered as a public service with commercial character.[2] The objective was to recover as much of the total costs as possible from the beneficiaries of the service. In addition, wastewater collection and treatment, which was previously considered as imposed on citizens for the sake of public health protection and then funded through local taxation, was considered as a service rendered, so it was added on top of the water bills. The price to pay for water services then progressively doubled. In Europe, WSS services became clearly more costly after the adoption of the Urban Waste Water Directive (EC 271/91).[3] All of these evolutions brought the increase in water consumption by various users to a halt and to a reversal: industry decreased first but, soon, residential customers also adopted water conservation measures. The decrease in potable water demand is now a well-known phenomenon in many European cities, and beyond a certain point, it does embarrass water supply operators, who rely on water bills to cover their fixed costs: less water sold means less revenues, while the infrastructure imposes a long-run fixed cost. The consequence of demand decrease is often unit price increase. And, in turn, overall price increases tend to hit poor families in the persistent economic crisis society faces. The social dimension of sustainability is now on the political agenda in several countries: opponents to "water as an economic good" are powerful enough to bring elected politicians to question full cost pricing, in particular, concerning the emerging category of the "water-poor" in developed countries. Many attempts are made in various countries to address this issue.

[2] In France, administrative law distinguishes public services that render a service to some beneficiaries, who then should pay for their use through billing, and other public services, called administrative, the cost of which is borne by citizens through taxation.

[3] The European Council Directive 91/271/EEC concerning urban wastewater treatment was adopted on 21 May 1991. Its objective is to protect the environment from the adverse effects of urban wastewater discharges and discharges from certain industrial sectors. See: http://ec.europa.eu/environment/water/water-urbanwaste/index_en.html

The Organisation for Economic Co-operation and Development conducted a survey on social issues related to water pricing (OECD 2003). Among other things, the OECD reviewed various methods used to cover bills in arrears or to support bills of the poorest families in various countries. The OECD usually supports full cost pricing of water services and its commercial or private law status. But, more recently, it has admitted the necessity of some taxation mechanisms and, of course, transfers from donors, to alleviate the impact of water tariffs on customers (the 3Ts approach).[4] The 2003 report is a landmark in the recognition of the social dimension of sustainable WSS services management. Most of the report is devoted to the affordability of water services. Indeed, water prices rose drastically in the 1990s, and this trend is estimated to continue, so that the social issue will necessarily remain. The OECD taskforce tried to develop an indicator, called macro-affordability, based on the ratio of average water charges to the mean aggregated household revenue or to the mean aggregated household expenses. It also developed an indicator of micro-affordability, this time looking at the impact of water expenses on various income groups, family sizes, and regions. To support its objectives, OECD justified metering as the basis of economic rationalization:

> The trade-off between efficiency and equity objectives in the provision of household water services typically occurs when moving from an unmeasured to a metered charging structure, when rebalancing tariffs away from fixed charges toward volumetric charges, and when increasing fees and tariffs toward full cost pricing. There is considerable experience in OECD countries with policy measures to address water affordability for vulnerable groups, while attempting to make water pricing reveal the full economic and environmental costs of water services. (OECD 2003, p. 12)

According to OECD, supporting measures for the poorest families can be grouped into two broad categories: those supporting revenues of targeted households, and preferential tariffs. The first group of measures includes social subsidies, vouchers, fractioned payments, and debt forgiveness. In the second group, preferential tariffs are meant to keep water bills below a certain fraction of revenue (e.g., 4 %). They include keeping water charges under a threshold, and increasing block tariffs.

In other words, for cases in which water service costs should be covered mainly though billing, one could either lower the bills of some category of users, at the expense of other users (or of the self-financing capacity of the service provider) or support the poor by helping them financially to pay their bill as the other customers. To provide a complete picture of actual situations, we also want to present cases in which part of the full cost of WSS services provision will not be covered by the beneficiaries, but by general budgets (i.e., taxation of citizens). One must indeed remember that if a service lowers the bills for all customers, automatically the percentage of families paying more than 3 % of their revenues for water will decrease. However, at present, water supply utilities seldom test the real redistributive impact of their ongoing tariff or of a proposed tariff change. Authorities in charge usually have their preferred solution to address the issue, and they publicize their choice so as to gain political legitimacy, but they frequently fail to see that the issue is more

[4] The 3Ts are tariffs, transfers, and taxes.

complex than they think and that tariffs may have counterintuitive effects. We discuss this and propose a simple tool to evaluate redistributive effects of tariffs at the end of this chapter.

18.3 Reducing the Water Bills for Targeted Populations

The first possibility is to offer some kind of bill reduction to some customers like disabled, retired, unemployed, living on benefits, etc. However, this section only applies to utilities that charge their services through some volumetric measure. It does not apply for instance to the majority of British water users who remain unmetered to this day or to Dutch wastewater collection, which is paid via real estate taxes, etc. In other developed countries or cities (France, Germany, Boston, and, recently, New York and Chicago), metering has been applied, but only at the property level (i.e., with only one meter per building). The collective bill is then allocated between residents on various bases: apartment surface area is quite common, but submetering is also frequent. In those cases, it is more difficult for water authorities or operators to subsidize targeted residents' water bills.

18.3.1 Rebates on Water Bills

The OECD 2003 report mentions various examples of rebates on water bills. In Australia, a system of identification (ID) cards giving the right to reductions (called concessions) was developed. It initially targeted single-family house owners who had become modest retired occupants but was extended in various ways by the various federated states that are in charge of social issues. Victoria's initiatives resulted in reducing the average water bill by a quarter, addressing up to 30 % of households. Additionally, in the same state, a special subsidy fund has been set up to support WSS subscribers undergoing unexpected difficulties (job loss, divorce, costly illness, and, eventually, consequence of an internal water leak). Demands are treated on a case-by-case basis, and in 2001, they included 12/10,000 customers.

Other studies (e.g., Smets 2003) indicate various possibilities to reduce bills altogether or wave part or all water bills in arrears, on the basis of a special fund generated by the operator for all water bills. A tiny increase in the per m^3 price can generate a substantial funding system for that purpose. In Belgian Wallonia, for instance, a little more than 1 eurocent per m^3 is added onto water bills to generate a fund to support the "water-poor" program. It represents 0.3 % of the total average bill and helps support around 10,000 customers per year (of an estimated total of 120,000 water-poor customers), with an average support of 200 €/customer (AquaWal 2009). In France, there is a support system designed at the county level, called solidarity fund for housing (FSL), in which public housing managers, county council social services, electricity, gas, and water utilities give various amounts of money to allow waving either unpaid rents or charges or bills. This is done on a case-by-case basis and is mainly concerned with temporary difficulties (not structural ones). In Scotland, water bills are systematically capped at 3 % of household's income. One difficulty with all these support schemes is the information needed by the operator or authority providing the service to identify the beneficiaries. In the United States, for instance, an electricity bill support system remains unused by most potential beneficiaries who are not aware or not willing to claim support. Usually these support systems require the use of other existing databanks on people who are poor or having difficulties (e.g., social services in local authorities, family benefits (CAF)[5] at the county level in France, national family fellowship in Brazil (bolsa familia), so as to reduce the cost of information. This gives an argument to WSS management associations like AWWA[6] in the United States, promoting a general support mechanism "outside the bill" (AWWA 2004). In developing countries, it remains particularly difficult to bring support to the significant fraction of the population that is not connected directly and has to carry water away from a well or a stand pipe. Water ends up far more expensive for them, usually.

England and Wales offer a special case, since water companies are not allowed to disconnect customers with bills in arrears. This has generated a dramatic increase of bills in arrears and on the duration of these arrears. Negotiations result in other solutions, like frequent (weekly or fortnight) billing, reduced water pressure (flow restrictors), and even prepaid water meters (working with coins or chip cards). The latter have, however, been banned. Conversely, their use was upheld by the Constitutional Court of South Africa.

Olivier Coutard (1999) proposes a typology of three groups of water indebtedness treatment:

1. No water consumption reduction, no bill waving, and household remains fully responsible. The company accepts some delay in payment by spreading the bill in arrears over time or changing frequency of billing.
2. No water consumption reduction, but rebates offered on the tariff or on the total bill.
3. No rebates on bills, but water consumption reduction.

New York City offers an interesting example for the third group above: under pressure by the Environmental Protection Agency (EPA) to improve the efficiency of wastewater treatment, in order to reduce excess water in drains, the city decided to introduce metering, but only at the property level. In many cases, poor leak control in condominiums would result in much higher water charges when shifting from the previous rate system to the volumetric payment. The city then offered to keep the bill at previous level during 2 years, giving some time for building residents to track and control leaks and replace inefficient appliances. Additionally, for specific places in which residents were identified as "poor," the city would bring subsidies to support leak control investments.

---

[5] Caisses d'allocations familiales—family benefits fund (benefits for low-income families with two children or more).

[6] American WaterWorks Association, the most important association of drinking water providers.

18.3.2 Increasing Block Tariffs

A different approach to redistribute water costs among users is to use increasing block tariffs (IBTs). Although the OECD acknowledged that some metering, plus IBTs, may have regressive effects on large, poor families, it claimed that "the design of increasing block tariffs can be adjusted in several ways to make the sizes and prices of tariff blocks deliver the intended distributive effects" (OECD 2003).

The first rationale for introducing volumetric payment of water and, additionally, increasing block tariffs (IBTs) is efficiency in use and demand management. But there is another argument: equity. One can indeed argue that even if elasticity of consumption to price is small, and IBTs have complex consequences, they may still be justified in terms of utilities getting higher revenues from users who generate a costly peak demand. Additionally, on moral grounds, most people support that water wasters should pay: metering and IBTs would then firstly be advocated in terms of consumer justice. But some also consider social justice: initial cheaper volumes would make water less expensive for the poor. And, indeed, in several studies, elasticity of water consumption to revenues is higher than to prices (see also the chapter by Barr and Ash in this volume).

In Brazil, most state water companies (CESB)[7] and many municipal or private water suppliers offer a cheaper initial volume of water to targeted populations, combining a rebate system and IBTs: typically, families identified as poor (e.g., receiving only one wage under the social minimum wage level), or on benefits (bolsa familia, state or municipal social support, etc.), or living in small homes (less than 60 m^2), or consuming little electricity (less than 200 kwh/month), will get an important rebate on the first 10 m^3/month (eventually 15). Beneficiaries must prove their eligibility every year or so (Britto 2015).

Some researchers challenge this claim of redistribution in favor of the poor, in particular in a developing country context. Boland and Whittington (2000) think that "this type of tariff deserves more careful attention. Even at first glance, the consensus appears somewhat curious because, although IBT structures were first designed in industrial countries by providing revenue-neutral cross-subsidies, only a small minority of water companies in countries like the United States now use them. Water and sanitation conditions may help explain the fact that IBTs are increasingly popular in developing countries, but this is not obvious. In many cities in developing countries, most poor households do not have private metered connections to the water distribution system, and thus IBTs do not help them" (Boland and Whittington 2000, pp. 215–216).

After careful examination, they conclude that "IBTs introduce inefficiency, inequity, complexity, lack of transparency, instability, and forecasting difficulties.... Every claimed advantage of an IBT can be achieved with a simpler and more efficient tariff design: a uniform price with rebate" (ibid). They argue that rebates can be targeted to low-income customers, provided the information on who belongs to

[7] Companhias Estaduais de Saneamento Básico.

this category is available. Komives et al. (2005) also draw from their experience in developing countries that IBTs have, in fact, regressive effects, because poor households are often large ones, so their consumption ends up in the upper blocks. It is, for instance, the case in South Africa, where the government's decision held in 2001 to provide a basic amount of water, free of charge, to all citizens was translated into 6 free cubic meters per household (in fact per subscriber and then behind him sometimes many households) and per month (Smith 2012; Burger and Jansen 2014). This tariff was challenged in the court by Soweto residents, who previously had access to unlimited water from standpipes against the payment of rates. After a long battle, the amount of "free basic water" was set at 42 m^3/month, which jeopardizes the capacity of water services to cover their costs from the users.

However, one could argue that these conclusions may not be valid in developed countries. Indeed, in parts of Europe, almost all households are connected to water supply systems, so that issues identified by Boland and Whittington regarding charging for collective consumption (e.g., villages depending upon standpipes or connected subscribers reselling water to poorer neighbors) do not occur. Yet, in many European and American cities, the meter used for billing is a collective one, so that it is difficult to apply a progressive tariff without additional information, typically the number of persons or apartments behind a meter. Where metering is collective and indoor water use is both moderate and inelastic, IBTs may well end up as a useless complexity, at least in condominiums.

In Barcelona, the superimposition on water bills of the sewer charges, plus a levy for environmental protection, was to be compensated by a growing block tariff system. But since this tariff was designed per meter rather than per capita, large families in the suburbs suffered dramatic bill increases, and they went on bill strike and to court, where they won and forced the Catalan water agency to redesign the tariff (Domene and Sauri 2012). In Belgian Flanders, the desire to implement the spirit of the Rio 1992 right to water declaration led to introduction of a tariff with an initial free volume of 15 m^3/year/capita. The information of the number of people behind meters was available. Yet, an ex post study (van Humbeeck 2000) showed that the redistributive effects were paradoxically negative for the poorer families. Boston, Massachusetts, is a very interesting example of collective metering through smart meters and progressive tariffs, which apparently brought water consumption to be much better controlled. Residents and building managers provide information on the number of residents per meter to the Boston Water and Sewer Commission, which allows setting the blocks on a per capita basis. However, to this day no social impact study has been made, and it remains to be seen whether the success in demand management is due to the tariff or to the interaction between the utility and the customers using smart metering systems.

In France, IBTs appear as an attractive formula to support water conservation and consumer justice. However, in most cities, metering is performed collectively for condominiums. And protection of privacy led the courts to deem it illegal for a utility to know how many people live in a housing unit that is being metered. Additionally, large water companies like Veolia have been able to calculate the extra cost they face when they have to meter and bill each apartment separately, and the result is adverse: namely, the additional information obtained through individual billing of apartments is not worth the cost (Barraqué 2011). This is why Veolia usually prefers to support the income of customers or give rebates rather than adopting growing blocks (see below). Lyonnaise des Eaux, another water company, supports another solution: IBTs are implemented in condominiums with collective metering, so the company sends only one bill to the building managers, but each apartment pays the same fixed part as a single family, and then the tariff blocks are multiplied by the number of apartments in the building.[8] Then it is possible to experiment with the combination of increasing block rate tariffs with social rebates: in Dunkerque, Lyonnaise des Eaux, refined its IBT tariff: for a single-family house that receives a separate bill, the first block, up to 75 m^3/year, is supplied at 0.80 €/m^3, and for families on benefits (CMU-C[9] in French), the price goes down to 0.30 €/m^3. The second block, up to 200 m^3, costs 1.50 €/m^3 for every user (regular or on benefit families). Additional consumption above that threshold is billed at 2 €/m^3. There are no social rebates for upper blocks. As for condominiums, since it was considered illegal to use data on family sizes, instead of setting the blocks per capita, these figures are multiplied by the number of apartments behind a meter, irrespective of the number of residents in each apartment. It is left to the building managers, who have to pay the collective bills, to allocate their bill among the resident families. Some will use submeters, others will calculate the cost according to the surface area of the apartments. It remains to be seen how this social tariff will perform in terms of social redistribution: will managers in turn give a rebate to those resident families that are eligible for a first block with rebates?

18.4 Supporting the Income of Targeted Populations

Many utilities argue that the social dimension of water services should be handled separately or as AWWA (2004) suggested, "think outside the bill." In downtown areas in particular, where water is paid in addition to the rents, it is much easier for tenants to pay a fixed charge for their water every month with the rent than a randomly sent variable bill. And when they cannot pay, they may need global support for the rent and general charges rather than for water alone.

One option is to get water suppliers, as well as electricity or gas suppliers, to give a small percentage of their turnover to a social housing fund, as is the case in France. The fund operates at the county level, since county councils are in charge of social and sanitary affairs. One of the problems is that this funding can only help people who are temporarily unable to pay. It is more difficult to support people who are in need but do not receive bills directly.

[8] Typically in Libourne, a condominium with 100 flats would replace the collective meter subscription of 200 €/year and a uniform variable price, by a fixed part of 100×15 €/year, and a first block of 100×15 m^3 at the "essential good" price of 0.1 €/m^3, etc.

[9] CMU-C means *couverture maladie universelle complémentaire*: these families get full social security coverage.

So, the best argument of those who favor income support is that as long as the percentage of people who cannot afford the bill is small, there is no need to create a sophisticated tariff. Social services should use municipal general budgets, or earmarked funds, to solve bad debts cases. Another option is to identify poor water users and offer them vouchers. It is implemented in France through "personalized water cheques" (coupons). They are being used as an experiment in the largest water supply utility in France, the Paris suburban SEDIF. Typically, a family of four with a yearly income of 12,000 € may have an average water consumption of 120 m^3/year and then, if billed separately, pay 380 €/year (sewer service included), i.e., more than 3 % of its income. This family would then be entitled to a yearly water coupon of 100 €. For those who live in condominiums and do not receive a separate bill, the support is channeled through the social support services of the concerned municipality.

A similar scheme is used in Chile: no water tariff for poor initiated by utilities, which were privatized during the dictatorship era, but municipalities are rather left with the task of identifying the poor water consumers and support them financially from a national fund allocated between them by the central government (Pflieger 2008). Families receiving the largest subsidies must still pay 15 % of their bills. Municipalities have difficulties in identifying the eligible households, and many errors are reported (Britto 2015).

According to OECD, in Finland, water charges are included in housing rents and are eligible to some support. Apparently 7 % of the households are concerned, and the support cannot go above 80 % of eligible charges. There is a minimal charge that all households must pay.

In the United States, according to AWWA (2004), adopting a system of vouchers to support payment of water services is under discussion, but it is not simple. Indeed, these vouchers could interfere with other forms of social support. In particular, the related artificial income increase might cause poor families to lose their eligibility to general rent support. In the end, only a minority (13 %) seem to offer rebates. Some wave the fixed part, and others give a rebate on variable parts or have set up "lifeline" prices for minimal volumes. Social support then usually comes from outside the bill, when it exists.

In most European Union (EU) member states, support for the water-poor is primarily left to municipalities or local public authorities. This is the case in particular when WSS services are provided by commercial utilities that consider it not their role to get involved in social support.

Centralization of water regulation in England and Wales is conversely translated into a national system for water-poor support. All poor or "incapable to pay" households are on benefits, i.e., they receive a financial support indexed based on the cost of living. Before 1989, water rates were eventually covered directly by a social support system. After the privatization of water utilities in 1989, benefits were added as a supplement to cover water expenses. But the price increase that followed was much faster than inflation, so that after 8 years, this fraction of benefits corresponding to water represented only 69 % in real terms of what it was in 1989. This probably added to the general discontent of British citizens with full privatization of 1989.

18.5 Reducing the Bills for All and Reintroducing Taxation

Against the objective of the 1990s, several international institutions have de facto shelved the project of full cost recovery. It is obviously the case for developing countries in which initial infrastructure finance needs taxes and transfers to avoid unaffordable water bills or charges. In France, for instance, back in 1954 a special national fund was created through a piggyback tax levied on all water bills (1 cent on each m^3). The funds were doubled with a tax on horse races bets, and this provided up to €300 million per year. Money was allocated at the county level exclusively to support the extension of water systems in the countryside. This was necessary, since France remains a low-density country with scattered housing. Yet today above 99 % French homes are connected, and the country has to maintain up to 950,000 km of water pipes in the long run. It will probably not be possible to do it solely with an increase in billing. In neighboring Italy, a long tradition to fund infrastructure through general local budgets supported by government grants results in the lowest average water price among EU member states, and it turns out impossible to stop the subsidy system to avoid having high water prices. This is politically impossible and would be in any country. Conversely, such low water prices result in a quasi absence of a specific water-poor issue.

Typically, in Portugal, in 1974 when the country returned to democracy, connections to WSS services were no better than in Brazil, despite the huge difference in rural-urban migrations. But the country did set up a dedicated national water company to channel the important subsidies coming from EU's cohesion funds. The resulting "public-public partnerships" allowed the country to catch up with the richer member states at an affordable price for water users.

The Netherlands offers a fascinating case at the other end of WSS services costs and turnover (on the high side): the Dutch pay potable water through volumetric water bills to a commercial utility owned by a mix of municipalities and provinces; they pay wastewater collection to their municipality through housing and land taxes, and they pay for sewage treatment to the institution in charge, which is one of the famous water boards, historically created to protect against seawater flooding and river flooding, and to drain the lowlands. Payment is by family: each family pays for three members irrespective of the number of family members, except single persons, who pay for one. Overall, this financing system spreads the high cost resulting from living lower than sea level in three different tiers and makes it both more affordable and acceptable. Paying for wastewater services through local rates is usually redistributive in favor of the poorer households, who live in less valuable homes, and the family tariff clearly favors large families. Yet there is no available analysis of the de facto redistributive effect of the global system. Water bills are sent individually to each household, even in condominiums, so it would be possible to develop redistributive formulas. But the Dutch are reluctant to do this. Water bills have to be paid, and in the rare cases when they are not, social services get involved. For wastewater, which is covered by taxes, families under a certain level of revenue are exempted. And the water boards exceptionally give rebates to poor people or to students. The overall philosophy is not to discuss water charges redistribution too much, but rather to keep the various charges paid by the population as low as possible. Hence, the benchmarking of water supply utilities, and water boards, systematized in recent years.

Many of the above remarks on indirect support of water-poor people through taxation apply to the majority of English water users who pay by rates (i.e., with some redistribution between rich and poor households via renting values).

In France, water price increases, due to the implementation of European directives (in particular Urban Waste Water Directive, EC 91/271), have led to the development of a national debate. Even the lobby of water supply companies and national representation have discussed the possibility of reducing the total bill through removing some of the elements from the bill. For instance, it has been advocated that under French institutional setup, public services with economic character should be covered by their beneficiaries, and, conversely, economic intervention of the public sector with no service rendered should be taxed and not charged through bills. Water supply has always been considered as a service, so it has been billed; sewage collection initially was mandated, so it was considered as a tax (housing/land tax) until 1967. Then it was considered that since all urban citizens were connected to sewers, one could consider wastewater collection as a service, so it could be transferred on water bills and paid by volumes. This allowed adding up the levies to be paid to the Agences de l'eau (water agencies) onto the water bills. However, the pollution discharge levy was originally used to fund the construction of sewage treatment plants, and these do not render a service to the sewer-served population but to downstream riparians of the water body. Therefore, there is a rationale to remove this part of water charges from bills and transfer it to local taxation. This would automatically reduce the percentage of people paying more than 3 % of their revenues to get water services.

18.6 Improving the Assessment of Water Tariff Redistributive Effects

The water-poor debate calls for a more important involvement of water utilities in the social dimension of water charges and a general reflection on the distributive effects of tariff levels and structures. This is still a relatively unexplored territory. Now that the water-poor notion is acknowledged, utilities are more or less forced to find alternative ways to address the situation.

In developing countries, the largest social issue is linked with the relatively low level of connections to water services: non- or poorly connected households are not really known by the authorities and the operators, so they may end up paying much more for the same quantity of water, or else they have illegal connections that jeopardize the reliability of the networks. Paradoxical situations also occur: in Rio de Janeiro, for instance, "favelas" (slums) get some form of collective water at reduced prices, because they are favelas. But in the outer periphery of this immense metropolis, there are thousands of families that are poorer than favela residents and who are not registered as poor and thus do not receive support. Any project to improve social access to water must then go through a real field survey and through residents' involvement in the issue.

But it seems to be the case also in developed countries: the level of connections is very high, but authorities and operators in fact know very little about what people do with tap water, and what is going on beyond the meter. For lack of socioeconomic analyses, they tend to indulge in simplistic considerations, like "water is a market good, so the rich can buy more than the poor, so if we charge the high level consumers more, there will be a redistributive effect in favor of the poor." Unfortunately it is not so simple, and real field assessments need to be done, starting with the distributive effect of fixed parts that are quite frequently used.

In a project funded by the French National Research Agency,[10] our partnership[11] decided to build upon the seminal approach developed by Fitch and Price (2002): their indicator for water affordability is the percentage of people who pay more than 3 % of their income on WSS services. This indicator can be supplemented by another one: those who pay less than 1 %. In turn, these two indicators can be calculated for a three-dimensional matrix, with deciles of income, number of persons behind the payer, and a proxy for water consumption habits (thrifty-average-hedonist). A similar approach has been used by Rajah and Smith (1993) in the United Kingdom and by P. Van Humbeeck (2000) in Belgian Flanders. It would be very useful for water utilities and authorities to use such tools to simulate and anticipate the potential impact of tariff changes on various categories of water users in practice. But additionally, this tool should calculate how much money is left at the end of the year in the cash flow of the operator to fund the long-term maintenance of the infrastructure.[12] The overall idea is that, since WSS services are a fixed cost industry, with frequent mandate for operators to balance costs with revenues from users, there is a zero sum game between various categories of customers and the operators' interests. This tool also allows for better assessing the real impact of a tariff change.

Thus, we have discovered that some new tariffs based on increasing blocks were not favorable to lower income populations, for the very reason (but counterintuitive for many decision-makers) that the additional administration costs of such tariffs led to everybody paying more to the operator. This is why one must first recommend to keep the water pricing simple, so as to reduce the transaction costs associated with the tariff. Social objectives are in the end better handled outside the tariff through income support for the poor. In particular, water poverty in extreme cases

[10] Within a sustainable cities program, this project dealt with the sustainability of water services in large cities. See http://eau3e.hypotheses.org

[11] The project was coordinated by B. Barraqué and involved seven partners, including Marielle Montginoul, and the public water supply utility of Paris.

[12] Such a tool is being developed by Ms Marie Tsanga Tabi in the Strasbourg research laboratory GESTE in ENGEES (Ecole Nationale du Génie de l'Eau et de l'Environnement). It is presented in the project's blog: http://eau3e.hypotheses.org

(homeless people) requires solutions completely out of the redistribution debate: in France, and we suppose also in other developed countries, one can find "water associative houses," in which deprived people can wash, clean their clothes, get support for other needs, and recover minimal dignity.

References

AquaWal. (2009). *Etude relative à l'impact sur les usagers des réformes en matière de tarification de l'eau et à l'estimation de l'emploi généré par le cycle anthropique de l'eau en Wallonie*. Report for the Walloon region, Belgium. http://www.aquawal.be/servlet/Repository/rapport-final.pdf?ID=1377

AWWA (American Water Works Association). (2004). *Thinking outside the bill: A utility manager's guide to assisting low-income water customers*. Denver: AWWA.

Barraqué, B. (2011). Is individual metering socially sustainable? The case of multifamily housing in France. *Water Alternatives, 4*(2), 223–244.

Boland, J. J., & Whittington, D. (2000). The political economy of water tariff design in developing countries: Increasing block tariffs versus uniform price with rebate. In A. Dinar (Ed.), *The political economy of water pricing reforms* (pp. 215–235). New York: Oxford University Press.

Britto, A. L. (2015). Políticas tarifárias e promoção do acesso ao saneamento no Brasil. In B. Barraqué & A. Britto (Eds.), *Gestão Sustentável das Águas em Áreas Metropolitanas Brasileiras e Francesas: novas questões relacionadas ao acesso universal aos serviços de saneamento.ed*. São Paulo: Annablume.

Burger, C., & Jansen, A. (2014). Increasing block tariff structures as a water subsidy mechanism in South Africa: An exploratory analysis. *Development Southern Africa, 31*(4), 553–562.

Coutard, O. (1999). L'accès des ménages à faible revenu aux services d'eau et d'énergie en France et en Grande-Bretagne. *Flux Cahiers scientifiques internationaux Réseaux et territoires, 36–37*, 7–15.

Domene, E., & Sauri, D. (2012). Water, public responsibility and equity: The Barcelona "water war" of the 1990s. In B. Barraqué (Ed.), *Urban water conflicts* (pp. 33–37). Paris: IHP-UNESCO and Taylor & Francis.

Fitch, M., & Price, H. (2002). *Water poverty in England and Wales*. Published jointly with the Centre for Utility Consumer Law under the auspices of the Public Utilities Access Forum (PUAF).

Komives, K., Foster, V., Halpern, J., Wodon, Q., with support from Roohi Abdullah (2005). *Water, electricity and the poor: Who benefits from electricity subsidies*? Washington, DC: The World Bank.

OECD. (2003). *Social issues in the provision and pricing of water services*. Paris: OECD.

Pflieger, G. (2008). Achieving universal access to drinking water and sanitation networks in Santiago de Chile: An historical analysis 1970–1995. *Journal of Urban Technology, 15–1*, 19–51.

Rajah, N., & Smith, S. (1993). Distributional aspects of household water charges. *Fiscal Studies, 14*(3), 86–108. doi:10.1111/j.1475-5890.1993.tb00488.x.

Smets, H. (2003). *La solidarité pour l'eau potable—Aspects économiques*. Paris: Académie de l'eau.

Smith, L. (2012). Conflict vs cooperation between the state and civil society: A water-demand management comparison between Cape Town and Johannesburg, South Africa. In B. Barraqué (Ed.), *Urban water conflicts* (Urban water, Vol. 8, pp. 147–168). Paris: IHP-UNESCO and Taylor & Francis.

van Humbeeck, P. (2000). The distributive effects of water price reform on households in the Flanders region of Belgium. In A. Dinar (Ed.), *The political economy of water pricing reforms* (pp. 279–295). New York: Oxford University Press.

Chapter 19
Sustainable Water Rate Design at the Western Municipal Water District: The Art of Revenue Recovery, Water Use Efficiency, and Customer Equity

Tim Barr and Tom Ash

Abstract Water providers in the United States have experienced years of revenue loss from lower water use/sales. The decline in water use has been caused by water restrictions, extended economic recession, and continued water conservation programs. When water users become more efficient, traditional rate designs cannot recognize and accommodate water conservation and/or a decline in water sales without a financial hardship to the agency and ultimately the end user. The impact of traditional water rate design when water is saved is a financial, political, and public credibility problem for water providers. However, a group of agencies in California have implemented rate structures that accurately reflect the costs of water and water service, recognize customer-by-customer water use efficiency, and also provide a strong economic signal as to the future or environmental costs of water. These agencies have experienced accurate and stable revenue recovery, increased customer awareness, and have seen more conservation (user behavior change) without a negative economic impact on the agency. The rate structure is referred to as "water budget-based rates" or, more accurately, as a "sustainable" rate design. This chapter will describe the evolution and the philosophy of a "sustainable" rate structure at the Western Municipal Water District and provide a glimpse into the agency motivation, design, and the impacts on finances and water efficiency.

Keywords California • Revenue loss • Water rate design • Agency credibility • Consumer satisfaction

T. Barr (✉)
Western Municipal Water District, 14205 Meridian Parkway, Riverside, CA 92508, USA
e-mail: tbarr@wmwd.com

T. Ash
Inland Empire Water Utility, 6075 Kimball Ave, Chino, CA 91708, USA
e-mail: tom.ash27@outlook.com

A. Dinar et al. (eds.), *Water Pricing Experiences and Innovations*,
Global Issues in Water Policy 9, DOI 10.1007/978-3-319-16465-6_19

19.1 Introduction

Water providers (agencies) in California have experienced years of revenue loss from lower water use/sales. Water agencies are now required by state legislation to reduce per capita water use to manage limited water supplies, protect state and regional watersheds, and meet the needs of world's eighth largest economy, which includes water for significant agricultural production, water for high-tech commerce, and a reliable water source for 38 million people. At the same time, water providers must accurately recover the costs of water and water services, be fair and equitable to water consumers, and protect local and regional ecosystems. These often competing goals and the current economic policy instruments (water rate structures) create significant financial, practical, and political challenges for water providers.

From 1990 to 2010, overall per capita water use has declined in California by 24 %, from 232 gal per day in 1990 to 178 gal per day in 2010 (Public Policy Institute 2014). The need to reduce water demand is expected to continue in California, based on legislation, climate change, and expected population growth. The common response by agencies to the predictable revenue/conservation conundrum is consistent water rate increases for all users. The traditional rate design, particularly where water conservation is necessary, creates a pattern of raising water rates to balance budgets after consumers have been asked to save water and responded successfully. This presents a difficult public relations position for the public water provider. The general public sentiment is "You asked us to save water, and then you raise water rates…we need to vote you out of office."

At a National Water Rates Summit in 2012, sponsored by the Alliance for Water Efficiency (AWE 2012), rate experts reported that "Partly due to successful water conservation programs, improved water-saving fixtures and technology, and a number of other factors, both water sales and water-related revenues are falling on a national level. Most importantly, how can [agencies] meet these costs while still encouraging much-needed conservation efforts?"

However, a small group of water agencies in California have designed and implemented a water rate stratagem that reflects the true costs of water and water service, provides adequate social equity measures, recognizes customer water use efficiency, and yet provides a strong economic signal for efficient water use based on the environmental or marginal costs of water (Wikipedia). This water rate is called an "allocation-based" rate or "water budget" rate and will be the subject of this chapter. The agencies that implemented water budget rate designs have experienced accurate and stable revenue recovery with less water being sold. The same agencies experience more educated and responsible customers, in terms of efficient water use. Some agencies cite 85 %+ customer satisfaction (The Farrell Group 1996) since implementing their allocation-based rate structure. With revenue, conservation, and customer acceptance, a sophisticated water budget or allocation-based rate stratagem may be more accurately referred to as a "sustainable" water rate design, maintaining financial stability or the agency, recognizing customer conservation efforts, and keeping more water for the environment.

This "conservation conundrum," the need to reduce demand yet maintain fiscal responsibility and customer equity, provides the backdrop for this investigation describing the design and deployment of a "sustainable" rate stratagem and why and how the Western Municipal Water District (WMWD) changed its rate stratagem to meet the agency, consumer, and environmental needs for the agency. The description will include an initial investigation into the first water budget rate structure, the process to change the water rate stratagem at WMWD, and key lessons learned with the impact of the new economic policy instrument.

The Western Municipal Water District's new rate stratagem aligned the district's core economic policy instrument in an innovative approach with water supplies, climate, social equity, local public policy, and state legislation. The district implemented the new rate structure in 2011. The WMWD process shows how this agency addressed the common challenges of revenue, conservation, and consumer equity with a sophisticated, yet personalized and flexible economic tool, the water use of individualized water budgets, referred to here as a "sustainable" water rate design.

19.2 The Agency

Western Municipal Water District (WMWD) is a California "special district" (California Special Districts Association 1954) established to provide water and sewer services to users in a 527-square-mile service area. Western Municipal serves 24,000 retail customers and provides wholesale water and services to eight retail areas agencies. WMWD was formed in 1954 as a member of the Metropolitan Water District of Southern California (MWD) to facilitate the delivery of Colorado River water to vast acres of citrus groves and to a rapidly growing residential and commercial region of Southern California. To expand on water supplies to meet consumer needs, WMWD added California State Water Project (SWP) water, via the regional wholesale agency, MWD, to their water supply portfolio in 1979. Approximately 75 % of water for WMWD users is imported, with the remaining water coming from local sources. Today, WMWD serves 24,000 accounts, 90 % of which are residential users.

WMWD is located in the hot inland valley of Southern California, where the climate is predominately a mid-desert environment with hot and dry conditions. The average annual rainfall is 10 in., and the average annual evapotranspiration (ET) is 61 in. (California Department of Water Resources, California Irrigation Management Information System 1984–2014).

With escalating limitations on the major sources of imported water, WMWD has implemented a wide range of innovative water conservation programs. Up to 2011, WMWD had utilized a common water rate design, built around two charges on the water bill. The rate design consisted of a fixed "service" charge, recovering 25 % of fixed costs on the fixed service charge, and imbedding the remaining fixed costs into a flat variable charge (volumetric rate) for water. The variable charge included a majority of fixed costs as well as the cost of water, energy, and treatment. The water is sold to agricultural, commercial, institutional, and residential users.

19.3 Rate Change Goals and Investigation by Western Municipal Water District

The challenge for WMWD revolved around how to reduce demand for the management of limited water supplies, maintain revenue sufficiency, and treat each end user equitably. Traditional rate designs, such as flat rates and uniform or fixed tier rates, have not been adept at revenue recovery and water conservation at the same time. Nor have traditional rates sent a clear and equitable message to the consumer of their individual water use efficiency, a prime goal depicted in various levels of California legislation.

WMWD officials sought to design a water rate that could deliver two important practical and political goals, increased water efficiency across the customer base, and a rate structure that would be considered fair and equitable by end users.

19.4 Examination of the First Water Budget Rate Structure

In 2008, WMWD began evaluating its current rate structure, particularly with regard to meeting state conservation legislation and customer equity. The prospect of changing the water rate structure required staff to investigate options, including what rate styles were working for agencies, the costs to change or update the billing system, conducting a financial study of the cost of service, the customer level data required to develop a more sophisticated rate structure, and the public outreach to constituents as a requirement in California under Proposition 218 (Proposition 18 ballot initiative).

The model for a "sustainable" rate structure was the Irvine Ranch Water District (IRWD; www.irwd.com), located in coastal Southern California. That agency implemented a "water budget" tiered rate structure in 1991, during a 6-year drought, partly due to a loss of revenue from drought restrictions and as a recognition of the need for long-term water use efficiency. The IRWD experience and results were the foundation for conducting educational workshops for the WMWD board and staff and have been the starting point in terms of education for other agencies looking to solve the revenue/conservation/public relations problem many agencies face.

An early study of the first water budget rate structures, conducted by the Metropolitan Water District of Southern California in 1997 (Metropolitan District of Southern California 1997), validated the success of the early innovators of water budget rate design and noted findings that included water budget-based rates identified as one of the most promising programs to achieve cost-effective water savings, community acceptance, and water agency support; the rate design offered flexibility with regard to customer needs, weather, and customer feedback; a great degree of satisfaction was found from agency staff and elected officials with the water budget rate structures as an effective and fair water conservation tool; the practical hurdles and cost of implementing water budget rate structures were more than outweighed by the benefits; the two primary agencies evaluated in the study showed a decrease in water use of 37 % and 35 %, respectively, within the first 4 years of water budget rate structure implementation; and elected officials often stressed the increased perception of fairness and a reduction in customer complaints after moving to water budget-based rates (ibid, Findings, Page V).

Within 5 years, the first water budget rate structure returned a 45 % decrease in the average water use, per account, and held steady for the next decade. Just as important for the agencies was the increased and stable recovery of fixed costs within the first year of the water budget rate structure (Source: IRWD Finance Dept.; January 2002).

The first water budget-based rate structure showed a distinct impact on landscape water use. Landscape water users methodically changed water waste behaviors, improved irrigation efficiency and management, and maintained efficiency to avoid paying "over-allocation" water use penalties. Dedicated irrigation water use declined by 55 % in 7 years and has held steady into 2014 with the continued use of a water budget rate structure by the water provider (Source: Joe Berg, MWDOC, personal call, May, 2014).

The sustained success of the first water budget rate structure suggested that stable revenues, conservation, and customer acceptance were possible to achieve for water providers. With real agency experience, the American Water Works Association (AWWA) produced a 2008 study of agencies with water budget style rates, titled *Water Budgets and Rate Structures*: *Innovative Management Tools*. The report confirmed the water savings potential of this rate design but also noted that fairness for consumers was a "fundamental benefit" identified by agencies and that the "additional complexity of customer specific water budgets was more than outweighed by the increased customer acceptance…" (Mayer et al. 2008).

By 2008, the time of initial water rate structure investigation by WMWD, the AWWA M1 Manual, *Principles of Water Rates, Fees and Charges* included information on water budget rates describing, "A water-budget rate structure is a form of increasing block rates that require the utility to set specific standards for what is considered efficient water use. Other rate designs do not directly account for efficiency of use a specific customer bill." Other specifics cited by the AWWA M1 Manual chapter on water budget rates also made reference to the rate structure qualities, including (1) water budgets are established for each individual consumer (metered account); (2) consumers with usage above their efficiency budget pay a significantly higher rate for the inefficient usage; (3) a water budget is based on an individual account characteristics, such as number of residents and size of irrigated area outside; and (4) water budgets may vary from billing period to billing period, depending upon the daily weather, input directly into the billing system.

WMWD officials, equipped with the compelling success of early adopters to water budget-based rates, took the next step, educating the elected officials of the issues and the potential for change.

19.5 Changing the Water Rate Stratagem at Western Municipal Water District: Meet Legislative Goals and Establish Equitable Public Policy

19.5.1 Educating Agency Officials

A fundamental first step in adapting a rate structure is to educate officials with the responsibility of establishing agency policies, particularly when related to water rates with the potential to change costs to constituents. Educational workshops with the district's elected officials and staff provided details on how water budget rates were designed and why they worked as reported by other agencies for the water provider and the end user. The prime objective for WMWD officials was to increase conservation and be fair to constituents.

The careful internal education process in turn led to an approval by WMWD elected officials to move to the new rate stratagem and to designate the tasks required to change the rate structure. The foremost goals of WMWD officials were "more conservation and customer equity," along with maintaining a low financial risk with lower water sales, due to increased water use efficiency (Personal interview with Kevin Mascaro). In late 2008, WMWD changed the economic policy instrument (the water rate structure) and became a full-fledged project for the WMWD Water Resources Department in coordination with all other district departments, including a strategic timeline for implementation.

19.5.2 Agency Water Use Efficiency Responsibility

New state legislation, the Water Conservation Bill of 2009 (SBX7x7) (State of California 2009),[1] established a timeline and methodologies for all California water providers to reduce water demand by 20 %. The legislation is called the "20 × 2020" (a 20 % per capita water use reduction by the year 2020). The water use efficiency goal established by this legislation is directly linked to the ability of water providers to apply for and receive state grants and low-interest loans used for water supply, water quality, and water infrastructure projects, creating a strong incentive for agencies to implement strategies for reducing water demand. This legislation drives the state requirement for water use efficiency and may also aggravate the common

[1] State of California Water Conservation Bill of 2009 (SBX7-7) requires a statewide 20 % reduction in urban per capita water use by 2020. Urban retail water suppliers must determine their base per capita water use and develop water use reduction targets using one of four specified methods: Option 1: 80 % of baseline per capita daily water use; Option 2: Sum of specified performance standards (55 gpd inside and 70 % of ETo outside); Option 3: 95 % of DWR Hydrologic Region target; and Option 4: A flexible alternative designed to adjust to local circumstances developed by the California Dept. of Water Resources.

problems seen by agencies related to revenue stability, conservation, and public relations when water is saved and rates are increased as a result.

19.5.3 Water User Responsibility

Yet, water agencies don't use water, the agency constituents use the water. While WMWD did not need to raise water rates, due to revenue loss from conservation, it did need to find an approach to reduce per capita demand and increase conservation programs and outreach to those who wasted water. This legislation required agencies to reduce demand that provided WMWD with the ability to educate customers about the users' role in conserving limited water resources. As an equitable social policy, the district uses the fact that state law establishes what is an efficient use of water, indoor and outdoor, relative to family size, weather, and parcel size, and would be using an economic policy instrument that places the responsibility for the costs of conservation on users who waste water. This approach to customer education and public relations is unique among water providers and is made possible through the deployment of "water budget" rate design.

19.5.4 Agency Political Reality

Providing safe and reliable water to all consumers is the prime responsibility of the agency. There is also the political reality that water must be priced in a fair manner across a widely varied customer base. In total, this means the agency must produce high-quality water, set prices that recover costs, and then also drive consumers to use water as efficiently as possible. Meeting these various political and practical needs is the objective of any rate structure. WMWD was able to determine that a well-designed water budget-based rate structure could deal with the multiple challenges agencies face when attempting to balance revenue, conservation, and public acceptance in a defensible and transparent manner, important to political stability in the service area.

19.5.5 Testing the Water Budget Concept with Water Users

One of the first steps taken by WMWD was to "test" the concept of water budget rates. It did that within the context of a broad customer survey in early 2010. Contained in the survey were two key questions related to the concept of water budget rates. They were: (a) Is it important to reward water use efficiency? (b) Is it important to penalize water waste? The customer responses to these questions are displayed in Fig. 19.1.

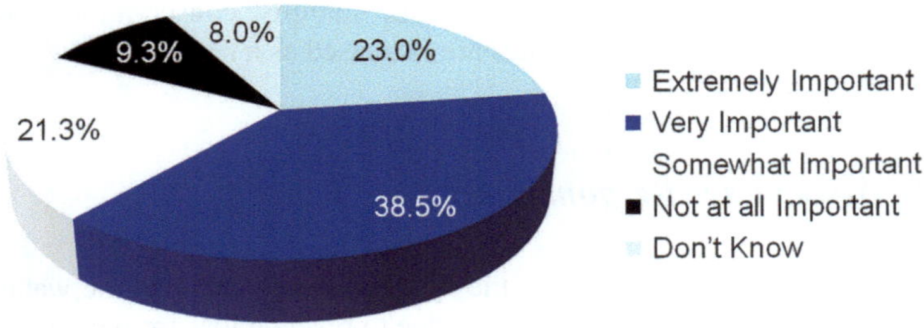

Fig. 19.1 How important is it to reward water use efficiency by homes and businesses and to penalize water waste (e.g., with higher water rates for waste)?

With 82.8 % of 1,700 survey respondents (7.7 % of the total customer base) claiming that "rewarding efficient use" and "penalizing water waste" was "important," a basic concept of water budget rates was validated for district officials. This step gave WMWD additional confidence that constituents would accept a change in the water rate structure if it was perceived as logical and fair (re: reward efficiency and penalize water waste).

19.6 The Timeline for Changing the Water Rate Structure Based on Tasks, Elections, and Seasons

Every water provider is different and has its own skill sets, staffing levels, billing systems, political climate, and available customer data. Some agencies have moved to sustainable rates within 6 months of approval. WMWD recognized that data collection and a billing system upgrade would take more time for implementation. However, the district viewed the extended timeline as one in which the sophisticated tasks could be well managed; customer outreach/education could be planned to insure accuracy, transparency, and fairness; and that there would be no overlap of a rate structure change with elections and other water projects. The timeline for implementing a more sophisticated rate structure was considered to be a key planning component by WMWD to insure its success in the public hearing approval process (Proposition 218) approved by the voters.

The project was broken down into sequential steps that involved all district departments, led by water resources (project management), finance (financial study, rate modeling), billing (billing system upgrade), engineering (parcel level data), public relations (customer outreach), and customer service (staff training, customer data verification, and ongoing administration) (Fig. 19.2).

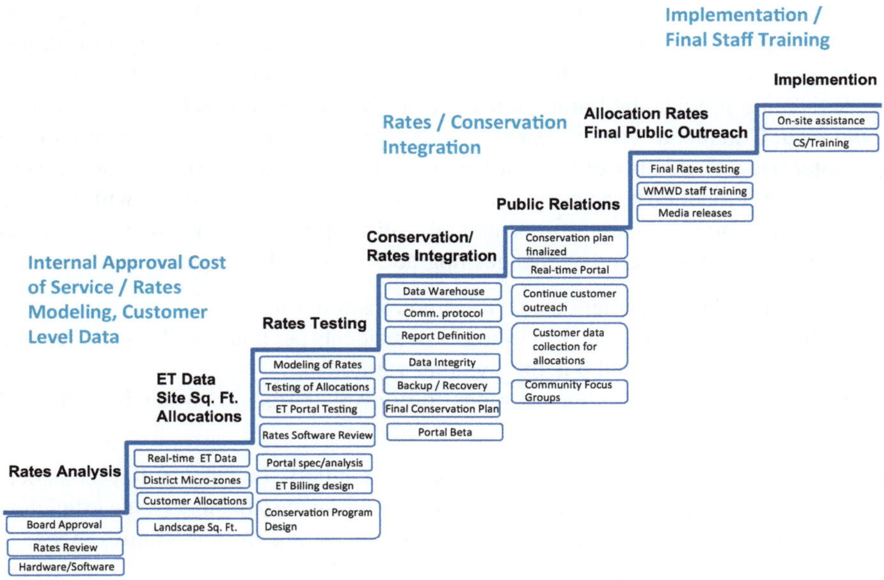

Fig. 19.2 Outline for water rate change tasks and timeline (Source: WMWD Water Resources Department)

19.7 Tasks Required to Develop and Implement Sustainable Water Rates

A change to a sophisticated and personalized water budget rate structure requires the recognition that such a water budget rate structure changes the agency, changes the relationship between the agency and the water users, and changes the way people think about their own water use. While the tasks are significant and a commitment by all levels of the agency is important, the 1997 Metropolitan Water District report evaluating water budget rates found “Many staff members began as skeptics, but after surmounting the practical hurdles of implementation came to believe that the benefits more than matched the costs incurred” (Metropolitan Water District of Southern California, Findings, pg. V).

The basic tasks for creating water budget rates overlap and are coordinated within the given timeline the agency identifies and fall within four major categories, including:

Legal: Include the agencies legal counsel on all aspects of the planning and implementation of a change in rate structure policy and conduct the rate evaluation, design, and public outreach with regard to meeting all legal requirements necessary, including creating an "administrative record" that includes the equations, philosophy, and policy-making details of the rate structure, including allocations, financial modeling, and public noticing. In California, Proposition 218 and other legislation provide the foundation and justifications for rate setting in the state.

Financial: Conduct a "cost of service" study; conduct a "rate" study (incorporating water budgets, number of tiers, tier break points, fixed, and variable cost recovery options); model cost recovery and varied rate structure scenarios (with varying fixed and variable charges, tiers, and allocations per customer group for internal evaluation); and determine changes/upgrades to the billing system software:

- Input of daily weather data (ET, evapotranspiration).
- Input of customer level data (number of residents per household, square footage of irrigated area, and potentially other variables).
- Make water bill design upgrades to convey charges to water budgets and individualized allocations.

Conservation: Establish water use efficiency standards (agency policy considerations), recognize/incorporate legislative requirements (that meet state efficiency standards), and evaluate/design conservation programs to support customers:

- Indoor water use efficiency.
- Outdoor water use efficiency.
- Commercial water use efficiency.
- Rebate programs.
- Evaluate/upgrade district website for ongoing customer education and communication.

Customer outreach and education: Outline a public education plan that includes key messages, timeline, and a multiple outreach approach; identify key stakeholder groups for one-on-one and small group education; utilize billing inserts, the local press, and special mailings informing customers of a prospective change in the water rate design; and conduct public hearings as required in the agency jurisdiction (constituent public hearing and voting process required in California) (Fig. 19.3).

19.8 The Western Municipal Water District Water Budget Rate Structure

The result of WMWD's water budget rate study project created a five-tiered, individualized rate structure. The example of the equation for the billing system for residential accounts is:

$$\text{Site water budget} = (\text{\# of residents})(60\text{gpd})(\text{days of service}) + (\text{ET})(\text{SF})(.8)(.625)(\text{days of service})(\text{DF})$$

Fig. 19.3 Images of WMWD public outreach and education (Source: Western Municipal Water District Community Affairs Dept., 2011)

where:

- # of residents recognizes family size.
- 60 gpd relates to State legislation (SBX7-7).
- ET = local daily evapotranspiration (from State legislation; AB 1881).
- SF = the size of the irrigated landscape area on the parcel.
- 0.8 = state landscape efficiency factor.
- 0.625 = conversion factor from inches to CCF (WMWD water billing units) (Note: 1 CCF = 28.3 l).
- Days of service = monthly bills may vary in the days of service; this insures customers receive an accurate allocation for their situation, including the varying days of service.
- DF = drought factor, used if/when there are water supply restrictions imposed by regional and/or state authorities (Fig. 19.4).

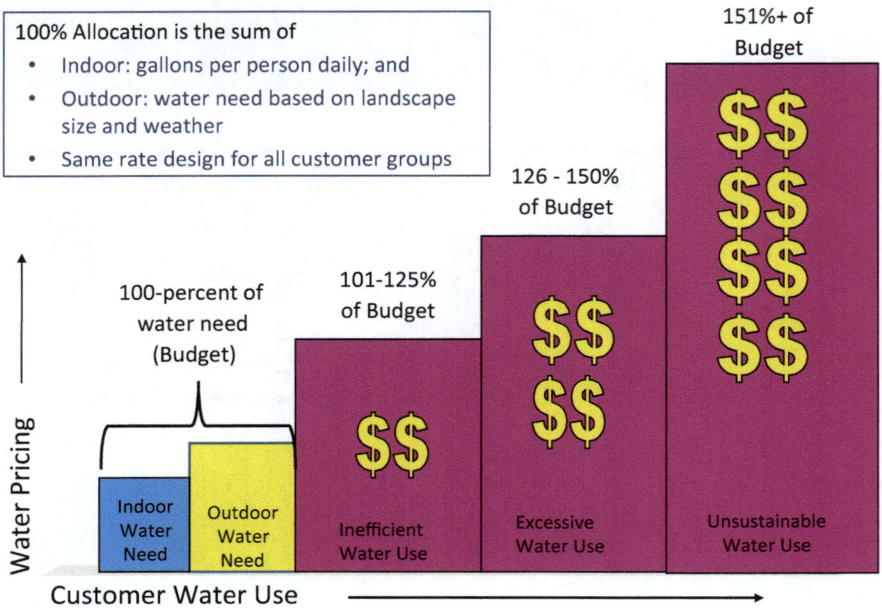

Fig. 19.4 Example of water budget rate "tiers" (width, height, and descriptive name related to use) (Source: Western Municipal Water District Public Meeting presentation, 4/2011)

19.9 Rate Structure Modeling

In Fig. 19.5, the WMWD financial consultants, using the recommended allocations from state guidelines, found that the majority of WMWD users would meet their individualized water budget, based on their current (2010) water use history.

The financial rate study found that most WMWD accounts were already efficient, as per the state efficiency standards, and were an important finding for the foundation for public education and outreach. Knowing which customers were already efficient would also assist the district's water use efficiency staff to target consumers who needed assistance to eliminate water waste and lower their water bills, a win-win for the customer and the agency. The financial modeling also revealed that a water bill of an efficient user would slightly decrease with a water budget rate design when compared with the existing rate structure (Figs. 19.6 and 19.7).

Important for increasing water efficiency was the ability to send a clear economic message that if an account exceeded its water budget allocation, the water bill would increase significantly. The result of water waste was directly communicated via a high water bill (and a corresponding tier name, such as "unsustainable" use).

With a customer survey supportive of the water budget concept in the mind of customers (re: reward efficiency and penalize water waste) and financial modeling showing most users were efficient, WMWD took a recommended five-tiered water

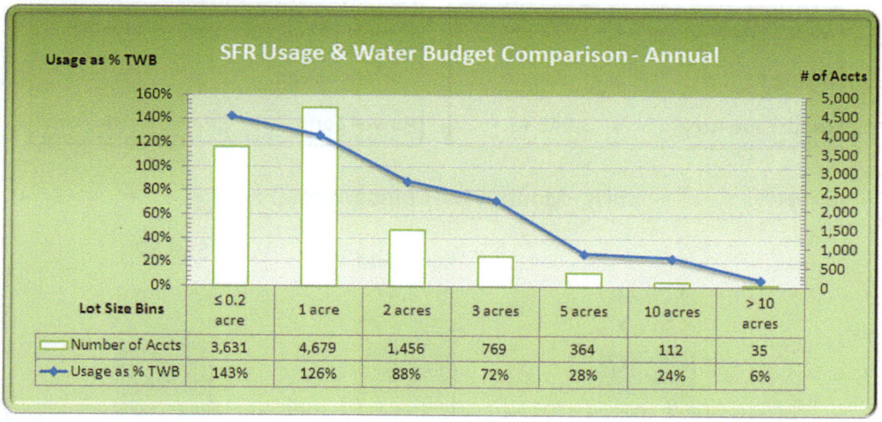

Fig. 19.5 Evaluation of WMWD customer water use compared to water budgets (Source: Western Municipal Water District/Raftelis Financial, Inc. 10/2010)

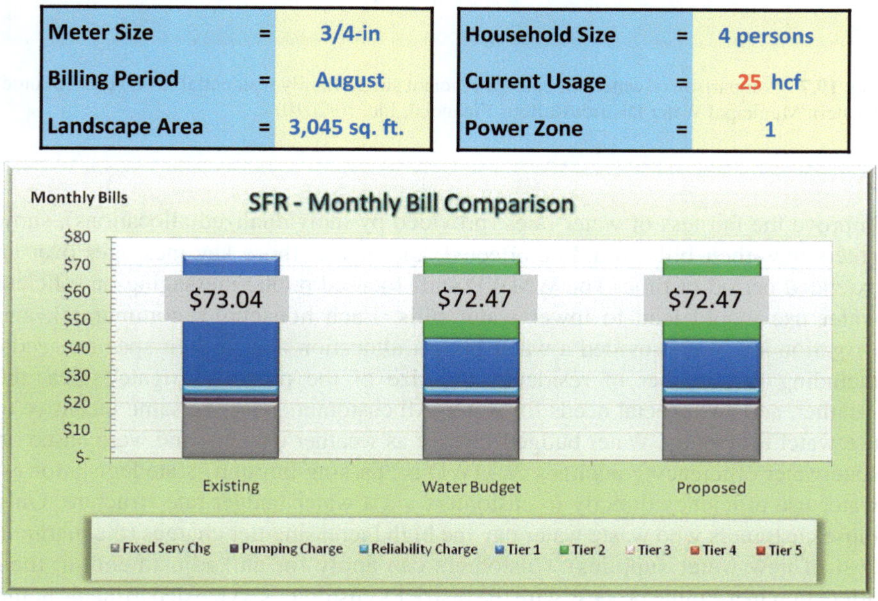

Fig. 19.6 Comparison of water bill cost for an average single-family residential (SFR) customer (Source: Western Municipal Water District/Raftelis Financial, Inc. 10/2010)

budget rate structure to constituents in mid-2011. During the required public hearing process (Proposition 218), only 400 ballots out of 23,000, or 1.7 %, voted not to approve the rate structure change. The overwhelming positive sentiment seen in the public voting was due, in the eyes of the agency, to the ability to significantly

Meter Size	=	3/4-in	Household Size	=	4 persons
Billing Period	=	August	Current Usage	=	19 hcf
Landscape Area	=	3,045 sq. ft.	Power Zone	=	1

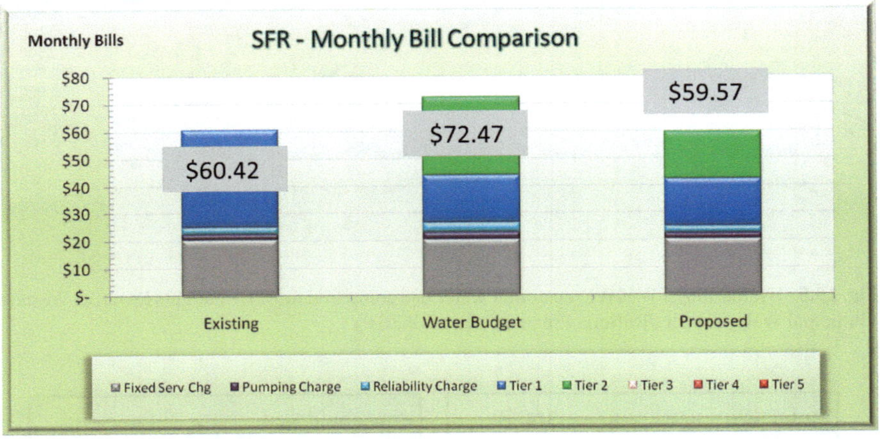

Fig. 19.7 Comparison of water bill cost for efficient single-family residential (SFR) user (Source: Western Municipal Water District/Raftelis Financial, Inc. 10/2010)

improve the fairness of water rates (provided by individualized allocations), show voters how their bills would be affected, and deliver those key messages over an extended period of time. The WMWD staff focused public messaging on efficient water use would lead to lower water bills. Each household, commercial, and irrigation site was provided a water budget allocation that fit their specific needs, including the number of residents, the size of the property/irrigated area, the weather, and any special needs for water. All customers have the same incentive to use water efficiently. Water budgets change as weather changes and were based on state water efficiency guidelines. WMWD is "passing through" state legislation on water use efficiency directly to customers via a water budget rate structure. Only those customers who waste water pay the high increasing tier charges (the marginal cost of new water supplies). Customers can apply for an "adjustment" to their allocation in a simple process with the district's customer service department should conditions change on their site. Any "excess" water revenue (from high-tier, over-allocation water use) would be directly funneled into customer conservation programs for consumer use.

19.10 The Impact on Water Use Efficiency with the Implementation of Water Budget Rates

In the first quarter of implementation of water budget rates, WMWD calculated the reduction of water waste at 28 %, or the demand reduction in the three over-allocation use penalty tiers, as shown in Fig. 19.8. The implementation of water budget rates had a rapid and dramatic impact on consumer water use behavior, particularly of users who wasted water, the intended targets of any "conservation"-oriented rate structure.

In calendar year 2012, the first full year of the WMWD water budget rate structure, WMWD found that customers meeting allocations rose from 63 % (shown in financial rate modeling) to 83 % (in actual use; the costs of implementing the water budget rate structure were recovered within 6 months, and revenue stability of fixed revenues was maintained, while less water was sold and delivered to consumers (Figs. 19.9 and 19.10).

In calendar year 2013, the second year of water budget rates, WMWD fully met its operations and maintenance fixed cost obligations; WMWD received $1.9 million in "penalty" revenue specifically earmarked for increased conservation programs and used to offset wholesale water rate increases for all customers; and 85 % of customers met their water budget allocations.

Meter Size	=	3/4-in	Household Size	=	4 persons
Billing Period	=	August	Current Usage	=	31 hcf
Landscape Area	=	3,045 sq. ft.	Power Zone	=	1

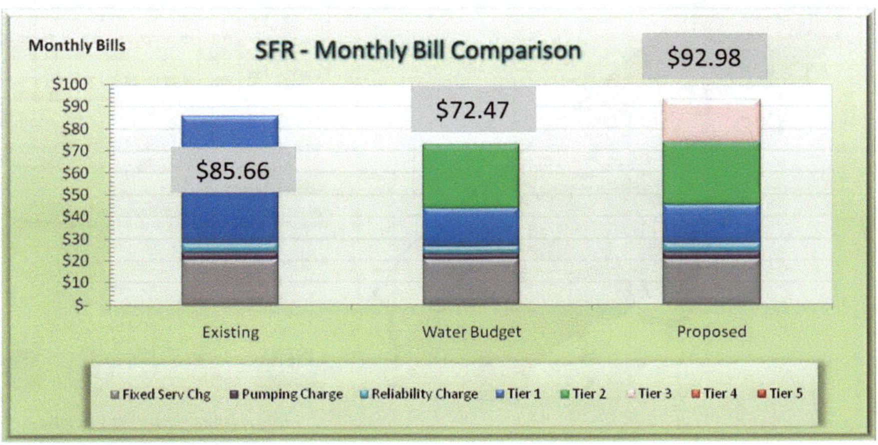

Fig. 19.8 Comparison of water bill cost for a high single-family residential (SFR) water user

Water Volume

PERIOD	TOTAL DEMAND	TIER 3	TIER 4	TIER 5	TOTAL PENALTY
1/12 – 3/12	4,660af	274af	172af	459af	905af
1/13 – 3/13	3,363af	146af	76af	193af	415af
	- 28%	- 47%	- 56%	- 58%	- 54%

- Tier 3 – Inefficient Water Use
- Tier 4 – Wasteful Water Use
- Tier 5 – Unsustainable Water Use

Fig. 19.9 Results in water use efficiency from the WMWD water budget rate implementation; reduction in use from one similar time period to the next year's same time period, weather normalized

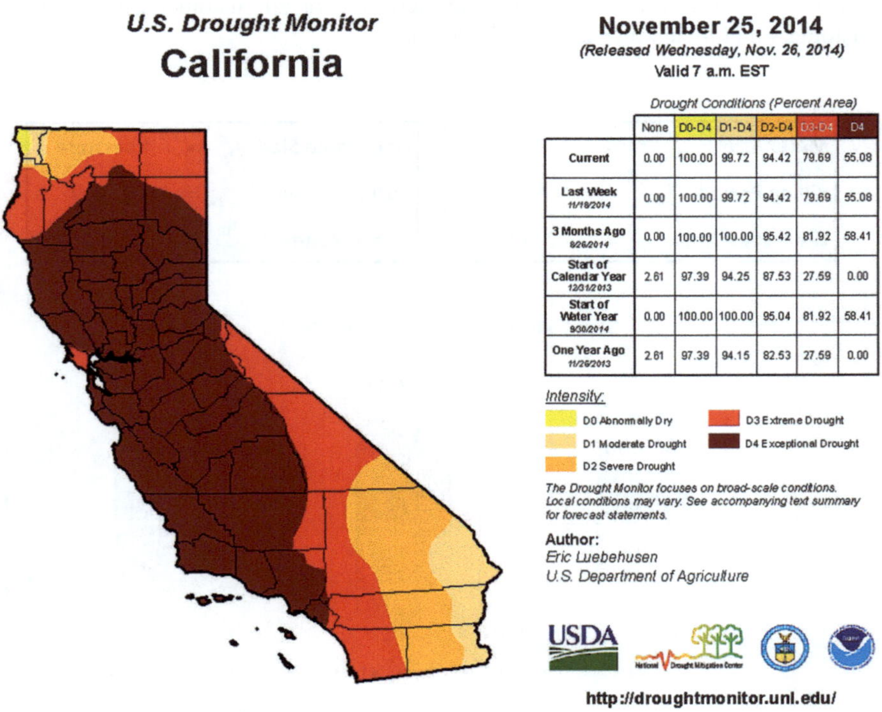

Fig. 19.10 Level of California drought (Source: US Drought Monitor: http://droughtmonitor.unl.edu/)

Other findings by WMWD with the implementation of water budget rates include:

Flexibility: Customers may "adjust" allocations as their water circumstances change (such as additional household members, medical or large animals, as per district policy). Individual allocations increase and decrease as weather changes. As well, should state water efficiency guidelines become stricter, reducing individual customer allocations can be done in the billing software without a "change" to the overall rate structure design.

Transparency: The costs of service (fixed charges) are more accurately attributed on the water bill for customers to see and the agency to recover independent of water sales. The cost of water, including the environmental (marginal cost of wasting water), is clearly displayed on the water bill in the form of "over-allocation use" or wasteful penalty charges. The relative water efficiency level of each customer is clearly displayed, with a corresponding economic message to conserve, each billing period (monthly).

Customer services: The need for customer service increases, in terms of supplying accurate information, directing customers to conservation programs to save water and lower water bills, and with the verification of customer level allocation data (re: number of residents, size of irrigated area). After an initial increase in customer service call volume in 2012, customer service call volume has returned to pre-rate structure change levels. Increased customer service ability, and the knowledge of individual account water budgets, is believed to be the key to having an informed constituency and overall customer acceptance of agency projects.

Drought response tool: The water budget rate structure provides a more equitable response to drought. The years 2011–2014 have been significantly dry years, with 2013–2014 being recorded as the third driest in history. WMWD is now able to use the water budget rate structure to lower individual allocations as a drought response (The State of California mandated emergency drought restrictions in July 2014), as opposed applying traditional drought restrictions that many users find inequitable.

19.11 Conclusions from the Implementation of Water Budget/Sustainable Water Rates by Western Municipal Water District

The implementation of a water budget rate structure by the Western Municipal Water District was intended and designed to deliver a wide range of new tools and benefits to both the customer and the agency, including (1) meeting state requirements for water use efficiency (AB 1881, SBX7-7); (2) establishing customer-by-customer allocations and identifying who is efficient and who wastes water; (3) providing a clear identification of water wasters (over-allocation users) and directing those users to conservation programs; (4) creating a new source of funding for efficiency programs, paid only by those who waste water; (5) directly linking customer efficiency with the rate charged for water (perceived fairness by customers); (6) directly linking the cost of buying more expensive water to meet higher demands and/or the cost of taking more water from the environment and end user behavior to meet wasteful use; (7) upgrading the billing system and increasing customer service (a certain cost, but also a practical and political benefit); (7) motivating customers to be more responsible with their water use on their own property, which WMWD views as the most cost-effective method of conservation; (8) providing more conservation program incentives and support (paid only by customers who waste water); (9) making water bills the main and most consistent educational tool the agency has with customers; and (10) having a method to measure water use efficiency and better manage local and imported water resources today and into the future.

19.11.1 Lessons Learned

While Western Municipal Water District was not the first agency to implement a sustainable rate structure, it is an "innovator" with the addition of a "drought factor" to the water budget billing system equations and with the level of modeling, financial, and political consideration included in the overall water rate change process. For example, the ability to "equitably" apply drought restrictions is considered a significant advancement to the ability of agencies, particularly in regions that experience drought and/or water shortages, to use the one communication tool that all customers see, that water bill.

Other lessons learned include (1) educate the elected officials and staff early and often; (2) include all the agency departments in the investigation and development of a new rate structure; (3) the vast majority of consumers do not know what an efficient use of water is (an individualized water budget presented on the monthly water bill provides this education or target); (4) consumers will make a rapid change given the proper economic signals, in this case the monthly water bill, water budget, and corresponding price for water waste; (5) conduct a financial and rate structure study and determine the most appropriate final rate design and charges BEFORE taking information out to the public; (6) accuracy of data (whether parcel by parcel irrigated area, number of residents, etc.) is very important to constituents and contributes to the credibility of the public water provider to be equitable and use a more sophisticated rate structure; (7) do not cut corners on collecting and testing data for accuracy; (8) recognize the importance to separate the recovery of fixed costs from water sales as much as can be politically feasible (this quickly results in more stable revenue recovery regardless of the fluctuations in water demands); (9) commitment by the agency at all levels is key to the success of a rate structure change, particularly from the policy-makers on down from top officials; (10) strategically planned public outreach is vital, punctuated with facts and simple messages that are logical to constituents; and (11) it is important to tell the real story of the costs of water and water service, the need to protect the environment by using water efficiently, and that every user has their own responsibility on their property in the process. The Western Municipal Water District director of finance cites today that "I was the biggest skeptic of water budget rates and changing the rate structure. Now, I am the biggest fan of water budget rates" (Personal interview 2013).

19.12 Epilogue

On July 15, 2014, the state of California Water Resource Control Board issued an emergency regulation for statewide urban water conservation, Resolution No. 2014-0038, due to severe drought. Included in the resolution is a "variance" for water providers that, as an alternative to prescribed emergency regulations and actions expected to be taken by agencies, the use of "allocation-based rate structures" that satisfies requirements may be approved, in conjunction with other measures, as it achieves a level of conservation that would be superior to that achieved by implementing limitations on outdoor irrigation (State Water Resources Control Board 2014: 7).

The recognition by the state of California, especially in times of severe drought, that the concept and history of the effectiveness of "allocation-based," "water budget," or "sustainable" water rate structures to send clear economic and environment signals to end users is testament to the logic, the linkage to other state legislation, and the success of such rate designs. Western Municipal Water District has freely shared its experiences in the investigation, design, and implementation of sustainable water budget-based rates. It is the desire of the district that this review of WMWD's process will assist any agency to ask itself the right questions, investigate successful solutions, and think outside of the traditional water rate structure box.

Note: for more information on sustainable water rates go to:

www.sustainablewaterrates.com
http://www.youtube.com/watch?v=qgLoHA4cer4
www.wmwd.com

References

Beecher, J. & Chestnut, T. *Declining water sales and utility revenue: A framework for understanding and adapting*. National water rates summit, Racine, Wisconsin, August 2012, Alliance for Water Efficiency.

California Department of Water Resources, California Irrigation Management Information System (CIMIS). (1984–2014). 30-year average rainfall and evapotranspiration rate. www.cimis.water.ca.gov

California Special Districts Association (CSDA). http://www.csda.net/special-districts/, Western Municipal established in 1954.

Mayer et al. (2008). Water Budget Rates — Innovative Management Tools. *Am Water Works Assoc J*, 117–131.

Metropolitan Water District of Southern California. (1997, September). *Evaluation of water budget based rate structures* (A&N Technical Report).

Personal interview, March 2013
Personal interview with WMWD's Kevin Mascaro, Director of Finance and Customer Service, July, 2014)
Proposition 218, California State Ballot Initiative. (1996).
Public Policy Institute website. (2014, September). http://www.ppic.org/main/publication_show.asp?i=1108, http://www.ppic.org/content/pubs/report/R_114EH2R.pdf. Mount, Freeman & Lund
Special districts are a form of local government created by a local community to meet a specific need. When residents or landowners want new services or higher levels of existing services, they can form a district to pay for and administer them. Nearly 85% of California's special districts perform a single function such as sewage, water, fire protection, pest abatement or cemetery management.
State Water Resources Control Board, Resolution No. 2014-0038, Emergency regulation for statewide urban water conservation, July 15, 2014, pg. 7 (2); www.swrcb.ca.gov
Steinberg, D. State of California Water Conservation Bill of 2009 (SBX7-7). Senate Bill No.7.
The Farrell Group. (1996, December). Irvine Ranch water district, voice of the customer study.
Water Conservation Bill of 2009 (Senate Bill X7-7, 20x2020); Association of California Water Agencies; http://www.acwd.org/DocumentCenter/View/23
Wikipedia. http://en.wikipedia.org/wiki/Marginal_cost. In economics and finance, marginal cost is the change in the total cost that arises when the quantity produced has an increment by unit. That is, it is the cost of producing one more unit of a good.

Chapter 20
Pricing Urban Water Services in the Developing World: The Case of Guayaquil, Ecuador

Abel Mejía, Jose Luis Santos, Daniel Rivera, and Germán E. Uzcátegui

Abstract Guayaquil, the largest city in Ecuador, offers a pragmatic and successful approach to pricing urban water services in the developing world. This chapter discusses the underlying principles and lessons learned to finance operations and investments from tariff revenue and subsidies, under a 30-year concession contract awarded in 2001. It reviews price-adjustment mechanisms to account for inflation and meet investment and service targets. It presents strategies followed to cover financing shortfalls to meet poverty and environmental goals. Finally, it summarizes strategic recommendations for other cities of the world.

Keywords Ecuador • Stormwater • Drainage • Public-private partnership • Concession

20.1 Introduction

Until February 1996, Guayaquil's water supply services were provided by a provincial water utility (Empresa Provincial de Agua Potable de Guayas), comprising 11 municipalities. Sewerage services were provided by a municipal utility (Empresa

A. Mejía (✉)
CAF Latin America Development Bank,
6024 Stonehenge Place, Rockville, MD 20852, USA
e-mail: abel.consultant@gmail.com

J.L. Santos
ECAPAG Empresa Cantonal de Agua Potable y Alcantarillado de Guayaquil,
Avenida Francisco de Orellana s/n y Miguel Alcivar, Guayaquil, Republica del Ecuador

D. Rivera
CAF Latin America Development Bank, Calle 50 y Calle Santo Domingo Edificio Global Bank Piso 40, Oficina 4001, CP 0832 Ciudad de Panamá, Panamá

G.E. Uzcátegui
UCAB Universidad Católica Andres Bello, Avenida Principal Las Esmeraldas Edificio Santa Marta, Apartamento 301-A La Tahona, Baruta Estado Miranda,
Caracas CP 1083, Venezuela

A. Dinar et al. (eds.), *Water Pricing Experiences and Innovations*,
Global Issues in Water Policy 9, DOI 10.1007/978-3-319-16465-6_20

Municipal de Alcantarillado de Guayaquil). These two utilities were merged into a new company (Empresa Cantonal de Agua Potable y Alcantarillado de Guayaquil—ECAPAG), incorporated under public law, with independent administrative, financial, and operating authority. The new company was established with a broad legal mandate to allow transferring its operational and investment responsibilities to the private sector.

ECAPAG inherited a low-performing system with 70 % of water losses, 55 % of past-due invoices, and four times the number of employees needed to manage both systems. Although the water supply system had a sufficient production of potable water, 40 % of the population (mostly low-income residents) received only 3 % of the system output. A number of treatment plants and pumping stations were inoperable, due to poor maintenance (Interamerican Development Bank 2001). The sewerage network had multiple breaks and cross connections with the stormwater drainage network. At the time, capital investments to update the systems to increase storage of drinking water and rehabilitate and expand the distribution network, as well as to conduct a major overhaul of the sewerage system, were estimated at US$700 million (Interamerican Development Bank 2005).

On October 2000, the national government authorized ECAPAG to delegate the provision of water and sanitation services of the city of Guayaquil to the private sector, through a concession contract. It also converted ECAPAG into the regulator of the contract (República del Ecuador 2000). Consequently, a 30-year concession contract to provide water supply, sewerage, and stormwater services was signed on April 2001 between ECAPAG on behalf of the city of Guayaquil and INTERAGUA as the concessionaire.

The concession contract guarantees INTERAGUA the exclusive right to provide and charge for water supply and sewerage services within a defined perimeter. According to the contract, ECAPAG remains responsible for services and owner of the assets affected by the concession. The operational perimeter comprised an estimated population of about three million in 2001, which was expected to grow to almost four million by the end of the contract in 2031 (UN Habitat 2009).

From the outset, a key policy decision was to establish clear and realistic goals for the first 5-year term of the concession. It included the installation of 55,238 water and 55,238 sewerage connections, equivalent to about 10 % of the population. A new connection is accounted for when it is registered in the commercial cadaster following preestablished standards, and it is externally audited. The contract established the obligation of the concessionaire to prepare a master plan every 5 years, starting in the sixth year of the concession, to achieve service goals on water supply, sewerage, and stormwater drainage (ECAPAG 2005).

To recover operational expenses and investments, and to have an adequate return on capital, INTERAGUA is entitled to tariffs applied to water supply and sewerage services, stormwater drainage fees associated with property taxes, a portion of taxes charged to telephone services in the city, and ad hoc contributions to investments that can be directly allocated to beneficiaries of improvements (ECAPAG 2006).

20.2 Financing Investments in Guayaquil

Revenue to finance investments and operations originates from water and sewerage tariffs, municipal subsidies, subsidized loans from the national governments, and a surcharge to finance urban drainage. To illustrate the composition of financing, the contribution to investment from different sources for year 2013 is shown in Table 20.1.

Investment programs for each 5-year period are guided by a master plan, starting from the sixth year of the concession. Operating costs are also updated every 5 years, but the contract allows for interim reviews on exceptional circumstances. In addition, to account for inflation, tariffs are indexed and quarterly adjusted. Investments from year 6 to year 30 of the concession were estimated at US$ 520 million, in constant terms at the date of the signing of the contract.

A key feature of the contract consisted of establishing a financial contribution to investments from service tariffs from the beginning of the contract. It was set at US$ 102 million for each of the 5-year periods until the end of the contract; but the distribution from year to year can be adjusted to better fit financial flows. For period 2011–2016, tariff revenue to finance annual investments was set at US$ 20.5 million.

Another interesting component of the contract is a cross subsidy from telecommunications to the water supply and sanitation sector, which is sizeable for the largest cities of the country. This subsidy is supposed to be transferred to finance investments under the concession contract (ECAPAG 2006). However, from year 2010, the municipality fixed the transfer at US$ 30 million per year.

Investments are also financed through subsidized credit lines of the government of Ecuador to the water supply and sanitation sector. In 2009, the government launched an ambitious investment program to reach 95 % water supply and 95 % sewerage coverage, respectively, by year 2019. The national development bank (Banco del Estado—BEDE) executes this program with subsidized interest rates and a grant component of nearly 50 % of the investment. To be eligible for this subsidy, a public institution should execute investments (BEDE 2014). To take advantage of such a generous program, the municipality of Guayaquil decided to request EMAPAG[1] to request the loans and execute investments, which were under the contractual responsibility of INTERAGUA.

Table 20.1 Sources of financing for investments 2013

Concepts	US$/year (million)
Water tariffs	20.5
Municipal subsidy	30.0
Drainage surcharge	3.5
Total	54.0

Source: EMAPAG (2014a)

[1]The name ECAPAG was changed to EMAPAG to reflect the municipal ownership of the company.

Table 20.2 Surcharge to finance stormwater investments

Monthly water consumption (m^3)	CEM (US$/month)
From 1 to 15	0.25
From 16 to 30	0.50
From 31 to 60	1.80
From 61 to 100	4.00
From 101 to 300	6.00
From 301 to 2,500	15.00
From 2,501 to 5,000	50.00
Larger than 5,001	120.00

Source: EMAPAG (2014a)

These investments are tendered by EMAPAG under engineering designs developed by INTERAGUA, which is also responsible for supervising the implementation of investments to ensure quality. The use of the line of credit managed by BEDE is considered equivalent to anticipating investments required from the concessionaire. Payment of the obligations under the BEDE loan is honored with the proceeds of the US$30 million municipal subsidy that was mentioned above. The impact of such investments should be neutral to the concessionaire, and as such, it is reflected in the corresponding 5-year reviews of operational costs and profitability.

The drainage surcharge (Contribución Especial de Mejoras—CEM) is a surcharge to finance stormwater investments included in the water bill. Since the concession contract did not provide for any other contributions for stormwater drainage investments, INTERAGUA and EMAPAG agreed to generate revenue for this purpose, according to the water consumption schedule in Table 20.2.

To maintain the financial equilibrium of the concession, cost increases or changes in the investment program should be financed from sources that are different from tariff revenue, like municipal subsidies, while ensuring neutrality to income and profits of the concessionaire.

Finally, if INTERAGUA fails to comply with the investment program, it will be penalized by EMAPAG, and the value corresponding to the unrealized investments are to be deposited by INTERAGUA in a trust account (Trust for Recovery of Investment), which is jointly managed by EMAPAG and INTERAGUA, and either one of them is entitled to build the missing investment.

20.3 Pricing Urban Water

The tariff offered by INTERAGUA to win the concession contract was 0.23 US dollars per cubic meter of drinking water supplied and metered at an individual connection. Sewerage services are calculated at 80 % of the water bill. However, in the meantime, Ecuador adopted the US dollar as its currency, which had an effect on the proposed cost structure of the concession. It led to an adjustment of the tariff to 0.403 US$/$m^3$, at the beginning of the contract, in August of 2001.

The financial model to calculate water prices is applied from the second 5-year period of the concession, and it is based on the agreements reached between the utility and the regulator on the basis of the diagnostic and investment planning included in the master plan for the period. During the first 5-year period, the tariff was fixed, and the contract was awarded to the bidder offering the largest number of connections of water supply and sewerage, under the quality and standards established in the contract.

The financial model calculates the reference rate, which is a unit value per cubic meter of drinking water, which ensures the recovery of the full costs of services under normal conditions. The reference rate that has been applied by December of each year from the beginning of the concession is shown in Table 20.3. Variable rates are computed from the reference rate by applying adjustment factors in such a way that rates for larger consumers cross-subsidize smaller consumers. Sewerage connections pay 80 % of the variable charge of the water supply. Stormwater service is charged in water bills as a fixed amount related to levels of water consumption.

The tariff structure does not differentiate between categories of users, only ranges of consumption. The two lowest ranks of consumption correspond to less than 30 m^3/month, comprising about 80 % of all accounts, and they are cross-subsidized by the large consumers.

Tariffs are reviewed every 5 years, but there is a quarterly adjustment formula to account for inflation, based on the consumer index (Ip), the cost of electric energy (E), and staff costs (Ru), as follows:

$$\mathrm{Tr} = \mathrm{FACT} \cdot \mathrm{Tr}_0$$

where Tr is the domestic water tariff for the next quarter, Tr_0 is the domestic tariff at the beginning of each 5-year term, and FACT is an adjustment factor calculated with the following formula:

$$\mathrm{FACT} = 0.60 \cdot \frac{\mathrm{Ip}_1}{\mathrm{Ip}_0} + 0.20. \frac{E_1}{E_0} + 0.20 \cdot \frac{\mathrm{Ru}_1}{\mathrm{Ru}_0}$$

The periodic review of tariffs and other charges is directly associated with the expansion and service goals for the next 5-year term. To facilitate the decision-making process, EMAPAG and INTERAGUA should follow a well-defined and time-bound process to validate the approval process of a new tariff.

To initiate this process, the regulator (EMAPAG) prepares terms of reference containing the minimum requirements to conduct a study to jointly review with the concessionaire (INTERAGUA) the investment program and service and efficiency goals for the next 5-year term. The study is prepared by an independent consultant, hired by INTERAGUA, to determine the tariffs that would be required to finance the operational and investment costs to achieve the proposed service goals. The study should consider affordability criteria for low-income users.

Table 20.3 Reference rate (US$/$m^3$ of metered water supply

Year	2001	2002	2003	2004	2005	2006	2007	2008	2009	2010	2011	2012	2013
Tariff US$/$m^3$	0.467	0.512	0.538	0.541	0.572	0.471	0.481	0.524	0.535	0.544	0.563	0.574	0.585

Source: EMAPAG (2014a)

The contract has practical mechanisms to resolve disputes about the tariff-setting process and other possible controversies. Specifically, it establishes deadlines for review and response by each one of the parties. Disputes should be addressed in direct negotiations, followed by a formal mediation process. If mediation is not successful in a period of 30 days, disputes are sent to arbitration following the rules of the International Chamber of Commerce of Paris at the time of the controversy. The arbitration itself should take place in Miami, Florida, and it will be conducted in Spanish following the laws of Ecuador. The arbitration tribunal consists of three members: one is designated by each party, and the third by the president of the International Chamber of Commerce. The decision of the tribunal is final and cannot be appealed.

In 2001, to protect the contract against the risks of expropriation, war, and civil disturbances of the investment made by the winner of the concession contract, the Multilateral Investment Guarantee Agency of the World Bank (MIGA) provided an US$ 18 million guarantee (The World Bank 2006). It also covers a performance bond guaranteeing the concessionaire's successful management, expansion, and operation of the water services. The bond was posted by the concessionaire in accordance with the 30-year term of the contract awarded by the government.

The methodology to review tariffs and other service charges is then conducted in four successive stages: (1) for each of the 5 years of the concession period, the total costs in constant US dollars will be projected; (2) for each of the 5 years of the period, the total revenues in constant US dollars will be projected; (3) a calculation will be made of the real change in income required during the period; and (4) modifications are applied to the various components of the structure of prices and tariffs until equilibrium is reached. Modifications are calculated by adjusting the reference tariff and the collection rate, multiplied by the cubic meters of water invoiced, until an equilibrium of total costs is reached.

20.3.1 First Stage

Total costs, CT_i, for each year should be the sum of the costs of operation and maintenance, taxes and levies, depreciation charges, and the cost of equity applied to the net value of investment in that year. These costs are calculated according to the following formula:

$$\mathrm{CT}_i = \mathrm{O\,\&\,M}_t + \mathrm{IMP}_t + \mathrm{Depr}_t + \left(\mathrm{CC}_t \cdot \mathrm{IN}_t\right)$$

Operation and maintenance (O&M) costs mean all the operational costs associated with the provision of the service, including the regulatory fee, except for the bad debts, which will not be considered at the total costs for the revision of the rates. These expenses will be calculated according to projections made by the independent consultant.

Taxes and levies (IMP) refer to the net of all taxes and levies to which the concessionaire is legally bound. Depreciation charges (Depr) refer to the application of a rate of depreciation to finished assets and works in process as provided in the contract. Net investments (IN) mean the sum of the book value at the date of revision of all the investments made and in process net depreciation, plus the value of the new investments programmed for the 5-year term, in constant, and net of the depreciation projected for the period.

The amount of investments to be carried out for each year of the period will be based on studies of the independent consultant using costs obtained from national and international benchmarks of projects of similar characteristics, and the cost of capital (CC) is calculated according to the following formula:

$$CC_t = 0.65 \cdot CD(1-\tau) + 0.35 \cdot CAC$$

where CD is the cost of debt in Ecuador expressed in US dollars, defined as the weighted average cost of the debt to the concessionaire in the last 3 years, τ is the tax rate on earnings in Ecuador at the time of the tariff review, and CAC is the cost of equity to an operator of water, which is calculated as follows:

$$CAC = TSR_E + \beta \cdot PRM \cdot \frac{TSR_E}{TSR_{USA}}$$

where TSR_E is the risk-free rate in Ecuador, which is equivalent to the average rate of sovereign debt of Ecuador in the preceding 12 months; TSR_{USA} is the risk-free rate in the United States of America, which is equal to the average of 10-year bond of the American Treasury; PRM is the risk premium of the capital market in the United States, equivalent to the average equity risk premium[2]; and β is the arithmetic average of unleveraged and leveraged coefficients of the concession as established in the contract (debt equal to 65 % and equity equal to 35 %) for the previous year, as established for water companies in the United Kingdom whose shares are traded on the London Stock Exchange and form FTWATR index, published by Reuters.

20.3.2 Second Stage

An independent firm hired under terms of reference and approved by the regulator estimates the total revenue that would be raised by the concessionaire for the next 5-year term, based on the tariff at the date of review, considering a collection rate greater than 80 %. INTERAGUA, however, should present another study supporting its estimate of the collection rate, which is prepared by an external auditing company that should analyze the performance of the concessionaire in the last

[2] It is the yearly average of Intermediate Horizon Equity Risk Premium, as published by the Yearbook, Valuation Edition of Ibbotson Associates (table A-2). This table reflects the risk-free in the US utilizing indexes published by S&P 500.

2 years to improve the collection rate as part of their business. This should include efforts and procedures to improve the collection rate.

20.3.3 Third Stage

Once the total costs and revenues for the next 5-year term are projected, the change in income required in the first year of the period to ensure a flow of revenue whose present value equals the present value of total costs is calculated by applying the following formula, to keep the economic-financial balance of the concessionaire.[3]

$$\Delta I = \sum_{t=1}^{5} \frac{\mathrm{CT}_t}{(1+r)^t} - \sum_{t=1}^{5} \frac{I_t}{(1+r)^t}$$

where CT_t is the total costs for year t (it is the total revenue for year t), r is the discount rate (equivalent to CC), and ΔI is the change of the present value of the income required for the period. The change in present value of revenue is made in the first year of the period. During the remainder of the period, the price of water is adjusted on a quarterly basis, according to the formula of automatic adjustment provided in the regulation of the water rate structure. EMAPAG is entitled to modify the water rate structure on the occasion of the periodic review, to ensure its progressive alignment with the goals contained in the contract.

20.3.4 Fourth Stage

The last step of the periodic review is translating to the water rate structure the change in present value of revenue[4] required for the 5-year term. This will be determined by equating the revenue estimated for the period with the revenue generated, multiplying each one of the charges of the rate structure by the corresponding consumption of water, plus the product of each of the fixed charges by the corresponding number of users, plus other revenue. This is accomplished by applying the following formula:

$$\sum_{t=1}^{5} \frac{I_t}{(1+r)_t} + \Delta I = \mathrm{TC} \cdot \sum_{t=1}^{5} \frac{\sum_{m=1}^{N} c_n v_{nt} + \sum_{m=1}^{n} x_m u_{mt} + \sum_{p=1}^{P} a_p w_{pt} + O_t}{(1+r)^t}$$

[3] The profit of the concessionaire included in the total cost is calculated with the weighted average cost of capital (WACC) applied to net investments at t period ($\mathrm{CC}_t \times \mathrm{IN}_t$).

[4] Revenue refers to the amount billed and not income after discounting uncollected revenue.

where It is the revenue projected for year t, r is the discount rate (equivalent to CC), ΔI is the change in present value of revenue required for the whole period, TC is the collection rate projected for the entire period, c is the variable charge of the tariff structure, v is the volume corresponding to each charge of the rate structure, N is the number of variable charges in the water rate structure, x is the fixed charge in the water rate structure, u represents the users (customers) corresponding to each charge, M is the number of different fixed charges, a is the charge for sanitary sewers, w is the volume invoiced for sanitary sewerage services, P is the number of charges for sanitary sewerage, and O represents charges for other services.

20.4 Extraordinary Price Reviews

Extraordinary reviews of water tariffs and other charges would be allowed in the event that the concessionaire experiences a variation of significant magnitude in their total costs, relative to what was planned during the last periodic review of rates and other charges for services. It will also consider a significant variation of demand and the potential impact of environmental legislation.

The extraordinary review will only be allowed if the variation in total costs occurs for any of the following reasons or events: expansion systems not considered in the program for the 5-year period, change in tax regime, change in the legislation governing the quality of drinking water or wastewater, and force majeure or unforeseen circumstances, provided that damage was not averted by the concessionaire using due diligence and care required by the circumstances.

The variation of the total costs is considered significant when the sum of the annual present value of these costs (until the next periodic review date) exceed more or less 10 % of the sum of the present value of the annual total costs projected at the beginning of the period and up to the same date of periodic review, according to the following formula:

$$\sum_{t=t_0}^{T} \frac{\Delta \mathrm{CT}_t}{(1+r)^t} > 0.10 \cdot \sum_{t=t_0}^{T} \frac{\mathrm{CTP}_t}{(1+r)^t}$$

where $\Delta \mathrm{CT}t$ is the change in total cost from year t, CTPt is the total cost projected at the beginning of the period to year t, r is the discount rate (equivalent to CC), t_0 is the current year, and T is the year for the next periodic review.

Once EMAPAG approves the circumstances that would allow for a special review, a tariff modification will be implemented that has the same present value as the variation in total costs until the next periodic review. In case the real change in revenue is negative, instead of a reduction in the rate, EMAPAG may deduct this amount from the total costs of the concessionaire in the following periodic or extraordinary revision of rates. The change will be calculated using the following formula:

Table 20.4 Tariff structure as of December 2013

Variable charge			Fixed charge	
Interval of consumption m³	Number of accounts	Tariff US$/m³	Diameter	US$/month
0–15	267.326	0.3	1/2″	1.28
16–30	140.575	0.45	3/4″	8.57
31–60	55.533	0.64	1″	22.04
61–100	8.783	0.83	1 1/2″	36.73
101–300	3.745	0.92	2″	36.73
301–2,500	923	1.4	3″	61.23
2,501–5,000	71	1.78	4″	183.67
5,000 or more	38	2.89	6″ or more	244.9
Total	476.904			
Reference rate by December 2013 of metered drinking water: 0.585 US$/m³				
Sewerage tariff is equivalent to 80 % of the variable charge for water supply				

Source: EMAPAG EP (2014b)

$$\Delta I = \sum_{t=t_0}^{T} \frac{\Delta \text{CT}_t}{(1+r)^t}$$

ΔI: change in present value of revenue for the remainder of the period
ΔCT_i: change in total costs for the year t for accepted causes and magnitude
T: year of the next periodic review
r: discount rate (equivalent to CC)

Table 20.4 presents the reference rate and the tariff structure as of December 2013 (EMAPAG EP 2014b).

20.5 Achievements After 12 Years of Concession

During the period that elapsed since the award of the concession in 2001 through December 31, 2013, water supply, sewerage, and stormwater services in the city of Guayaquil had improved significantly (Mejia et al. 2013). The key underlining policy principle behind this successful experience is that tariffs and financial resources have been made available on a timely basis to ensure the financial equilibrium of the concession. Pricing policies have been implemented through well-defined procedures and methodologies for tariff setting with a review every 5 years, which is based on a master plan. The master plan supports a 5-year investment program associated with realistic quality and performance goals for the provision of each service.

The most relevant operational outcomes during this period are the following: (1) expanding the coverage of drinking water from 63 % in 2001 to 100 % in the year 2012, considering the limit of urban area established by the municipality of Guayaquil, and expanding sewerage coverage from 50 % in 2001 to 90 % in 2013; (2) monitoring compliance to water quality standards (certified laboratories collect samples of raw water and potable water in the distribution network and in households, all according to procedures established in the contract); (3) continuity of water supply (the central and south zones of the city that received water for a few hours a day now receive it 24 h a day); and (4) improved operational flexibility of the water distribution network and better control of water leakages by creating 81 macro and 783 micro sectors. It also helps to control pressures and improve efficiency of electricity consumption. However, while water leaks measured as non-invoiced water have decreased from 72 % in 2001 to 58.3 % in 2013, there is still much to do to reach the level of best utilities in the region, averaging 25 %(EMAPAG EP 2014a, b).

20.6 Lessons Learned

Guayaquil demonstrates that public-private partnerships represent a valid public policy option to improve the provision of urban water services in the developing world. It also shows that water supply, sanitation, and stormwater services can be successfully integrated in one single contract with one operator. Moreover, policy reforms can be implemented on a difficult country environment if there is strong resolve and leadership of local authorities. To be successful, however, countries need several concurrent conditions; among them are well-designed contracts and the adoption of sound pricing principles. The main lessons for success can be extracted from this case, including:

1. Commitment to policy reforms. Guayaquil enjoys a strong leadership at the municipal level with enviable consistency for more than two decades, which include galvanized social support and political commitment to improve water supply and sanitation services.
2. A well-designed contract. The design of the contract reflects a realistic balance of risks between the public and the private sector, including well-defined procedures to set and review pricing of services and resolve disputes quickly and effectible. In addition, it provided adequate checks and balances to account for noncompliance issues and sovereign risks. The contract is pragmatic to establish realistic goals and investments, particularly for the first 5-year period of the concession.
3. A strong and capable regulator. EMAPAG is technically competent and has an adequate budget to operate. It is a single utility regulator with a special status within the municipal government of Guayaquil. It manages a professional and constructive relationship with the concessionaire, which has allowed to successfully resolve differences and impasses.
4. Long-term planning. Reliance on a comprehensive master plan for water supply, sewerage, wastewater treatment and disposal, and stormwater provides a diagnosis and the analytical basis for short-, medium-, and long-term investments and actions.

References

Banco del Estado. (2014). Banco del Estado. In B. d. Estado (Ed.), *Recuperado el Mayo de 2014*, de Banco del Estado: http://bancoestado.com

ECAPAG. (2005). *Historia de ECAPAG. Recuperado el 21 de March de 2014*, de ECAPAG: web.archive.com

ECAPAG. (24 de agosto de 2006). Acta de Sesión Ordinaria de Directorio de ECAPAG No. 23/206.

EMAPAG EP. (2014a). *Rendición de Cuentas Ejercicio Fiscal 2013*. Obtenido de EMAPAG: http://www.emapag-ep.gob.ec

EMAPAG-EP. (2014b). *Dirección de Regulación y Control Técnico. Informe de Labores*. Guayaquil: EMAPAG-EP.

IADB. (11 de Abril de 2001). *Contrato de Concesión de los Servicios Públicos de Agua Potable y Saneamiento de la Ciudad de Guayaquil, celebrado entre la Empresa Cantonal de Agua Potable y de Alcantarillado de Guayaquil (ECAPAG) e International Water Services (Guayaquil) INTERAGUA C. LTDA*. Washington.

INTERAGUA. (2006). *Plan Maestro de Agua Potable, Alcantarillad Sanitario y Alcantarillado Pluvial*. Guayaquil: Unpublished.

INTERAGUA. (2012). *Ajuste del Plan Maestro Agua Potable; Alcantarillado Sanitario y Alcantarillado Pluvial*. Tomo I. Guayaquil: Inédito.

Interamerican Development Bank. (2005). *Concesión de Agua y Alcantarillado: Guayaquil*. Project Completion Report. Washington: Unpublished.

Mejia, A., Nucete, M., & Quintero, A. (2013). *Análisis Cualitativo de Diez Proyectos y Programas, y Propuesta de Mejoras Operacionales*. Caracas: Unpublished.

República del Ecuador. (2000). Decreto Ejecutivo No. 872. Registro Oficial No. 186, 11.

The World Bank. (2006). Project appraisal on a proposed adaptable loan in the amount of US$ 48 million to the Republic o Ecuador for the second rural an small towns water supply and sanitation project (PAGUAS II) Report No: 36579-EC. Washington: The World Bank.

UN HABITAT. (2009). *Global report on human settlements*. London: Earthscan.

Chapter 21
Price for Domestic Water Supply: An Innovative Method Developed for the Tucano Aquifer in the State of Bahia, Brazil

Raymundo José Santos Garrido

Abstract This chapter analyzes the pricing of bulk water extracted from the Tucano aquifer in the semiarid region of Bahia, Brazil, using the optimizing economic behavior agent model. The starting point is a set of demand and supply equations on groundwater that is pumped from the aquifer and used for domestic supply. The main goal of this chapter is to offer bulk water tariff levels through a methodology especially adequate to a region that, due to the scarcity of this natural resource as well as the level of poverty that characterizes the region, demands more and more application of mechanisms that contribute to the efficiency of its use, while ensuring adequate prices to be paid by poor families.

Keywords Brazil • Conveyance cost • Economic prices • Price elasticity of demand • Tucano aquifer

21.1 Introduction

This chapter addresses the issue of bulk water pricing in the region of the Tucano aquifer for multiple uses, considering the domestic use as a priority, which is a practical application, based on the process of price optimization developed in a previous research by the Federal University of the State of Bahia (UFBA).[1]

The work leading to this chapter is part of the Project "Water Resources Economics" founded by "Grupo de Estudos de Relações Interssetoriais – GERI" at the Federal University of Bahia – Brazil. Various components of the chapter were included in various reports in Portuguese by that Project.

[1] Professor José Carrera-Fernandez developed the price optimization methodology of bulk water source applied in this chapter.

R.J.S. Garrido (✉)
Alameda Umbuzeiros, 347, Caminho das Árvores, Salvador City,
Estado da Bahia CEP 41.820-680, Brazil
e-mail: raymundojosegarrido@gmail.com

A. Dinar et al. (eds.), *Water Pricing Experiences and Innovations*,
Global Issues in Water Policy 9, DOI 10.1007/978-3-319-16465-6_21

The objective is the use of bulk water from the Tucano aquifer, taking advantage of the elements from an enterprise conceived and executed by the water resource and environmental engineering company of the state of Bahia (Companhia de Engenharia da Bahia—CERB).[2] This enterprise consists of the abstraction, reserving, and adduction to delivery points of bulk water in a semiarid perimeter in the state of Bahia that was formed by municipalities that receive water from the Tucano aquifer. This chapter analyzes the tariffication[3] of bulk water for Phase 1 of the Northeast Block of the mentioned perimeter, which was recently built and is already operating.

In using the elements of the enterprise mentioned earlier, the author proceeded by adapting a hypothetical simulation of it being operated by a private company, which introduced profit in the exploration of the activity. The intention of this hypothesis is not to compare public and private management in water projects but rather to present a reality of mixed companies that work with the water supply, gain profit, and demonstrate annual results that are made available to the public.

The scenario assumed here corresponds to the part of the region identified in Fig. 21.1, which represents the region where the enterprise is located and the part of it that is subject to the pricing exercise. This part was the objective of Phase 1 of the Northeast Block of the aquifer region. It includes municipalities, whose territories overlap a fraction of the Tucano aquifer. The hydrogeological formation of the Tucano aquifer is of sedimentary nature, which means that the underground reservoir has an abundant storage and excellent quality. Furthermore, it has the advantage of being protected due to the depth to which the water percolates. The depth in the case of this aquifer is almost always over 200 m. The water would not need any treatment for domestic use, only the simple disinfection that is recommended by the health authorities.

Phase 1 encompasses the municipalities of Banzaê, Cícero Dantas, Fátima, Heliópolis, Adustina, and Paripiranga, the latter of which borders the state of Sergipe. From hereafter, Phase 1 will be referred to as "Tucano Project" or "supply system" or "system" or simply "enterprise."

The innovative character of this enterprise proceeds from various aspects of its concept. Initially, the greatest outstanding fact is that CERB, the owner of the enterprise, is anticipating the availability of a high-quality water infrastructure for the development of a region that, besides lacking this natural resource due to its being a part of the semiarid region, is immersed in a poor area of low levels of HDI (human development index).[4]

---

[2] The author expresses his gratitude toward the water resource and environmental engineering company of the state of Bahia (CERB) for the opportunity of using the engineering knowledge of the project for the purpose of setting prices.

[3] The term tariff is used in the place of public price to differentiate the water that is simply abstracted (object of public price) from water that, besides being abstracted, is reserved and adducted to delivery points by means of a service offered by the projects' enterprise (object of the tariff).

[4] The Northeast is the poorest region of Brazil and has been subject of successive plans for sustainable development. The "Projeto Áridas" (SEPLAN – PR, 1994) was one that dealt with broader scope the issue of water resources.

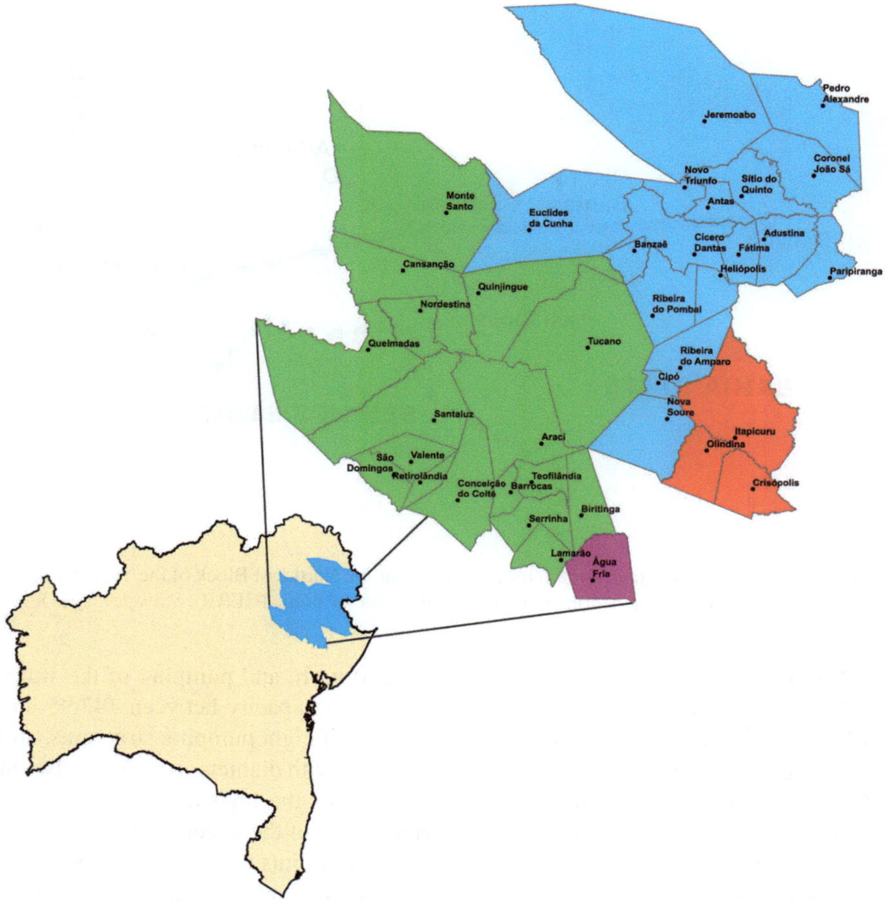

Fig. 21.1 Location of the water supply enterprise when it is at its final configuration (Source: Environmental Engineering Company of the State of Bahia (CERB). Salvador. 2011)

21.2 Brief Notes on the Enterprise Project

The bulk water supply project in the specified region (Phase 1 of the Northeast Block) is formed by hydraulic engineering work aimed to produce, at the end of 30 years of the plan, an outflow of 1,600 m^3/h (m^3/h) by means of eight wells operating 18 h a day. Its main elements are a groundwater abstraction structure by means of the mentioned eight wells that will be operating at the final configuration of the enterprise, in which four will be in the municipality of Banzaê and four in the municipality of Cícero Dantas. The abstracted water will be conducted by an aqueduct that goes from Banzaê to Paripiranga, crossing the territories of the municipalities of Cícero Dantas and Fátima and diverting in two directions with deviation points near Fátima, one to the northeast, toward Adustina, and the other to the southeast, toward Heliópolis. Figure 21.2 illustrates this path.

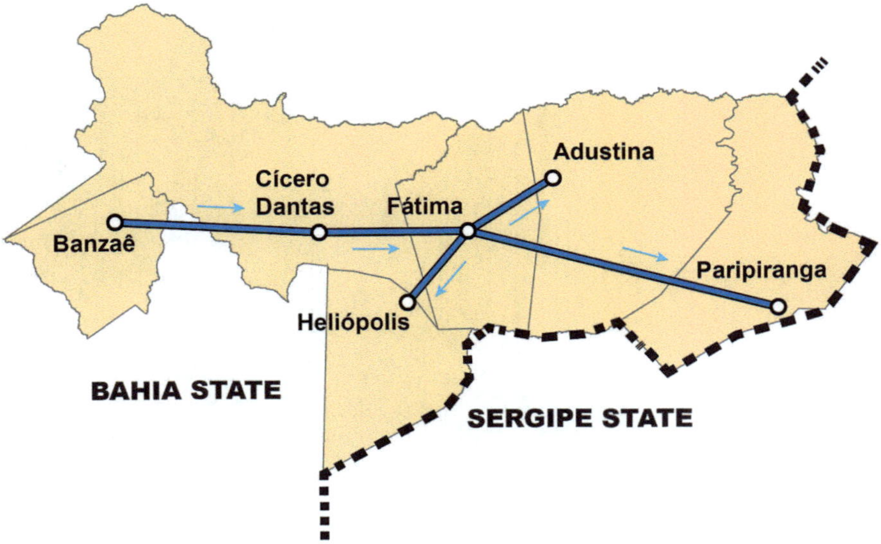

Fig. 21.2 Path of the aqueduct system for the Phase 1 of the Northeast Block of the Tucano Project (Source: Environmental Engineering Company of the State of Bahia (CERB). Salvador. 2011)

The well outflow varies between 150 and 250 m^3/h, and pumping of the water will be done by submerged pumps with a potential capacity between 147.95 and 246.58 horsepower (HP). The system is composed of eight pumping structures, and the total extension of the aqueducts is of 161.04 km, with diameters varying between 50 and 600 millimeters (mm). Its path accompanies the highways of the region, contributing significantly to cost-effective maintenance of the system. Interconnecting ramifications will be built to connect water delivery points to the distribution reservoirs that are located in rural area settlements. This is an additional characteristic that distinguishes the project, by allowing a capillarity in which the system can attend to demands of water users dispersed in rural areas, facilitating small farmers' access to water. The connecting ramifications are not a part of the project, since their purpose is to take water to what we have called the "delivery points," from which the users collect, treat, and transport to its final destination. In the case of urban distribution, the user is the water company that holds the water supply concession that will sell the service to the final user. In the case of rural distribution, the users will agglutinate in order to construct and operate the so connecting ramifications.

The storage system groups of reservoirs that will be located in the production centers are the following: the "lung" reservoirs located in strategic points of the aqueduct; the "delivery point" reservoirs, responsible for supplying a stretch of 10 km in each direction of the aqueducts in rural areas; and, finally, the treated water distribution reservoirs that are responsible for the water supply to different locations. The treated water distribution reservoirs are not a part of the project, which is essentially bulk water. The apparatus to treat and distribute water is part of the system served by the Tucano Project being, however, a responsibility of the urban supply company that is one of the users of the enterprise. In that respect, the Tucano Project is the wholesale enterprise, and the urban supply companies are the retailers. Finally, water flow gauges and shut-off valves for maintenance are added to the structures, among other complements of the engineering project.

Water will be used for domestic and agricultural purposes, as well as for animals' needs. The agroindustry that will use this water is small scaled, and irrigation, although recommended for semiarid regions in which soils are fertile, is not a part of this project due to the strong outflows it requires, which cannot be supplied.

21.3 Pricing Methodology

As already mentioned, the focus of this chapter is the pricing of water from the project, based on the economic optimization criteria. There are several methodologies for bulk water pricing (Seroa Da Motta 1988). Based on economic theory, the method applied results from the optimization study of an indirect function of social well-being, with the differentiation of prices derived from price elasticity of the demand for each type of water use, subject to the condition of providing, with the resources of this charge, the necessary funds to operate and maintain the enterprise.

This method has an advantage that, besides recuperating costs, it additionally prioritizes economic and distributional effects. The main variables used are operational and maintenance costs of the enterprise and those related to water outflow that will be produced and distributed in the enterprise's life cycle.

The method that was developed by Carrera-Fernandez[5] departs from the "second best" theory (Lypsei and Lancaster 1956–1957) and is based on the maximization of the difference between social benefits and costs and the minimization of distributive impacts on the economy, subject to the request to cover all the operational costs of the enterprise. The starting point is the recognition of the existence of a function of indirect utility of social well-being:

$$U = U\left(p, M\right)$$

$$\text{with}\frac{\partial U}{\partial p} < 0 \text{and} \frac{\partial U}{\partial M} > 0$$

Conditioned on the budget constraint of society,

$$M\left(p\right) = \sum_j p_j q_j\left(p\right) - \sum_j c_j\left[q_j\left(p\right)\right]$$

where:

"p" is the vector of economic prices.
"M" is the community's income, which depends on the vector of economic prices.

---

[5] In his reasoning, Carrera-Fernandez takes into account the proposition of Baumol, W. and Bradford, D. (1970) that prices that deviate from marginal cost may be required for an optimal allocation of resources, even in the absence of externalities.

"q_j" is the water outflow used and therefore the subject for charge.
"c_j" is the operational and maintenance cost of the system of production and delivery of bulk water.

First-order conditions under the given restriction of the indirect utility function yield

$$\frac{\partial U}{\partial p_j}+\lambda\left[p_j\left(\frac{\partial q_j}{\partial p_j}\right)+q_j-\left(\frac{\partial c_j}{\partial q_j}\right)\left(\frac{\partial q_j}{\partial p_j}\right)\right]=0,\forall j$$

where λ is the Lagrange multiplier that corresponds to the marginal use of income.

Based on Roy's identity $\left[\left(\frac{\partial U}{\partial p_j}\right)/\left(\frac{\partial U}{\partial M}\right)\right]=-q_j$ and with a few algebraic manipulations, this expression evolves to

$$-q_j\left(\frac{\partial U}{\partial M}\right)+\lambda q_j+\lambda q_j\left\{\left[pj-\left(\partial c_j\right)/\left(\partial q_j\right)\right]/pj\right\}\varepsilon_j=0,\forall j$$

Considering that the price elasticity of the demand of the good j is given by

$$\varepsilon_j=\left(\frac{\partial q_j}{\partial p_j}\right)\left(\frac{p_j}{q_j}\right)<0,$$

the previous equation may be presented in the following manner:

$$\left(p_j-CMg_j\right)/p_j=\varphi\left(\frac{1}{\left|\varepsilon_j\right|}\right),\forall j$$

where $\varphi=\frac{1-\left(\frac{\partial U}{\partial M}\right)}{\lambda}$ corresponds to the relative difference between marginal benefits and costs (Carrera-Fernandez 2001). From there, results show that the percentile price variation of water in the "j'th" use, in relation to its marginal cost, is inversely proportional to the price elasticity of demand.

In practical terms, the conditioned optimization can be represented by the following equation systems:

$$\begin{cases}\frac{\left(p_j-CMg_j\right)}{p_j}=\varphi\left(\frac{1}{\left|\varepsilon_j\right|}\right),\\ \sum_j p_j^* q_i \quad C=0\end{cases}$$

where:

p_j^* is the optimum price, the unknown of the problem, of water in the type of use j.
CMg_j is the marginal management cost in use "j."
φ is a proportionality constant that corresponds to the difference between the benefits and marginal costs of the project.
$|\varepsilon_j|$ is the price elasticity of demand for water in use "j."
x_j is the water outflow demanded in the basin based on investments made.
C is the total cost of management that in the case of the Tucano Project, it excludes the most part of amortization of investments needed according to the explanation in this text.

The practical application of the instruments above to the enterprise case being studied is presented below, and the price levels produced are supposed to signal water users to a behavior of efficient use of this natural resource, while promoting the internalization of social costs, reflecting the true opportunity cost for each water use in the scenario it is inserted in the enterprise, and assuring the financial sustainability of the project.

The following two sections present surveys performed by the project on outflow demand and cost figures needed for tariff calculation.

21.4 Water Demands

As mentioned before, the Tucano Project waters are supposed to be directed toward domestic use, water for animals, and small-scale agricultural industry. The demands have been estimated on an annual basis for the next 30 years, which is the lifetime of the project, and are presented per water use and totals in the first years of a series of 5 years and in the last year of the last series of 5 years in Table 21.1. The whole annual evolution of the complete series of 30 years of the lifetime of the project is not presented, due to lack of space in this text. However, in the last line of Table 21.1,

Table 21.1 Evolution of annual demand noting the first of every 5 years and the thirtieth year of the economic lifetime of the enterprise (m^3/year)

Annual period	Domestic use	Water for animals	Agric. industry	Total
1	988,128	296,175	29,565	1,314,000
6	2,593,836	777,460	77,608	3,449,250
11	3,952,512	1,184,702	118,260	5,256,000
16	5,249,430	1,573,432	157,064	6,980,625
21	6,916.896	2,073,229	206,955	9,198,000
26	7,905,024	2,369,404	236,520	10,512,000
30	8,893,152	2,665,580	266,085	11,826,000
Total (30 Years)	129,022,755	38,672,511	3,860,388	171,555,655

Source: Environmental Engineering Company of the State of Bahia (CERB). Salvador. 2011

the total demand per use and the grand total for the complete series of the 30 years of the project's horizon are mentioned. There might be small discrepancies in the summation of the values in several tables of this chapter due to truncation, when appropriate, of the values after the decimal point.

The demographic growth rate used to determine the evolution of demand for domestic supply was 1.5 % per year. In terms of water consumption indexes used to determine demand, the rate of 124 l/inhabitant/day (l/inhab/day) was used, which resulted from the unit consumptions of the following scales of population conglomerates[6]:

Populations of 1,000 inhabitants at the most	80 l/inhab/day
Populations between 1,000 and 2,000 inhabitants	100 l/inhab/day
Populations between 2,000 and 4,000 inhabitants	120 l/inhab/day
Populations above 4,000 inhabitants	150 l/inhab/day

In the case of watering animals, they were grouped in large and small scales. The large-scale animals include cattle, horses, and mules. The small-scale animals include swine, goats, and sheep. The large-scale animals present a usage index of 75 l/day of water and the small-scale use 17.50 l/day, average indexes that absorb the effects of seasonal changes in a year.

Finally, a limit was established for water use in agriculture so that it would not exceed 10 % of the demand of water for animals. This percentage resulted from a project criterion by the owner of the enterprise to prefer domestic supply and water for animals, since these are priority uses of bulk water in situations of scarcity. This opens opportunities for small farmers to generate opportunities for small enterprises.

The total volume demanded in the realm of the 30 years of operation will be, as indicated in Table 21.1, 171,555,655 m^3. In order to meet this demand, the owner of the project enterprise considered a loss of 20 % over the total volume produced during the 30 years of operation. This increased the total volume produced to 214,452,289 m^3, according to an annual schedule to accompany the evolution of demand. This schedule is presented in Table 21.2 that, also for lack of space, has only the figures of the first

Table 21.2 Demand outflows and outflows to be produced (m^3/year)

Year	Demand outflow	Produced outflow
1st	1,314,000	1,642,500
6th	3,449,250	4,311,562
11th	5,256,000	6,570,000
16th	6,980,625	8,725,781
21st	9,198,000	11,497,500
26th	10,512,000	13,140,000
30th	11,826,000	14,782,500
Total (30 years)	171,555,655.22	214,444,569

Source: Environmental Engineering Company of the State of Bahia (CERB). Salvador. 2011

[6] Criteria constants of the executive project of CERB.

of every 5 years and the last year of the complete series of 30 years of operation of the project.

The outflows to be produced are the result of the well operation, which will vary from two—each operating 8 h a day (this will be the case of the first year)—to eight wells, each operating 18 h a day (the case of the 30th year).

The detailed schedule of well operation is presented in the text. The immediately following section presents the cost estimates of the enterprise, an essential element to form levels of tariffs.

21.5 Costs

The total costs of a hydraulic engineering enterprise include the amortization of the investment and the costs of operation and maintenance (O&M). However, in the case of the Tucano Project, the annual amortization of the investment in the system of storage and conveyance, US$34.09 × 10^6, was not included in the tariff calculation. The only amortization considered in the calculation was related to the construction and installation of wells, resulting in a much lower total cost. The first reason for this exclusion resides in the source of these funds, which with the exception of the investment in wells,[7] were passed on by the state of Bahia in the context of the growth acceleration program (PAC) of the federal government as a grant. Second, but not less important, the investment resources were excluded from the tariff formation formula, due to a tradition in the sanitation sector, the main water user of this enterprise. This tradition is not only Brazilian but a current practice in some Latin American countries. In some countries, it is only partially adopted, that is, a small part of investment is included in the price for use of the water and the complement is cleared.

The total costs were calculated, based on the sum of the following classes of expenses: (1) labor costs, (2) vehicle rentals, (3) power supply for the operational system, (4) supervision, (5) maintenance, (6) well investment amortization, and (7) fiscal obligations.

As a starting point, the total costs were estimated for a period of a fiscal year in order to include seasonal variations, since water use depends on temperature. The demand for water is calculated annually. Additionally, prices were estimated in a detailed manner for the first year of the enterprise and projected to each of the following 30 subsequent years, based on the expected variation. This variation is measured by the tendency of the indexes inherent to each class of cost. In this section, only the calculations for the first year are presented in order to clarify the budget criteria. This presentation is done separately for each of the expense classes according to the plan of accounts of the enterprise company. The costs of the other years will be presented in a consolidated manner and presented only for the first of every 5 years of the economic lifetime of the enterprise and for the last year of the last 5 years. The next subsections present the cost composition of the first year, per expense class, accompanied by the projection criteria indication of that class.

[7] The installation of wells corresponds to investments made by CERB's own resources.

21.5.1 Labor Costs and Its Implications

The labor costs correspond to salaries of the professional team in charge of O&M of the enterprise, including labor union costs (Table 21.3). This team is composed of 20 professionals and will probably increase its size over the lifetime of the enterprise in order to meet the increase in workload due to the amount of demand for water already shown in Table 21.1.

The increase in team size, after exhaustive evaluations of the owning company of the enterprise, implied that after the fifth year of operation, the inclusion of an electrician, an electric mechanic, and an electric technician maintains the group of professionals in this new dimension during the temporal horizon of the project.

The projection of these costs for the next 29 years of the lifetime of the project was made, based on the programmed team increases mentioned above and the application of an annual average representation of the last collective agreements signed by CERB with various relevant labor unions. The result is presented in Table 21.4.

It may be noted that preference was given for the inflation target, instead of the tendency outlined by the broad consumer price index (IPCA), used by the Central Bank to control inflation.[8] This is due to the official declarations by the federal government that the Brazilian economy will make restitution of inflation toward this target among a series of microeconomic measures that have been adopted, besides the macroeconomic key measure for the same reason that has been administered by the SELIC interest rate in a higher level than what has been currently used.[9]

Table 21.3 Labor costs and its implications (US$)

Job position	Quantity	Unit cost includ. labor union costs	Subtotal
Supervising engineer	1	6,987.07	6,987
Administrative assistant	1	1,228.57	1,228
Unit operator	6	1,047.93	6,287
Electric technician	1	3,721.57	3,721
Electric mechanic	1	3,721.57	3,721
Electrician	1	1,948.54	1,948
Assistant electric mechanic	2	731.50	1,463
Watchman	1	1,342.58	1,342
Night watchman	1	1,611.61	1,611
Plumber	1	1,655.17	1,655
Driver	1	1,998.50	1,998
Helper	3	731.50	2,194
Monthly total	–	0,00	34,160
Annual total	12 months	34,160.29/month	409,923

Source: Environmental Engineering Company of the State of Bahia (CERB). Salvador. 2011

[8] www.bcb.gov.br. Sistema de Metas de Inflação (série completa – 1999–2013).

[9] 7.25 % per year.

Table 21.4 Labor costs during the temporal horizon of the project

Year	(US$/year)
1st	409,923
6th	641,989
11th	788,617
16th	968,734
21st	1,189,990
25th	1,461,779
30th	1,723,268
Total (30 years)	29,756,100

Source: Auxiliary calculations to the work. Includes labor union costs and implications

21.5.2 Vehicle Rental

The cost of vehicle rentals for the first year of operation corresponds to US$130,363.64 per year, according to the data shown in Table 21.5.

The cost projection of vehicle rentals was made, based on the subindex IPCA services for transportation, at 10.59 % per year, between 2004 and 2011, according to a research by the Central Bank,[10] which resulted in the consolidated totals that are shown in Table 21.6.

21.5.2.1 Electric Energy of the Operational System

The cost of electricity in the first year of operation was determined by tariff simulations. The electrical equipment forecasted for the project totals 3,083.80 horsepower (HP), corresponding to 2,333.12 kW.

The installed power serves as a base for fixed cost calculation of power, which corresponds to the reserve of demand that the power company should assure to the project. The variable part of power costs depends on the consumption, which is proportional to the intensity of use of the wells, the elevation stations, and the water production centers. The calculation for the first year presented a synthesized result (Table 21.7), which reveals a preponderance of the production centers for power consumption, due to the small number of wells (only two units) and its number of hours in operation also being low——only 8 h per well.

In the last year of the project, however, all eight wells will be operating at 18 h/day each, reflecting a greater sacrifice in terms of cost. This sacrifice results from the abstraction of groundwater at depths of over 200 m. The cost of electric energy for the 30th year shows the prevalence of work to impound water from the wells to the elevators and the water production centers according to the figures in Table 21.8.

---

[10] (op.cit.)

Table 21.5 Vehicle rental costs (US$)

Type	Quantity	Unit cost	Subtotal
Small utility vehicle	1	1,090.91	1,090
Medium-size utility w/extended cab	1	2,954.55	2,954
Munck truck	1	5,454.55	5,454
Motorcycle 125 cc	3	454.55	1,363
Monthly total	–	0.00	10,863
Annual total	12	10,863.64	130,363

Source: Environmental Engineering Company of the State of Bahia (CERB). Salvador. 2011

Table 21.6 Vehicle rental costs during the temporal horizon of the project

Year	(US$/year)
1st	130,363
6th	172,408
11th	228,013
16th	301,551
21th	398,808
25th	527,431
30th	659,610
Total (30 years)	9,863,898

Source: Auxiliary Calculations to this research

Table 21.7 Electric energy cost in the first year

Equipment	Cost (US$/year)	Participation (%)
Wells	115,587	14.65
Elevators	211,134	26.76
Production centers	462,271	58.59
Total	788,994	100.00

Source: Calculations based on no. 971/2010 Resolution of ANEEL

Table 21.8 Cost of electric energy in the 30th year

Equipment	Cost (US$/year)	Participation (%)
Wells	1,523,654	56.64
Elevators	761,827	28.32
Production centers	404,720	15.04
Total	2,690,202	100.00

Source: Calculations based on Number 971/2010 Resolution of ANEEL

Table 21.9 Electric energy cost evolution during the temporal horizon of the project

Períod	Cost (US$/year)
Year 1	821,816
Year 6	1,160,235
Year 11	1,594,173
Year 16	2,225,021
Year 21	3,084,460
Year 26	3,953,734
Year 30	5,187,374
Total (30 years)	73,201,348

Source: Calculations elaborated based on Number 971/2010 Resolution of ANEEL

The figures in Tables 21.7 and 21.8 were obtained by a tariff simulator that includes all the technical information on the number and capacity of pumps in operation, elevators, and accompanying equipment of the electric energy supply of the system, offering all of the electric power tariff possibilities, and already selecting the lowest cost, adopted in this chapter.

The future energy cost projection, between the second and 30th year of the enterprise's lifetime, including the two extreme years, includes the average IGP-M[11] variation combined with factor X and monitored by ANEEL with the goal of passing on to consumers the productivity gains of the concessionaries due to the expansion of the electric energy market (Central Bank of Brazil 2012a, b).

Based on these criteria, the results in Table 21.9 were calculated and present the evolution of the electric energy's annual cost. In this enterprise, the cost of electric energy is of greatest share, corresponding to 38.4 % of the total operational cost of the system, and increases over time by the amortization of the wells.

This is mainly due to the fact that water is abstracted from such a depth along with successive high conveyance costs.

21.5.3 *Supervision*

The supervision services of the water supply system include renting an office and the costs associated with the administrative activities. Table 21.10 presents in detail the different components included in such costs in the first year of operation of the enterprise.

The supervision services referred to here are strictly local, that is, they refer to the direct supervision for the functioning of the enterprise. Besides this local supervision, the company owner of the enterprise carries out the central administration whose annual cost attributed to the project is US$32,908/year. The central administration corresponds to an indirect supervision, and the costs should be added to the

[11] General price index.

Table 21.10 Supervision costs in the first year of operation

Items	Month or monthly amount	Unitary cost (US$)	Subtotal (US$)
Office rental	1	136.36	136
Electric energy bill	1	136.36	136
Phone and internet	1	386.36	386
Office supplies	Budgeted allowance	–	113
Cleaning supplies	Budgeted allowance	–	45
Mobility	Budgeted allowance	–	454
Fuel and maintenance	Budgeted allowance	–	1,159
Monthly total	–		2,431
Annual total	12 months	2,431.82/month	29,181

Source: Environmental Engineering Company of the State of Bahia (CERB). Salvador. 2011

operational costs of the enterprise. This way, the total cost with supervision, direct and indirect, is equal to the sum of US$29,181 and U$32,908, respectively, in the first year of the enterprise. For the other years, the supervision costs were estimated based on the proportionality in relation to other costs of the system, that is, the same proportion was maintained for each subsequent year, the same proportion found for the first year of operation of the enterprise.

21.5.4 Maintenance

The maintenance costs, like the supervision costs, result from rental services for this purpose, such as electric energy and communication, office supplies and cleaning supplies, and fuel. Table 21.11 presents the components of the maintenance costs for the functioning of the system.

A scrutiny of Table 21.11 suggests that besides the components mentioned above, there are a few expenses for outsourced plumbing, electromechanical, and patrimonial services, to mention a few.

The estimate of these outsourced services figures was based on experience of the enterprise's company. The projection of the evolution over the 30 years was determined based on the percentage of increase of 5.75 % per year, an average observed value in the period of 2004–2011 for the subgroup IPCA services, according to a study conducted by the Central Bank. The future figures are presented in Table 21.12.

21.5.5 Well Amortization

According to the comment made in the "costs" section, the investment in works of the project was based on the grant allocation from the federal government. In the case of the enterprise being studied, however, the costs of the wells were taken care

Table 21.11 Maintenance costs

Specified items	Month or monthly amount	Unit cost (US$)	Subtotal (US$)
Office rental	1	181.82	181
Electric energy bill	1	68.18	68
Telephone and Internet	1	181.82	181
Office supplies	Budgeted allowance	–	45
Cleaning supplies	Budgeted allowance	–	90
Fuel and maintenance	Budgeted allowance	–	1,954
Outsourced plumbing	Budgeted allowance	–	10,000
Outsourced electromechanical services	Budgeted allowance	–	1,136
Monthly total	–		13,659
Annual total	12 months	13,659.09/month	163,909

Source: Environmental Engineering Company of the State of Bahia (CERB). Salvador. 2011

Table 21.12 Evolution of maintenance costs during a temporal horizon of the project

Year	Cost (US$/year)
1	163,909
6	238,450
11	315,354
16	417,062
21	551,573
26	729,466
30	912,276
Total (30 years)	13,550,371

Source: Auxiliary calculations to this research

of by the enterprise itself. The cost of each well, after drilling, finishing, the filter, the prefilter (gravel), and cementation, as well as the pumping installation, totaled US$428.18 thousand. The execution program of the eight wells of the system, based on other previous experiments (Lopes 1996), was established in the following manner: two wells will be initiated in the first year of operation of the system; another well after the sixth year, totaling three wells; another well after the ninth year, resulting in four wells; another well after the 13th year, a total of five wells; another well after the 17th year, totaling six wells operating; another well after the 20th year, so that there are seven wells in operation; and the last well after the 25th year, completing a series of eight wells, and the system reaches its final configuration.

The operational regime of the wells oscillates between 8 and 18 daily hours. The lowest limit was established based on initial demands that may be satisfied with only two wells under the regime of 8 h a day, and the highest limit was obtained by geological studies conducted by CERB that indicated the need for the remaining aquifer, in relation to each well in "rest" for 6 h a day. These operation time limits took into account the production of water needed to attend demand and considering the

Table 21.13 Evolution of costs for the amortization of wells

Period	Cost (R$/year)
Year 1	62,800
Year 6	124,581
Year 11	219,681
Year 16	363,166
Year 21	672,411
Year 26	1,016,316
Year 30	1,271,014
Total (30 years)	14,634,180

Source: Auxiliary calculations to this research

losses of the project (20 %), as well as the average outflow of each well (200 m^3/h), resulting in the projected quantity of eight wells. Finally, the amortization of the wells was calculated linearly to recuperate investment in 25 years, according to estimates made by the federal revenue office.[12] It may be observed that six of the wells will start operating after the sixth year of operation of the system; the revenue of the enterprise will not amortize the costs of all wells during the life of the project. In other words, these other six wells will only have their costs totally recuperated after the 30-year period, as they complete 25 years of operation. The evolution of these costs with the amortization of the wells is presented in Table 21.13.

21.5.6 Profit

The profit of the enterprise was set as a percentage of the costs, at 10 %. This value was based on the last two balances available from the water and sanitation company of Bahia's (Embasa's) website that showed a profit of 10.62 % in 2010 and 3.75 % in 2011.[13] Profit rates of other states were not considered so that the study would remain adherent to the reality of the state of Bahia.

21.5.7 Fiscal Obligations

The fiscal obligations were left for the last of all costs, due to legislation that does not allow the incidence of profit over obligations. The sanitation companies incur, in the development of their mission, the following tributes: the PIS (1.65 % overbilling), the COFINS (7.6 % overbilling), the income tax (15 % of profit that exceeds the

[12] Empresa Baiana de Águas e Saneamento – Embasa (2012). Relatório do Exercício de 2011. Salvador.

[13] Ministério da Fazenda (1998). Instrução Normativa da Secretaria da Receita Federal no 162. Brasília.

Table 21.14 Costs for the charge for use of the water

Year	Volume captured (m^3)	Public charge (US$/$m^3$)	Total charge (US$/year)
1	1,368,662	0.00455	6,221
6	3,592,738	0.00475	17,065
11	5,474,649	0.00496	27,174
16	7,271,019	0.00519	37,713
21	9,580,636	0.00542	51,931
25	10,949,299	0.00566	62,017
30	12,317,961	0.00566	69,770
Total (30 years)	214,452,289	–	1,128,167

Source: Calculations for the research

value of US$109,090), and the contribution over liquid profit, CSLL (9 %). A charge is also paid in the state for the use of groundwater that is a property of the state.

The charge for the use of water resources is still not implemented in the region of the Tucano aquifer. However, since the wells are subject to a right to use (grant), the implementation of a charge will be a natural consequence, due to legislation. There is still an established practice in Bahia, in terms of public charges for the use of water resources, except in the case of tariffs for administration of supply reservoirs. By virtue of these circumstances, the charge adopted is US$4.50 per 10^6 m^3, which has been used in watersheds in Brazil and being increased every 5 years, based on the inflation rate presently used in the country. It should be considered also that the charge begins on the first year of operation of the project, so that the costs of the project may be according to the established law. Table 21.14 presents an estimate of annual costs with the charge for use of water resources. The use of five decimals is necessary in this case, due to the high volume, which combines the charge with the low-level public price, so that a hundredth makes a difference, more or less, in the total charge.

In terms of the readjustment of costs with the charge along this period, the established inflationary rate percentage was adopted. As an example, the watersheds of the Piracicaba-Jundiaí-Capivari (PCJ), which are considered the best performing, in terms of water resource management in Brazil, shared their experience of the use of inflationary rate. However, the national water agency (ANA) insisted on an index to mark this readjustment. In any case, since the inflation rate reflects the desire to obtain a determined level of IPCA at the end of each year and the watershed management collects the charge, the inflation rate results in an indicator that is close to the actual costs established by the watershed (basin) committee.

21.5.8 Panoramic View of the Costs and the Producer's Profit

The summary of the costs of the project plus the profit is shown in Table 21.15 with the same criteria adopted in this chapter of showing only the first of every 5 years and the last of the temporal horizon of the project.

Table 21.15 Annual operational cost estimate increased by the amortization of wells and profit for the first of every 5 years and the 30th year of the economic lifetime of the enterprise (US$)

Expense	Year 1	Year 6	Year 11	Year 16	Year 21	Year 26	Year 30	Total (30 years)
Labor cost	409,923	641,989	788,617	968,734	1,189,990	1,461,779	1,723,268	29,756,100
Vehicle rent	130,363	172,408	228,013	301,551	398,808	527,431	659,610	9,863,898
Elec. energ.	821,816	1.160,235	1,594,173	2,225,021	3,084,460	3,953,734	5,187,374	73,201,348
Supervision	62,090	87,247	122,598	172,272	242,072	340,154	446,540	5,907,485
Maintenance	163,909	238,450	315,354	417,062	551,573	729,466	912,276	13,550,371
Wells[a]	28,545	56,627	99,855	165,075	305,641	461,962	577,733	6,651,900
Profit	161,664	235,695	314,861	424,971	577,254	747,452	950,680	13,893,110
Fiscal works	194,040	298,385	408,480	558,089	764,640	989,687	1,254,115	18,184,257
Annual total	1,972,353	2.891,040	3,871,954	5,232,779	7,114,442	9,211,668	11,711,601	171,008,471

Source: Calculations for the research

[a]Amortization

Combining the set of figures referred to in Table 21.15 and the total volume of water produced (at the end of Sect. 21.4), it may be concluded that the average sale price of water produced by the Tucano Project is US\$0.79/m^3. Embasa's website shows that the lowest unit price charged for potable water is US\$0.80/m^3, which almost coincides with the cost of water of the Tucano Project. We are comparing bulk water, in the case of the Tucano, to treated water, in the case of Embasa for its consumers. However, if not for the aqueduct impurities, the quality of the Tucano Project water would be far superior to the potable water mentioned above, and its treatment costs are much less (only chlorination) to remove impurities. These considerations explain the goals of the project.

Table 21.15 shows that only two components contribute 70.24 % to the total costs: "electric energy" and "labor cost." This means that in administering the enterprise, greater control should be exerted over these two cost components, since any savings that is obtained by either one or both of these imply a significant reduction in operation and maintenance costs.

The next step consists of determining the unit prices in which water may be delivered to the user in each defined use by the enterprise, using the price methodology that forms prices that induce efficiency and, at the same time, are connected to distribution criteria and subject to the additional condition of attempting to cover all costs throughout the entire scope of the project.

21.6 Tariff Calculation

According to the theoretical framework presented earlier, the optimum tariffs will result from the solution of the following system of third-degree equations:

$$(p_{ad} - CMg) / p_{ad} = \varphi / |\varepsilon_{ad}|$$

$$(p_{ag} - CMg) / p_{ag} = \varphi / | \varepsilon_{ag}$$

$$(p_{da} - CMg) / p_{da} = \varphi / | \varepsilon_{da} |$$

$$p_{ad} V_{ad} + p_{ag} V_{ag} + p_{da} V_{da} - C_{tot} = 0,$$

The unknowns of the system above are the tariffs for each use, represented by p_{ad}, p_{da}, and p_{ag} for the domestic supply, water for animals, and agricultural industry, besides the factor φ, already defined conceptually as the difference between the benefits and marginal costs.

The marginal cost is the incremental cost that corresponds to the additional cost as the water supply of the project is expanded by an additional production of a cubic meter, independent of the use of the water. In this case, the following formula was used:

$$CMg^{LT} = \frac{\sum_{t=0}^{T} \frac{I_t + R_t}{(1+\rho)^t}}{\sum_{t=0}^{T} \frac{q_t}{(1+\rho)^t}}$$

where:

CM_g^{LT} is the long-term marginal cost.
I_t is the amortization of the investment in year "*t*."
R_t represents the cost increment for operation and maintenance in year "*t*," including the cost of water resource management.
q_t is the annual incremental impounding of bulk water.
ρ is the capital opportunity cost (social discount rate).
T is the planning horizon of the project.

The definition of the capital opportunity cost is relatively complex, due to the multiplicity of proposals that have been discussed in the past. The World Bank, for example, indicates a rate that should reflect the marginal cost of the investment in each country, submitting the issue to an evaluation of the rate in which the economic agents choose to postpone consumption. In spite of this, the World Bank usually recommends a rate of 12 % for the economic and financial evaluation of the projects it supports. The equality between the two rates—economic and financial—seems unreasonable, due to the different purposes each is used for.

In any case, the social discount rate should be adopted at a lower level than the market rates, since the government and society have a lot more means to delude the risk than the private economic agent. The average interest rate on credit to a legal entity in Brazil from 2012 was 13.90 % per year.[14] Therefore, a social discount rate of 10.00 % a year was adopted, which seems reasonable for the present economic analysis of projects. Thus the marginal cost to be considered is $\rho = 0.10$.

Based on the figures and levels of outflows available and using the abovementioned social discount rate, the marginal cost obtained was US$0.82252 through the application of the expression presented above.

The price elasticities of demand for each use, extracted from a study elaborated by the Secretariat of Water Resources of the Ministry of the Environment for the Watershed of the Vaza-Barris River (Carrera-Fernandez 1999), which is a part of the Tucano aquifer, are:

$|\varepsilon_{ad}| = 0.13$
$|\varepsilon_{da}| = 0.19$
$|\varepsilon_{ag}| = 1.01$

[14] Central Bank of Brazil (2013). Política Monetária e Operações de Crédito do Sistema Financeiro Nacional. Nota para a Imprensa. Brasília.

The price elasticities for domestic supply and agricultural industry were calculated by Carrera-Fernandez (2002) in an application study of the watershed of the Vaza-Barris River. In the study by Carrera-Fernandez (2002), the price elasticity for water for animals was established in the same methodology as in the Secretariat of Water Resources (1999).

The total cost of operation of the system, including the amortization of wells and the profit of the enterprise is US$171,008,471. The volumes to be produced by the system in its 30 years of operation are the following:

V_{ad} = 161.269.605 m^3
V_{da} = 48,347,894 m^3
V_{ag} = 4,834,789 m^3

The division of the total cost by the total volume of 214,452,289 m^3 (sum of the volumes of the three uses referred to above) results in an average cost of US$0.79742/m^3. We will return to this average cost later in the chapter with a few more comments.

With the data already presented, the system of equations that allows the determination of tariffs may be written in the following manner:

$$\left(p_{ad} - CMg\right) / \, p_{ad} = \varphi / 0.13$$

$$\left(p_{da} - CMg\right) / \, p_{da} = \varphi / 0.19$$

$$\left(p_{ag} - CMg\right) / \, p_{ag} = \varphi / 1.01$$

$$\left(161{,}269{,}605 \times p_{ad}\right) + \left(48{,}347{,}894 \times p_{da}\right) + \left(4{,}834{,}789 \times p_{da}\right) \quad 171{,}008{,}471 = 0,$$

One of the solutions of the system above offers the following result:

p_{ad} = US$0.79/m^3
p_{da} = US$0.80/m^3
p_{ag} = US$0.82/m^3

Two of the tariff levels are above the average cost, and one is below. These relative values suggest a cross subsidy among sectors that altogether attempt to cover the cost budget of the system in totality of the 30 years of operation. In other words, the balance is only "cleared" at the end of the analysis period of the operation of the system. The release of these tariff levels with a schematic representation of the long-term cost curves is presented in Fig. 21.3.

As mentioned earlier, the suggested tariffs contribute to the efficient use of water, besides considering the payment capacity of each user class and promoting the coverage of all costs associated with the use of the Tucano Project water. In terms of the coverage of all of the costs, it is important to repeat that this condition is materialized along the temporal horizon of the project and not necessarily in each of the years. This means that a few years may be deficient on a revenue-versus-cost basis,

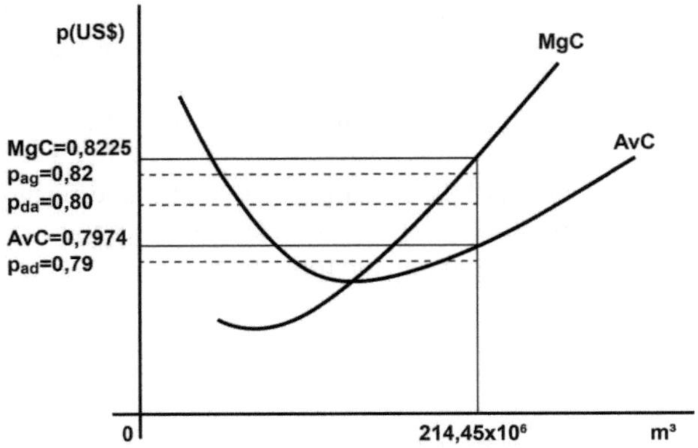

Fig. 21.3 Tariff and cost levels of the system

while others may present a surplus. In the analysis of the enterprise, a deficit may be observed up to the eighth year, and after the eighth year, it shows surpluses.

The eighth year corresponds to a payback of the project. This project demonstrates the need for the owner to have reserve funds in order to cover the deficits of the first 8 years. The surplus will then be returned between the ninth and 30th years. This measure is reasonable in social projects, as is the case of water supply in regions that are located in poverty areas.

21.7 Additional Considerations

The enterprise being studied regarding water tariffication has characteristics of a structuring project as it organizes the production and delivery of the most valuable good (water) for the population in a semiarid region. This condition in itself makes it necessary to analyze the price formation connected with use efficiency criteria of this natural resource as it operates in order to minimize the distributive impacts on the economy and recuperate costs. The fact that the prices calculated by an optimization method induce economic efficiency in water use allows the water company to sell drinking water at even lower prices, contributing to the fight against poverty, which is, as mentioned, a characteristic of the Tucano aquifer region. This present study opens the way also to proceed with a broader evaluation of the distribution of the water from the Tucano aquifer, considering not only the six municipalities that are a part of it but also broadening the research to all of the regions inscribed in the polygon shown in Fig. 21.1. Certainly, the change in scale will lead to the reduction of average prices (economies of scale) and benefits to the society.

The tariffs calculated by the optimization process serve also as a benchmark in discussions at the state water resource council on public price for bulk water abstracted from hydrogeological bodies of the Tucano. This reference is important as public prices for the use of water are established in Brazil, as a result of criteria that are not necessarily based on sound methods and postulates of economic theory.

In this sense, the rich experience CERB has opens the way to proceed with a broader study that encompasses the extension of the Tucano aquifer in conjunction with the basins in which it includes. This will allow bulk water supply and demand considerations to use the same optimizing approaches and may define price levels for efficient use of water in the region, whose physiography presents, predominantly, the effects of prolonged droughts.

References

ANEEL. (2010). *Tarifas de energia elétrica*. Brasília: SRE/ANEEL.

Baumol, W., & Bradford, D. (1970). Optimal departures from marginal cost pricing. *American Economic Review, 60*, 265–283.

Carrera-Fernandez, J. (1999). *Estudo de cobrança pelo uso da água na bacia hidrográfica do Vaza-Barris*. Relatório Final. Secretaria de Recursos Hídricos do Ministério do Meio Ambiente (SRH/MMA). Brasília: 1999.

Carrera-Fernandez, J. (2001). *Curso básico de microeconomia*. Edufba. Salvador.

Carrera-Fernandez, J. (2002). *Updating the water pricing system for Vaza Barris River basin*. Text for discussion in the Water Resources Secretariat Seminar. Brasília. 2002.

Central Bank of Brazil. (2012a). *Administrated prices* (Paper). Brasília.

Central Bank of Brazil. (2012b). *The dynamics of the prices of services*. Brasília.

Central Bank of Brazil. (2013). *Política Monetária e Operações de Crédito do Sistema Financeiro Nacional*. Nota para a Imprensa. Brasília.

Empresa Baiana De Águas E Saneamento—Embasa. (2012). *Relatório do Exercício de 2011*.Salvador.

Lopes, A. B. (1996). Estudo de viabilidade econômica de sistemas de bombeamento d'água em regiões rurais. In *Congresso Brasileiro de Energia: II Seminário Latino Americano de Energia*. Anais. Rio de Janeiro: UFRJ.

Lypsei, R. G. & Lancaster, K. J. (1956–1957). The general theory of the second best. *Review of Economic Studies*, 24, 11–32.

Ministério Da Fazenda. (1998). *Instrução Normativa da Secretaria da Receita Federal nº 162*. Brasília.

Normative Instruction of the Federal Revenue Secretariat no. 162/1998.

SEPLAN-PR—Secretaria de Planejamento, Orçamento e Coordenação da Presidência da República. (1994). *Projeto Áridas—uma estratégia de desenvolvimento sustentável para o Nordeste*. Brasília: SEPLAN-PR, 1994.

Seroa Da Motta, R. (1998). *Utilização de critérios econômicos para a valorização da água no Brasil*. IPEA. Texto para Discussão n. 556, 1998.

Chapter 22
Pricing for Reclaimed Water in Valencia, Spain: Externalities and Cost Recovery

Francesc Hernández-Sancho, Maria Molinos-Senante, and Ramón Sala-Garrido

Abstract The cost of reclaimed water and the tariffs paid by water users illustrate that the principle of cost recovery is not met in the majority of water reuse projects. However, such projects may also generate positive externalities, contributing to improved welfare of the entire society. This chapter describes the case of the Valencia region of Spain, referring to agreements among water stakeholders. It also includes a proposal of pricing for reclaimed water to be implemented in this area as a pilot case in order to develop a framework for costs and financial, institutional, and social arrangements for water reuse projects. A two-part tariff with a combination of a decreasing and increasing rate structure is proposed. This experience will help water associations and water companies to focus on new water reuse projects and opportunities they introduce. The chapter also explains why the cost-recovery principle is not met for almost all water reuse projects and identifies the major constraints hindering the implementation of this economic principle.

Keywords Valencia • Reclaimed water • Water stakeholders • Cost-benefit analysis • Operation and maintenance cost

22.1 Introduction

It is well known that water reuse is beneficial, not only because it enables water resources to be recovered but also because it reduces environmental impacts. However, it is also equally true that economic variables, such as costs and the price of treated water, affect its use, particularly in comparison to other available resources.

F. Hernández-Sancho (✉)
Department of Applied Economics II, Faculty of Economics, University of Valencia, Campus dels Tarongers Av. dels Tarongers, s/n 46022 Valencia, Spain
e-mail: francesc.hernandez@uv.es

M. Molinos-Senante • R. Sala-Garrido
Department of Mathematics for Economics, Faculty of Economics, University of Valencia, Campus dels Tarongers Av. dels Tarongers, s/n 46022 Valencia, Spain

A. Dinar et al. (eds.), *Water Pricing Experiences and Innovations*, Global Issues in Water Policy 9, DOI 10.1007/978-3-319-16465-6_22

The strict application of the cost-recovery principle would suggest that many water reuse projects would not be commissioned if the costs were only to be paid by private users. However, in this context, intervention by the public administration in water reuse projects can be justified, since they generate positive externalities that improve social welfare. In this respect, governments may choose to fund and maintain these types of projects.

In this sense, an in-depth analysis of the local water resources is required in order to establish a profitability threshold in technical and economic terms that makes water reuse an attractive option for various sectors, such as agriculture, industry, and aquifer recharging. This analysis should consider both the costs involved in the various water supply alternatives (Onkal and Demir 2006) and the profit margins, social and environmental impacts, and guaranteeing quantity and quality of water supply, among other aspects.

Any given analysis of the potential of water reuse in a particular region and for a series of specific uses requires an extensive knowledge of cost structure (Asano 1991, 1998; Sipala et al. 2005; Iglesias et al. 2010). Once the necessary quality parameters have been defined, existing alternatives (mainly conventional resources and desalination) should be studied. It is important to know the structure of the costs linked to alternative water resources. When calculating the structure, not only technical costs should be taken into consideration but also the value of externalities and opportunity costs (Hernández et al. 2006). Once the structure of costs is known, price-fixing mechanisms should be studied in order to assess the possibilities of reclaimed water demand. In this sense, the role of the various incentives for using reclaimed water could be assessed (Renzetti 2003).

A methodology based on cost-effectiveness and cost-benefit analysis would be advisable in order to assess the potential of water reuse as an alternative to conventional resources in each type of possible use in a particular region. This methodology provides valuable information in a simple way in order to optimize the management of available water resources in water-scarce regions. It is also important to value the potential of the water reuse market, particularly in agriculture, by comparing the various relevant cost structures to the conventional alternatives. The economic power and interest of users in treated water distribution systems must also be considered. The study of water demand is also relevant in order to assess the possibilities of reclaimed water usage accordingly.

With the aim to develop a comparative analysis of the costs associated with each alternative, the following issues should be taken into account: energy consumption for pumping and distributing water, well maintenance and piping for conventional resources; and personnel costs, maintenance, chemical products, and energy for reclaimed water. In the case of treated water, we should also consider that the structure of costs varies, depending on the size of the plant. As a result, determining a profitability threshold would give us the minimal size of a plant at which competitive treated water use is guaranteed.

In order to analyze the potential of the existing resources from treatment plants in a given region, we should assess, first, the resources available for reuse. Second, we should assess the potentially available resources, that is, those that could be obtained from existing plants, but after implementing suitable treatment systems. Finally, we should assess the resources that could be obtained from newly built plants (Angelakis et al. 1999). In all cases, the cost per cubic meter of reclaimed water at the outlet of the plant would have to be taken into account, together with the costs stemming from transport via piping to the final destination (Richard 1998). This exhaustive knowledge of the costs associated to the treatment and the water reuse is considered a basic requirement when valuing the true potential for reclaimed water in any territorial area.

In the context of resource management, the objective would be to optimize water resources (current and potentially available) in order to satisfy efficiently the various types of demand. From a supply point of view, it is necessary to consider both existing resources and those potentially available, either through new water reuse facilities, through desalination, or from other types of infrastructure. Potable water is often subsidized. If the principle of cost recovery were implemented both in the potable water industry and water reuse industry and reflected in their respective tariff system, the competitiveness of regenerated water may be significantly improved. From a demand perspective, it is important to analyze current uses, quality requirements in each case, possibilities of savings, as well as the forecasts of potential new requirements, the seasonal nature of demand, and potential uses.

In terms of the conventional resources available at present, it is necessary to analyze the price paid for using both surface and groundwater in the considered area. In this sense, it is important to take into account the availability of resources, their quality, whether or not there are aquifers, their price of extraction, etc. In relation to water resources available in the future, it is important to know the series of direct or indirect benefits and the disadvantages that stem from the use of reclaimed water. Furthermore, appropriate information concerning potential demand would also be required, in order to assess the possibilities of reclaimed water use.

Any pricing strategy will need to fit within the broad industrial, agricultural, and environmental policy setting of each case study. In this context, the successful implementation of pricing reform is often embedded in large reform processes (Dinar 2000). From the perspective of integrated water resources management, the analysis of water reuse economics should take into account the cost of regenerated water (and sometimes the benefits), as well as the costs of alternative water supply options, such as drinking water, desalination water, or storm water. Hence, it is possible to determine a ranking of cost-effective solutions for guaranteeing water demand. If the tariffs of regenerated water must be increased to meet the principle of cost recovery, then the price of drinking water must also be increased to achieve the same aim and prevent growth in total water consumption. A more transparent full-cost pricing of all water sources is required. In this sense, a higher cost for drinking water (full-cost tariff) could be a factor driving some utilities into developing or expanding their regenerated water programs.

22.2 Pricing and Economic Feasibility of Water Reuse Projects

According to their economics, worldwide experiences on water reuse can be classified in two groups, namely, projects in which water is not chargeable and projects in which water is chargeable.

Projects in which water is not chargeable usually refer to those of a public good nature, in which water is used for recharging aquifers, restoring water bodies, or public garden watering. For example, the Water Services Association of Australia (2005) reported several water reuse experiences in which regenerated water is not charged at all. In Mediterranean river basins, water reuse is being expected to achieve the good ecological status of water bodies required by the Water Framework Directive (CHJ 2012). Although these projects do not generate income from a market point of view, they create a number of significant positive externalities benefiting the entire society. Hence, to justify their economic feasibility, the cost-benefit analysis (CBA) must involve not only internal but also external (positive externality) benefits.

On the other hand, projects in which water is chargeable are characterized by the supply of regenerated water to private users. The regulations allow different water uses, and they define quality levels of water for each use. In this context, it is well known that investment and O&M costs of regenerating water vary greatly, depending on the quality of the required water (Iglesias et al. 2010). In most of these projects, water regeneration systems are designed ad hoc, based on the requirements of the private users.

To recover the cost of regenerating the water, the tariff of the water must be equal to or greater than the cost associated with such processes. According to Molinos-Senante et al. (2013), and from a static point of view, the selling price of regenerated water (SPRW) should be:

$$\mathrm{SPRW} \geq \frac{\left(\mathrm{IC} + \mathrm{OMC} + \mathrm{FC} + T\right)}{\mathrm{AVW}} \tag{22.1}$$

where SPRW is the selling price of regenerated water ($\$/m^3$); IC is the investment cost (\$); OMC is the operational and maintenance cost (\$); FC is the financial cost (\$); T are taxes in (\$); and AVW is the annual volume of regenerated water (m^3).

As shown in Eq. (22.1), the economics of achieving full cost recovery in water reuse projects is well known. However, real water reuse projects rarely charge full cost recovery to water users, but, in most cases, some degree of subsidy is needed (Molinos-Senante et al. 2013). Italy and Israel are two examples of this policy. On the one hand, Italy promotes water reuse through the Legislative Decree 152/2006, which orders discount tariffs for industrial users of regenerated water. In Israel, the state pays a non-negligible fraction of the total costs of water reuse projects (Fine et al. 2006).

With this information, the obvious question is why the cost-recovery principle is not met for most of the water reuse projects. In the context of water pricing in water-scarce regions, there are three main desirable objectives, namely, pricing for water demand management, pricing to promote the use of regenerated water, and pricing for cost recovery. The simultaneous achievement of all of these three objectives is almost impossible. Radcliffe (2003) analyzed an Australian case study in which regenerated water was priced at $0.28/m^3, while drinking water was also $0.28/m^3. These tariffs are a good incentive to promote the use of regenerated water, but neither to recover the full cost of the project nor to reduce water demand. In the United States, the American Water Works Association (2008) reported that for 42 % of utilities, it is more important to promote the use of regenerated water than to recover the full cost of the project.

The lack of an integrated water resources management (IWRM) perspective is one of the main limitations of meeting the full cost-recovery principle in water reuse projects. According to Tsagarakis and Georgantzis (2003), the willingness to use regenerated water by farmers depends on the price differential between conventional and reclaimed water. In this sense, in almost all water reuse projects, the tariff for regenerated water ranges from 0 to 25 % of drinking water. AWWA (2008) reported that only five US water reuse projects achieved 75–100 % of drinking water rates. It should be noted that the low price for drinking water is due to, in most cases, being subsidized.

From a policy perspective, to meet the principle of cost recovery, not only the tariffs of regenerated water must be increased but also the price of drinking water to achieve the same aim. In this sense, the full-cost tariff of drinking water might be a driving factor to enhance the development of water reuse programs.

22.3 The Case of the Valencia Region

The Valencia region is located on the Mediterranean coast of Spain. Due to water-scarcity problems and public awareness about this issue, the Valencia region is one of the most advanced Spanish regions in the reuse of water.

The aim of the regional government is to define water reclamation facilities and infrastructure needed to increase the current level of water reuse, up to 70 % of wastewater treated. The quality of reclaimed water is especially important and always has to fulfill the requirements of the Spanish Royal Decree 1620/2007 about water reuse. To achieve this aim is essential to facilitate agreements between users that make economically and functionally feasible water reuse projects.

Because neither Spain nor the Valencia region has a market for reclaimed water, it is very difficult setting a price for this product. Hence, in order to calculate the recovery of costs, it is assumed that for each cubic meter, the costs of reclaiming the water should be equal to the maximum selling price of the water. In this sense, the question is who should or can afford the cost of implementing a water reuse project (Hernández et al. 2006). It should be taken into account that oftentimes reclaimed water is used simultaneously for economic and environmental purposes, such as environmental flows or crop irrigation. An example of that use for reclaimed water is the Pinedo WWTP, located in Valencia, which reuses 78 hm^3/year for the irrigation of crops and for environmental restoration of Albufera Natural Park.

In order to increase the use of reclaimed water, the Valencia Water Authority is promoting agreements among municipalities, managers of WWTPs, and local farmers. The agreements are based on the infrastructure required for water reuse, which is provided by the government, either state or regional, while farmers transfer their water rights so conventional water can be used for urban users. The public investment in water reuse projects is justified by the positive impacts for the society taking into account the fact that all stakeholders will benefit from water reuse. Usually, the irrigation communities are favorable to these agreements as long as they do not have to pay an extra cost and the proper irrigation of crops is ensured.

An example of these agreements is the case of the water reuse project of the Marina Baixa district (Alicante). It is based on an exchange of water among the "Consorcio de Aguas de la Marina Baixa" and farmers, which guarantees the adequacy of resources in the district. Irrigators agree not to use the conventional water, as it is reserved for the urban users of the coastal municipalities of the district. In return, the "consortium" is committed to supply to irrigators reclaimed water of adequate quality to meet the needs of irrigation. It is an integrated system whose equilibrium is ensured through the transfer of economic resources though two procedures. The first one is direct contributions of the consortium to the budgets of the irrigation communities, while the second one is based on the payment of energy bills and infrastructure maintenance of the irrigation communities.

When user agreements based on the exchange of water are not feasible due to the scarcity of conventional water resources, the price of the reclaimed water should be based on the cost of regenerating the water. Figure 22.1 shows water treatment in

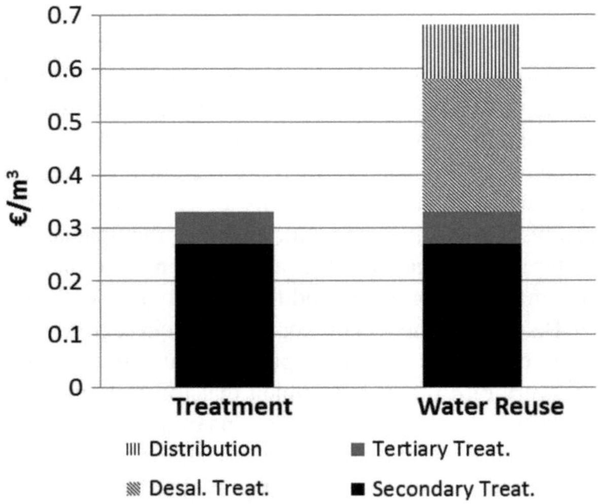

Fig. 22.1 Cost of water treatment and water reuse in Valencia region (Source: EPSAR (Valencia Public Entity for Wastewater Treatment))

the Valencia region. It should be noted that the treatment costs to fulfill the quality requirements of the European Directive 91/271 are covered by the so-called sanitation levy, which is paid by all urban water users. The extra costs needed to get reclaimed water suitable for agriculture should be paid by the users, i.e., by the farmers. However, such costs are usually considered by farmers as too high, especially if investment costs are also considered. For this reason, the participation of the government is required to cover a percentage of the costs, as long as environmental and social benefits from reclaimed water are identified. An example of this scheme has been developed in the water reuse project from Pinedo WWTP, in which a part of the reclaimed water flow is used to improve the ecological status of Albufera Lake. In this case, irrigators only pay pumping costs, which are around 0.006–0.012 €/m^3. The low price of reclaimed water was set in order to promote the use of reclaimed water by irrigators.

While water reclamation costs should be covered by the users, it should be noted that these costs are highly variable, depending on several factors such as the characteristics of the wastewater to be treated and the use and quality required for the reclaimed water. Because different wastewater treatments involve significant cost differences, several prices should be set. Energy consumption is different among wastewater treatment technologies; therefore, special attention should be paid to energy costs. On average, investment costs associated with a physical-chemical treatment range from 20 to 30 €/m^3, while operating costs range from 0.02 to 0.03 €/m^3.

As is known, microfiltration (MF), ultrafiltration (UF), reverse osmosis (RO), and reversible electrodialysis (RED) processes are increasingly incorporated into wastewater treatment and water reclamation. Their investment costs range from 200 to 400 €/m^3, while their operating costs are from 0.05 to 0.09 €/m^3. According to data obtained from the EPSAR (Valencia public entity for wastewater treatment) in the region of Valencia, wastewater treatment costs are 0.27 €/m^3, which are distributed as follows: personnel costs (40 %), energy costs (19 %), waste management costs (16 %), maintenance costs (12 %), reagents costs (7 %), and other costs, such as laboratory equipment, gardening, etc. (6 %).

In setting the price that users have to pay for the reclaimed water, not only costs associated with the treatment and reclamation of water should be considered. There are other costs, such as monitoring costs, administrative costs, pumping costs, and updated network costs, that must also be considered. Moreover, sometimes, wastewater has high salt concentration and an RO process is needed to reclaim the wastewater.

For example, the water reclamation plants located on Rincon de Leon and Benidorm (Alicante) have implemented RO processes with an associated cost of 0.26 €/m^3. Another foundation for setting the price of water reuse is the price of the drinking water. Hence, some WWTP companies set prices for reclaimed water as a percentage of the price of drinking water. This procedure is very useful in promoting water reuse, but it does not allow recovery of the full cost of the water reuse project.

22.3.1 Proposal for Setting Price to Reclaimed Water in Valencia Region

The aim of this section is to develop a case study, based on the water reuse tariff proposed by Molinos-Senante et al. (2013). To deal with the objectives of controlling water demand and encouraging water reuse and cost recovery, a two-part tariff consisting of a fixed charge and a volumetric charge was proposed.

On the one hand, a water reuse project includes fixed costs related to the system for regenerating the water, as well as distribution network costs. Hence, the fixed charge should be designed to ensure that all fixed costs are covered. On the other hand, volumetric charges include operating and maintenance costs, pumping costs, and monitoring water quality costs (i.e., all volume-related costs). Within the volumetric charge, Molinos-Senante et al. (2013) proposed a combination of decreasing and increasing rates, which are given by the following expression Eq. 22.2:

$$\begin{aligned} &\text{Declining charge}: \mathrm{VC} = e^{-K_1 Q} \ \text{ if } \ Q < Q' \\ &\text{Increasing charge}: \mathrm{VC} = e^{K_2 Q} \ \text{ if } \ Q \geq Q' \end{aligned} \tag{22.2}$$

where VC is the variable cost of the regenerated water per volumetric unit ($\$/m^3$), K_1 and K_2 are positive parameters, Q is the volume of regenerated water consumed (m^3/month), and Q' is the volume of regenerated water on which it is considered that larger water consumption is due to a low price (m^3/month) ($Q' > Q > 0$).

Initially, when a water reuse project is developed, the structure of the volumetric charge should be declining, since the aim is to break down social prejudices of using regenerated water. In other words, the aim is to promote water reuse. However, when a certain volume of regenerated water is used (Q') and basic needs are covered, there is no need for promoting the consumption of regenerated water. To control water demand, the low price of the water should be avoided. Therefore, when the volume of regenerated water is larger than Q', it is more appropriate pricing water following an increasing volumetric rate. The estimation of Q' should be performed for each individual water reuse project taking into account several factors, such as policy considerations, price of alternative water sources, equity, affordability, etc.

Integrating the fixed and volumetric water rates (Eq. 22.3), the total cost of the regenerated water is defined as follows (see Eq. 22.3):

$$\begin{aligned} &C = \mathrm{FC} + e^{-K_1 Q} + e^{K_2 Q} \\ &\text{If } \ Q < Q', K_1 > 0 \text{ and } K_2 = 0 \\ &\text{If } \ Q \geq Q', K_1 = 0 \text{ and } K_2 > 0 \end{aligned} \tag{22.3}$$

where C is the total cost of the regenerated water per volumetric unit ($\$/m^3$) and FC is the fixed cost of the regenerated water per volumetric unit ($\$/m^3$).

Based on total cost water rate shown in Eq. 22.3, a hypothetical case study was simulated. It involves a water reuse project in which the volume of regenerated water might range from a minimum of 10 m^3/month to a maximum of 1,000 m^3/month. Taking into account infrastructure costs, it was estimated that the fixed charge should be around 0.4 $ per cubic meter of regenerated water. Regarding the volumetric charge, the first step was to estimate the Q' value (i.e., the volume of water in which the rate of the volumetric charge changes from declining to increasing). Based on the use of the regenerated water and in the availability of alternative water sources, in this hypothetical case study, it was assumed that Q' was 500 m^3/month.

Subsequently, it estimated the price of the regenerated water for three scenarios: (1) decreasing volumetric rate, (2) increasing volumetric rate, and (3) declining-increasing volumetric rate. Figure 22.2 shows the price of the regenerated water, expressed in $ per cubic meter for each scenario defined.

If the volumetric rate is based only on a declining rate, the price of the regenerated water would range from \$0.50/m^3 to \$0.40/m^3. This means that for a large volume of regenerated water, the price of the water will be basically determined by the fixed costs. This approach would be very useful for promoting the use of regenerated water but contrary to control water demand. On the other hand, if an increasing rate is applied, the minimum price of the water (taken into account fixed and variable charges) will be \$0.41/m^3. This price is for a consumption of 190 m^3/month. The increasing volumetric rate creates barriers for promoting water reuse among small water users, since they have to pay more for each cubic meter of regenerated water. This rate is appropriate for reducing water consumption, since large water users (1,000 m^3/month) have to pay around \$0.59/m^3. The advantages of the decreasing-increasing volumetric rate can be derived from Fig. 22.2.

From a volume of water of 10–500 m^3/month, the price of the regenerated water exponentially decreases. This fact is especially important for small farmers, who cannot take advantages of economies of scale. For the case study, in particular, the

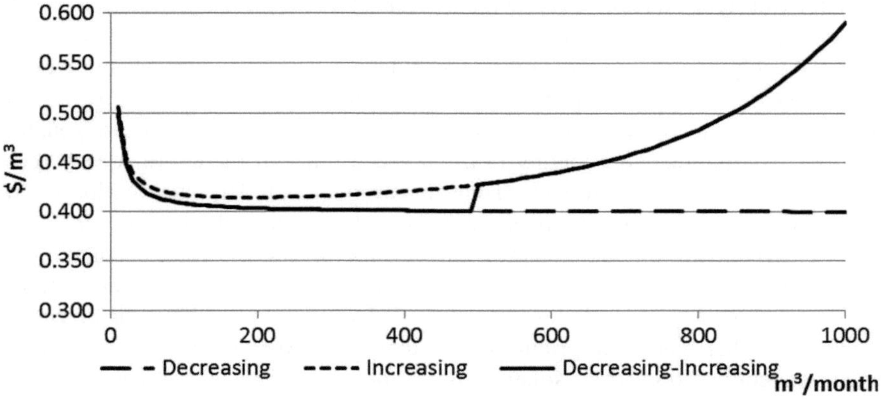

Fig. 22.2 Price of the regenerated water (volumetric charges) in \$/m^3 for the three scenarios

minimum price of the regenerated water will be around $0.40/m^3. From this volume of water consumption (Q'), the main aim of water pricing is not to promote water reuse but to control water demand, since it was assumed that basic water needs are covered. Hence, as shown in Fig. 22.2, the price of the water increases exponentially. The aim is to avoid the wasting of water associated with its low price.

The empirical application developed shows the usefulness of implementing a declining-increasing water rate to simultaneously promote water reuse, meet full cost recovery, and control water demand.

22.4 Conclusions

Increasing water shortage due to climate change leads arid and semiarid regions worldwide to reusing reclaimed water for several purposes. Three aspects have been identified as key points to improve the use of reclaimed water: (i) the increase in the quantity of wastewater treated motivated by new regulations, (ii) technical improvements in water regeneration systems for producing high-quality water at affordable costs, and (iii) the institutional and societal context focus on water reuse regulations.

Although the objectives of water reuse are very desirable, there are some challenges to be addressed. In doing so, information about current costs of water reuse projects, tariffs, and subsidy arrangements, as well as the overall acceptance and issues of raising awareness, should be investigated. While undertaking a water reuse project is fully justified in terms of objectives, it is not always possible to defray costs by charging tariffs. Moreover, who should pay for water reuse projects? The regenerated water in these projects is supplied to private users. The regulations allow different water uses, such as golf course irrigation and industrial use, although agricultural irrigation is the most widespread use. The economics of these projects differ significantly from the previous projects, because the water reuse systems are generally designed ad hoc in line with the characteristics of the private users.

Taking into account the water regeneration costs and the tariffs paid by water users, it is obvious that in most cases, some degree of subsidy is needed to recover the full costs. It has been illustrated that until now, the principle of cost recovery is not met in almost all water reuse projects. However, such projects may also generate positive externalities contributing to improved welfare of the entire society (e.g., concerning health, environment, and water availability). The intervention of the public administration in water reuse projects could be justified for the generation of these externalities. Although the principle of full cost recovery price accounting for environmental externalities represents an ambitious and long-term goal, it is necessary to start introducing policies and mechanisms aimed to facilitate this objective.

On the other hand, any pricing policy that encourages the reuse of reclaimed water cannot be adopted in isolation. From the point of view of integrated water resource management, it is essential to act holistically on water prices from all sources. The same economic principles should be applied to all water sources so that they "compete" on equal terms (level playing fields). Moreover, awareness campaigns, education, and dissemination of results from previous experiences are needed to help change attitudes and encourage water reuse.

The Valencia Water Authority is promoting agreements among municipalities, managers of WWTPs, and local farmers in order to increase the use of reclaimed water in the Valencia region. In the context of these agreements, the infrastructure required for water reuse projects is provided by the authorities, while farmers transfer their water rights on conventional water to the municipalities. For cases in which the exchange of water is not possible due to scarcity reasons, a price mechanism for reclaimed water is applied. This price should be based on the cost of regenerating the water. In order to promote the feasible use of this nonconventional water, a two-part tariff based on a decreasing-increasing rate for volumetric charge has been proposed in this chapter, using data from Valencia region. It represents a good alternative to promote the use of regenerated water without increasing the total amount of water consumed.

References

Angelakis, A. N., Marecos do Monte, M. H. F., Bontoux, L., & Asano, T. (1999). The status of wastewater reuse practice in the Mediterranean basin: Need for guidelines. *Water Research, 33*(10), 2201–2217. Elsevier.

Asano, T. (1991). Planning and implementation for water reuse projects. *Water Science and Technology, 24*(9), 1–10.

Asano, T. (Ed.). (1998). *Wastewater reclamation and reuse* (Water Quality Management Library, Vol. 10). Lancaster: Technomie Publishing.

AWWA (American Water Works Association). (2008). *Water reuse rates and charges*. Survey results. Available at http://www.awwa.org. Accessed 17 June 2014.

CHJ (Jucar Water Authority). (2012). *Water reuse for the restoration of the Albufera Lake*. Available at http://www.chj.es/es-es/medioambiente/albufera/. Accessed 17 June 2014 (In Spanish).

Dinar, A. (2000). Political economy of water pricing reforms. In A. Dinar (Ed.), *The political economy of water pricing reforms* (pp. 1–25). Oxford: Oxford University Press.

Fine, P., Halperin, R., & Hadas, E. (2006). Economic considerations for wastewater upgrading alternatives: An Israeli test case. *Journal of Environmental Management, 78*, 163–169.

Hernández, F., Urkiaga, A., De las Fuentes, L., Bis, B., Chiru, E., Balazs, B., & Wintgens, T. (2006). Feasibility studies for water reuse projects: An economical approach. *Desalination, 187*, 253–261.

Iglesias, R., Ortega, E., Batanero, G., & Quintas, L. (2010). Water reuse in Spain: Data overview and costs estimation of suitable treatment trains. *Desalination, 263*, 1–10.

Molinos-Senante, M., Hernandez-Sancho, F., & Sala-Garrido, R. (2013). Tariffs and cost recovery in water reuse. *Water Resources Management, 27*(6), 1797–1808.

Onkal, G., & Demir, I. (2006). Cost analysis of alternative methods for wastewater handling in small communities. *Journal of Environmental Management, 79*, 357–363.

Radcliffe, J. (2003, November). *Water: Second/third time around*. ATSE 2003 symposium—water the Australian Dilemma (pp. 17–18).

Renzetti, S. (2003). *Full cost accounting for water supply and sewage treatment: A case study of the Niagara region, Canada*. World Bank's Water Resources Management Group on Economic Instrument. http://worldbank.org

Richard, D. (1998). The cost of wastewater reclamation and reuse. In T. Asano (Ed.), *Wastewater reclamation and reuse: Water quality management library* (Vol. 10, pp. 1335–1396). Florida: CRC Press LLC.

Sipala, S., Mancini, G., & Vagliasindi, F. G. A. (2005). Development of a web-based tool for the calculation of costs of different wastewater treatment and reuse scenarios. *Water Science and Technology: Water Supply, 3*, 89–96.

Tsagarakis, K. P., & Georgantzis, N. (2003). The role of information on farmers' willingness to use recycled water for irrigation. *Water Science and Technology: Water Supply, 3*(4), 105–113.

WSSA. (2005). *WSAA facts 2005. Pricing for recycled water* (Occasional Paper No. 12). Water Services Association of Australia.

Chapter 23
Pricing Municipal Water and Wastewater Services in Developing Countries: Are Utilities Making Progress Toward Sustainability?

Caroline van den Berg

Abstract This chapter uses data from the International Benchmarking Network for Water and Sanitation Utilities (IBNET) to assess the progress utilities are making toward financial sustainability while ensuring that the services remain affordable. The analysis finds that many utilities are only recovering operation and maintenance (O&M) costs and fall far short of achieving full financial cost recovery. The level of (O&M) cost recovery has, on average, barely changed between 2006 and 2011. Over the same period, with incomes rising in many parts of the world, water supply and wastewater services have become more affordable. This suggests that many utilities may be able to increase water prices. The analysis also shows that the differences in utilities' performance between and within countries are large. Better understanding of how the differences in costs affect revenue sufficiency and affordability is needed to improve pricing policies.

Keywords Developing countries • Wastewater • Cost recovery • Subsidies • Benchmarking

23.1 Introduction

Partly as a result of the Dublin principles issued in 1992 that emphasized water as an economic good, sector professionals in the 1990s increasingly discussed the need for cost recovery of water supply and wastewater services. Water pricing was seen as an important tool to ensure that water users would use the resource more efficiently. Especially in the municipal water sector, cost recovery became the main goal for ensuring that utilities can provide water supply and wastewater services in

---

The opinions reflected in this chapter are the opinions of the author and should not be contributed to the World Bank Group, its Executive Board of Directors or any of its member countries.

C. van den Berg (✉)
The World Bank, MSN J-6-601 1818 H Street NW, Washington, DC 20433, USA
e-mail: cvandenberg@worldbank.org

A. Dinar et al. (eds.), *Water Pricing Experiences and Innovations*,
Global Issues in Water Policy 9, DOI 10.1007/978-3-319-16465-6_23

a financially sustainable manner. Yet, the lack of financial cost recovery in many utilities has led to low-quality service that, in turn, results in underinvestment, lack of access to services, low reliability of services, and low quality of water supply and wastewater services (with increasing pollution loads in water bodies). The low quality of service results in reduced benefits for users, thereby diminishing their willingness to pay for water supply and/or wastewater services (OECD 2009).

At the time of the formulation of the Dublin principles, there was also a major concern about the affordability of water supply and wastewater services. In 2010, the United Nations (UN) General Assembly, through Resolution 64/292, explicitly recognized the human right to water and sanitation and acknowledged that clean drinking water and sanitation are essential to the realization of all human rights (UN General Assembly 2010). In 2012, 748 million people worldwide still lacked access to safe water supplies (WHO-UNICEF 2014).

Up to now, the debate on pricing and affordability has not yet been settled, and water pricing has not become any less relevant. With water becoming increasingly scarce in many regions as a result of a growing population, rising incomes, and increasing water pollution and climate change, pricing remains an important tool to manage scarce resources more efficiently. Whittington (2003) notes several reasons for the lack of consensus on water tariffs. The first is disagreement on the objectives of pricing, as water pricing decisions affect policymakers' objectives often in competing ways. Rogers et al. (2002) identify a list of objectives for water pricing. Three main criteria are efficiency, cost recovery, and affordability.[1] It is clear that utilities must recover their costs; otherwise, they will not be able to continue providing services. Tariffs are the most common tool to do so, but they can also serve other goals besides raising revenues to cover the financial costs of the service. Tariffs can also be used to ensure that water is efficiently used when prices accurately reflect water's economic value, and tariffs can provide information to consumers about their consumption decisions. Another important objective is that water and wastewater utilities must provide services that are affordable to consumers. Water pricing alone cannot realize these three objectives simultaneously (European Environment Agency 2013). Second, because of lack of empirical work, there is disagreement on what effect the different water prices have on consumer behavior. The largest meta-analysis carried out in industrial countries covers 64 studies over a 40-year time span (Dalhuizen et al. 2003). Nauges and Whittington (2010) undertook a meta-analysis of 11 studies covering a 20-year time period in developing countries. In view of the large number of utilities all over the world, the wide variety in geographic, hydrological, economic, social, institutional, political, and cultural differences and changes over time and space, this number of studies is indeed very small.[2]

Affordability is a concern, as lack of access to improved water and sanitation services can have significant public health impacts. The challenge with affordability is how to define the concept. What is a minimum level of use? There are minimum

---

[1] Rogers et al. (2002) identify also objectives such as equity, fairness, transparency, simplicity, or administrative ease.

[2] Whittington (2003) also mentions a third reason, namely, the lack of a market test for different water tariff structures, as consumers have limited options to reject inappropriate tariff structures.

standards for water use set at 50 l per capita per day (lcd) to ensure that most water needs are met (consumption; personal hygiene, including laundry and bathing) and health concerns are mostly addressed (WHO 2003). Yet, that same report also mentions 20 lcd as the minimum needed for consumption and basic personal hygiene and 100 lcd or more for optimal access. Moreover, income data are rarely available at the utility level, and hence proxies like gross domestic product per capita are used to calculate affordability. This income proxy does not take into account income distribution (Zetland and Gasson 2013), and it also does not adjust for the fact that many utilities in developing countries do not provide universal service—with the poor more likely to be excluded from the service.

In this chapter, we will look exclusively at how balancing the principles of managing water as a social and economic good has worked out globally in striking the right balance between different pricing policy objectives. Due to lack of data on the economic cost of water and wastewater services, we focus on the two goals of revenue sufficiency and affordability. By presenting data from about 1,800 utilities in mostly developing countries, we add some empirical work to the debate on revenue sufficiency and affordability.

The chapter is organized as follows. We use the database of the International Benchmarking Network for Water and Sanitation Utilities (IBNET).[3] This database covers performance data from utilities all over the world and will be described in Section 23.2. In Sect. 23.3, we look at the progress made in achieving cost recovery as an economic good, whereas in Sect. 23.4, we look at how much progress has been made in making water and wastewater services affordable to all. Section 23.5 looks into the subsidies that are provided in water and wastewater provision. We conclude in the last section.

23.2 IBNET Database

For this empirical analysis, we use a unique database that collects data on the financial and operational performance of water utilities, namely, the International Benchmarking Network for Water and Sanitation Utilities (IBNET) at www.ib-net.org. More details on the IBNET database can be found in Appendix 23.1. IBNET guarantees that the data used in this analysis are relatively homogeneous and comparable across countries. For each country, we have annual data on different aspects of the utilities' operational and financial performance. As of 2010, the IBNET database contains information on performance from 1,861 water utilities, serving nearly 513 million people, in 12,480 cities and towns. This is approximately 14 % of the total population of all households with piped water access in the world. The database represents the equivalent of more than US$40 billion in annual utility revenue

[3] IBNET is being developed by the World Bank with the objective to provide comparative international benchmark performance information that can inform utilities and policymakers on how to improve service delivery (for more information, see www.ib-net.org).

in 2010. The utilities represented in the database employ nearly 623,000 utility staff. The database compromises municipal water and/or wastewater utilities, regional and national utilities; these utilities differ widely in scope and size.

Participation in IBNET is voluntary. This may affect the sample, as it is possible that better-performing utilities are more likely to report results than those that are poor performers. IBNET works with national utility associations, regulatory agencies, and, in some cases, ministries that collect data on utility performance. Hence, in many countries, data are collected from a range of utilities that are likely to include better- and worse-performing utilities. The voluntary character may also result in changes in the database over time, as sometimes utilities participating in one round of data collection do not necessarily participate in a next round of data collection. In addition, the data collection process takes time, as the data is collected once annual reporting has taken place, while the cost of data collection does not allow for annual updates. In this analysis, we use data from 2006 to 2011 when the number of utilities in the database is relatively stable, and the sample size does not differ significantly year by year.[4]

Information is collected from countries in all the world's regions. However, data from countries in Africa and South Asia, which tend to be categorized as low income, are less well represented for the year 2010, as data collection has not yet been completed. Data collection is especially complicated in South Asia where utilities are often still municipal departments, and reporting in such departments is less well established.

In the past years, economic development has been rapid, especially in emerging countries. The average per capita nominal gross domestic product (GDP) increased rapidly from US$4,937 in 2006 to US$9,567 in 2011. This rapid economic growth has had major consequences, as it has reduced the proportion of people living in low-income countries, as many countries have moved to higher-income categories. As a result, the income classification in the IBNET sample has changed significantly, too.

23.3 Cost Recovery

Definition of cost recovery. In this section, we will measure in how far a sample of utilities around the world is able to achieve the goal of (financial) cost recovery. Cost recovery is defined as in how far the revenues generated by the utility cover the costs of supplying water and wastewater services. In many cases, the costs to be considered are the operation and maintenance costs (which for the largest part are made up of labor costs, energy costs, and chemical costs) that enable the utility to continue to provide water and/or wastewater services to existing customers. In addition, the capital costs should be recovered to ensure that the longer run assets can be replaced. These two cost categories would make up the full financial cost of the

[4] The IBNET database has been growing over time. It started with a small number of utilities in 1996 and has been growing to 1,861 utilities in 2010.

Table 23.1 Operating cost coverage between 2006 and 2011

Indicator	2006	2007	2008	2009	2010	2011
Median OCCR	1.09	1.11	1.10	1.15	1.09	1.09
Average OCCR	1.17	1.22	1.21	1.22	1.15	1.15
Standard deviation	0.52	0.58	0.56	0.59	0.54	0.58
Number of utilities reporting	1,447	1,420	1,494	1,449	1,664	1,429
Top 5 %	2.17	2.40	2.28	2.35	2.15	2.26
Top quartile	1.38	1.39	1.40	1.45	1.40	1.38
Bottom quartile	0.88	0.91	0.90	0.89	0.86	0.83
Bottom 5 %	0.46	0.47	0.42	0.37	0.31	0.30

Source: IBNET database
Note: *OCCR* operating cost coverage ratio. The 2011 data collection cycle is not yet fully completed

service. There are, however, complications[5] with collecting data on especially the capital cost of the service. The cost of depreciation and the capital costs can differ significantly between countries, because of accounting practices and the patterns of allocating ownership of assets and economic risk (European Environment Agency 2013). In addition, in countries in which the infrastructure was built (many) decades ago, the historical capital costs may be relatively small and insufficient to replace assets over time.

We will use the operating cost coverage ratio[6] (OCCR) as an indicator for revenue sufficiency. This ratio measures how far the total revenues billed to customers cover the operation and maintenance costs of water and wastewater services. This is certainly a limited definition of cost recovery, as no capital costs will be specifically included in this definition.[7] Using the OCCR will help us to determine whether in the short run the utility can service its existing customers.

Short-term cost recovery. As can be seen in Table 23.1, the OCCR has remained at a level that just covers the operation and maintenance costs of the service—with revenues covering about 109 % of operation and maintenance costs in 2010. The standard deviation has increased over time, which means that the gap between better- and worse-performing utilities is increasing. The top 25 % of utilities are able to cover operation and maintenance costs and a small portion of its capital costs with the median OCCR for this income group being 1.40 in 2010. The bottom 25 % of utilities in the sample are not able to cover their operation and maintenance costs, and they have a median OCCR of 0.86 in 2010. For the bottom 25 % utilities, the OCCR has continuously declined since the financial and fuel crisis of 2007 and 2008. The worst-performing utilities, those that are in the bottom 5 % of the sample,

[5] There are more complications, such as the costs of supplying the services efficiently (European Environment Agency 2013).

[6] It should also be noted that, although covering costs through revenues is important, it does not necessarily mean that cash needs are met (WHO 1989). Cash flow needs are not similar to that of revenue needs, but are of critical importance, as they ensure that the utility can keep on providing the services.

[7] Capital costs make up the bulk of the cost of water and wastewater services, due to the high capital intensity of water and wastewater infrastructure.

were only able to cover 31 % of their operation and maintenance costs in 2010. These bottom 5 % of utilities saw the most rapid decline in their ability to cover operation and maintenance costs between 2006 and 2011.

As can be expected, utilities in high-income countries are much better able to cover their operation and maintenance costs than utilities in other countries. A typical utility in a high-income country had an OCCR of 1.42 over the period between 2006 and 2011, as can be seen in Table 23.2. Yet, an OCCR of 1.42 is unlikely to cover the full financial cost of the service. A typical utility in a low-income country has an OCCR of only 1.09. Interestingly, the lowest levels of short-term cost recovery are registered in utilities in lower middle-income countries, which typically are not able to cover their operation and maintenance costs (OCCR is 0.99), while upper middle-income countries have an OCCR more or less similar to that of low-income countries. Table 23.2 shows there is a wide variety in behaviors between utilities in the different income country categories. Utilities in high-income countries show better performance, but there are equally well-performing utilities in low- and middle-income countries. Apart from the variance between countries, we also observe a large variance within countries. In Brazil, for instance, the IBNET database finds that the OCCR varies between 0.33 and lower for the bottom 5 % of utilities and 2.03 and higher for the top 5 %. In high-income countries, the variance between utilities may be less, but still can be very large. According to the IBNET database, the OCCR of the top 5 % performing utilities is three times higher than that of the bottom 5 % in Australia. One important factor for these differences is that operation and maintenance costs can be highly variable, due to local circumstances (distance to water source, type of technology to produce water and quality of water sources, and corresponding treatment needs among others), as already mentioned by other authors, such as Whittington et al. (2008) and OECD (2010).

When disaggregating revenue sufficiency, the major reason for the lower levels of cost recovery in middle-income countries is linked to the provision of wastewater services. When countries become more affluent, the services that utilities provide tend to change. In low-income countries, utilities usually provide drinking water

Table 23.2 Operating cost coverage ratio between 2006 and 2011 by income status

Indicator	Low-income countries	Lower middle-income countries	Upper middle-income countries	High-income countries
Median OCCR	1.09	0.99	1.12	1.42
Mean OCCR	1.29	1.05	1.19	1.56
Standard deviation	0.68	0.55	0.52	0.58
Top 5 %	2.71	2.10	2.12	2.74
Top quartile	1.53	1.24	1.39	1.77
Bottom quartile	0.86	0.74	0.92	1.18
Bottom 5 %	0.45	0.30	0.39	0.90

Source: IBNET database
Note: *OCCR* operating cost coverage ratio, *Prelim* preliminary. The 2011 data collection cycle is not yet fully completed

services, and, if they provide wastewater services, these services tend to be limited in scope. In the period between 2006 and 2011, about 64 % of the population in low-income countries had access to water supply services, but only a paltry 27 % had access to some form of wastewater collection. In high-income countries, water and wastewater coverage is almost universal, whereas most people are having access to wastewater collection services while they are also provided with drinking water and wastewater treatment at often increasingly high standards—as standards tend to increase with economic development.

Operation and maintenance cost. The lack of progress in achieving higher levels of revenue sufficiency may give an impression of stagnation. But a disaggregation of the OCCR shows that operation and maintenance costs per cubic meter of water and wastewater sold have increased rapidly since 2006 (see Table 23.3). The reasons for this increase in operation and maintenance costs are manyfold, but an important one is that countries have grown increasingly richer. The number of low-income countries has declined rapidly over the past decade, with two effects: a significant change in the scope and quality of water and wastewater services provided and increases in overall price levels.

Nominal operation and maintenance costs increased from US$0.50 in 2006 to US$0.75 in 2010. This is a cost increase of 50 % over the 5-year period. At the same time, the World Bank Indicator database showed that world inflation over the same period was almost 22 %, resulting in a real increase in operation and maintenance costs of about 23 %. The standard deviation for this indicator has increased over the period, which means that the gap in operation and maintenance costs between utilities has increased. Those with the lowest operation and maintenance costs saw their costs grow most rapidly, as can be seen in Table 23.3. The "top" quartile saw a nominal cost increase of 57 %, compared to nominal cost increase of 33 % in the bottom quartile.

Table 23.4 shows median O&M cost per cubic meter of US$0.26 in low-income counties. These costs increase to US$1.68 in high-income countries. Yet, the same table shows that the differences within the various country income groups are very

Table 23.3 Operation and maintenance cost per cubic meter of water and wastewater sold (current USD) in between 2006 and 2011

Indicator	2006	2007	2008	2009	2010	2011
Median O&M cost	0.50	0.58	0.60	0.62	0.75	0.70
Average O&M cost	0.65	0.75	0.80	0.79	0.88	0.82
Standard deviation	0.53	0.58	0.65	0.62	0.61	0.57
Number of utilities reporting	1,264	1,468	1,381	1,415	1,565	1,304
Top 5 %[a]	0.11	0.12	0.12	0.13	0.16	0.12
Top quartile	0.28	0.32	0.33	0.34	0.44	0.40
Bottom quartile	0.87	0.99	1.03	1.05	1.16	1.12
Bottom 5 %	1.72	1.91	2.20	2.05	2.12	1.89

Source: IBNET database

Note: The 2011 data collection cycle is not yet complete

[a]The very low operation and maintenance costs are likely to reflect large subsidy flows that push down the cost of service or incomplete accounting of the cost of service

Table 23.4 Operation and maintenance costs per cubic meter of water and wastewater sold (current USD) between 2006 and 2011 by income status

Indicator	Low-income countries	Lower middle-income countries	Upper middle-income countries	High-income countries
Median O&M costs	0.26	0.43	0.68	1.68
Mean average O&M costs	0.50	0.61	0.79	1.71
Standard deviation	0.54	0.53	0.50	0.76
Top 5 %	0.06	0.11	0.20	0.47
Top quartile	0.12	0.23	0.43	1.18
Bottom quartile	0.72	0.81	1.02	2.22
Bottom 5 %	1.56	1.75	1.78	3.02

Source: IBNET database
Note: The 2011 data collection cycle is not yet complete

large. As shown elsewhere, the variance of utilities within and between countries is very large and is for a large part due to the very local factors that affect the cost of service. Nauges and van den Berg (2008) noted a similar result from a set of case studies with cost structures of water and sewerage utilities varying significantly within and between countries and over time.

The estimation of an operation and maintenance cost (per cubic meter of water sold) function provides insight on variables influencing these costs. The operation and maintenance costs and its explaining variables were specified in natural logarithm terms, so that the coefficients would measure the elasticity of the operation and maintenance costs. Danilenko et al. (2014) show that operation and maintenance costs are essentially driven by changes in staff costs, per capita water production, the scope of services that are provided (the utility provides water and/or wastewater services), and gross national income (GNI) per capita (Table 23.5). In most utilities, staff costs make up the largest proportion of costs, and they are mostly fixed in nature. In 2010, the IBNET database found that about 40 % of the operating costs were made up of labor costs. The higher the staff cost per employee, the higher the O&M costs, whereas the lower the staff productivity (i.e., the higher the number of staff per 1,000 people served), the higher the O&M costs. Water production has a negative impact on operation and maintenance costs, because of economies of scale: the higher the water production, the lower the operation and maintenance costs. Wastewater coverage pushes operation and maintenance costs up. The level of economic development tends to have a positive effect on the operation and maintenance costs: the higher the income levels in the country, the higher the operation and maintenance costs are[8] (Nauges and van den Berg 2010). Operation and maintenance costs may therefore have an upward bias and will increase over time, as the scope and quality of the services increase when countries become more affluent. In addition, the effect of climate change may further push costs upward, as this will result in many places in water scarcity, and the subsequent higher costs of providing water through the use of more expensive technologies

[8] Labor costs tend to increase with economic growth and so do water production levels.

Table 23.5 The drivers of operation and maintenance costs of water and wastewater services

Variable = LN_OMCOST	Coefficient	Standard error	T-value	P-value
LN_staff productivity	0.412	0.015	27.18	0.000
LN_cost per employee	0.589	0.020	29.26	0.000
LN_water production (lcd)	−0.627	0.024	−21.28	0.000
SEWCOV_DUMMY	0.604	0.144	3.35	0.000
LN_nonrevenue water (cum/km/day)	0.048	0.027	29.14	0.000
LN_metering	0.007	0.005	1.30	0.195
LN_pipe breaks per km	0.032	0,017	6.50	
LN_gross national income per capita	0.315	0.005	18.49	0.000
Constant	−3.92		−26.87	0.000
Number of observations		2,100		
R-square adjusted		0.7385		
F-test		742.10		

Table 23.6 Average revenues per cubic meter of water and wastewater sold (USD) between 2006 and 2011

Indicator	2006	2007	2008	2009	2010	2011
Median average revenues	0.54	0.66	0.69	0.70	0.81	0.72
Mean average revenues	0.74	0.87	0.95	0.94	1.00	0.91
Standard deviation	0.64	0.74	0.88	0.82	0.81	0.75
Number of utilities reporting	1,290	1,498	1,400	1,480	1,567	1,299
Top 5 %	1.99	2.34	2.93	2.74	2.62	2.25
Top quartile	0.97	1.15	1.19	1.23	1.34	1.23
Bottom quartile	0.29	0.33	0.34	0.36	0.41	0.35
Bottom 5 %	0.10	0.13	0.16	0.14	0.14	0.13

Source: IBNET database
Note: The 2011 data collection cycle is not yet complete

(desalination, water reuse) (Freebairn 2008), whereas the risks associated with more extreme weather events may result in additional costs to climate proof the water and wastewater infrastructure.

Operating revenues. Average revenues per cubic meter of water sold have increased significantly over time (Table 23.6). In 2006, the typical utility had an average tariff per cubic meter of water sold of US$0.54, which increased to US$0.81 in 2010. This amounts to a nominal US dollar increase of 50 % and, when using world inflation data over the same period, a real increase of 23 %. As shown, the variance of utilities within countries tends to be larger than the variance between countries and is, for a large part, due to the very local factors that affect the cost of services and the subsequent willingness to charge and pay for services.

Average revenues per cubic meter of water and wastewater sold are about US$0.31 in low-income countries, increasing to US$2.43 in high-income countries

Table 23.7 Average revenues per cubic meter of water sold (current USD) between 2006 and 2011 by income status

Indicator	Low-income countries	Lower middle-income countries	Upper middle-income countries	High-income countries
Median average revenues	0.31	0.39	0.80	2.43
Mean average revenues	0.47	0.55	0.92	2.49
Standard deviation	0.41	0.51	0.61	1.10
Number of utilities reporting	818	2,023	5,074	619
Top 5 %	1.33	1.64	2.09	4.57
Top quartile	0.64	0.67	1.23	3.19
Bottom quartile	0.19	0.24	0.48	1.73
Bottom 5%	0.07	0.11	0.18	0.67

Source: IBNET database
Note: The 2011 data collection cycle is not yet complete

Table 23.8 Universal metering—descriptive statistics

Variable	Utilities with universal metering	Utilities without universal metering	T-value
Average revenues for WWW per cum	1.11	0.82	−14.32[a]
Unit operation and maintenance costs for WWW per cum	0.88	0.75	−8.44[a]
OCCR	1.35	1.15	−14.38[a]

[a]Significant at 1 % level

(Table 23.7). The table also shows large variances within the groupings, and these may be linked to differences in cost structures and differences in pricing policies.

Ideally, each customer should pay for its use, as that will provide incentives to use water and wastewater services more efficiently and more equitable. In the IBNET sample, metering levels are relatively high, but universal metering is much less common, with 35 % of the utilities stating that they meter all their customers. Yet, the effect of metering is important. Utilities that have universal metering tend to have higher water prices than those that do not. Universal water metering also has a positive effect on the operation and maintenance costs, but, in general, the benefits associated with water metering tend to exceed the costs. As a result, the OCCR is higher in those utilities in which metering is universal (Table 23.8).

Pricing is an important means to reduce water consumption (Dalhuizen et al. 2003), and increase revenues (as residential water demand is mostly inelastic). Yet, the effect of higher water prices and lower water consumption on the operation and maintenance costs can be perverse in certain circumstances. In high-income countries, for instance, the effect of pricing may result in lower utility revenues (see also chapter by Barr and Ash in this volume). This effect was demonstrated in Germany, where the combination of a high dependence on volumetric metering (with a small proportion of fixed charges) and high tariffs resulted in a decline in utility revenues (European Environment Agency 2013).

23.4 Affordability

One of the reasons why many utilities do not charge the full cost of service is that they are often worried that consumers cannot pay for the service. In 2010, the UN General Assembly declared clean drinking water and sanitation as a human right, with states being called to provide safe, clean, accessible, and affordable drinking water and sanitation for all.

In this section, we will determine how affordable water and wastewater services are. Affordability is measured as the ratio between average revenue per capita to GNI per capita (expressed in percentage). This indicator is a proxy for affordability, because although we can measure the average revenue per capita, the income indicator is by necessity a proxy variable. Most utilities are municipality based, and income data at this level is rarely available. By using this proxy, however, the actual affordability may be overestimated as most utilities serve urban populations, which tend to have higher average incomes than rural populations, whereas the GNI per capita is a national average. Yet, at the same time, an average number does conceal the effects of income inequality, while many utilities still do not provide universal access to services (average access to piped water was 85 % in 2010). In general, those not connected to the service tend to be disproportionally poor (Komives et al. 2005).

Median affordability has improved significantly for those that have access to piped water services (Table 23.9). In 2006, 0.86 % of income was spent on water and wastewater services, compared to only 0.59 % in 2010. The standard deviation is decreasing over time, which means that the variation in affordability performance between utilities is decreasing. A general used affordability rule relies on international limits; 3–5 % of household income is an often-quoted figure of what house-

Table 23.9 Affordability in percentage of GNI between 2006 and 2011

Indicator	2006	2007	2008	2009	2010	2011
Median affordability	0.86	1.00	0.92	0.76	0.59	0.55
Average affordability	1.14	1.27	1.11	0.84	0.73	0.71
Standard deviation	1.01	1.00	0.89	0.73	0.59	0.59
Number of utilities reporting	1,437	1,521	1,476	1,600	1,633	1,383
Top 5 %	0.23	0.28	0.24	0.15	0.12	0.10
Top quartile	0.56	0.65	0.56	0.39	0.36	0.31
Bottom quartile	1.38	1.54	1.36	1.03	0.92	0.92
Bottom 5 %	3.04	3.16	2.80	2.25	1.84	1.82

Source: IBNET database
Note: The 2011 data collection cycle is not yet complete

holds should spend on their water and wastewater services (OECD 2009). Based on the IBNET sample, the data above shows that in the typical utility, water and wastewater services are very affordable for the majority of households (with median affordability significantly below the earlier mentioned threshold of 3–5 % of income). The table shows that the services have also become more affordable over time. However, it should also be noted that income levels can vary widely within the service areas of utilities, and, hence, serving poor people still may be an issue in some utilities, even when the service in the typical utility is very affordable.

Consumers served by utilities in low-income countries spent more of their income on water and/or wastewater services than consumers from utilities in middle-income and high-income countries (Table 23.10). In 2010, median affordability for households in low-income countries was 1.48 %, compared to 0.78 % in lower middle-income countries, and 0.66 % in upper middle-income countries, but 0.85 % in high-income countries. In general, affordability by income group seems to show a U-shape form, with utility consumers in low-income countries paying significantly more than consumers in middle-income countries. Yet, utility consumers in middle-income countries pay less for the water and/or wastewater services than utility customers in high-income countries.

The differences in median affordability within the income groups can be very large (Table 23.10). In the low-income country group, for instance, the difference in affordability ranges from 0.19 % of GNI per capita for the top 5 % of utilities to 4.62 % of the bottom 5 % of utilities. Banerjee et al. (2008) show that tariffs in Africa are significantly higher than those in South Asia, suggesting that approaches to water pricing may be very different.

Summarizing, the typical utility provides services that are eminently affordable for those that have access to piped services. This result, however, does not take into account that many households that have access to piped services benefit from cross-subsidies.

Table 23.10 Affordability as percentage of GNI per capita between 2006 and 2011 by income status

Indicator	Low-income countries	Lower middle-income countries	Upper middle-income countries	High-income countries
Median affordability	1.48	0.78	0.66	0.85
Mean affordability	1.81	1.04	0.81	0.93
Standard deviation	1.43	0.90	0.64	0.45
Number of utilities reporting	818	2,229	5,336	667
Top 5 %	0.19	0.19	0.14	0.03
Top quartile	0.86	0.45	0.40	0.64
Bottom quartile	2.39	1.31	1.05	1.13
Bottom 5 %	4.62	2.75	1.95	1.71

Source: IBNET database
Note: The 2011 data collection cycle is not yet complete

Cross-subsidies. In many utilities, subsidies to residential consumers are (partially) provided through cross-subsidies, where certain consumer categories (mostly commercial and industrial water users) subsidize residential consumers. The IBNET database provides some details on the level of cross-subsidies within utilities, but the data is not complete because only one-third (or less) of the utilities provide this information. In 2010, the median utility charged industrial users up to 1.98 times more per cubic meter of water than they charged residential users (Table 23.11). Yet, the average cross-subsidy rates are much higher and show the wide variation in behavior between utilities in using cross-subsidies. The financial crisis of 2008 resulted in an uptake of cross-subsidies as the median cross-subsidy rate increased in 2008 and has not yet returned to precrisis levels. The increased use of cross-subsidies in utilities reporting the data suggests that when faced with the need for higher revenues, many utilities have tried to reduce the impact on residential consumers by putting more of the burden on nonresidential water users.

Cross-subsidies are especially prevalent in poorer countries (Table 23.12). The typical utility in a lower-income country has an industrial tariff that is three times the residential tariffs. There is also wide variation within the different income

Table 23.11 Cross-subsidy levels, 2006–2011 (ratio of industrial to residential tariff)

Indicator	2006	2007	2008	2009	2010	2011
Median cross-subsidy rate	1.89	1.69	1.99	1.99	1.98	1.96
Average cross-subsidy	4.22	3.24	3.98	4.02	4.21	3.80
Standard deviation	6.04	4.37	5.39	5.63	5.54	5.24
Number of utilities reporting	553	691	574	589	464	371

Source: IBNET database
Note: The 2011 data collection cycle is not yet complete

Table 23.12 Cross-subsidy levels between 2006 and 2011 by income status

Indicator	Low-income countries	Lower middle-income countries	Upper middle-income countries	High-income countries
Median cross-subsidy rate	2.37	3.37	1.32	1.37
Median cross-subsidy rate	4.11	5.86	2.32	2.27
Standard deviation	4.99	6.75	3.58	2.76
Number of utilities reporting	489	1,184	1,336	233
Top 5 %	0.67	0.94	0.65	0.90
Top quartile	1.31	1.68	0.97	1.05
Bottom quartile	4.52	6.76	2.16	2.46
Bottom 5 %	13.91	20.89	6.95	5.72

Source: IBNET database
Note: *Prelim* preliminary. The 2011 data collection cycle is not yet complete

Table 23.13 Collection periods, 2006–2011 (number of days)

Indicator	2006	2007	2008	2009	2010	2011
Median collection period	96	88	82	75	70	66
Average collection period	157	142	139	125	116	122
Standard deviation	181	158	175	154	153	161
Number of utilities reporting	1,179	1,244	1,174	1,237	1,155	975

Source: IBNET database
Note: The 2011 data collection cycle is not yet complete

groupings, with high standard deviations. Cross-subsidies tend to be more common in certain regions, especially countries in Eastern Europe, Central Asia, and the Middle East.

High industrial water tariffs do not automatically translate into more revenues per cubic meter sold, as industrial consumers react to tariffs (just like other types of consumers), and their demand for water tends to be more price elastic than that of residential water users (Dalhuizen et al. 2003). When the ratio of industrial to residential tariffs is getting too large, the effect may be that utilities see their revenues reduced. A rapid analysis shows that when the cross-subsidy rate is more than 2, overall water consumption decreases, and the share of residential consumers in total water consumption increases, whereas that of industrial consumers declines. The effect is that the OCCR is statistically significantly lower in utilities with high levels of cross-subsidies. In the IBNET database, utilities with a cross-subsidy rate of more than 2 have an OCCR of 1.14, compared to 1.28 for those utilities with cross-subsidy rates below 2. Too much cross-subsidies can result in industrial water users leaving the piped network and, thus, reducing the basis on which cross-subsidies are being determined. Hence, there is likely to be an optimal level for cross-subsidies.

Collection efficiency. An indirect measure of whether the services are affordable to customers is whether utilities are able to collect the revenues that it charges to its customers. Liquidity maintenance is of utmost importance for utilities (WHO 1989), as all cash needs to be covered. If revenues are billed but not collected, these cash needs will not be recovered. Low collection efficiencies are often linked to a lack of willingness to pay from consumers, although the lack of interest from the utilities to collect their unpaid bill revenues may also play a role, especially when average water and wastewater tariffs are very low, and/or access to public funds is very easy. As services have become more affordable, collection efficiency has improved. In 2010, the median collection period had decreased to 70 days (from 96 days in 2006), as shown in Table 23.13.

23.5 Subsidies in Water and Wastewater Provision

The typical utility is able to cover its operation and maintenance costs of the service, but barely, and hence in most utilities revenues are insufficient to cover the full financial costs of the services. Hence, this subsidy flows to the utilities are large.

In some cases, the subsidy flows may be explicit with governments paying for the capital cost of the service and in a smaller number of utilities by providing subsidies for recovering operation and maintenance costs. Yet, there are also often hidden subsidies provided, where low quality of services—often the result of postponing maintenance—may be used by utilities to deal with an insufficient revenue base (Saavalainen and ten Berge 2006). Since the financial crisis of 2008, maintenance costs have been crowded out by increases in labor and energy costs and suggest that these hidden subsidies have become more important. In some countries, utilities also may benefit from general subsidy policies, most notable energy subsidies that push down the cost of the service.

Whittington et al. (2008) estimate that the economic cost of conventional water and wastewater infrastructure is US$2.50 per cubic meter[9] (at a discount rate of 10 %[10] in 2006 prices), assuming very low opportunity cost of raw water and very limited externalities associated with the discharge of treated wastewater.[11] If we exclude any externalities from the US$2.50 and adjust them to 2010 prices, this would result in a US$2.85 per cubic meter of water sold, with about US$1.14 for water and US$1.71 for wastewater. If we note that the typical utility is able to charge costs of US$0.75 in 2010, the size of the subsidy for a cubic meter of water and wastewater sold is very significant, and it would imply that fixed (capital) costs make up 74 % of the total costs of the service.

There is significant scope for reducing the current size of the subsidies, especially as the costs of water and wastewater continue to rise, and long-term trends suggest that this is going to be the case in the future (Freebairn 2008). In many countries, especially higher-income countries, the infrastructure assets are at the end of their lifetime, and, hence, large replacement investments will be needed to ensure the continuity of the service (American Society of Civil Engineers 2011). Because of the long lifetime of the assets (up to 100 years for certain infrastructure assets), and accounting practices in many countries, this will result in a sharp increase in costs, especially when utilities strive to cover a larger part of their full financial costs through to their customers. Moreover, water stress is increasing all over the world, while demand for water is still increasing. This will result in an increasing dependence on more expensive forms of water supply (desalination, recycling) and will result in an increasing marginal cost curve for water. Climate change and the corresponding adverse weather shocks will add to the costs, as utilities will have to safeguard their supplies and deal with increased supply risks.

[9] It should be noted that these estimates are highly variable. In places, for instance, where water has to be hauled over long distances, or where water scarcity is a reality, the costs can be significantly higher.

[10] This assumes that, for utilities, financing is available against international market rates, whereas it also assumes that there are no particular country risks included.

[11] There is some trade-off between wastewater treatment costs and externalities associated with discharge of treated wastewater. The higher the levels of wastewater treatment, the lower the externalities associated with the discharge of treated wastewater.

The first route of improving cost recovery levels is by charging for water and wastewater. The analysis above shows that water and wastewater services are very affordable, even in many poor countries. In 95 % of the sample, the affordability in 2010 was less than 1.8 % of income (as can be seen in Table 23.9)—significantly below the threshold of 3–5 % that is often used to measure the affordability of the services. As services are becoming rapidly more affordable, there is scope to increase tariffs in many utilities.

Another way to improve cost recovery is by increasing the efficiency with which services are provided. There is scope for improvements in operations, as an analysis of the IBNET team shows. There are wide variations in utilities' access rates to services, staff productivity, nonrevenue water, and energy efficiency (Danilenko et al. 2014). Some progress has been made with reductions in nonrevenue water and increases in staff productivity. Despite staff productivity gains, the total share of labor in the operating costs has nonetheless increased. The share of energy in total operation and maintenance costs saw a similar trend; the share of energy has increased with any energy efficiency gains nullified by higher fuel prices. The increase in both labor and energy share makes it likely that maintenance has been put on hold, as the financial crisis hit utilities. The slow progress in improving efficiency in utility operations is linked to the fact that fixed costs make up the bulk of costs in water and wastewater systems. The data from Whittington et al. (2008), and IBNET, show that typically operation and maintenance costs may make up only 26 % of total financial costs. In such systems, the impact of efficiency improvements in operation and maintenance can be important, but is by definition relatively limited.

An important element that often is overlooked in the discussion on efficiency is designing the systems in such a way that they are more efficient once they start operating. In case the system is already in place, this may look like a forgone option. However, having the right (water and/or wastewater infrastructure) design is critical in the delivery of water and/or wastewater services. With the total number of people not yet served by piped water running around three billion (UNICEF-WHO JMP 2014), this is not merely a theoretical discussion. Many of the water and/or wastewater services are coming at the end of their economic lifetime requiring this infrastructure to be replaced; optimizing water and wastewater systems is of major importance. Water systems are laid out using specific design criteria for, among others, water demand, design horizons, peak factors, material use, pressure requirements, and water (and wastewater) quality standards. The more stringent these design criteria are set, the higher the cost of the services, but also often the more difficult to operate and maintain these systems, as most of the costs are fixed. Hence, the need to think through the trade-offs between design standards, system costs (both full financial and economic costs), customer's willingness to pay, and fiscal impacts is important. Whittington et al. (2008) conclude that in certain circumstances, conventional network technologies may not produce the economic benefits to cover the costs associated with these investments, and the use of alternative technologies should be considered.

23.6 Conclusions and Policy Implications

The typical utility in a developing country is able to cover its operation and maintenance costs of the service, however, progress in improving the levels of cost recovery between 2006 and 2011 has been minimal. This lack of progress in recovering the financial cost of service through utility revenues is for a large part the result of the particular characteristics of water and wastewater infrastructure. Two of the most important characteristics are the capital intensity of the infrastructure and the ever-changing nature of what constitutes water and wastewater services. The capital intensity of water and wastewater infrastructure ensures that a large part of the total cost of water and wastewater is made up of fixed costs. The second complication is that what makes up water and wastewater services is not fixed over time. Hence, the service is consistently evolving. The early stages of providing these services are mainly focused on drinking water provision. Yet, when piped water access rises, increasing flows of wastewater need to be collected, treated, and safely disposed of. This trend of the evolving nature of water and wastewater services is likely to continue in the future, as water quality standards are continuously scaled up. It is likely that in the future, costs will increase even more as water scarcity becomes more widespread (Freebairn 2008), whereas increases in the occurrence of weather shocks and the need to address these shocks will further add to the cost of providing services.

While the costs of providing water and wastewater services have been increasing both in nominal and real terms in developing countries, the cost increases have been matched by increases in tariffs. Increases in cross-subsidies have, in some cases, dampened the effect of rising tariffs for residential water users. Many governments, which are often responsible for tariff setting in utilities, have been less than willing to support significant residential tariff increases, and, as a result, services have become increasingly affordable for most residential customers.

The combination of higher future costs of water and the current levels of affordability means that in many utilities, there is ample scope for balancing the goals of revenue sufficiency and affordability—more in favor of the former as government subsidies will otherwise need to increase rapidly. The high dependence on subsidies raises large questions about the equity of such policies that disadvantage those not yet connected to piped network services (in most developing countries, those not connected to the piped network services are disproportionally poorer citizens and/or disadvantage future generations).

The variance of utilities within and between countries is very large. This is for a large part because water and wastewater services are locally provided. These local factors can vary widely between utilities and include factors such as the distance to the water source and the effect on the cost to store and transport water, the quality of the water source, and the need for treatment, among others. These cost differences will affect the subsequent willingness to charge and pay for services. The implication of this large variance in performance is that specific local circumstances have a major impact on revenue sufficiency and affordability. Policymakers will need to understand how the variation in the costs of water and wastewater services affects the balance between the objectives of revenue sufficiency and affordability[12] in their pricing policies. This will require much more information and research than is currently available.

Finally, as the fixed cost in water and wastewater service provision is so large, the design standards under which the services are constructed will determine the long-term cost of the services. Hence, it is important to undertake a proper analysis when investment decisions are made to ensure that the benefits and costs of such investments are properly analyzed, because the financial and social implications of such investment decisions will be felt for many decades.

Appendix 23.1: IBNET Data

IBNET collects data on financial and operational utility performance that can be grouped in the following categories:

- *Utility information*: This data includes basic details on the utility providing the data, including the type of legal status of the utility, the extent of private sector participation.
- *Service area*: Each utility serves a certain service area for water and/or wastewater services.
- *Staff data*: Data on the number of employees by service provided.
- *Data on water supply services*: This is the largest category of data that is collected and refers to the population served with water, connections (including connections with water meters) and length of network, and water production and consumption (by customer category). These data also include some data on the quality of the service provided, such as duration of supply, pipe breaks, and water quality tests.
- *Data on wastewater services*: This section includes data on the population served with wastewater services (collection and treatment), connections and length of sewer network, and wastewater collected and treated (by customer category). These data also include some data on the quality of the service provided, such as sewer blockages.
- *Financial information*: This section includes data on the billed revenues and collected cash income by customer category. This also includes data on the operation and maintenance costs of the water and/or wastewater services by cost categories. Other financial information, such as fixed assets, accounts receivables, and funding sources are also being included in the data collection sheets.

[12] And any other policy objectives that are to be included, such as economic efficiency, transparency, etc.

- *Tariff information*: This section in the data collection tool includes information on the fixed charges and connection charges, whereas it also documents the water bill for a household with a minimum level of consumption (set at 6 m^3 per month).

These data translate into a large set of indicators that are being calculated on the basis of the input data described above.

The quality of the IBNET database (Danilenko et al. 2014) depends on the quality of the data submitted by individual utilities, utility associations, regulatory agencies, ministries, and others. The quality of the data is variable and reflects for a large part the reporter's performance. Some of the data comes from sources that have excellent quality assurance procedures (as in the case of regulatory data) and others that follow less sound procedures. The need for rigorous quality assurance procedures has to be balanced against the need to avoid discouraging potentially valuable data sources from participating. The IBNET team checks the data and removes the data that cannot be verified. For the data used in this analysis, the bottom and top 1 % of data was taken out of the dataset.

References

American Society of Civil Engineers. (2011). *Failure to act: The economic impact of current investment trends in water and wastewater infrastructure*. American Society of Civil Engineers, Washington, DC.

Banerjee, S., Foster, V., Ying, Y., Skilling, H., & Wodon, Q. (2008). *Cost recovery, equity, and efficiency in water tariffs: Evidence from African utilities* (Africa Infrastructure Diagnostic Survey, Working Paper 7).

Dalhuizen, J., Raymond, M., Florax, J. G. M., de Groot, H. L. F., & Nijkamp, P. (2003). Price and income elasticities of residential water demand: A meta-analysis. *Land Economics, 79*, 292–308.

Danilenko, A., van den Berg, C., Macheve, B., & Moffitt, L. J. (2014). *The IBNET water supply and sanitation blue book 2014: The international benchmarking network for water and sanitation utilities databook*. World Bank Group, Washington, DC.

European Environment Agency. (2013). *Assessment of cost recovery through pricing of water* (Technical report 16/2013). Copenhagen: European Environment Agency.

Freebairn, J. (2008). Some emerging issues in urban water supply and pricing. *Economic Papers (the Economic Society of Australia), 27*, 184–193.

International Benchmarking Network for Water and Sanitation Utilities (IBNET) at www.ib-net.org

Komives, K., Foster, V., Halpern, J., & Wodon, Q. (2005). *Water, electricity and the poor: Who benefits from utility subsidies?* Washington, DC: World Bank.

Nauges, C., & van den Berg, C. (2008). Economies of density, scale and scope in the water supply and sewerage sector: A study of four developing and transition economies. *Journal of Regulatory Economics, 34*, 144–163.

Nauges, C., & van den Berg, C. (2010). Heterogeneity in the cost structure of water and sanitation services: A cross-country comparison of conditions for scale economies. *Oxford Development Studies, 38*, 199–217.

Nauges, C., & Whittington, D. (2010). Estimation of water demand in developing countries: An overview. *World Bank Research Observer, 25*, 263–294.

OECD. (2009). *Managing water for all: An OECD perspective on pricing and financing*. Paris: OECD.

OECD. (2010). *Pricing water resources and water and sanitation services*. Paris: OECD Publishing.

Rogers, P., de Silva, R., & Bhatia, R. (2002). Water is an economic good: How to use prices to promote equity, efficiency and sustainability. *Water Policy, 4*, 1–17.

Saavalainen, T. & ten Berge, J. (2006). *Quasi-fiscal deficits and energy conditionality in selected CIS countries* (IMF Research Working Paper 06/43). Washington, DC.

UN General Assembly. (2010). *Resolution 64/292: The human right to water and sanitation*. http://www.un.org/ga/search/view_doc.asp?symbol=A/RES/64/292

Whittington, D. (2003). Municipal water pricing and tariff design: A reform agenda for south Asia. *Water Policy, 5*, 61–76.

Whittington, D., Hanemann, C. S., & Jeuland, M. (2008). The challenge of improving water and sanitation services in less developed countries. *Foundations and Trends in Microeconomics, 4*, 469–609.

WHO. (1989). *Managerial and financial principles for water supply and sanitation agencies*. Report of the fourth consultation on institutional development working group on cost recovery. Geneva, 21–25 Nov 1988.

WHO. (2003). *Domestic water quality, service levels and health*. Geneva: World Health Organization.

WHO and UNICEF Joint Monitoring Program. (2014). *Progress on drinking water and sanitation: Update for 2014*. World Health Organization and UNICEF, Geneva.

Zetland, D., & Gasson, C. (2013). A global survey of urban water tariffs: Are they sustainable, efficient and fair? *International Journal for Water Resources Development, 29*, 1–16.

Index

A. Dinar et al. (eds.), *Water Pricing Experiences and Innovations*, Global Issues in Water Policy 9, DOI 10.1007/978-3-319-16465-6

J

L

M

S

T

MIX
Papier aus verantwortungsvollen Quellen
Paper from responsible sources
FSC® C105338

If you have any concerns about our products,
you can contact us on
ProductSafety@springernature.com

In case Publisher is established outside the EU,
the EU authorized representative is:
Springer Nature Customer Service Center GmbH
Europaplatz 3, 69115 Heidelberg, Germany

Printed by Libri Plureos GmbH
in Hamburg, Germany